LONDON MATHEMATICAL SOCIETY LECTURE NOTE SERIES

Managing Editor: Professor N.J. Hitchin, Mathematical Institute,
University of Oxford, 24–29 St Giles, Oxford OX1 3LB, United Kingdom

The titles below are available from booksellers, or from Cambridge University Press at www.cambridge.org.

161 Lectures on block theory, BURKHARD KÜLSHAMMER
163 Topics in varieties of group representations, S.M. VOVSI
164 Quasi-symmetric designs, M.S. SHRIKANDE & S.S. SANE
166 Surveys in combinatorics, 1991, A.D. KEEDWELL (ed)
168 Representations of algebras, H. TACHIKAWA & S. BRENNER (eds)
169 Boolean function complexity, M.S. PATERSON (ed)
170 Manifolds with singularities and the Adams-Novikov spectral sequence, B. BOTVINNIK
171 Squares, A.R. RAJWADE
172 Algebraic varieties, GEORGE R. KEMPF
173 Discrete groups and geometry, W.J. HARVEY & C. MACLACHLAN (eds)
174 Lectures on mechanics, J.E. MARSDEN
175 Adams memorial symposium on algebraic topology 1, N. RAY & G. WALKER (eds)
176 Adams memorial symposium on algebraic topology 2, N. RAY & G. WALKER (eds)
177 Applications of categories in computer science, M. FOURMAN, P. JOHNSTONE & A. PITTS (eds)
178 Lower K- and L-theory, A. RANICKI
179 Complex projective geometry, G. ELLINGSRUD *et al*
180 Lectures on ergodic theory and Pesin theory on compact manifolds, M. POLLICOTT
181 Geometric group theory I, G.A. NIBLO & M.A. ROLLER (eds)
182 Geometric group theory II, G.A. NIBLO & M.A. ROLLER (eds)
183 Shintani zeta functions, A. YUKIE
184 Arithmetical functions, W. SCHWARZ & J. SPILKER
185 Representations of solvable groups, O. MANZ & T.R. WOLF
186 Complexity: knots, colourings and counting, D.J.A. WELSH
187 Surveys in combinatorics, 1993, K. WALKER (ed)
188 Local analysis for the odd order theorem, H. BENDER & G. GLAUBERMAN
189 Locally presentable and accessible categories, J. ADAMEK & J. ROSICKY
190 Polynomial invariants of finite groups, D.J. BENSON
191 Finite geometry and combinatorics, F. DE CLERCK *et al*
192 Symplectic geometry, D. SALAMON (ed)
194 Independent random variables and rearrangement invariant spaces, M. BRAVERMAN
195 Arithmetic of blowup algebras, WOLMER VASCONCELOS
196 Microlocal analysis for differential operators, A. GRIGIS & J. SJÖSTRAND
197 Two-dimensional homotopy and combinatorial group theory, C. HOG-ANGELONI *et al*
198 The algebraic characterization of geometric 4-manifolds, J.A. HILLMAN
199 Invariant potential theory in the unit ball of $\mathbf{C}^n$, MANFRED STOLL
200 The Grothendieck theory of dessins d'enfant, L. SCHNEPS (ed)
201 Singularities, JEAN-PAUL BRASSELET (ed)
202 The technique of pseudodifferential operators, H.O. CORDES
203 Hochschild cohomology of von Neumann algebras, A. SINCLAIR & R. SMITH
204 Combinatorial and geometric group theory, A.J. DUNCAN, N.D. GILBERT & J. HOWIE (eds)
205 Ergodic theory and its connections with harmonic analysis, K. PETERSEN & I. SALAMA (eds)
207 Groups of Lie type and their geometries, W.M. KANTOR & L. DI MARTINO (eds)
208 Vector bundles in algebraic geometry, N.J. HITCHIN, P. NEWSTEAD & W.M. OXBURY (eds)
209 Arithmetic of diagonal hypersurfaces over infite fields, F.Q. GOUVEA & N. YUI
210 Hilbert C*-modules, E.C. LANCE
211 Groups 93 Galway / St Andrews I, C.M. CAMPBELL *et al* (eds)
212 Groups 93 Galway / St Andrews II, C.M. CAMPBELL *et al* (eds)
214 Generalised Euler-Jacobi inversion formula and asymptotics beyond all orders, V. KOWALENKO *et al*
215 Number theory 1992–93, S. DAVID (ed)
216 Stochastic partial differential equations, A. ETHERIDGE (ed)
217 Quadratic forms with applications to algebraic geometry and topology, A. PFISTER
218 Surveys in combinatorics, 1995, PETER ROWLINSON (ed)
220 Algebraic set theory, A. JOYAL & I. MOERDIJK
221 Harmonic approximation, S.J. GARDINER
222 Advances in linear logic, J.-Y. GIRARD, Y. LAFONT & L. REGNIER (eds)
223 Analytic semigroups and semilinear initial boundary value problems, KAZUAKI TAIRA
224 Computability, enumerability, unsolvability, S.B. COOPER, T.A. SLAMAN & S.S. WAINER (eds)
225 A mathematical introduction to string theory, S. ALBEVERIO, J. JOST, S. PAYCHA, S. SCARLATTI
226 Novikov conjectures, index theorems and rigidity I, S. FERRY, A. RANICKI & J. ROSENBERG (eds)
227 Novikov conjectures, index theorems and rigidity II, S. FERRY, A. RANICKI & J. ROSENBERG (eds)
228 Ergodic theory of $\mathbf{Z}_d$ actions, M. POLLICOTT & K. SCHMIDT (eds)
229 Ergodicity for infinite dimensional systems, G. DA PRATO & J. ZABCZYK
230 Prolegomena to a middlebrow arithmetic of curves of genus 2, J.W.S. CASSELS & E.V. FLYNN
231 Semigroup theory and its applications, K.H. HOFMANN & M.W. MISLOVE (eds)
232 The descriptive set theory of Polish group actions, H. BECKER & A.S. KECHRIS
233 Finite fields and applications, S.COHEN & H. NIEDERREITER (eds)
234 Introduction to subfactors, V. JONES & V.S. SUNDER
235 Number theory 1993–94, S. DAVID (ed)
236 The James forest, H. FETTER & B. GAMBOA DE BUEN
237 Sieve methods, exponential sums, and their applications in number theory, G.R.H. GREAVES *et al*
238 Representation theory and algebraic geometry, A. MARTSINKOVSKY & G. TODOROV (eds)
239 Clifford algebras and spinors, P. LOUNESTO
240 Stable groups, FRANK O. WAGNER

241 Surveys in combinatorics, 1997, R.A. BAILEY (ed)
242 Geometric Galois actions I, L. SCHNEPS & P. LOCHAK (eds)
243 Geometric Galois actions II, L. SCHNEPS & P. LOCHAK (eds)
244 Model theory of groups and automorphism groups, D. EVANS (ed)
245 Geometry, combinatorial designs and related structures, J.W.P. HIRSCHFELD *et al*
246 p-Automorphisms of finite p-groups, E.I. KHUKHRO
247 Analytic number theory, Y. MOTOHASHI (ed)
248 Tame topology and o-minimal structures, LOU VAN DEN DRIES
249 The atlas of finite groups: ten years on, ROBERT CURTIS & ROBERT WILSON (eds)
250 Characters and blocks of finite groups, G. NAVARRO
251 Gröbner bases and applications, B. BUCHBERGER & F. WINKLER (eds)
252 Geometry and cohomology in group theory, P. KROPHOLLER, G. NIBLO, R. STÖHR (eds)
253 The q-Schur algebra, S. DONKIN
254 Galois representations in arithmetic algebraic geometry, A.J. SCHOLL & R.L. TAYLOR (eds)
255 Symmetries and integrability of difference equations, P.A. CLARKSON & F.W. NIJHOFF (eds)
256 Aspects of Galois theory, HELMUT VÖLKLEIN *et al*
257 An introduction to noncommutative differential geometry and its physical applications 2ed, J. MADORE
258 Sets and proofs, S.B. COOPER & J. TRUSS (eds)
259 Models and computability, S.B. COOPER & J. TRUSS (eds)
260 Groups St Andrews 1997 in Bath, I, C.M. CAMPBELL *et al*
261 Groups St Andrews 1997 in Bath, II, C.M. CAMPBELL *et al*
263 Singularity theory, BILL BRUCE & DAVID MOND (eds)
264 New trends in algebraic geometry, K. HULEK, F. CATANESE, C. PETERS & M. REID (eds)
265 Elliptic curves in cryptography, I. BLAKE, G. SEROUSSI & N. SMART
267 Surveys in combinatorics, 1999, J.D. LAMB & D.A. PREECE (eds)
268 Spectral asymptotics in the semi-classical limit, M. DIMASSI & J. SJÖSTRAND
269 Ergodic theory and topological dynamics, M.B. BEKKA & M. MAYER
270 Analysis on Lie Groups, N.T. VAROPOULOS & S. MUSTAPHA
271 Singular perturbations of differential operators, S. ALBEVERIO & P. KURASOV
272 Character theory for the odd order function, T. PETERFALVI
273 Spectral theory and geometry, E.B. DAVIES & Y. SAFAROV (eds)
274 The Mandelbrot set, theme and variations, TAN LEI (ed)
275 Computational and geometric aspects of modern algebra, M. D. ATKINSON *et al* (eds)
276 Singularities of plane curves, E. CASAS-ALVERO
277 Descriptive set theory and dynamical systems, M. FOREMAN *et al* (eds)
278 Global attractors in abstract parabolic problems, J.W. CHOLEWA & T. DLOTKO
279 Topics in symbolic dynamics and applications, F. BLANCHARD, A. MAASS & A. NOGUEIRA (eds)
280 Characters and Automorphism Groups of Compact Riemann Surfaces, T. BREUER
281 Explicit birational geometry of 3-folds, ALESSIO CORTI & MILES REID (eds)
282 Auslander-Buchweitz approximations of equivariant modules, M. HASHIMOTO
283 Nonlinear elasticity, R. OGDEN & Y. FU (eds)
284 Foundations of computational mathematics, R. DEVORE, A. ISERLES & E. SÜLI (eds)
285 Rational points on curves over finite fields: Theory and Applications, H. NIEDERREITER & C. XING
286 Clifford algebras and spinors 2nd edn, P. LOUNESTO
287 Topics on Riemann surfaces and Fuchsian groups, E. BUJALANCE, A. F. COSTA & E. MARTINEZ (eds)
288 Surveys in combinatorics, 2001, J. W. P. HIRSCHFELD (ed)
289 Aspects of Sobolev-type inequalities, L. SALOFF-COSTE
290 Quantum groups and Lie theory, A. PRESSLEY (ed)
291 Tits buildings and the model theory of groups, K. TENT (ed)
292 A quantum groups primer, S. MAJID
293 Second order partial differential equations in Hilbert spaces, G. DA PRATO & J. ZABCZYK
294 Introduction to operator space theory, G. PISIER
295 Geometry and integrability, L. MASON & Y. NUTKU (eds)
296 Lectures on invariant theory, I. DOLGACHEV
297 The homotopy category of simply connected 4-manifolds, H.-J. BAUES
298 Higher operads, higher categories, T. LEINSTER
299 Kleinian groups and hyperbolic 3-manifolds, Y. KOMORI, V. MARKOVIC & C. SERIES (eds)
300 Introduction to Möbius differential geometry, U. HERTRICH-JEROMIN
301 Stable modules and the D(2)-problem, F. E. A. JOHNSON
302 Discrete and continuous nonlinear Schrödinger systems, M. ABLOWITZ, B. PRINARI & A.D. TRUBATCH
303 Number theory and algebraic geometry, MILES REID & ALEXEI SKOROBOGATOV (eds)
304 Groups St Andrews 2001 in Oxford vol. 1, C. M. CAMPBELL, E. F. ROBERTSON & G. C. SMITH (eds)
305 Groups St Andrews 2001 in Oxford vol. 2, C. M. CAMPBELL, E. F. ROBERTSON & G. C. SMITH (eds)
306 Geometric Mechanics and Symmetry, J. MONTALDI & T. RATIU (eds)
307 Surveys in combinatorics 2003, C. D. WENSLEY (ed)
308 Topology, Geometry and Quantum Field Theory, U. TILLMANN (ed)
309 Corings and comdules, TOMASZ BRZEZINSKI & ROBERT WISBAUER
310 Topics in dynamics and ergodic theory, SERGEY BEZUGLYI & SERGIY KOLYADA (eds)
311 Groups, T.W. MÜLLER
312 Foundations of computational mathematics, Minneapolis 2002, GELIPE CUCKER *et al* (eds)
313 Transcendantal Aspects of Algebraic Cycles, S. MÜLLER-STACH & C. PETERS (eds)
314 Spectral Generalizations of Line Graphs, D CVETKOVIC, P. ROWLINSON & S. SIMIC
315 Structured ring spectra, A. BAKER & B. RICHTER (eds)
317 Advances in elliptic curve cryptography, I.F. BLAKE, G. SEROUSSI & N. SMART
318 Perturbation of the boundary in boundary-value problems, DAN HENRY
319 Double Affine Hecke Algebras, I. CHEREDNIK
321 Surveys in Modern Mathematics, V. PRASOLOV & Y. ILYASHENKO (eds)
322 Recent perspectives in random matrix theory and number theory, F. MEZZADRI & N. C. SNAITH (eds)
323 Poisson geometry, deformation quantisation and group representations, S. GUTT, J. RAWNSLEY & D. STERNHEIMER (eds)
327 Surveys in Combinatorics 2005, B.S. WEBB (ed)

London Mathematical Society Lecture Note Series: 335

Lectures on the Combinatorics of Free Probability

Alexandru Nica
University of Waterloo, Ontario

Roland Speicher
Queen's University, Ontario

CAMBRIDGE
UNIVERSITY PRESS

CAMBRIDGE UNIVERSITY PRESS
Cambridge, New York, Melbourne, Madrid, Cape Town,
Singapore, São Paulo, Delhi, Mexico City

Cambridge University Press
The Edinburgh Building, Cambridge CB2 8RU, UK

Published in the United States of America by Cambridge University Press, New York

www.cambridge.org
Information on this title: www.cambridge.org/9780521858526

First published 2006

A catalogue record for this publication is available from the British Library

Library of Congress Cataloguing in Publication Data

ISBN 978-0-521-85852-6 Paperback

Dedicated to Anisoara and Betina

Contents

Introduction xiii

Part 1. Basic concepts 1

Lecture 1. Non-commutative probability spaces and distributions 3
- Non-commutative probability spaces 3
- $*$-distributions (case of normal elements) 7
- $*$-distributions (general case) 13
- Exercises 15

Lecture 2. A case study of non-normal distribution 19
- Description of the example 19
- Dyck paths 22
- The distribution of $a + a^*$ 26
- Using the Cauchy transform 30
- Exercises 33

Lecture 3. C^*-probability spaces 35
- Functional calculus in a C^*-algebra 35
- C^*-probability spaces 39
- $*$-distribution, norm and spectrum for a normal element 43
- Exercises 46

Lecture 4. Non-commutative joint distributions 49
- Joint distributions 49
- Joint $*$-distributions 53
- Joint $*$-distributions and isomorphism 55
- Exercises 59

Lecture 5. Definition and basic properties of free independence 63
- The classical situation: tensor independence 63
- Definition of free independence 64
- The example of a free product of groups 66
- Free independence and joint moments 69
- Some basic properties of free independence 71
- Are there other universal product constructions? 75

Exercises 78

Lecture 6. Free product of $*$-probability spaces 81
Free product of unital algebras 81
Free product of non-commutative probability spaces 84
Free product of $*$-probability spaces 86
Exercises 92

Lecture 7. Free product of C^*–probability spaces 95
The GNS representation 95
Free product of C^*-probability spaces 99
Example: semicircular systems and the full Fock space 102
Exercises 109

Part 2. Cumulants 113

Lecture 8. Motivation: free central limit theorem 115
Convergence in distribution 115
General central limit theorem 117
Classical central limit theorem 120
Free central limit theorem 121
The multi-dimensional case 125
Conclusion and outlook 131
Exercises 132

Lecture 9. Basic combinatorics I: non-crossing partitions 135
Non-crossing partitions of an ordered set 135
The lattice structure of $NC(n)$ 144
The factorization of intervals in NC 148
Exercises 153

Lecture 10. Basic combinatorics II: Möbius inversion 155
Convolution in the framework of a poset 155
Möbius inversion in a lattice 160
The Möbius function of NC 162
Multiplicative functions on NC 164
Functional equation for convolution with μ_n 168
Exercises 171

Lecture 11. Free cumulants: definition and basic properties 173
Multiplicative functionals on NC 173
Definition of free cumulants 175
Products as arguments 178
Free independence and free cumulants 182

Cumulants of random variables 185
Example: semicircular and circular elements 187
Even elements 188
Appendix: classical cumulants 190
Exercises 193

Lecture 12. Sums of free random variables 195
Free convolution 195
Analytic calculation of free convolution 200
Proof of the free central limit theorem via $\mathcal{R}$-transform 202
Free Poisson distribution 203
Compound free Poisson distribution 206
Exercises 208

Lecture 13. More about limit theorems and infinitely divisible distributions 211
Limit theorem for triangular arrays 211
Cumulants of operators on Fock space 214
Infinitely divisible distributions 215
Conditionally positive definite sequences 216
Characterization of infinitely divisible distributions 220
Exercises 221

Lecture 14. Products of free random variables 223
Multiplicative free convolution 223
Combinatorial description of free multiplication 225
Compression by a free projection 228
Convolution semigroups $(\mu^{\boxplus t})_{t\geq 1}$ 231
Compression by a free family of matrix units 233
Exercises 236

Lecture 15. R-diagonal elements 237
Motivation: cumulants of Haar unitary elements 237
Definition of R-diagonal elements 240
Special realizations of tracial R-diagonal elements 245
Product of two free even elements 249
The free anti-commutator of even elements 251
Powers of R-diagonal elements 253
Exercises 254

Part 3. Transforms and models 257

Lecture 16. The R-transform 259
The multi-variable R-transform 259

The functional equation for the R-transform 265
More about the one-dimensional case 269
Exercises 272

Lecture 17. The operation of boxed convolution 273
The definition of boxed convolution, and its motivation 273
Basic properties of boxed convolution 275
Radial series 277
The Möbius series and its use 280
Exercises 285

Lecture 18. More on the one-dimensional boxed convolution 287
Relation to multiplicative functions on NC 287
The S-transform 293
Exercises 300

Lecture 19. The free commutator 303
Free commutators of even elements 303
Free commutators in the general case 310
The cancelation phenomenon 314
Exercises 317

Lecture 20. R-cyclic matrices 321
Definition and examples of R-cyclic matrices 321
The convolution formula for an R-cyclic matrix 324
R-cyclic families of matrices 329
Applications of the convolution formula 331
Exercises 335

Lecture 21. The full Fock space model for the R-transform 339
Description of the Fock space model 339
An application: revisiting free compressions 346
Exercises 356

Lecture 22. Gaussian random matrices 359
Moments of Gaussian random variables 359
Random matrices in general 361
Selfadjoint Gaussian random matrices and genus expansion 363
Asymptotic free independence for several independent Gaussian random matrices 368
Asymptotic free independence between Gaussian random matrices and constant matrices 371

Lecture 23. Unitary random matrices 379
Haar unitary random matrices 379

The length function on permutations 381
Asymptotic freeness for Haar unitary random matrices 384
Asymptotic freeness between randomly rotated constant matrices 385
Embedding of non-crossing partitions into permutations 390
Exercises 393

Notes and comments 395

References 405

Index 411

Introduction

Free probability theory is a quite recent theory, bringing together many different fields of mathematics, for example operator algebras, random matrices, combinatorics, or representation theory of symmetric groups. So it has a lot to offer to various mathematical communities, and interest in free probability has steadily increased in recent years.

However, this diversity of the field also has the consequence that it is considered hard to access for a beginner. Most of the literature on free probability consists of a mixture of operator algebraic and probabilistic notions and arguments, interwoven with random matrices and combinatorics.

Whereas more advanced operator algebraic or probabilistic expertise might indeed be necessary for a deeper appreciation of special applications in the respective fields, the basic core of the theory, however, can be mostly freed from this and it is possible to give a fairly elementary introduction to the main notions, ideas and problems of free probability theory. The present lectures are intended to provide such an introduction.

Our main emphasis will be on the combinatorial side of free probability. Even when stripped from analytical structure, the main features of free independence are still present; moreover, even on this more combinatorial level it is important to organize all relevant information about the considered variables in the right way. Anyone who has tried to perform computations of joint distributions for non-commuting variables will probably agree that they tend to be horribly messy if done in a naive way. One of the main goals of the book is to show how such computations can be vastly simplified by appropriately relying on a suitable combinatorial structure – the lattices of non-crossing partitions. The combinatorial development starts from the standard theory of Möbius inversion on non-crossing partitions, but has its own specific flavor – one arrives to a theory of free or non-crossing cumulants or, in an alternative approach, one talks about R-transforms for non-commutative random variables.

While writing this book, there were two kinds of readers that we had primarily in mind:

(a) a reader with background in operator algebras or probability who wants to see the more advanced "tools of the trade" on the combinatorial side of free probability;
(b) a reader with background from algebraic combinatorics who wants to get acquainted with a field (and a possible source of interesting problems) where non-trivial combinatorial tools are used.

We wrote our lectures by trying to accommodate readers from both these categories. The result is a fairly elementary exposition, which should be accessible to a beginning graduate student or even to a strong senior undergraduate student.

Free probability also has applications outside of mathematics, in particular in electrical engineering. Our exposition should also be useful for readers with engineering background, who have seen the use of R- or S-transform techniques in applications, for example in wireless communications, and who want to learn more about the underlying theory.

We emphasize that the presentation style used throughout the book is a detailed one, making the material largely self-contained, and only rarely requiring that other textbooks or research papers are consulted. The basic units of this book are called "lectures." They were written following the idea that the material contained in one of them should be suitable for being presented in one class of a first-year graduate course. (We have in mind a class of 90 minutes, where the instructor presents the essential points of the lecture, and leaves a number of things for individual study.)

While the emphasis is on combinatorial aspects, we still felt that we must give an introduction of how the general framework of free probability comes about. Also, we felt that the flavor of the theory will be better conveyed if we show, with moderation and within a self-contained exposition, how analytical arguments can be interwoven with the combinatorial ones. However, it should be understood that in the analytical respects, this book is only an appetizer and an invitation to further reading. In particular, the analytical framework used for illustrations is exclusively that of a C^*-probability space. The reader should be aware that some of the most significant applications of free probability to operator algebras take place in the more elaborate framework of W^*-probability spaces; but going to W^*-structures (or in other words, to von Neumann algebra theory) did not seem possible within the detailed, self-contained style of the book, and within the given page limits.

A consequence of the frugality of the analytic aspects covered by the book is that we do not discuss free entropy and free Fisher information, and how free cumulants can be used in some cases to perform free information calculations. Free entropy is currently one of the main directions of development in free probability; for an overview of the topic see the recent survey by Voiculescu [**85**].

Coming to things that are not covered by the book we must also say, with regret, that we only consider free independence over the complex field. The combinatorial ideas of free probability have a far-reaching extension to the situation when free independence is considered over an algebra $\mathcal{B}$ (instead of just $\mathbb{C}$) – the reader interested in this direction is referred to the memoir [**73**].

References to the literature are not made in the body of the lectures, but are collected in the "Notes and comments" section at the end of the book. The literature on free probability is growing at an explosive rate, and, with due apologies, we felt it is beyond our limits to even try to provide an exhaustive bibliography. We have followed the line of only citing the research work which is presented in the lectures, or is very directly connected to it. For a more complete image of work in this field, the reader can consult the survey papers indicated at the beginning of the "Notes and comments" section.

So, to summarize, from one point of view this is a research monograph, presenting the current state of the combinatorial facet of free probability. At the same time it is an introduction to the field – one which is, we hope, friendly and self-contained. Finally, the book is written with the specific purpose of being used for teaching a course. We hope this will be a contribution towards making free probability appear more often as a topic for a graduate course, and we look forward to hearing from other people how following these lectures has worked for them.

Finally we would like to mention that the idea of writing this book came from a sequence of lectures which we gave at the Henri Poincaré Institute in Paris, during a special semester on free probability and operator spaces hosted by the institute in Fall 1999. Time has flown quickly since then, but we hope it is not too late to thank the Poincaré Institute, and particularly the organizers of that special semester – Philippe Biane, Gilles Pisier, and Dan Voiculescu – for the great environment they offered us, and for the opportunity of getting started on this project.

Part 1

Basic concepts

LECTURE 1

Non-commutative probability spaces and distributions

Since we are interested in the combinatorial aspects of free probability, we will focus on a framework which is stripped of its analytical structure (i.e. where we ignore the metric or topological structure of the spaces involved). The reason for the existence of this monograph is that even so (without analytical structure), the phenomenon of free independence is rich enough to be worth studying. The interesting combinatorial features of this phenomenon come from the fact that *we will allow the algebras of random variables to be non-commutative.* This certainly means that we have to consider a generalized concept of "random variable" (since in the usual meaning of the concept, where a random variable is a function on a probability space, the algebras of random variables would have to be commutative).

Non-commutative probability spaces

DEFINITION 1.1. (1) A **non-commutative probability space** $(\mathcal{A}, \varphi)$ consists of a unital algebra $\mathcal{A}$ over $\mathbb{C}$ and a unital linear functional

$$\varphi : \mathcal{A} \to \mathbb{C}; \qquad \varphi(1_{\mathcal{A}}) = 1.$$

The elements $a \in \mathcal{A}$ are called **non-commutative random variables** in $(\mathcal{A}, \varphi)$. Usually, we will skip the adjective "non-commutative" and just talk about "random variables $a \in \mathcal{A}$."

An additional property which we will sometimes impose on the linear functional φ is that it is a **trace**, i.e. it has the property that

$$\varphi(ab) = \varphi(ba), \quad \forall\, a, b \in \mathcal{A}.$$

When this happens, we say that the non-commutative probability space $(\mathcal{A}, \varphi)$ is **tracial.**

(2) In the framework of part (1) of the definition, suppose that $\mathcal{A}$ is a $*$-algebra, i.e. that $\mathcal{A}$ is also endowed with an antilinear $*$-operation $\mathcal{A} \ni a \mapsto a^* \in \mathcal{A}$, such that $(a^*)^* = a$ and $(ab)^* = b^*a^*$ for all $a, b \in \mathcal{A}$. If we have that

$$\varphi(a^*a) \geq 0, \quad \forall\, a \in \mathcal{A},$$

then we say that the functional φ is **positive** and we will call $(\mathcal{A}, \varphi)$ a **$*$-probability space**.

(3) In the framework of a $*$-probability space we can talk about:

• **selfadjoint** random variables, i.e. elements $a \in \mathcal{A}$ with the property that $a = a^*$;

• **unitary** random variables, i.e. elements $u \in \mathcal{A}$ with the property that $u^*u = uu^* = 1$;

• **normal** random variables, i.e. elements $a \in \mathcal{A}$ with the property that $a^*a = aa^*$.

In these lectures we will be mostly interested in $*$-probability spaces, since this is the framework which provides us with a multitude of exciting examples. However, plain non-commutative probability spaces are also useful, because sometimes we encounter arguments relying solely on the linear and multiplicative structure of the algebra involved – these arguments are more easily understood when the $*$-operation is ignored (even if it happened that the algebra had a $*$-operation on it).

REMARKS 1.2. Let $(\mathcal{A}, \varphi)$ be a $*$-probability space.

(1) The functional φ is selfadjoint, i.e. it has the property that

$$\varphi(a^*) = \overline{\varphi(a)}, \quad \forall a \in \mathcal{A}.$$

Indeed, since every $a \in \mathcal{A}$ can be written uniquely in the form $a = x+iy$ where $x, y \in \mathcal{A}$ are selfadjoint, the latter equation is immediately seen to be equivalent to the fact that $\varphi(x) \in \mathbb{R}$ for every selfadjoint element $x \in \mathcal{A}$. This in turn is implied by the positivity of φ and the fact that every selfadjoint element $x \in \mathcal{A}$ can be written in the form $x = a^*a - b^*b$ for some $a, b \in \mathcal{A}$ (take e.g. $a = (x + 1_{\mathcal{A}})/2$, $b = (x - 1_{\mathcal{A}})/2$).

(2) Another consequence of the positivity of φ is that we have:

$$|\varphi(b^*a)|^2 \leq \varphi(a^*a)\varphi(b^*b), \quad \forall\, a, b \in \mathcal{A}. \tag{1.1}$$

The inequality (1.1) is commonly called the **Cauchy–Schwarz inequality** for the functional φ. It is proved in exactly the same way as the usual Cauchy–Schwarz inequality (see Exercise 1.21 at the end of the lecture).

(3) If an element $a \in \mathcal{A}$ is such that $\varphi(a^*a) = 0$, then the Cauchy–Schwarz inequality (1.1) implies that $\varphi(ba) = 0$ for all $b \in \mathcal{A}$ (hence a is in a certain sense a degenerate element for the functional φ). We will use the term "faithful" for the situation when no such degenerate elements exist, except for $a = 0$. That is, we make the following definition.

DEFINITION 1.3. Let $(\mathcal{A}, \varphi)$ be a $*$-probability space. If we have the implication:

$$a \in \mathcal{A},\ \varphi(a^*a) = 0 \ \Rightarrow\ a = 0,$$

then we say that the functional φ is **faithful**.

EXAMPLES 1.4. (1) Let $(\Omega, \mathcal{Q}, P)$ be a probability space in the classical sense, i.e. Ω is a set, $\mathcal{Q}$ is a σ-field of measurable subsets of Ω and $P : \mathcal{Q} \to [0,1]$ is a probability measure. Let $\mathcal{A} = L^\infty(\Omega, P)$, and let φ be defined by

$$\varphi(a) = \int_\Omega a(\omega)\, dP(\omega), \quad a \in \mathcal{A}.$$

Then $(\mathcal{A}, \varphi)$ is a $*$-probability space (the $*$-operation on $\mathcal{A}$ is the operation of complex-conjugating a complex-valued function). The random variables appearing in this example are thus genuine random variables in the sense of "usual" probability theory.

The reader could object at this point that the example presented in the preceding paragraph only deals with genuine random variables that are bounded, and thus misses for instance the most important random variables from usual probability – those having a Gaussian distribution. We can overcome this problem by replacing the algebra $L^\infty(\Omega, P)$ with:

$$L^{\infty-}(\Omega, P) := \bigcap_{1 \leq p < \infty} L^p(\Omega, P).$$

That is, we can make $\mathcal{A}$ become the algebra of genuine random variables which have finite moments of all orders. (The fact that $L^{\infty-}(\Omega, P)$ is indeed closed under multiplication follows by an immediate application of the Cauchy–Schwarz inequality in $L^2(\Omega, P)$ – cf. Exercise 1.22 at the end of the lecture.) In this enlarged version, our algebra of random variables will then contain the Gaussian ones.

Of course, one could also point out that in classical probability there are important cases of random variables which do not have moments of all orders. These ones, unfortunately, are beyond the scope of the present set of lectures – we cannot catch them in the framework of Definition 1.1.

(2) Let d be a positive integer, let $M_d(\mathbb{C})$ be the algebra of $d \times d$ complex matrices with usual matrix multiplication, and let $\mathrm{tr} : M_d(\mathbb{C}) \to \mathbb{C}$ be the normalized trace,

$$\mathrm{tr}(a) = \frac{1}{d} \cdot \sum_{i=1}^{d} \alpha_{ii} \qquad \text{for} \qquad a = (\alpha_{ij})_{i,j=1}^{d} \in M_d(\mathbb{C}). \tag{1.2}$$

Then $(M_d(\mathbb{C}), \mathrm{tr})$ is a $*$-probability space (where the $*$-operation is given by taking both the transpose of the matrix and the complex conjugate of the entries).

(3) The above examples (1) and (2) can be "put together" into one example where the algebra consists of all the $d \times d$ matrices over $L^{\infty-}(\Omega, P)$:

$$\mathcal{A} = M_d(\, L^{\infty-}(\Omega, P)\,),$$

and the functional φ on it is

$$\varphi(a) := \int \mathrm{tr}(a(\omega))\, dP(\omega), \quad a \in \mathcal{A}.$$

The non-commutative random variables obtained here are thus **random matrices** over $(\Omega, \mathcal{Q}, P)$. (Observe that this example is obtained by starting with the space in Example 1.4.1 and by performing the $d \times d$ matrix construction described in Exercise 1.23.) We will elaborate more on random matrix examples later (cf. Lectures 22 and 23).

(4) Let G be a group, and let $\mathbb{C}G$ denote its **group algebra**. That is, $\mathbb{C}G$ is a complex vector space having a basis indexed by the elements of G, and where the operations of multiplication and $*$-operation are defined in the natural way:

$$\mathbb{C}G := \Big\{ \sum_{g \in G} \alpha_g g \mid \alpha_g \in \mathbb{C}, \text{ only finitely many } \alpha_g \neq 0 \Big\},$$

with

$$\Big(\sum \alpha_g g \Big) \cdot \Big(\sum \beta_h h \Big) := \sum_{g,h} \alpha_g \beta_h (gh) = \sum_{k \in G} \Big(\sum_{g,h:\, gh=k} \alpha_g \beta_h \Big) k,$$

and

$$\Big(\sum \alpha_g g \Big)^* := \sum \bar{\alpha}_g g^{-1}.$$

Let e be the unit element of G. The functional $\tau_G : \mathbb{C}G \to \mathbb{C}$ defined by the formula

$$\tau_G \Big(\sum \alpha_g g \Big) := \alpha_e$$

is called the canonical trace on $\mathbb{C}G$. Then $(\mathbb{C}G, \tau_G)$ is a $*$-probability space. It is easily verified that τ_G is indeed a trace (in the sense of Definition 1.1.1) and is faithful (in the sense of Definition 1.3).

(5) Let $\mathcal{H}$ be a Hilbert space and let $B(\mathcal{H})$ be the algebra of all bounded linear operators on $\mathcal{H}$. This is a $*$-algebra, where the adjoint a^* of an operator $a \in B(\mathcal{H})$ is uniquely determined by the fact that

$$\langle a\xi, \eta \rangle = \langle \xi, a^* \eta \rangle, \quad \forall\, \xi, \eta \in \mathcal{H}.$$

Suppose that $\mathcal{A}$ is a unital $*$-subalgebra of $B(\mathcal{H})$ and that $\xi_o \in \mathcal{H}$ is a vector of norm one ($||\xi_o|| := \langle \xi_o, \xi_o \rangle^{1/2} = 1$). Then we get an example of $*$-probability space $(\mathcal{A}, \varphi)$, where $\varphi : \mathcal{A} \to \mathbb{C}$ is defined by:

$$\varphi(a) := \langle a\xi_o, \xi_o \rangle, \quad a \in \mathcal{A}. \tag{1.3}$$

A linear functional as defined in (1.3) is usually called a **vector-state** (on the algebra of operators $\mathcal{A}$).

EXERCISE 1.5. (1) Verify that in each of the examples described in 1.4, the functional considered as part of the definition of the $*$-probability space is indeed positive.

(2) Show that in Examples 1.4.1–1.4.4, the functional considered as part of the definition of the $*$-probability space is a faithful trace.

DEFINITION 1.6. (1) A **morphism** between two $*$-probability spaces $(\mathcal{A}, \varphi)$ and $(\mathcal{B}, \psi)$ is a unital $*$-algebra homomorphism $\Phi : \mathcal{A} \to \mathcal{B}$ with the property that $\psi \circ \Phi = \varphi$.

(2) In the case when $(\mathcal{B}, \psi)$ is a $*$-probability space of the special kind discussed in Example 1.4.5, we will refer to a morphism Φ from $(\mathcal{A}, \varphi)$ to $(\mathcal{B}, \psi)$ as a **representation** of $(\mathcal{A}, \varphi)$. So, to be precise, giving a representation of $(\mathcal{A}, \varphi)$ amounts to giving a triple $(\mathcal{H}, \Phi, \xi_o)$ where $\mathcal{H}$ is a Hilbert space, $\Phi : \mathcal{A} \to B(\mathcal{H})$ is a unital $*$-homomorphism, and $\xi_o \in \mathcal{H}$ is a vector of norm one, such that $\varphi(a) = \langle \Phi(a)\xi_o, \xi_o \rangle$ for all $a \in \mathcal{A}$.

REMARK 1.7. The $*$-probability spaces appearing in Examples 1.4.1, 1.4.2 and 1.4.4 have natural representations, on Hilbert spaces related to how the algebras of random variables were constructed – see Exercise 1.25 at the end of the lecture.

$*$-distributions (case of normal elements)

A fundamental concept in the statistical study of random variables is that of distribution of a random variable. In the framework of a $*$-probability space $(\mathcal{A}, \varphi)$, the appropriate concept to consider is the $*$-*distribution* of an element $a \in \mathcal{A}$. Roughly speaking, the $*$-distribution of a has to be some "standardized" way of reading the values of φ on the unital $*$-subalgebra generated by a.

We start the discussion of $*$-distributions with the simpler case when $a \in \mathcal{A}$ is normal (i.e. is such that $a^*a = aa^*$). In this case the unital $*$-algebra generated by a is

$$\mathcal{A} := \text{span}\{a^k(a^*)^l \mid k, l \geq 0\}; \tag{1.4}$$

the job of the $*$-distribution of a must thus be to keep track of the values $\varphi(\, a^k(a^*)^l\,)$, where k and l run in $\mathbb{N} \cup \{0\}$. The kind of object which does this job and which we prefer to have whenever possible is a compactly supported probability measure on $\mathbb{C}$.

DEFINITION 1.8. Let $(\mathcal{A}, \varphi)$ be a $*$-probability space and let a be a normal element of $\mathcal{A}$. If there exists a compactly supported probability measure μ on $\mathbb{C}$ such that

$$\int z^k \bar{z}^l \, d\mu(z) = \varphi(a^k (a^*)^l), \qquad \text{for every } k, l \in \mathbb{N}, \tag{1.5}$$

then this μ is uniquely determined and we will call the probability measure μ the **$*$-distribution** of a.

REMARKS 1.9. (1) The fact that a compactly supported probability measure μ on $\mathbb{C}$ is uniquely determined by how it integrates functions of the form $z \mapsto z^k \bar{z}^l$ with $k, l \in \mathbb{N}$ is an immediate consequence of the Stone–Weierstrass theorem. Or more precisely: due to Stone–Weierstrass, μ is determined as a linear functional on the space $C(K)$ of complex-valued continuous functions on K, where K is the support of μ; it is then well known that this in turn determines μ uniquely.

(2) It is not said that every normal element in a $*$-probability space has to have a $*$-distribution in the sense defined above. But this turns out to be true in a good number of important examples. Actually, this is *always* true when we look at $*$-probability spaces which have a representation on a Hilbert space, in the sense of the above Definition 1.6 (see Corollary 3.14 in Lecture 3); and civilized examples do have representations on Hilbert spaces – see Lecture 7.

REMARK 1.10. *(The case of a selfadjoint element)*
Let $(\mathcal{A}, \varphi)$ be a $*$-probability space, and let a be a selfadjoint element of $\mathcal{A}$ (that is, we have $a = a^*$, which implies in particular that a is normal). Suppose that a has $*$-distribution μ, in the sense of Definition 1.8. Then μ is supported in $\mathbb{R}$. Indeed, we have

$$\begin{aligned} \int_{\mathbb{C}} | \, z - \bar{z} \, |^2 \, d\mu(z) &= \int_{\mathbb{C}} (z - \bar{z})(\bar{z} - z) \, d\mu(z) \\ &= \int_{\mathbb{C}} 2z\bar{z} - z^2 - \bar{z}^2 \, d\mu(z) \\ &= 2\varphi(aa^*) - \varphi(a^2) - \varphi(\,(a^*)^2\,) = 0. \end{aligned}$$

Since $z \mapsto | \, z - \bar{z} \, |^2$ is a continuous non-negative function, we must have that $z - \bar{z}$ vanishes on the support $\text{supp}(\mu)$ of our measure, and hence:

$$\text{supp}(\mu) \subset \{z \in \mathbb{C} \mid z = \bar{z}\} = \mathbb{R}.$$

So in this case μ is really a measure on $\mathbb{R}$, and Equation (1.5) is better written in this case as

$$\int t^p \, d\mu(t) = \varphi(a^p), \qquad \forall\, p \in \mathbb{N}. \tag{1.6}$$

Conversely, suppose that we have a compactly supported measure μ on $\mathbb{R}$ such that (1.6) holds. Then clearly μ is the $*$-distribution of a in the sense of Definition 1.8 (because $\int z^k \bar{z}^l \, d\mu(z)$ becomes $\int t^{k+l} \, d\mu(t)$, while $\varphi(\, a^k (a^*)^l \,)$ becomes $\varphi(a^{k+l})$).

The conclusion of this discussion is that for a selfadjoint element $a \in \mathcal{A}$ it would be more appropriate to talk about the **distribution** of a (rather than talking about its $*$-distribution); this is defined as a compactly supported measure on $\mathbb{R}$ such that (1.6) holds. But there is actually no harm in treating a as a general normal element, and in looking for its $*$-distribution, since in the end we arrive at the same result.

EXAMPLES 1.11. (1) Consider the framework of Example 1.4.1, where the algebra of random variables is $L^\infty(\Omega, P)$. Let a be an element in $\mathcal{A}$; in other words, a is a bounded measurable function, $a : \Omega \to \mathbb{C}$. Let us consider the probability measure ν on $\mathbb{C}$ which is called "the distribution of a" in usual probability; this is defined by

$$\nu(E) \;=\; P(\,\{\omega \in \Omega : a(\omega) \in E\}\,), \quad E \subset \mathbb{C} \text{ Borel set.} \tag{1.7}$$

Note that ν is compactly supported. More precisely, if we choose a positive r such that $|a(\omega)| \leq r$, $\forall\, \omega \in \Omega$, then it is clear that ν is supported in the closed disc centered at 0 and of radius r.

Now, a is a normal element of $\mathcal{A}$ (all the elements of $\mathcal{A}$ are normal, since $\mathcal{A}$ is commutative). So it makes sense to place a in the framework of Definition 1.8. We will show that the above measure ν is exactly the $*$-distribution of a in this framework.

Indeed, Equation (1.7) can be read as

$$\int_{\mathbb{C}} f(z) \, d\nu(z) \;=\; \int_{\Omega} f(\,a(\omega)\,) \, dP(\omega), \tag{1.8}$$

where f is the characteristic function of the set E. By going through the usual process of taking linear combinations of characteristic functions, and then doing approximations of a bounded measurable function by step functions, we see that Equation (1.8) actually holds for every bounded measurable function $f : \mathbb{C} \to \mathbb{C}$. (The details of this are left to the reader.) Finally, let k, l be arbitrary non-negative integers, and let $r > 0$ be such that $|a(\omega)| \leq r$ for every $\omega \in \Omega$. Consider a bounded measurable function $f : \mathbb{C} \to \mathbb{C}$ such that $f(z) = z^k \bar{z}^l$ for every $z \in \mathbb{C}$ having $|z| \le r$. Since ν is supported in the closed disc of radius r centered at 0, it follows that

$$\int_{\mathbb{C}} f(z)\, d\nu(z) = \int_{\mathbb{C}} z^k \bar{z}^l \, d\nu(z),$$

and, consequently, that

$$\int_{\Omega} f(a(\omega))\, dP(\omega) = \int_{\Omega} a(\omega)^k \overline{a(\omega)}^{\,l} \, dP(\omega) = \varphi(a^k (a^*)^l).$$

Thus for this particular choice of f, Equation (1.8) gives us that

$$\int_{\mathbb{C}} z^k \bar{z}^l \, d\nu(z) = \varphi(a^k (a^*)^l),$$

and this is precisely (1.5), implying that ν is the $*$-distribution of a in the sense of Definition 1.8.

(2) Consider the framework of Example 1.4.2, and let $a \in M_d(\mathbb{C})$ be a normal matrix. Let $\lambda_1, \ldots, \lambda_d$ be the eigenvalues of a, counted with multiplicities. By diagonalizing a we find that

$$\mathrm{tr}(a^k (a^*)^l) = \frac{1}{d} \sum_{i=1}^{d} \lambda_i^k \bar{\lambda}_i^l, \qquad k, l \in \mathbb{N}.$$

The latter quantity can obviously be written as $\int z^k \bar{z}^l \, d\mu(z)$, where

$$\mu := \frac{1}{d} \sum_{i=1}^{d} \delta_{\lambda_i} \tag{1.9}$$

(δ_λ stands here for the Dirac mass at $\lambda \in \mathbb{C}$). Thus it follows that a has a $*$-distribution μ, which is described by Equation (1.9). Usually this μ is called the **eigenvalue distribution** of the matrix a.

One can consider the question of how to generalize the above fact to the framework of random matrices (as in Example 1.4.3). It can be shown that the formula which appears in place of (1.9) in this case is

$$\mu := \frac{1}{d} \sum_{i=1}^{d} \int_{\Omega} \delta_{\lambda_i(\omega)} \, dP(\omega), \tag{1.10}$$

where $a = a^* \in M_d(L^{\infty-}(\Omega, P))$, and where $\lambda_1(\omega) \le \cdots \le \lambda_d(\omega)$ are the eigenvalues of $a(\omega)$, $\omega \in \Omega$. (Strictly speaking, Equation (1.10) requires an extension of the framework used in Definition 1.8, since the resulting averaged eigenvalue distribution μ will generally not have compact support. See Lecture 22 for more details about this.)

Our next example will be in connection to a special kind of element in a $*$-probability space, called a Haar unitary.

DEFINITION 1.12. Let $(\mathcal{A}, \varphi)$ be a $*$-probability space.

(1) An element $u \in \mathcal{A}$ is said to be a **Haar unitary** if it is a unitary (i.e. if $uu^* = u^*u = 1$) and if

$$\varphi(u^k) = 0, \quad \forall\, k \in \mathbb{Z} \setminus \{0\}. \tag{1.11}$$

(2) Let p be a positive integer. An element $u \in \mathcal{A}$ is said to be a ***p*-Haar unitary** if it is a unitary, if $u^p = 1$, and if

$$\varphi(u^k) = 0, \qquad \text{for all } k \in \mathbb{Z} \text{ such that } p \text{ does not divide } k. \tag{1.12}$$

REMARKS 1.13. (1) The name "Haar unitary" comes from the fact that if u is a Haar unitary in a $*$-probability space, then the normalized Lebesgue measure (also called "Haar measure") on the circle serves as $*$-distribution for u. Indeed, for every $k, l \in \mathbb{N} \cup \{0\}$ we have

$$\varphi(u^k(u^*)^l) = \varphi(u^{k-l}) = \begin{cases} 0 & \text{if } k \neq l \\ 1 & \text{if } k = l, \end{cases}$$

and the computation of the integral

$$\int_{\mathbb{T}} z^k \bar{z}^l \, dz = \int_0^{2\pi} e^{i(k-l)t} \, \frac{dt}{2\pi}$$

(where $\mathbb{T} = \{z \in \mathbb{C} \mid |z| = 1\}$ and dz is the normalized Haar measure on $\mathbb{T}$) gives exactly the same thing.

(2) Haar unitaries appear naturally in the framework of Example 1.4.4. Indeed, if g is any element of infinite order in the group G, then one can verify immediately that g viewed as an element of the $*$-probability space $(\mathbb{C}G, \tau_G)$ is a Haar unitary.

(3) The p-Haar unitaries also appear naturally in the framework of Example 1.4.4 – an element of order p in G becomes a p-Haar unitary when viewed in $(\mathbb{C}G, \tau_G)$. It is immediately verified that a p-Haar unitary has $*$-distribution

$$\mu = \frac{1}{p} \sum_{j=1}^{p} \delta_{\lambda_j}, \tag{1.13}$$

where $\lambda_1, \ldots, \lambda_p \in \mathbb{C}$ are the roots of order p of unity.

EXAMPLE 1.14. Let $(\mathcal{A}, \varphi)$ be a $*$-probability space, and let $u \in \mathcal{A}$ be a Haar unitary. We consider the selfadjoint element $u + u^* \in \mathcal{A}$, and we would like to answer the following questions.

(1) Does $u + u^*$ have a $*$-distribution?

(2) Suppose that $u + u^*$ does have a $*$-distribution μ. Then, as observed in Remark 1.10, μ is a probability measure on $\mathbb{R}$, and satisfies

Equation (1.6). Do we have some "nice" formula for the moments $\int t^k \, d\mu(t)$ which appear in Equation (1.6)?

Let us note that the second question is actually very easy. Indeed, this question really asks for the values $\varphi(\,(u+u^*)^k\,)$, $k \in \mathbb{N}$, which are easily derived from Equation (1.11). We argue like this: due to the fact that u and u^* commute, we can expand

$$(u+u^*)^k = \sum_{j=0}^{k} \binom{k}{j} u^j (u^*)^{k-j}.$$

Then we use the fact that $u^* = u^{-1}$ and we apply φ to both sides of the latter equation, to obtain:

$$\varphi((u+u^*)^k) = \sum_{j=0}^{k} \binom{k}{j} \varphi(u^{2j-k}).$$

It only remains to take (1.11) into account, in order to get that:

$$\int t^k \, d\mu(t) = \begin{cases} 0 & \text{if } k \text{ is odd} \\ \binom{k}{k/2} & \text{if } k \text{ is even.} \end{cases} \tag{1.14}$$

This is the answer to the second question.

Now we could treat the first question as the problem of finding a compactly supported probability measure μ on $\mathbb{R}$ which has moments as described by Equation (1.14). This is feasible, but somewhat cumbersome. It is more convenient to forget for the moment the calculation done in the preceding paragraph, and attack question (1) directly, by only using the fact that we know the $*$-distribution of u. (The distribution of $u+u^*$ has to be obtainable from the $*$-distribution of u!) We go like this:

$$\begin{aligned}
\varphi((u+u^*)^k) &= \sum_{j=0}^{k} \binom{k}{j} \varphi(u^j (u^*)^{k-j}) \\
&= \sum_{j=0}^{k} \binom{k}{j} \int_{\mathbb{T}} z^j \bar{z}^{k-j} \, dz \qquad \text{(by Remark 1.13.1)} \\
&= \int_{\mathbb{T}} (z+\bar{z})^k \, dz \\
&= \frac{1}{2\pi} \int_{-\pi}^{\pi} (e^{it}+e^{-it})^k \, dt \\
&= \frac{1}{2\pi} \int_{-\pi}^{\pi} (2\cos t)^k \, dt
\end{aligned}$$

$$= \frac{1}{2\pi}\Big(\int_{-\pi}^{0} (2\cos t)^k \, dt \; + \; \int_{0}^{\pi} (2\cos t)^k \, dt\Big)$$
$$= \frac{1}{\pi}\int_0^{\pi} (2\cos t)^k \, dt.$$

For the last integral obtained above, our goal is not to compute it explicitly (this would only yield a more complicated derivation of Equation (1.14)), but to rewrite it in the form $\int t^k \rho(t)\, dt$, where ρ is an appropriate density. This is achieved by the change of variable

$$2\cos t = r, \qquad dt = d(\arccos(r/2)) = -dr/\sqrt{4-r^2},$$

which gives us that

$$\frac{1}{\pi}\int_0^{\pi} (2\cos t)^k \, dt = \frac{1}{\pi}\int_{-2}^{2} r^k \frac{dr}{\sqrt{4-r^2}}.$$

In this way we obtain that

$$\varphi\big((u+u^*)^k\big) = \int_{\mathbb{R}} t^k \rho(t)\, dt, \qquad k \geq 0, \tag{1.15}$$

where $\rho(t)$ is the so-called "arcsine density on $[-2,2]$":

$$\rho(t) = \begin{cases} \frac{1}{\pi\sqrt{4-t^2}} & \text{if } |t| < 2 \\ 0 & \text{if } |t| \geq 2. \end{cases} \tag{1.16}$$

So, as a solution to the first question of this example, we find that the distribution of $u + u^*$ is the arcsine law.

$*$-distributions (general case)

Let us now consider the concept of $*$-distribution for an arbitrary (not necessarily normal) element a in a $*$-probability space $(\mathcal{A}, \varphi)$. The unital $*$-subalgebra of $\mathcal{A}$ generated by a is

$$\mathcal{A}_o = \text{span}\{a^{\varepsilon(1)} \cdots a^{\varepsilon(k)} \mid k \geq 0,\ \varepsilon(1), \ldots, \varepsilon(k) \in \{1, *\}\ \}, \tag{1.17}$$

i.e. it is the linear span of all the "words" that one can make by using the "letters" a and a^*. The values of φ on such words are usually referred to under the name of $*$-*moments.*

DEFINITION 1.15. Let a be a random variable in a $*$-probability space $(\mathcal{A}, \varphi)$. An expression of the form

$$\varphi(a^{\varepsilon(1)} \cdots a^{\varepsilon(k)}), \quad \text{with } k \geq 0 \text{ and } \varepsilon(1), \ldots, \varepsilon(k) \in \{1, *\}, \tag{1.18}$$

is called a **$*$-moment** of a.

So in this case the $*$-distribution of a must be a device which keeps track of its $*$-moments. Unlike in the case of normal elements, there is no handy analytic structure which does this. As a consequence, we will have to define the $*$-distribution of a as a purely algebraic object.

NOTATION 1.16. We denote by $\mathbb{C}\langle X, X^*\rangle$ the unital algebra which is freely generated by two non-commuting indeterminates X and X^*. More concretely, $\mathbb{C}\langle X, X^*\rangle$ can be described as follows. The monomials of the form $X^{\varepsilon(1)} \cdots X^{\varepsilon(k)}$, where $k \geq 0$ and $\varepsilon(1), \ldots, \varepsilon(k) \in \{1, *\}$, give a linear basis for $\mathbb{C}\langle X, X^*\rangle$, and the multiplication of two such monomials is done by juxtaposition. $\mathbb{C}\langle X, X^*\rangle$ has a natural $*$-operation, determined by the requirement that the $*$-operation applied to X gives X^*.

DEFINITION 1.17. Let a be a random variable in a $*$-probability space $(\mathcal{A}, \varphi)$. The **$*$-distribution** of a is the linear functional

$$\mu : \mathbb{C}\langle X, X^*\rangle \to \mathbb{C}$$

determined by the fact that:

$$\mu(X^{\varepsilon(1)} \cdots X^{\varepsilon(k)}) = \varphi(a^{\varepsilon(1)} \cdots a^{\varepsilon(k)}), \tag{1.19}$$

for every $k \geq 0$ and all $\varepsilon(1), \ldots, \varepsilon(k) \in \{1, *\}$.

REMARKS 1.18. (1) The advantage of the formal definition made above is that even when we consider random variables which live in different $*$-probability spaces, the corresponding $*$-distributions are all defined on the same space $\mathbb{C}\langle X, X^*\rangle$ (and hence can be more easily compared with each other).

(2) Definition 1.17 will apply to a even if a happens to be normal. In this case the functional μ of (1.19) could actually be factored through the more familiar commutative algebra $\mathbb{C}[X, X^*]$ of polynomials in two commuting indeterminates. But this would not bring much benefit to the subsequent presentation. (In fact there are places where we will *have to* consider all the possible words in a and a^* despite knowing a to be normal – see e.g. the computations shown in the section on Haar unitaries of Lecture 15.) So it will be easier to consistently use $\mathbb{C}\langle X, X^*\rangle$ throughout these notes.

(3) If a is a normal element of a $*$-probability space, then the $*$-distribution of a is now defined twice, in Definition 1.8 and in Definition 1.17. When there is a risk of ambiguity, we will distinguish between the two versions of the definition by calling them *"$*$-distribution in analytic sense"* and respectively *"$*$-distribution in algebraic sense."*

DEFINITION 1.19. Let $(\mathcal{A}, \varphi)$ be a $*$-probability space, and let a be a selfadjoint element of $\mathcal{A}$. In this case, the $*$-moments from (1.18) are just the numbers $\varphi(a^k)$, $k \geq 0$, and they are simply called **moments** of a. Following the standard terminology from classical probability, the first moment $\varphi(a)$ is also called the **mean** of a, while the quantity

$$\mathrm{Var}(a) := \varphi(a^2) - \varphi(a)^2$$

is called the **variance** of a.

REMARK 1.20. We would like next to introduce an important example of $*$-distribution, which is in some sense a non-normal counterpart of the Haar unitary; and moreover, we would like to show how the analog of the questions treated in Example 1.14 can be pursued for this non-normal example. The discussion will be longer than that for the Haar unitary (precisely because we do not have an analytic $*$-distribution to start from), and will be the object of the next lecture.

Exercises

EXERCISE 1.21. (1) Let $(\mathcal{A}, \varphi)$ be a $*$-probability space, and let a, b be elements of $\mathcal{A}$. By examining the quadratic function

$$t \mapsto \varphi(\,(a - tb)^*(a - tb)\,), \quad t \in \mathbb{R},$$

prove that

$$\Big(\mathrm{Re}\varphi(b^*a)\Big)^2 \leq \varphi(a^*a)\varphi(b^*b).$$

(2) Prove the Cauchy–Schwarz inequality which was stated in Remark 1.2.2.

EXERCISE 1.22. Let $(\Omega, \mathcal{Q}, P)$ be a probability space, and consider the space of functions

$$L^{\infty-}(\Omega, P) := \bigcap_{1 \leq p < \infty} L^p(\Omega, P)$$

(as in Example 1.4.1).

(1) Prove that the spaces intersected on the right-hand side of the above equation form a decreasing family (that is, $L^p(\Omega, P) \supset L^q(\Omega, P)$ for $p \leq q$).

(2) Observe that $L^{\infty-}(\Omega, P)$ could also be defined as $\cap_p L^p(\Omega, P)$ with p running in $\mathbb{N} \setminus \{0\}$. Or equivalently, observe that $L^{\infty-}(\Omega, P)$ could be defined as the algebra of complex random variables on Ω which have finite moments of all orders.

(3) Prove that $L^{\infty-}(\Omega, P)$ is closed under multiplication.
[Hint for part (3): use the Cauchy–Schwarz inequality in $L^2(\Omega, P)$.]

EXERCISE 1.23. Let $(\mathcal{A}, \varphi)$ be a $*$-probability space and let d be a positive integer. Let $M_d(\mathcal{A})$ be the space of $d \times d$ matrices over $\mathcal{A}$,

$$M_d(\mathcal{A}) := \{(a_{ij})_{i,j=1}^d \mid a_{ij} \in \mathcal{A} \text{ for } 1 \le i, j \le d\}.$$

On $M_d(\mathcal{A})$ we can define canonically a $*$-operation by

$$(\,(a_{ij})_{i,j=1}^d\,)^* =: (b_{ij})_{i,j=1}^d,$$

where $b_{ij} := a_{ji}^*$ for $1 \le i, j \le d$; thus $M_d(\mathcal{A})$ becomes a $*$-algebra. Then consider the linear functional $\varphi_d : M_d(\mathcal{A}) \to \mathbb{C}$ defined by

$$\varphi_d(A) = \frac{1}{d}\sum_{i=1}^{d} \varphi(a_{ii}), \quad \text{for } A = (a_{ij})_{i,j=1}^d \in M_d(\mathcal{A}).$$

Note that $M_d(\mathcal{A})$ is canonically isomorphic to $M_d(\mathbb{C}) \otimes \mathcal{A}$, and that under this isomorphism φ_d corresponds to $\mathrm{tr} \otimes \varphi$.

(1) Verify that $(M_d(\mathcal{A}), \varphi_d)$ is a $*$-probability space.

(2) Show that if the $*$-probability space $(\mathcal{A}, \varphi)$ is tracial, then so is $(M_d(\mathcal{A}), \varphi_d)$.

(3) Show that if the functional φ is faithful, then so is φ_d.

EXERCISE 1.24. Let $(\mathcal{A}, \varphi)$ and $(\mathcal{B}, \psi)$ be $*$-probability spaces, and suppose that φ is faithful. Let Φ be a morphism between $(\mathcal{A}, \varphi)$ and $(\mathcal{B}, \psi)$. Prove that Φ is one-to-one.

EXERCISE 1.25. (1) Consider the $*$-probability space discussed in Example 1.4.1. Write a representation of this $*$-probability space, living on the Hilbert space $L^2(\Omega, \mathcal{Q}, P)$.

(2) Consider the $*$-probability space discussed in Example 1.4.2. Write a representation of this $*$-probability space, living on the Hilbert space $\mathbb{C}^{d^2}$.

(3) Consider the $*$-probability space discussed in Example 1.4.4. Write a representation of this $*$-probability space, living on the Hilbert space $l^2(G) := \{\xi : G \to \mathbb{C} \mid \sum_{g \in G} |\xi_g|^2 < \infty\}$.

EXERCISE 1.26. Let $(\mathcal{A}, \varphi)$ be a $*$-probability space, let a be a normal element of $\mathcal{A}$, and suppose that a has $*$-distribution μ in analytic sense (i.e. in the sense of Definition 1.8).

(1) Let $P : \mathbb{C} \to \mathbb{C}$ be a polynomial in z and $\bar{z}$, and let ν be the probability measure on $\mathbb{C}$ defined by:

$$\nu(E) := \mu(\,P^{-1}(E)\,), \text{ for } E \subset \mathbb{C} \text{ Borel set.}$$

Show that ν is compactly supported and that the normal element $b := P(a, a^*) \in \mathcal{A}$ has $*$-distribution ν.

(2) By using the result in part (1), describe the $*$-distributions of the following elements: i) a^*; ii) $a+\alpha$, where α is an arbitrary complex number; iii) ra, where r is an arbitrary positive number.

EXERCISE 1.27. Do the analog of the first question treated in Example 1.14 for a p-Haar unitary.

LECTURE 2

A case study of non-normal distribution

In this lecture we study the example of the so-called "vacuum-state" on the $*$-algebra generated by the one-sided shift operator, and see how the important concept of semicircular random variable is connected to it.

Description of the example

NOTATION 2.1. Throughout the lecture we fix a $*$-probability space $(\mathcal{A}, \varphi)$ and an element $a \in \mathcal{A}$, such that:
(i) $a^*a = 1_{\mathcal{A}} \neq aa^*$;
(ii) a generates $\mathcal{A}$ as a $*$-algebra.

One refers to the condition $a^*a = 1_{\mathcal{A}}$ by saying that a is an **isometry**; since the above assumption (i) also requires that $aa^* \neq 1_{\mathcal{A}}$, one can rephrase it by saying that "a is a non-unitary isometry."

Some more assumptions made on a and $(\mathcal{A}, \varphi)$ will be stated after we observe the following simple consequence of (i) and (ii).

LEMMA 2.2. $\mathcal{A} = \text{span}\{a^m(a^*)^n \mid m, n \geq 0\}$.

PROOF. The condition $a^*a = 1_{\mathcal{A}}$ immediately implies that for every $m, n, p, q \geq 0$ we have:

$$\Big(a^m(a^*)^n\Big) \cdot \Big(a^p(a^*)^q\Big) = \begin{cases} a^{m+p-n}(a^*)^q & \text{if } n < p \\ a^m(a^*)^q & \text{if } n = p \\ a^m(a^*)^{n-p+q} & \text{if } n > p. \end{cases} \tag{2.1}$$

Since the family $\{a^m(a^*)^n \mid m, n \geq 0\}$ is, clearly, also closed under $*$-operation, it follows that its linear span has to be equal to the unital $*$-subalgebra of $\mathcal{A}$ generated by a. But this is all of $\mathcal{A}$, by (ii) of Notation 2.1. $\square$

NOTATION 2.3. In addition to what was stated in 2.1, we will make the following assumptions on a and $(\mathcal{A}, \varphi)$:
(iii) the elements $\{a^m(a^*)^n \mid m, n \geq 0\}$ are linearly independent;

(iv) the functional $\varphi : \mathcal{A} \to \mathbb{C}$ satisfies (and is determined by) the equation

$$\varphi(a^m (a^*)^n) = \begin{cases} 1 & \text{if } m = n = 0 \\ 0 & \text{otherwise,} \end{cases} \tag{2.2}$$

for $m, n \in \mathbb{N} \cup \{0\}$.

REMARK 2.4. As the reader may recognize, Equation (2.1) is connected to a semigroup structure on $(\mathbb{N} \cup \{0\})^2$, where the multiplication is defined by

$$(m,n) \cdot (p,q) = \begin{cases} (m+p-n, q) & \text{if } n < p \\ (m,q) & \text{if } n = p \\ (m, n-p+q) & \text{if } n > p. \end{cases} \tag{2.3}$$

This is called the *bicyclic semigroup*, and is a fundamental example in a class of semigroups with a well-developed theory, which are called "inverse semigroups." So from this perspective, the algebra $\mathcal{A}$ appearing in this example could be called "the semigroup algebra of the bicyclic semigroup."

REMARK 2.5. From another perspective, the algebra $\mathcal{A}$ is related to an important example from the theory of C^*-algebras, called the *Toeplitz algebra*, and obtained by completing $\mathcal{A}$ with respect to a suitable norm. Equivalently, the Toeplitz algebra can be defined as the closure in the norm-topology of $\pi(\mathcal{A}) \subset B(l^2)$, where $\pi : \mathcal{A} \to B(l^2)$ is the natural representation described in what follows.

Consider the Hilbert space $l^2 := l^2(\mathbb{N} \cup \{0\})$. The vectors of l^2 are thus of the form $\xi = (\alpha_k)_{k \geq 0}$, where the α_k are from $\mathbb{C}$ and have $\sum_{k=0}^{\infty} |\alpha_k|^2 < \infty$. The inner product of $\xi = (\alpha_k)_{k \geq 0}$ with $\eta = (\beta_k)_{k \geq 0}$ is

$$\langle \xi, \eta \rangle := \sum_{k=0}^{\infty} \alpha_k \bar{\beta}_k.$$

For every $n \geq 0$ we denote:

$$\xi_n := (0, 0, \ldots, 0, 1, 0, \ldots, 0, \ldots), \tag{2.4}$$

with the 1 occurring on component n. Then $\xi_0, \xi_1, \ldots, \xi_n, \ldots$ form an orthonormal basis for l^2.

Let $S \in B(l^2)$ be the **one-sided shift operator**, determined by the fact that

$$S\xi_n = \xi_{n+1}, \quad \forall\, n \geq 0.$$

Its adjoint S^* is determined by

$$S^*\xi_0 = 0 \qquad \text{and} \qquad S^*\xi_n = \xi_{n-1}, \quad \forall\, n \geq 1.$$

It is immediate that $S^*S = 1_{B(l^2)}$ (the identity operator on l^2), but $SS^* \neq 1_{B(l^2)}$.

Since $(a^m(a^*)^n)_{m,n\geq 0}$ form a linear basis in $\mathcal{A}$, we can define a linear map $\pi : \mathcal{A} \to B(l^2)$ by asking that:

$$\pi(a^m(a^*)^n) = S^m(S^*)^n, \quad \forall\, m,n \geq 0.$$

It is easily verified that π is a unital $*$-homomorphism. (The multiplicativity of π follows from the fact that, as a consequence of the relation $S^*S = 1_{B(l^2)}$, the product of two members of the family $(S^m(S^*)^n)_{m,n\geq 0}$ is described by the same rules as in Equation (2.1).)

Now, it is also easy to see that the operators $(S^m(S^*)^n)_{m,n\geq 0}$ are linearly independent (see Exercise 2.22 at the end of the lecture). This implies that the $*$-homomorphism π defined above is one-to-one, hence it actually gives us an identification between the algebra $\mathcal{A}$ fixed in Notation 2.1 and an algebra of operators on l^2.

Let $\varphi_0 : B(l^2) \to \mathbb{C}$ be the functional defined by

$$\varphi_0(T) = \langle T\xi_0, \xi_0\rangle, \quad T \in B(l^2), \tag{2.5}$$

where ξ_0 is the first vector of the canonical orthonormal basis considered in (2.4). If $m,n \in \mathbb{N}\cup\{0\}$ and $(m,n) \neq (0,0)$ then

$$\varphi_0(S^m(S^*)^n) = \langle S^m(S^*)^n\xi_0, \xi_0\rangle = \langle (S^*)^n\xi_0, (S^*)^m\xi_0\rangle,$$

which is equal to 0 because at least one of $(S^*)^m\xi_0$ and $(S^*)^n\xi_0$ is the zero-vector. Comparing this with (2.2) makes it clear that π is a morphism between $(\mathcal{A},\varphi)$ and $(B(l^2),\varphi_0)$, in the sense discussed in Definition 1.6 of Lecture 1. Or, in the sense of the same definition, (l^2,π,ξ_0) is a representation of $(\mathcal{A},\varphi)$ on the Hilbert space l^2.

As mentioned above, the closure $\mathcal{T}$ of $\pi(\mathcal{A})$ in the norm-topology of $B(l^2)$ is called the **Toeplitz algebra**. Moreover, the restriction to $\mathcal{T}$ of the functional φ_0 defined by Equation (2.5) is called *"the vacuum-state on the Toeplitz algebra"* (which is why, by a slight abuse of terminology, the $*$-algebraic example discussed throughout the lecture is also termed in that way).

REMARK 2.6. Our goal in this lecture is to look at the $*$-distribution of the non-normal element a which was fixed in Notation 2.1. But as the reader has surely noticed, the equation describing $\mathcal{A}$ in Lemma 2.2 is a repeat of Equation (1.4) from the discussion on normal elements, in Lecture 1. Does this indicate that we can treat a as if it was normal? It is instructive to take a second to notice that this is not the case. Indeed, the unique compactly supported probability measure on $\mathbb{C}$ which fits the $*$-moments in (2.2) is the Dirac mass δ_0 – so we would come to the unconvincing conclusion that a has the same $*$-distribution as the zero-element of $\mathcal{A}$.

The point here is that, besides the information given in (2.2), one must also understand the process (quite different from the case of normal elements) of reducing a word $a^{\varepsilon(1)} \cdots a^{\varepsilon(k)}$ to the form $a^m(a^*)^n$. Or at least, one should be able to understand how to distinguish the words $a^{\varepsilon(1)} \cdots a^{\varepsilon(k)}$ which reduce to $1_{\mathcal{A}}$ from those which reduce to something else. The latter question is best clarified by using a class of combinatorial objects called Dyck paths.

Dyck paths

DEFINITION 2.7. (1) We will use the term **NE-SE path** for a path in the lattice $\mathbb{Z}^2$ which starts at $(0,0)$ and makes steps either of the form $(1,1)$ ("North-East steps") or of the form $(1,-1)$ ("South-East steps").

(2) A **Dyck path** is a NE-SE path γ which ends on the x-axis, and never goes strictly below the x-axis. (That is, all the lattice points visited by γ are of the form (i,j) with $j \geq 0$, and the last of them is of the form $(k,0)$.)

REMARKS 2.8. (1) For a given positive integer k, the set of NE-SE paths with k steps is naturally identified with $\{-1,1\}^k$, by identifying a path γ with the sequence of ± 1s which appear as second components for the k steps of γ.

Concrete example: here is the NE-SE path of length 6 which corresponds to the 6-tuple $(1,-1,-1,1,-1,1)$.

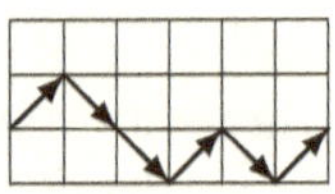

This path is not a Dyck path, because it goes twice under the x-axis.

(2) Let k be a positive integer, and consider the identification described above between the NE-SE paths with k steps and $\{-1,1\}^k$. It is immediately seen that a k-tuple $(\lambda_1, \ldots, \lambda_k)$ corresponds to a Dyck path if and only if

$$\begin{cases} \lambda_1 + \cdots + \lambda_j \geq 0, \quad \forall\, 1 \leq j < k, \\ \\ \lambda_1 + \cdots + \lambda_k = 0. \end{cases} \tag{2.6}$$

From the equality stated in (2.6) it is clear that Dyck paths with k steps can only exist when k is even.

Concrete examples: there are 5 Dyck paths with 6 steps. We draw them in the pictures below, and for each of them we indicate the corresponding tuple in $\{-1, 1\}^6$ (thus listing the 5 tuples in $\{-1, 1\}^6$ which satisfy (2.6)).

$(+1, +1, +1, -1, -1, -1)$

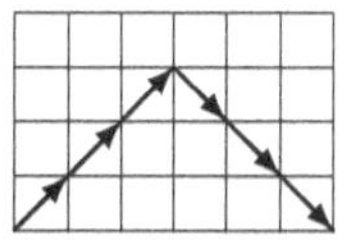

$(+1, +1, -1, +1, -1, -1)$

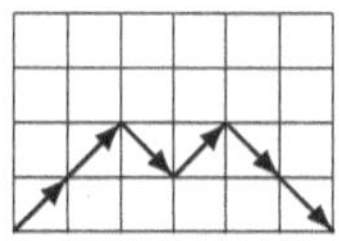

$(+1, +1, -1, -1, +1, -1)$

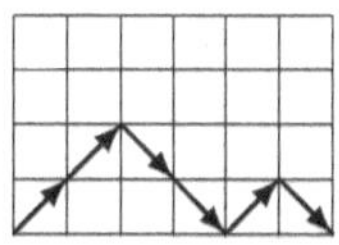

$(+1, -1, +1, +1, -1, -1)$

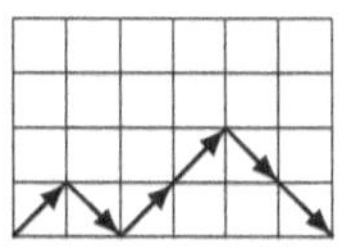

$(+1, -1, +1, -1, +1, -1)$

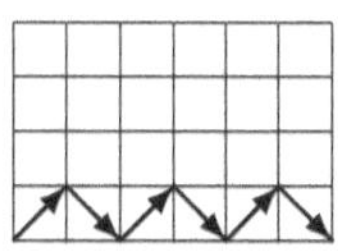

The Dyck paths can be enumerated by using a celebrated "reflection trick" of Desiré André, and turn out to be counted by the (even more celebrated) Catalan numbers.

NOTATION 2.9. For every integer $n \geq 0$ we will denote by C_n the nth **Catalan number**,

$$C_n := \frac{1}{n+1}\binom{2n}{n} = \frac{(2n)!}{n!(n+1)!} \tag{2.7}$$

(with the convention that $C_0 = 1$).

REMARK 2.10. An equivalent (and often used) way of introducing the Catalan numbers is via the following recurrence relation:

$$\begin{cases} C_0 = C_1 = 1 \\ \\ C_p = \sum_{j=1}^{p} C_{j-1}C_{p-j}, \quad p \geq 2. \end{cases} \tag{2.8}$$

It is not hard to see that the numbers defined by (2.7) do indeed satisfy the recurrence (2.8). One way of proving this fact can actually be read from the following discussion about the enumeration of Dyck paths (see the last paragraph in Remark 2.12).

PROPOSITION 2.11. *For every positive integer p, the number of Dyck paths with $2p$ steps is equal to the pth Catalan number C_p.*

PROOF. Let us first count all the NE-SE paths which end at a given point $(m, n) \in \mathbb{Z}^2$. A NE-SE path with u NE-steps and v SE-steps ends at $(u+v, u-v)$, so there are NE-SE paths arriving at (m, n) if and only if $(m, n) = (u + v, u - v)$ for some $u, v \in \mathbb{N} \cup \{0\}$ with $u + v > 0$; this happens if and only if $m > 0$, $|n| \leq m$, and m, n have the same parity. If the latter conditions are satisfied, then the NE-SE paths arriving at (m, n) are precisely those which have $(m+n)/2$ NE-steps and $(m-n)/2$ SE-steps. These paths are hence counted by the binomial coefficient $\binom{m}{(m+n)/2}$, because Remark 2.8.1 identifies them with the m-tuples in $\{-1, 1\}^m$ which have precisely $(m + n)/2$ components equal to 1.

In particular, it follows that the total number of NE-SE paths arriving at $(2p, 0)$ is $\binom{2p}{p}$.

We now look at the NE-SE paths arriving at $(2p, 0)$ which are *not* Dyck paths. Let us fix for the moment such a path, γ, and let $j \in \{1, \ldots, 2p - 1\}$ be minimal with the property that γ goes under the x-axis after j steps. Then γ is written as a juxtaposition of two paths, $\gamma = \gamma' \vee \gamma''$, where γ' goes from $(0, 0)$ to $(j, -1)$, and γ'' goes from $(j, -1)$ to $(2p, 0)$. Let $\widehat{\gamma''}$ be the reflection of γ'' in the horizontal line of equation $y = -1$; thus $\widehat{\gamma''}$ is a path from $(j, -1)$ to $(2p, -2)$. Then let us define $F(\gamma) := \gamma' \vee \widehat{\gamma''}$, a NE-SE path from $(0, 0)$ to $(2p, -2)$.

[Concrete example: suppose that $p = 10$ and that γ is the NE-SE path from $(0, 0)$ to $(20, 0)$ which appears drawn in bold-face fonts in the following picture. It is not a Dyck path, and the first time when it goes under the x-axis is after 5 steps. Thus for this example, the decomposition $\gamma = \gamma' \vee \gamma''$ described above looks as follows: γ' has 5 steps, going from $(0,0)$ to $(5,-1)$, and γ'' has 15 steps, going from $(5,-1)$ to $(20,0)$.

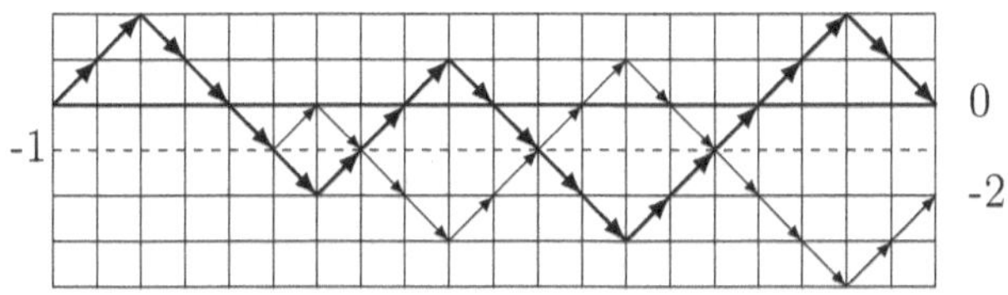

The reflection of γ'' in the horizontal line of equation $x = -1$ is shown in the above picture as a thinner line, going from $(5,-1)$ to $(20,-2)$. The path $F(\gamma)$ goes from $(0,0)$ to $(20,-2)$; it is obtained by pursuing the first five steps of γ, and then by continuing along the thinner line.]

So, the construction described in the preceding paragraph gives a map F from the set of NE-SE paths ending at $(2p,0)$ and which are not Dyck paths, to the set of all NE-SE paths ending at $(2p,-2)$. The map F is a bijection. Indeed, if β is a NE-SE path ending at $(2p,-2)$, then there has to be a minimal $j \in \{1,\ldots,2p-1\}$ such that β is at height $y = -1$ after j steps. Write $\beta = \beta' \vee \beta''$ with β' from $(0,0)$ to $(j,-1)$ and β'' from $(j,-1)$ to $(2p,-2)$, and let $\widehat{\beta''}$ be the reflection of β'' in the line $y = -1$; then $\gamma := \beta' \vee \widehat{\beta''}$ is the unique path in the domain of F which has $F(\gamma) = \beta$.

It follows that the number of NE-SE paths which end at $(2p,0)$ but are not Dyck paths is equal to the total number of NE-SE paths ending at $(2p,-2)$, which is $\binom{2p}{p-1}$. Finally, the number of Dyck paths with $2p$ steps is

$$\binom{2p}{p} - \binom{2p}{p-1} = \frac{1}{p+1}\binom{2p}{p} = C_p.$$

□

REMARK 2.12. Another approach to the enumeration of Dyck paths is obtained by making some simple remarks about the structure of such a path, which yield a recurrence relation. Let us call a Dyck path γ *irreducible* if it only touches the x-axis at its starting and ending points (but never in between them). For instance, out of the 5 Dyck paths pictured in Remark 2.8.2, 2 paths are irreducible and 3 are reducible.

Given an even integer $k \geq 2$. If γ is an irreducible Dyck path with k steps, then it is immediate that the k-tuple in $\{-1,1\}^k$ associated to γ is of the form $(1,\lambda_1,\ldots,\lambda_{k-2},-1)$, where $(\lambda_1,\ldots,\lambda_{k-2}) \in \{-1,1\}^{k-2}$

corresponds to a Dyck path with $k-2$ steps. Conversely, it is also immediate that if $(\lambda_1, \ldots, \lambda_{k-2}) \in \{-1,1\}^{k-2}$ corresponds to a Dyck path, then $(1, \lambda_1, \ldots, \lambda_{k-2}, -1) \in \{-1,1\}^k$ will correspond to an irreducible Dyck path with k steps. Thus the irreducible Dyck paths with k steps are in natural bijection with the set of all Dyck paths with $k-2$ steps.

On the other hand, suppose that γ is a reducible Dyck path with k steps, and that the first time when γ touches the x-axis following to its starting point is after j steps $(1 < j < k)$. Then γ splits as a juxtaposition $\gamma = \gamma' \vee \gamma''$, where γ' is an irreducible Dyck path with j steps and γ'' is a Dyck path with $k-j$ steps. Moreover, this decomposition is unique, if we insist that its first piece, γ', is irreducible.

For every $p \geq 1$, let then D_p denote the number of Dyck paths with $2p$ steps, and let D'_p be the number of irreducible Dyck paths with $2p$ steps. The observation made in the preceding paragraph gives us that

$$D_p = D'_1 D_{p-1} + D'_2 D_{p-2} + \cdots + D'_{p-1} D_1 + D'_p, \quad p \geq 2. \tag{2.9}$$

(Every term $D'_j D_{p-j}$ on the right-hand side of (2.9) counts the reducible Dyck paths with $2p$ steps which touch for the first time the x-axis after $2j$ steps.) The observation made one paragraph before the preceding one says that $D'_p = D_{p-1}$, $\forall\, p \geq 2$. This equality is also true for $p = 1$, if we make the convention to set $D_0 := 1$. So we get the recurrence

$$\begin{cases} D_0 = D_1 = 1 \\ D_p = \sum_{j=1}^{p} D_{j-1} D_{p-j}, \quad p \geq 2. \end{cases} \tag{2.10}$$

This is exactly (2.8), and shows that $D_p = C_p$, $\forall\, p \geq 1$.

The argument presented above can be viewed as an alternative proof of Proposition 2.11. On the other hand, since the derivation of (2.10) was made independently from Proposition 2.11, a reader who is not already familiar with Catalan numbers can view the above argument as a proof of the fact that the numbers introduced in Notation 2.9 do indeed satisfy the recurrence (2.8).

The distribution of $a + a^*$

We now return to the example of $(\mathcal{A}, \varphi)$ and $a \in \mathcal{A}$ introduced in Notations 2.1 and 2.3. The connection between the $*$-distribution of a and Dyck paths appears as follows.

PROPOSITION 2.13. *Let k be a positive integer, let $\varepsilon(1), \ldots, \varepsilon(k)$ be in $\{1, *\}$, and consider the monomial $a^{\varepsilon(1)} \cdots a^{\varepsilon(k)} \in \mathcal{A}$. Let us set*

$$\lambda_j := \begin{cases} 1 & \text{if } \varepsilon(j) = * \\ -1 & \text{if } \varepsilon(j) = 1 \end{cases} \quad \text{for } 1 \leq j \leq k, \tag{2.11}$$

and let us denote by γ the NE-SE path which corresponds to the tuple $(\lambda_1, \ldots, \lambda_k) \in \{-1, 1\}^k$. Then

$$\varphi(a^{\varepsilon(1)} \cdots a^{\varepsilon(k)}) = \begin{cases} 1 & \text{if } \gamma \text{ is a Dyck path} \\ 0 & \text{otherwise.} \end{cases} \tag{2.12}$$

PROOF. It is convenient to use the representation of $(\mathcal{A}, \varphi)$ discussed in Remark 2.5. With notations as in that remark, we write:

$$\begin{aligned} \varphi(a^{\varepsilon(1)} \cdots a^{\varepsilon(k)}) &= \langle S^{\varepsilon(1)} \cdots S^{\varepsilon(k)} \xi_0 \,, \xi_0 \rangle \\ &= \langle \xi_0 \,, (S^{\varepsilon(k)})^* \cdots (S^{\varepsilon(1)})^* \xi_0 \rangle. \end{aligned} \tag{2.13}$$

Applying successively the operators $(S^{\varepsilon(1)})^*, \ldots, (S^{\varepsilon(k)})^*$ to ξ_0 takes us either to a vector of the orthonormal basis $\{\xi_n \mid n \geq 0\}$ of l^2, or to the zero-vector. More precisely: by keeping track of how $\lambda_1, \ldots, \lambda_k$ were defined in Equation (2.11) in terms of $\varepsilon(1), \ldots, \varepsilon(k)$, the reader should have no difficulty to verify by induction on j, $1 \leq j \leq k$, that

$$(S^{\varepsilon(j)})^* \cdots (S^{\varepsilon(1)})^* \xi_0 = \begin{cases} \xi_{\lambda_1 + \cdots + \lambda_j} & \text{if } \lambda_1 \geq 0, \lambda_1 + \lambda_2 \geq 0, \\ & \quad \ldots, \lambda_1 + \cdots + \lambda_j \geq 0 \\ 0 & \text{otherwise.} \end{cases} \tag{2.14}$$

If we make $j = k$ in (2.14) and substitute this expression into (2.13), then we obtain:

$$\begin{aligned} \varphi(a^{\varepsilon(1)} \cdots a^{\varepsilon(k)}) &= \begin{cases} \langle \xi_0, \xi_{\lambda_1 + \cdots + \lambda_k} \rangle & \text{if } \sum_{i=1}^{j} \lambda_i \geq 0, \ 1 \leq j \leq k \\ 0 & \text{otherwise} \end{cases} \\ &= \begin{cases} 1 & \text{if } \sum_{i=1}^{j} \lambda_i \geq 0 \text{ for } 1 \leq j < k \\ & \text{and if } \sum_{i=1}^{k} \lambda_i = 0 \\ 0 & \text{otherwise} \end{cases} \\ &= \begin{cases} 1 & \text{if } \gamma \text{ is a Dyck path} \\ 0 & \text{otherwise} \end{cases} \end{aligned}$$

(where at the last equality sign we used Equation (2.6) of Remark 2.8). □

Let us next consider the selfadjoint element $a + a^* \in \mathcal{A}$, and ask the following two questions (identical to those asked in Example 1.14 of Lecture 1, in connection to Haar unitaries).

(1) Does $a+a^*$ have a $*$-distribution in analytic sense (as discussed in Definition 1.8) ?

(2) Suppose that $a+a^*$ does have a $*$-distribution μ. Then (as observed in Remark 1.10) μ is a compactly supported probability measure on $\mathbb{R}$, determined by the fact that

$$\int_{\mathbb{R}} t^k \, d\mu(t) = \varphi(\,(a+a^*)^k\,), \quad \forall\, k \geq 0.$$

Do we have some "nice" formula for the moments of μ (or in other words, for the values of $\varphi(\,(a+a^*)^k\,)$, $k \geq 0$) ?

We can derive the answer to the second question as an immediate consequence of Proposition 2.13.

COROLLARY 2.14. *If k is an odd positive integer, then*

$$\varphi(\,(a+a^*)^k\,) = 0.$$

If $k = 2p$ is an even positive integer, then

$$\varphi(\,(a+a^*)^k\,) = C_p,$$

where C_p is the pth Catalan number.

PROOF.

$$\begin{aligned}
\varphi(\,(a+a^*)^k\,) &= \varphi\Big(\sum_{\varepsilon(1),\ldots,\varepsilon(k)\in\{1,*\}} a^{\varepsilon(1)} \cdots a^{\varepsilon(k)} \Big) \\
&= \sum_{\varepsilon(1),\ldots,\varepsilon(k)\in\{1,*\}} \varphi(\, a^{\varepsilon(1)} \cdots a^{\varepsilon(k)}\,) \\
&= \sum_{\substack{\text{Dyck paths} \\ \text{with } k \text{ steps}}} 1 \qquad\qquad \text{(by Proposition 2.13).}
\end{aligned}$$

So $\varphi(\,(a+a^*)^k\,)$ is equal to the number of Dyck paths with k steps, and the result follows from Proposition 2.11. □

It remains to look at the first question asked above about $a+a^*$, that of finding (if it exists) a compactly supported probability measure μ on $\mathbb{R}$ which has moment of order k equal to $\varphi(\,(a+a^*)^k\,)$, $k \geq 0$. The answer to this question turns out to be the following.

PROPOSITION 2.15. *The distribution of $a+a^*$ in $(\mathcal{A},\varphi)$ is the measure $d\mu(t) = \frac{1}{2\pi}\sqrt{4-t^2}\,dt$ on the interval $[-2,2]$.*

PROOF. By taking into account Corollary 2.14, what we have to show is that

$$\int_{-2}^{2} t^k \sqrt{4-t^2}\, dt = \begin{cases} 0 & \text{if } k \text{ is odd} \\ \frac{2\pi}{p+1}\binom{2p}{p} & \text{if } k \text{ is even, } k = 2p. \end{cases} \tag{2.15}$$

The case of odd k is obvious, because in that case $t \mapsto t^k\sqrt{4-t^2}$ is an odd function. When k is even, $k = 2p$, we use the change of variable $t = 2\cos\theta$, $dt = -2\sin\theta\, d\theta$, with θ running from π to 0. We obtain that:

$$\int_{-2}^{2} t^{2p}\sqrt{4-t^2}\, dt = \int_0^{\pi} 2^{2p+2}\cos^{2p}\theta\, \sin^2\theta\, d\theta = 4^{p+1}(I_p - I_{p+1}),$$

where

$$I_p := \int_0^{\pi} \cos^{2p}\theta\, d\theta, \quad p \geq 0.$$

The integral I_p has already appeared in Example 1.14 of Lecture 1; in fact, if we combine Equation (1.14) of that example with the calculations following it (in the same example), we clearly obtain that

$$I_p = \frac{\pi}{4^p}\binom{2p}{p}, \quad p \geq 0,$$

and (2.15) quickly follows. □

Proposition 2.15 can be rephrased by saying that $a + a^*$ is a semicircular element of radius 2, in the sense of the next definition.

DEFINITION 2.16. Let $(\mathcal{A}, \varphi)$ be a $*$-probability space, let x be a selfadjoint element of $\mathcal{A}$ and let r be a positive number. If x has distribution (in analytical sense, as in Remark 1.10 of Lecture 1) equal to $\frac{2}{\pi r^2}\sqrt{r^2-t^2}\, dt$ on the interval $[-r, r]$, then we will say that x is a **semicircular element** of radius r.

REMARKS 2.17. (1) It is customary to talk about *semicircular* elements, despite the fact that the graph of a function of the form $[-r, r] \ni t \mapsto \frac{2}{\pi r^2}\sqrt{r^2-t^2}$ is not exactly a semicircle (but rather a semi-ellipse). Semicircular elements will play an important role in subsequent lectures – see e.g. Lecture 8. The semicircular distribution is also a fundamental object in random matrix theory; we will address this relation in Lecture 22.

(2) The semicircular elements of radius 2 are also called **standard semicircular**, due to the fact that they are normalized by the variance. Indeed, it is immediate that a semicircular element x of radius r has its variance $\mathrm{Var}(x) := \varphi(x^2) - \varphi(x)^2$ given by

$$\mathrm{Var}(x) = r^2/4.$$

(It is in fact customary to talk about semicircular elements in terms of their variance, rather than radius. Of course, the above equation shows that either radius or variance can be used, depending on the user's preference.)

(3) Strictly speaking, the above definition has only introduced the concept of a *centered* semicircular element; it is quite straightforward how to adjust it in order to define a "semicircular element of mean $m \in \mathbb{R}$ and radius $r > 0$," but this will not be needed in the sequel.

(4) The proof shown above for Proposition 2.15 was immediate, but not too illuminating, as it does not show how one arrives to consider the semicircular density in the first place. (It is easier to *verify* that the given density has the right moments, rather than *derive* what the density should be!) We will conclude the lecture by elaborating on this point. The object which we will use as an intermediate in order to derive μ from the knowledge of its moments is an analytic function in the upper half plane called the *Cauchy transform*.

Using the Cauchy transform

DEFINITION 2.18. Let μ be a probability measure on $\mathbb{R}$. The **Cauchy transform** of μ is the function G_μ defined on the upper half plane $\mathbb{C}^+ = \{s + it \mid s, t \in \mathbb{R},\ t > 0\}$ by the formula:

$$G_\mu(z) = \int_{\mathbb{R}} \frac{1}{z-t}\, d\mu(t), \quad z \in \mathbb{C}^+.$$

REMARKS 2.19. (1) It is easily verified that G_μ is analytic on $\mathbb{C}^+$ and that it takes values in $\mathbb{C}^- := \{s + it \mid s, t \in \mathbb{R},\ t < 0\}$.

(2) Suppose that μ is compactly supported, and let us denote $r := \sup\{|t| \mid t \in \operatorname{supp}(\mu)\}$. We then have the power series expansion:

$$G_\mu(z) = \sum_{n=0}^{\infty} \frac{\alpha_n}{z^{n+1}}, \quad |z| > r, \tag{2.16}$$

where $\alpha_n := \int_{\mathbb{R}} t^n\, d\mu(t)$ is the nth moment of μ, for $n \geq 0$. Indeed, for $|z| > r$ we can expand:

$$\frac{1}{z-t} = \sum_{n=0}^{\infty} \frac{t^n}{z^{n+1}}, \quad \forall\, t \in \operatorname{supp}(\mu).$$

The convergence of the latter series is uniform in $t \in \operatorname{supp}(\mu)$; hence we can integrate the series term by term against $d\mu(t)$, and (2.16) is obtained.

Note that the expansion (2.16) of G_μ around the point at infinity has as an obvious consequence the fact that

$$\lim_{z\in\mathbb{C}^+,\ |z|\to\infty} zG_\mu(z) = 1. \tag{2.17}$$

REMARK 2.20. The property of the Cauchy transform that we want to use is the following: there is an effective way of recovering the probability measure μ from its Cauchy transform G_μ, via the *Stieltjes inversion formula.* If we denote

$$h_\varepsilon(t) := -\frac{1}{\pi}\Im G_\mu(t+i\varepsilon), \quad \forall \varepsilon > 0,\ \forall\, t \in \mathbb{R} \tag{2.18}$$

(where "$\Im$" stands for the operation of taking the imaginary part of a complex number), then the Stieltjes inversion formula says that

$$d\mu(t) = \lim_{\varepsilon\to 0} h_\varepsilon(t)\, dt. \tag{2.19}$$

The latter limit is considered in the weak topology on the space of probability measures on $\mathbb{R}$, and thus amounts to the fact that

$$\int_{\mathbb{R}} f(t)\, d\mu(t) = \lim_{\varepsilon\to 0}\int_{\mathbb{R}} f(t)h_\varepsilon(t)\, dt, \tag{2.20}$$

for every bounded continuous function $f : \mathbb{R} \to \mathbb{C}$.

The fact that (2.19) holds is a consequence of the connection between the Cauchy transform and the family of functions $(P_\varepsilon)_{\varepsilon>0}$ defined by

$$P_\varepsilon(t) := \frac{1}{\pi}\frac{\varepsilon}{t^2+\varepsilon^2}, \quad \text{for } \varepsilon > 0 \text{ and } t \in \mathbb{R},$$

which forms the so-called "Poisson kernel on the upper half plane." For every $\varepsilon > 0$ and $t \in \mathbb{R}$ we have that

$$\begin{aligned} h_\varepsilon(t) &= -\frac{1}{\pi}\Im\int_{\mathbb{R}} \frac{1}{t+i\varepsilon-s}\, d\mu(s) \\ &= -\frac{1}{\pi}\Im\int_{\mathbb{R}} \frac{t-s-i\varepsilon}{(t-s)^2+\varepsilon^2}\, d\mu(s) \\ &= \frac{1}{\pi}\int_{\mathbb{R}} \frac{\varepsilon}{(t-s)^2+\varepsilon^2}\, d\mu(s) \\ &= \int_{\mathbb{R}} P_\varepsilon(t-s)\, d\mu(s). \end{aligned}$$

The last expression in the above sequence of equalities is called a *convolution* integral, and one of the fundamental properties of the Poisson kernel is that the h_ε given by such an integral will converge weakly to μ for $\varepsilon \to 0$.

Let us record explicitly how the Stieltjes inversion formula looks in the case when the Cauchy transform G_μ happens to have a continuous extension to $\mathbb{C}^+ \cup \mathbb{R}$. The values on $\mathbb{R}$ of this extension must of course be given by the function g obtained as

$$g(t) = \lim_{\varepsilon \to 0} G_\mu(t + i\varepsilon), \quad t \in \mathbb{R}. \tag{2.21}$$

It is immediate that in this case the measures $h_\varepsilon(t)\,dt$ converge weakly to $-\frac{1}{\pi}\Im g(t)\,dt$. Hence in this case the Stieltjes inversion formula is simply telling us that:

$$d\mu(t) = -\frac{1}{\pi}\Im g(t)\,dt, \tag{2.22}$$

with g defined as in (2.21).

Let us now look once more at the random variable a fixed at the beginning of the lecture, and see how we can use the Cauchy transform in order to derive the distribution of $a + a^*$ from the knowledge of its moments.

LEMMA 2.21. *Suppose that μ is a probability measure with compact support on $\mathbb{R}$ such that*

$$\int_{\mathbb{R}} t^k \, d\mu(t) = \begin{cases} 0 & \text{if } k \text{ is odd} \\ \frac{1}{p+1}\binom{2p}{p} & \text{if } k \text{ is even, } k = 2p. \end{cases}$$

Then the Cauchy transform of μ is

$$G_\mu(z) = \frac{z - \sqrt{z^2 - 4}}{2}, \quad z \in \mathbb{C}^+. \tag{2.23}$$

(Note: on the right-hand side of (2.23) we can view

$$\sqrt{z^2 - 4} := \sqrt{z - 2} \cdot \sqrt{z + 2},$$

where $z \mapsto \sqrt{z \pm 2}$ is analytic on $\mathbb{C} \setminus \{\mp 2 - it \mid t > 0\} \supset \mathbb{C}^+$, and is defined such that it gives the usual square root for $z \in \mathbb{R},\ z > 2$.)

PROOF. We know that for $|z|$ sufficiently large we have the series expansion (2.16), which becomes here

$$G_\mu(z) = \sum_{p=0}^{\infty} \frac{C_p}{z^{2p+1}},$$

with C_p the pth Catalan number. The recurrence relation (2.8) of the Catalan numbers and some elementary manipulations of power series then give us that:

$$\begin{aligned}
G_\mu(z) &= \frac{1}{z} + \sum_{p=1}^{\infty} \frac{1}{z^{2p+1}} \Big(\sum_{j=1}^{p} C_{j-1} C_{p-j} \Big) \\
&= \frac{1}{z} + \frac{1}{z} \sum_{p=1}^{\infty} \sum_{j=1}^{p} \frac{C_{j-1}}{z^{2j-1}} \cdot \frac{C_{p-j}}{z^{2(p-j)+1}} \\
&= \frac{1}{z} + \frac{1}{z} \sum_{j=1}^{\infty} \frac{C_{j-1}}{z^{2j-1}} \cdot \Big(\sum_{p=j}^{\infty} \frac{C_{p-j}}{z^{2(p-j)+1}} \Big) \\
&= \frac{1}{z} + \frac{1}{z} \sum_{j=1}^{\infty} \frac{C_{j-1}}{z^{2j-1}} \cdot G_\mu(z) \\
&= \frac{1}{z} + \frac{1}{z} G_\mu(z)^2.
\end{aligned}$$

It follows that G_μ satisfies the quadratic equation

$$G_\mu(z)^2 - zG_\mu(z) + 1 = 0, \quad z \in \mathbb{C}^+.$$

(The above computations only obtain this for a z such that $|z|$ is large enough, but after that the fulfilling of the quadratic equation is extended to all of $\mathbb{C}^+$ by analyticity.) By solving this quadratic equation we find that $G_\mu(z) = (z \pm \sqrt{z^2-4})/2$, and from the condition $\lim_{|z|\to\infty} zG_\mu(z) = 1$ we see that the "$-$" sign has to be chosen in the "$\pm$" of the quadratic formula. □

Finally, let us remark that the analytic function found in Equation (2.23) has a continuous extension to $\mathbb{C}^+ \cup \mathbb{R}$, where the extension acts on $\mathbb{R}$ by:

$$t \mapsto g(t) := \begin{cases} (t - i\sqrt{4-t^2})/2 & \text{if } |t| \le 2 \\ (t - \sqrt{t^2-4})/2 & \text{if } |t| > 2. \end{cases}$$

By taking the imaginary part of g, and by using the observation made at the end of Remark 2.20, we see why the semicircular density is the appropriate choice in the statement of Proposition 2.15.

Exercises

EXERCISE 2.22. Let $S \in B(l^2)$ be the shift operator considered in Remark 2.5, and let $\{\xi_n \mid n \ge 0\}$ be the orthonormal basis of l^2 considered in the same remark.

(1) Let $(m,n) \neq (0,0)$ be in $\big(\mathbb{N} \cup \{0\}\big)^2$. Based on the fact that $\langle S^m (S^*)^n \xi_n \,,\, \xi_m \rangle = 1$, show that

$$S^m (S^*)^n \notin \text{span}\{ S^k (S^*)^l \mid \text{ either } k > m, \text{ or } k = m \text{ and } l > n \}.$$

(2) By using the result in part (1) and the lexicographic order on $(\mathbb{N} \cup \{0\})^2$, prove that the operators $(S^m(S^*)^n)_{m,n\geq 0}$ form a linearly independent family in $B(l^2)$.

EXERCISE 2.23. Write a proof of Proposition 2.13 which only uses the framework introduced in Notations 2.1 and 2.3, and does not appeal to the representation of a as a shift operator.

EXERCISE 2.24. Re-derive the formula (1.16) from Example 1.14, by starting from Equation (1.14) of the same example and by using the Stieltjes inversion formula.

LECTURE 3

C^*-probability spaces

C^*-algebras provide a natural environment where non-commutative probabilistic ideas can be seen at work. In this lecture we provide some basic background for our readers who are not familiar with them. The emphasis will be on the concept of C^*-probability space and on the relations between spectrum and $*$-distribution for a normal element in a C^*-probability space.

The line followed by our sequence of lectures does not require any substantial C^*-algebra apparatus, and we hope it will be comprehensible to present the fairly few and elementary C^*-algebra facts which are needed, at the places where they appear. We will keep to minimum the number of statements which have to be accepted without proof – for instance in the present lecture the only such statement is that of Theorem 3.1, which collects some fundamental facts about the spectral theory of normal elements.

Functional calculus in a C^*-algebra

A **C^*-probability space** is a $*$-probability space $(\mathcal{A}, \varphi)$ where the $*$-algebra $\mathcal{A}$ is required to be a unital C^*-algebra. Being a unital **C^*-algebra** means that (in addition to being a unital $*$-algebra) $\mathcal{A}$ is endowed with a norm $||\cdot|| : \mathcal{A} \to [0, \infty)$ which makes it a complete normed vector space, and such that we have:

$$||ab|| \leq ||a|| \cdot ||b||, \quad \forall\, a, b \in \mathcal{A}; \tag{3.1}$$

$$||a^*a|| = ||a||^2, \quad \forall\, a \in \mathcal{A}. \tag{3.2}$$

Out of the very extensive theory of C^*-algebras we will only need some basic facts of spectral theory, which are reviewed in a concentrated way in the following theorem. Recall that if $\mathcal{A}$ is a unital C^*-algebra and if $a \in \mathcal{A}$, then the **spectrum** of a is the set

$$\mathrm{Sp}(a) = \{z \in \mathbb{C} \mid z1_{\mathcal{A}} - a \text{ is not invertible}\}.$$

THEOREM 3.1. *Let $\mathcal{A}$ be a unital C^*-algebra.*

(1) *For every $a \in \mathcal{A}$, $\mathrm{Sp}(a)$ is a non-empty compact subset of $\mathbb{C}$, contained in the disc $\{z \in \mathbb{C} \mid |z| \leq ||a||\}$.*

(2) *Let a be a normal element of $\mathcal{A}$, and consider the algebra $C(\mathrm{Sp}(a))$ of complex-valued continuous functions on* $\mathrm{Sp}(a)$. *There exists a map $\Phi : C(\mathrm{Sp}(a)) \to \mathcal{A}$ which has the following properties:*

(i) *Φ is a unital $*$-algebra homomorphism.*

(ii) $||\Phi(f)|| = ||f||_\infty$, $\forall\, f \in C(\mathrm{Sp}(a))$ *(where for $f \in C(\mathrm{Sp}(a))$ we define* $||f||_\infty := \sup\{|f(z)| \mid z \in \mathrm{Sp}(a)\}$*).*

(iii) *Denoting by $id : \mathrm{Sp}(a) \to \mathbb{C}$ the identity function $id(z) = z$, we have that $\Phi(id) = a$.*

REMARKS 3.2. Let $\mathcal{A}$ be a unital C^*-algebra, let a be a normal element of $\mathcal{A}$, and let $\Phi : C(\mathrm{Sp}(a)) \to \mathcal{A}$ have the properties (i), (ii) and (iii) listed in Theorem 3.1.2.

(1) Condition (ii) (together with the linearity part of (i)) implies that Φ is one-to-one. Hence in a certain sense, Φ provides us with a copy of the algebra $C(\mathrm{Sp}(a))$ which sits inside $\mathcal{A}$.

(2) Suppose that $p : \mathrm{Sp}(a) \to \mathbb{C}$ is a polynomial in z and $\bar{z}$, i.e. it is of the form

$$p(z) = \sum_{j,k=0}^{n} \alpha_{j,k} z^j \bar{z}^k, \quad z \in \mathrm{Sp}(a). \tag{3.3}$$

Then the properties (i) and (iii) of Φ immediately imply that

$$\Phi(p) = \sum_{j,k=0}^{n} \alpha_{j,k} a^j (a^*)^k. \tag{3.4}$$

(3) The preceding remark shows that the values of Φ on polynomials in z and $\bar{z}$ are uniquely determined. Since these polynomials are dense in $C(\mathrm{Sp}(a))$ with respect to uniform convergence, and since (by (i)+(ii)) Φ is continuous with respect to uniform convergence, it follows that the properties (i), (ii) and (iii) *determine Φ uniquely.*

(4) The name commonly used for Φ is **functional calculus with continuous functions** for the element a. A justification for this name is seen by looking at polynomials p such as the one appearing in Equation (3.3). Indeed, for such a p, the corresponding element $\Phi(p) \in \mathcal{A}$ (appearing in (3.4)) is what one naturally tends to denote as "$p(a)$." It is in fact customary to use the notation

$$\text{“}f(a)\text{” instead of “}\Phi(f)\text{”} \tag{3.5}$$

when f is an arbitrary continuous function on $\mathrm{Sp}(a)$ (not necessarily a polynomial in z and $\bar{z}$). The notation (3.5) will be consistently used in the remainder of this lecture.

REMARKS 3.3. Let $\mathcal{A}$ be a unital C^*-algebra. Theorem 3.1.2 contains in a concentrated way a good amount of information about the spectra of the normal elements of $\mathcal{A}$. We record here a few facts which are immediately implied by it. (Note: it is handy to record these facts as consequences of Theorem 3.1; but the reader should be warned that in a detailed development of basic C^*-algebra theory, some of these facts would be proved directly from the axioms, preceding the discussion about functional calculus.)

(1) If a is a normal element of $\mathcal{A}$, then

$$||a|| = ||a^*|| = \sup\{|z| \mid z \in \text{Sp}(a)\}. \tag{3.6}$$

This is seen by using (ii) of Theorem 3.1.2 for the functions id and $\overline{id}$ on Sp(a).

(2) If x is a selfadjoint element of $\mathcal{A}$ then Sp$(x) \subset \mathbb{R}$. Indeed, when we apply (ii) of Theorem 3.1.2 to the function $id - \overline{id}$ on Sp(x), we get that

$$||x - x^*|| = \sup\{|z - \bar{z}| \mid z \in \text{Sp}(x)\}. \tag{3.7}$$

The left-hand side of (3.7) is 0; hence so must be the right-hand side of (3.7), and this implies that Sp$(x) \subset \{z \in \mathbb{C} : z - \bar{z} = 0\} = \mathbb{R}$.

Conversely, if $x \in \mathcal{A}$ is normal and has Sp$(x) \subset \mathbb{R}$, then it follows that $x = x^*$; this is again by (3.7), where now we know that the right-hand side vanishes.

(3) If u is a unitary element of $\mathcal{A}$, then Sp$(u) \subset \mathbb{T} = \{z \in \mathbb{C} : |z| = 1\}$. And conversely, if $u \in \mathcal{A}$ is normal and has Sp $\subset \mathbb{T}$ then u has to be a unitary. The argument is the same as in part (2) of this remark, where now we use the equation:

$$||1 - u^*u|| = \sup\{|\, 1 - |z|^2 \,| \mid z \in \text{Sp}(u)\}.$$

The following statement is known as the "spectral mapping theorem."

THEOREM 3.4. *Let $\mathcal{A}$ be a unital C^*-algebra, let a be a normal element of $\mathcal{A}$, and let $f : \text{Sp}(a) \to \mathbb{C}$ be a continuous function. Then the element $f(a) \in \mathcal{A}$ (defined by functional calculus) has*

$$\text{Sp}(\, f(a)\,) = f(\,\text{Sp}(a)\,). \tag{3.8}$$

PROOF. By considering functions of the form $z \mapsto f(z) - \lambda$ on Sp(a) (where $\lambda \in \mathbb{C}$), one immediately sees that it suffices to prove the following statement. Let $\mathcal{A}$ and a be as above, and let $g : \text{Sp}(a) \to \mathbb{C}$ be a continuous function, then:

$$g(a) \text{ is invertible in } \mathcal{A} \Leftrightarrow 0 \notin g(\,\text{Sp}(a)\,). \tag{3.9}$$

The implication "$\Leftarrow$" in (3.9) is immediate: if $0 \notin g(\,\text{Sp}(a)\,)$, then one can define the continuous function $h = 1/g : \text{Sp}(a) \to \mathbb{C}$, and the properties of functional calculus imply that the element $h(a) \in \mathcal{A}$ is an inverse for $g(a)$.

In order to prove the implication "$\Rightarrow$" in (3.9), we proceed by contradiction. Assume that $g(a)$ is invertible in $\mathcal{A}$, but that nevertheless there exists $z_o \in \mathrm{Sp}(a)$ such that $g(z_o) = 0$. Let us pick a positive number $\alpha > ||g(a)^{-1}||$. Because of the fact that $g(z_o) = 0$, one can construct a function $h \in C(\mathrm{Sp}(a))$ such that $h(z_o) = \alpha$ while at the same time $||g \cdot h||_\infty \leq 1$. (Indeed, there exists $\varepsilon > 0$ such that $|g(z)| < 1/\alpha$ for all $z \in \mathrm{Sp}(a)$ with $|z - z_o| < \varepsilon$, and one can construct h with values in $[0, \alpha]$ and supported inside the disc of radius $\varepsilon/2$ centered at z_o. For instance $h(z) := \alpha \cdot \max(0, 1 - 2|z - z_o|/\varepsilon)$ will do.) From the properties of functional calculus it follows that the element $h(a) \in \mathcal{A}$ is such that its norm equals

$$||h(a)|| = ||h||_\infty \geq \alpha$$

while at the same time we have:

$$||g(a) \cdot h(a)|| = ||g \cdot h||_\infty \leq 1.$$

We then get that

$$\alpha \leq ||h(a)|| = ||g(a)^{-1} \cdot (g(a) \cdot h(a))|| \leq ||g(a)^{-1}|| \cdot ||g(a) \cdot h(a))|| < \alpha,$$

a contradiction. □

REMARK 3.5. Let $\mathcal{A}$ be a unital C^*-algebra. It is customary to define the set of **positive elements** of $\mathcal{A}$ as

$$\mathcal{A}^+ := \{p \in \mathcal{A} \mid p = p^* \text{ and } \mathrm{Sp}(p) \subset [0, \infty)\}. \tag{3.10}$$

It is fairly easy to show that

$$p, q \in \mathcal{A}^+, \ \alpha, \beta \in [0, \infty) \ \Rightarrow \ \alpha p + \beta q \in \mathcal{A}^+, \tag{3.11}$$

i.e. that $\mathcal{A}^+$ is a convex cone in the real vector space of selfadjoint elements of $\mathcal{A}$ – see Exercise 3.18 at the end of the lecture. Moreover, the cone $\mathcal{A}^+$ is "pointed," in the sense that $\mathcal{A}^+ \cap (-\mathcal{A}^+) = \{0\}$. (Or in other words: if a selfadjoint element $x \in \mathcal{A}$ is such that both x and $-x$ are in $\mathcal{A}^+$, then $x = 0$. This is indeed so, because $x, -x \in \mathcal{A}^+ \Rightarrow \mathrm{Sp}(x) \subset [0, \infty) \cap (-\infty, 0] = \{0\} \Rightarrow ||x|| = \sup\{|z| \mid z \in \mathrm{Sp}(x)\} = 0$.)

Note also that the spectral mapping theorem provides us with a rich supply of positive elements in $\mathcal{A}$. Indeed, if a is an arbitrary normal element of $\mathcal{A}$ and if $f : \mathrm{Sp}(a) \to [0, \infty)$ is a continuous function, then the element $f(a)$ is in $\mathcal{A}^+$ (it is selfadjoint because $f = \overline{f}$, and has spectrum in $[0, \infty)$ by Theorem 3.4).

Recall from Lecture 1 that a linear functional $\varphi : \mathcal{A} \to \mathbb{C}$ is declared to be positive when it satisfies the condition $\varphi(a^*a) \geq 0, \forall\, a \in \mathcal{A}$. This brings up the question of whether there is any relation between $\mathcal{A}^+$ and the set $\{a^*a \mid a \in \mathcal{A}\}$. It is quite convenient that these two sets actually coincide.

PROPOSITION 3.6. *Let $\mathcal{A}$ be a unital C^*-algebra, and consider the set $\mathcal{A}^+$ of positive elements of $\mathcal{A}$ (defined as in Equation (3.10) of the preceding remark). Then*

$$\mathcal{A}^+ = \{a^*a \mid a \in \mathcal{A}\}. \tag{3.12}$$

PROOF. "$\subset$" Let p be in $\mathcal{A}^+$, and define $a = f(p)$ (functional calculus) where f is the square root function on the spectrum of p. Then the properties of functional calculus immediately give us that $a = a^*$ (coming from $f = \bar{f}$) and that $a^*a = a^2 = (f^2)(p) = p$.

"$\supset$" Fix an $a \in \mathcal{A}$, for which we want to prove that $a^*a \in \mathcal{A}^+$. It is clear that a^*a is selfadjoint, the issue is to prove that $\text{Sp}(a^*a) \subset [0, \infty)$.

Consider the functions $f, g : \text{Sp}(a^*a) \to [0, \infty)$ defined by

$$f(t) := \max(0, t), \quad g(t) := \max(0, -t), \quad t \in \text{Sp}(a^*a),$$

and denote $f(a^*a) =: x$, $g(a^*a) =: y$. We have that $x, y \in \mathcal{A}^+$ (cf. the second paragraph of Remark 3.5). The properties of functional calculus also give us that

$$x - y = a^*a, \quad xy = yx = 0. \tag{3.13}$$

Consider now the element $b := ay \in \mathcal{A}$. We have (by direct calculation and by using (3.13)) that

$$b^*b = ya^*ay = y(x - y)y = -y^3.$$

Since $y \in \mathcal{A}^+$, it is immediate by functional calculus that $y^3 \in \mathcal{A}^+$; hence it follows that $b^*b \in -\mathcal{A}^+$. We leave it as an exercise to the meticulous reader to go through the details of why "$b^*b \in -\mathcal{A}^+$" implies "$b = 0$" – cf. Exercise 3.20 at the end of the lecture. Here we will assume that this is proved, and will finish the argument as follows:

$$\begin{aligned} y^3 = -b^*b = 0 \quad &\Rightarrow \quad \{t^3 \mid t \in \text{Sp}(y)\} = \text{Sp}(y^3) = \{0\} \\ &\Rightarrow \quad \text{Sp}(y) = \{0\} \\ &\Rightarrow \quad ||y|| = \sup\{|t| \mid t \in \text{Sp}(y)\} = 0. \end{aligned}$$

So we found that $y = 0$, and therefore $a^*a = x - y = x \in \mathcal{A}^+$. □

C^*-probability spaces

DEFINITION 3.7. A ***C^**-probability space** is a $*$-probability space $(\mathcal{A}, \varphi)$ where $\mathcal{A}$ is a unital C^*-algebra.

Let us note that in the C^*-framework, the expectation functional is automatically continuous. More precisely, we have the following.

PROPOSITION 3.8. *Let $(\mathcal{A}, \varphi)$ be a C^*-probability space. Then*

$$|\varphi(a)| \leq ||a||, \quad \forall\ a \in \mathcal{A}. \tag{3.14}$$

PROOF. We first pick an arbitrary element $p \in \mathcal{A}^+$. We know that $\varphi(p) \in [0, \infty)$ (by Proposition 3.6 and the positivity of φ). We claim that:

$$\varphi(p) \leq ||p||. \tag{3.15}$$

Indeed, we have (by Theorem 3.1.1 and Equation (3.10) of Remark 3.5) that

$$\mathrm{Sp}(p) \subset \{z \in \mathbb{C} \mid |z| \leq ||p||\} \cap [0, \infty) = [\,0, ||p||\,].$$

As a consequence, we can use functional calculus to define the element $b := (\,||p|| - p\,)^{1/2} \in \mathcal{A}$ (or more precisely, $b := f(p)$ where $f \in C(\,\mathrm{Sp}(p)\,)$ is defined by $f(t) = (\,||p|| - t\,)^{1/2}$, $t \in \mathrm{Sp}(p)$). It is immediate that $b = b^*$ and that $p + b^2 = ||p|| \cdot 1_{\mathcal{A}}$; therefore

$$||p|| - \varphi(p) = \varphi(b^*b) \geq 0,$$

and (3.15) is obtained.

Now for an arbitrary $a \in \mathcal{A}$ we have

$$\begin{aligned}
|\varphi(a)| &= |\varphi(1_{\mathcal{A}}^* \cdot a)| \\
&\leq \varphi(a^*a)^{1/2} && \text{(by Cauchy–Schwarz – cf. Lecture 1)} \\
&\leq ||a^*a||^{1/2} && \text{(by (3.15), where we take } p = a^*a) \\
&= ||a|| && \text{(by (3.2)).}
\end{aligned}$$

□

REMARK 3.9. The following partial converse of Proposition 3.8 is also true. Let $\mathcal{A}$ be a unital C^*-algebra. Let $\varphi : \mathcal{A} \to \mathbb{C}$ be a linear functional such that $|\varphi(a)| \leq ||a||$, $\forall\ a \in \mathcal{A}$, and such that $\varphi(1_{\mathcal{A}}) = 1$ (where $1_{\mathcal{A}}$ is the unit of $\mathcal{A}$). Then φ is positive, and hence $(\mathcal{A}, \varphi)$ is a C^*-probability space. See Exercise 3.21 at the end of the lecture.

EXAMPLE 3.10. Let Ω be a compact Hausdorff topological space, and let μ be a Radon probability measure on the Borel σ-algebra of Ω. (Asking the probability measure μ to be "a Radon measure" amounts to requesting that for every Borel set $A \subset \Omega$ one has

$$\mu(A) = \sup\{\mu(K) \mid K \subset A,\ \text{compact}\} = \inf\{\mu(D) \mid D \supset A,\ \text{open}\}.$$

In many natural situations – when Ω is a compact metric space, for instance – one has that every probability measure on the Borel σ-algebra of Ω is actually a Radon measure.)

Consider the algebra $\mathcal{A} = C(\Omega)$ of complex-valued continuous functions on Ω, and let $\varphi : \mathcal{A} \to \mathbb{C}$ be defined by

$$\varphi(f) = \int_\Omega f \, d\mu, \quad f \in \mathcal{A}. \tag{3.16}$$

Then $(\mathcal{A}, \varphi)$ is a C^*-probability space, and all the elements of $\mathcal{A}$ are normal. The functional calculus with continuous functions for an element $a \in \mathcal{A}$ is reduced in this case to performing a functional composition – see Exercise 3.22 below.

There are two important theorems in functional analysis which are worth recalling in connection to this example. First, a basic theorem of Riesz states that every positive linear functional on $C(\Omega)$ can be put in the form (3.16) for an appropriate Radon probability measure μ. Second, a theorem of Gelfand states that every commutative unital C^*-algebra $\mathcal{A}$ can be identified as $C(\Omega)$ for a suitable compact Hausdorff space Ω. Hence the example presented here is the "generic" one, as far as commutative C^*-probability spaces are concerned.

In non-commutative examples, C^*-algebras appear most frequently as $*$-subalgebras $\mathcal{A} \subset B(\mathcal{H})$ ($\mathcal{H}$ Hilbert space), such that $\mathcal{A}$ is closed in the norm-topology of $B(\mathcal{H})$. We present here the example of this kind which is the C^*-counterpart of Example 1.4.4 from Lecture 1.

EXAMPLE 3.11. Let G be a discrete group, and let $\lambda : G \to B(\,l^2(G)\,)$ be its **left regular representation**. This is defined by the formula

$$\lambda(g)\xi_h = \xi_{gh}, \quad \forall \, g, h \in G, \tag{3.17}$$

where $\{\xi_h \mid h \in G\}$ is the canonical orthonormal basis of $l^2(G)$. (That is, every $\lambda(g)$ is a unitary operator on $l^2(G)$, which permutes the orthonormal basis $\{\xi_h \mid h \in G\}$ according to the formula (3.17).) It is not hard to show that the operators $(\,\lambda(g)\,)_{g\in G}$ are linearly independent, and that their linear span is a unital $*$-algebra of $B(\,l^2(G)\,)$, isomorphic to the group algebra $\mathbb{C}G$ from Example 1.4.4. (See Exercise 3.24 below.) The closure in the norm-topology:

$$C^*_{\text{red}}(G) := \text{cl}\Big(\text{span}\{\lambda(g) \mid g \in G\} \Big)$$

is then a unital C^*-algebra of operators on $l^2(G)$; it is called the **reduced C^*-algebra** of the group G.

Let e be the unit of G and let ξ_e be the corresponding vector in the canonical basis of $l^2(G)$. Let τ be the vector-state defined by ξ_e on $C^*_{\text{red}}(G)$:

$$\tau(T) = \langle T\xi_e, \xi_e \rangle, \quad T \in C^*_{\text{red}}(G). \tag{3.18}$$

Then $(C^*_{\text{red}}(G), \tau)$ is an example of C^*-probability space.

Let us observe that when T is the image of $\sum_g \alpha_g g \in \mathbb{C}G$ via the canonical isomorphism $\mathbb{C}G \simeq \text{span}\{\lambda(g) \mid g \in G\} \subset C^*_{\text{red}}(G)$, then we get

$$\tau(T) = \big\langle \big(\sum_g \alpha_g \lambda(g)\big)\xi_e, \xi_e \big\rangle = \big\langle \sum_g \alpha_g \xi_g, \xi_e \big\rangle = \alpha_e.$$

So, via natural identifications, τ extends the trace τ_G on $\mathbb{C}G$ which appeared in Example 1.4.4. Thus, in a certain sense, $(C^*_{\text{red}}(G), \tau)$ is an upgrade of $(\mathbb{C}G, \tau_G)$ from the $*$-algebraic framework to the C^*-algebraic one.

Moreover, the C^*-probability space $(C^*_{\text{red}}(G), \tau)$ retains the pleasing features which we trust that the reader has verified (in the course of solving Exercise 1.5) for the canonical trace on $\mathbb{C}G$. That is, we have the following.

PROPOSITION 3.12. *In the framework of the preceding example, the functional τ is a faithful trace on $C^*_{\text{red}}(G)$.*

PROOF. The traciality of τ is immediate. Indeed, since τ is continuous (by Proposition 3.8) and since the linear span of the operators $\{\lambda(g) \mid g \in G\}$ is dense in $C^*_{\text{red}}(G)$, it suffices to check that

$$\tau(\,\lambda(g_1) \cdot \lambda(g_2)\,) = \tau(\,\lambda(g_2) \cdot \lambda(g_1)\,), \quad \forall\, g_1, g_2 \in G. \tag{3.19}$$

But (3.19) is obviously true – both sides are equal to 1 when $g_1 = g_2^{-1}$, and are equal to 0 otherwise.

In order to prove that τ is faithful on $C^*_{\text{red}}(G)$, it is convenient that (in addition to the left translation operators $\lambda(g)$) we look at *right translation operators* on $l^2(G)$. So, for every $g \in G$ let us consider the unitary operator $\rho(g)$ on $l^2(G)$ which permutes the canonical basis $(\xi_h)_{h\in G}$ according to the formula:

$$\rho(g)\xi_h = \xi_{hg^{-1}}, \quad h \in G.$$

Then $\rho : G \to B(\,l^2(G)\,)$ is a unitary representation, called the *right* regular representation of G. It is immediately verified that the left and the right translation operators commute with each other:

$$\rho(g)\lambda(g') = \lambda(g')\rho(g), \quad \forall\, g, g' \in G. \tag{3.20}$$

If in (3.20) we fix an element $g \in G$ and make linear combinations of the operators $\lambda(g')$, followed by approximations in norm, we obtain that

$$\rho(g)T = T\rho(g), \quad \forall\, g \in G, \ \ \forall\, T \in C^*_{\text{red}}(G). \tag{3.21}$$

Now, let $T \in C^*_{\text{red}}(G)$ be such that $\tau(T^*T) = 0$. Since

$$\tau(T^*T) = \langle T^*T\xi_e, \xi_e\rangle = ||T\xi_e||^2,$$

we thus have that $T\xi_e = 0$. But then for every $g \in G$ we find that

$$T\xi_g = T(\rho(g^{-1})\xi_e) = \rho(g^{-1})(T\xi_e) = \rho(g^{-1}) \cdot 0 = 0.$$

(The second equality follows by Equation (3.21).) So T vanishes on the orthonormal basis $(\xi_g)_{g\in G}$ of $l^2(G)$, and this implies that $T = 0$. □

∗-distribution, norm and spectrum for a normal element

PROPOSITION 3.13. *Let $(\mathcal{A}, \varphi)$ be a C^*-probability space, and let a be a normal element of $\mathcal{A}$ Then a has a ∗-distribution μ in analytic sense (as described in Definition 1.8). Moreover:*

(1) The support of μ is contained in the spectrum of a.

(2) For $f \in C(\mathrm{Sp}(a))$ we have the formula

$$\int f \, d\mu = \varphi(f(a)), \tag{3.22}$$

where on the right-hand side $f(a) \in \mathcal{A}$ is obtained by functional calculus, and on the left-hand side μ is viewed as a probability measure on $\mathrm{Sp}(a)$.

PROOF. Let $\Phi : C(\mathrm{Sp}(a)) \to \mathcal{A}$ be the functional calculus for a, as in Theorem 3.1.2 ($\Phi(f) = f(a)$, for $f \in C(\mathrm{Sp}(a))$). Then $\varphi \circ \Phi : C(\mathrm{Sp}(a)) \to \mathbb{C}$ is a positive linear functional, so by the theorem of Riesz mentioned in Example 3.10 there exists a probability measure μ on the Borel σ-algebra of $\mathrm{Sp}(a)$ such that

$$(\varphi \circ \Phi)(f) = \int f \, d\mu, \quad \forall \, f \in C(\mathrm{Sp}(a)). \tag{3.23}$$

If we set f in (3.23) to be of the form $f(z) = z^m \bar{z}^n$ for some $m, n \geq 0$, then $\Phi(f) = a^m(a^*)^n$ (cf. Remark 3.2.2), and (3.23) gives us that

$$\varphi(a^m(a^*)^n) = \int_{\mathrm{Sp}(a)} z^m \bar{z}^n \, d\mu(z), \quad \forall \, m, n \geq 0. \tag{3.24}$$

Of course, the measure μ of (3.23), (3.24) can also be viewed as a compactly supported measure on $\mathbb{C}$, with $\mathrm{supp}(\mu) \subset \mathrm{Sp}(a)$. In this interpretation, (3.24) tells us that μ is the ∗-distribution of a, in analytic sense, while (3.23) becomes (3.22). □

COROLLARY 3.14. *Let $(\mathcal{A}, \varphi)$ be a ∗-probability space. If $(\mathcal{A}, \varphi)$ admits a representation on a Hilbert space (in the sense of Definition 1.6), then every normal element of $\mathcal{A}$ has a ∗-distribution in analytic sense.*

PROOF. The existence of representations means in particular that we can find a C^*-probability space $(\mathcal{B}, \psi)$ and a unital $*$-homomorphism $\Phi : \mathcal{A} \to \mathcal{B}$ such that $\psi \circ \Phi = \varphi$. For every normal element $a \in \mathcal{A}$, it is clear that $b := \Phi(a)$ is a normal element of $\mathcal{B}$; hence, by Proposition 3.13, b has a $*$-distribution μ in analytic sense. But then for every $m, n \geq 0$ we can write:

$$\varphi(\, a^m(a^*)^n \,) = \psi(\Phi(\, a^m(a^*)^n \,)) = \psi(\, b^m(b^*)^n \,) = \int z^m \bar{z}^n \, d\mu(z),$$

which shows that μ is the $*$-distribution of a as well. □

In the rest of this section we look at some additional facts which can be derived for a C^*-probability space where the expectation is faithful.

PROPOSITION 3.15. *Let $(\mathcal{A}, \varphi)$ be a C^*-probability space such that φ is faithful. Let a be a normal element of $\mathcal{A}$, and let μ be the $*$-distribution of a in analytic sense. Then the support of μ is equal to* $\mathrm{Sp}(a)$.

PROOF. The inclusion "$\subset$" was observed in Proposition 3.13, so we only have to prove "$\supset$". Let us fix an element $\lambda \in \mathrm{Sp}(a)$, and assume by contradiction that $\lambda \notin \mathrm{supp}(\mu)$. Since $\mathbb{C} \setminus \mathrm{supp}(\mu)$ is an open set of μ-measure 0, it follows that we can find $r > 0$ such that $\mu(\, B(\lambda; r)\,) = 0$, where $B(\lambda; r) := \{z \in \mathbb{C} \mid |z - \lambda| < r\}$. Let $f : \mathrm{Sp}(a) \to [0, 1]$ be a continuous function such that $f(\lambda) = 1$ and such that $f(z) = 0$ for all $z \in \mathrm{Sp}(a)$ with $|z - \lambda| \geq r$ (e.g. $f(z) = \max(0, 1 - |\lambda - z|/r)$ will do); and let us define $b := f(a) \in \mathcal{A}$, by functional calculus. Property (ii) from Theorem 3.1.2 gives us that $||b|| = 1$, so in particular we know that $b \neq 0$. On the other hand we have that

$$\begin{aligned} \varphi(b^*b) &= \varphi(b^2) && \text{(since } f = \bar{f}\text{, which implies } b = b^*) \\ &= \int f^2 \, d\mu && \text{(since } b^2 = f^2(a)\text{, and by Prop. 3.13)} \\ &\leq \int_{B(\lambda;r)} 1 \, d\mu, \end{aligned}$$

with the last inequality holding because f^2 is bounded above by the characteristic function of $B(\lambda; r)$. We thus get

$$\varphi(b^*b) \leq \mu(\, B(\lambda; r)\,) = 0,$$

and this contradicts the faithfulness of φ. □

REMARK 3.16. The above proposition can be read as follows: if $(\mathcal{A}, \varphi)$ is a C^*-probability space such that φ is faithful, and if a is a normal element of $\mathcal{A}$, then knowledge of the $*$-distribution μ of a allows us to compute the spectrum of a, via the formula

$$\mathrm{Sp}(a) = \mathrm{supp}(\mu). \tag{3.25}$$

Note that knowledge of μ will then also give us the norm of a – indeed, from (3.25) and Equation (3.6) of Remark 3.3.1 it follows that

$$||a|| = \sup\{|z| \mid z \in \mathrm{supp}(\mu)\}. \tag{3.26}$$

The following proposition indicates another (more direct) way of computing the norm of a from combinatorial information on $*$-moments.

PROPOSITION 3.17. *Let $(\mathcal{A}, \varphi)$ be a C^*-probability space such that φ is faithful. For every $a \in \mathcal{A}$ (normal or not) we have that*

$$||a|| = \lim_{n\to\infty} \varphi(\,(a^*a)^n\,)^{1/2n}. \tag{3.27}$$

PROOF. Equivalently, we have to show that

$$||p|| = \lim_{n\to\infty} \varphi(p^n)^{1/n}, \tag{3.28}$$

where $p := a^*a \in \mathcal{A}^+$ and where we used the C^*-axiom (3.2). An immediate application of functional calculus shows that $p^n \in \mathcal{A}^+, \forall\, n \geq 1$; so the sequence appearing on the right-hand side of Equation (3.28) consists of non-negative numbers. Note also that for every $n \geq 1$ we have:

$$\begin{aligned} \varphi(p^n)^{1/n} &\leq (\,||p^n||\,)^{1/n} && \text{(by Proposition 3.8)} \\ &\leq (\,||p||^n\,)^{1/n} && \text{(by Equation (3.1))} \\ &= ||p||. \end{aligned}$$

So what we actually have to do is to fix an $\alpha \in (0, ||p||)$, and show that $\varphi(p^n)^{1/n} > \alpha$ if n is sufficiently large.

Now, we have that $\mathrm{Sp}(p) \subset [0, ||p||]$ (same argument as in the proof of Proposition 3.8). Moreover, from Remark 3.3.1 we infer that $||p|| \in \mathrm{Sp}(p)$. Let μ be the $*$-distribution of p, in analytic sense. Then $||p|| \in \mathrm{supp}(\mu)$ (by Proposition 3.15), and it follows that we have

$$\mu(\,[\beta, ||p||]\,) > 0, \quad \forall\, 0 < \beta < ||p||. \tag{3.29}$$

For the number $\alpha \in (0, ||p||)$ which was fixed above, let us choose a $\beta \in (\alpha, ||p||)$ (for instance $\beta = (\alpha + ||p||)/2$ will do). Then we can write, for every $n \geq 1$:

$$\varphi(p^n) = \int_{\mathrm{Sp}(p)} t^n \, d\mu(t) \geq \int_{\mathrm{Sp}(p)\cap[\beta,||p||]} t^n \, d\mu(t) \geq \beta^n \cdot \mu(\,[\beta, ||p||]\,).$$

(The first equality follows by Proposition 3.13.2.) Hence

$$\varphi(p^n)^{1/n} \geq \beta \cdot \mu(\,[\beta, ||p||]\,)^{1/n}, \quad \forall\, n \geq 1, \tag{3.30}$$

and the right-hand side of (3.30) exceeds α when n is sufficiently large (since (3.29) implies that $\mu(\,[\beta, ||p||]\,)^{1/n} \to 1$ as $n \to \infty$). □

Exercises

Exercises 3.18–3.20 fill in the details left during the discussion on positive elements of a C^*-algebra (cf. Remark 3.5, proof of Proposition 3.6).

EXERCISE 3.18. Let $\mathcal{A}$ be a unital C^*-algebra.

(1) By using functional calculus, prove that if x is a selfadjoint element of $\mathcal{A}$ and if $\alpha \in \mathbb{R}$ is such that $\alpha \geq ||x||$, then we have

$$||\alpha - x|| = \alpha - \inf(\,\mathrm{Sp}(x)\,).$$

(2) By using the formula found in part (1) of the exercise, prove that if x, y are selfadjoint elements of $\mathcal{A}$, then

$$\inf(\,\mathrm{Sp}(x+y)\,) \geq \inf(\,\mathrm{Sp}(x)\,) + \inf(\,\mathrm{Sp}(y)\,).$$

(3) Consider the set $\mathcal{A}^+$ of positive elements of $\mathcal{A}$ (defined as in Equation (3.10) of Remark 3.5). Prove that if $p, q \in \mathcal{A}^+$ and if $\alpha, \beta \in [0, \infty)$, then $\alpha p + \beta q \in \mathcal{A}^+$.

EXERCISE 3.19. Let $\mathcal{A}$ be a unital C^*-algebra and let a, b be elements of $\mathcal{A}$. Prove that $\mathrm{Sp}(ab) \cup \{0\} = \mathrm{Sp}(ba) \cup \{0\}$.
[This is a version of the exercise, usually given in a basic algebra course, which goes as follows: for a, b elements of a unital ring, prove that $1-ab$ is invertible if and only if $1 - ba$ is invertible.]

EXERCISE 3.20. Let $\mathcal{A}$ be a unital C^*-algebra, and let $b \in \mathcal{A}$ be such that $\mathrm{Sp}(b^*b) \subset (-\infty, 0]$. The goal of this exercise is to draw the conclusion that $b = 0$.

(1) Prove that $b^*b+bb^* \in -\mathcal{A}^+$ (where $\mathcal{A}^+$ is defined as in Equation (3.10) of Remark 3.5).
[Hint: One has $-b^*b \in \mathcal{A}^+$ by hypothesis and $-bb^* \in \mathcal{A}^+$ due to Exercise 3.19. Then use Exercise 3.18.]

(2) Let $x := (b+b^*)/2$ and $y := (b-b^*)/2i$ be the real and imaginary parts of b. Verify that $b^*b + bb^* = 2(x^2 + y^2)$, and conclude from there that $x^2 + y^2 \in \mathcal{A}^+ \cap (-\mathcal{A}^+)$.

(3) Prove that $b = 0$.

EXERCISE 3.21. (1) Let K be a non-empty compact subset of $[0, \infty)$, and consider the algebra $C(K)$ of complex-valued continuous functions on K. Suppose that $\varphi : C(K) \to \mathbb{C}$ is a linear functional such that $|\varphi(f)| \leq ||f||_\infty$, $\forall\, f \in C(K)$, and such that $\varphi(1_{C(K)}) = 1$ (where $1_{C(K)}$ is the function constantly equal to 1). Let h be the function in $C(K)$ defined by $h(t) = t$, for $t \in K$. Prove that $\varphi(h) \geq 0$.
[Hint: In order to verify that $\varphi(h) \in \mathbb{R}$ look at functions of the form $h + i\alpha 1_{C(K)}$, $\alpha \in \mathbb{R}$. Then in order to verify that $\varphi(h) \geq 0$ look at functions of the form $h - \alpha 1_{C(K)}$, $\alpha \in [0, \infty)$.]

(2) Let $\mathcal{A}$ be a unital C^*-algebra. Let $\varphi : \mathcal{A} \to \mathbb{C}$ be a linear functional such that $|\varphi(a)| \leq ||a||$, $\forall\, a \in \mathcal{A}$, and such that $\varphi(1_\mathcal{A}) = 1$. Prove that φ is a positive functional, and hence that $(\mathcal{A}, \varphi)$ is a C^*-probability space.
[Hint: Given $p \in \mathcal{A}^+$, observe that the inequality $\varphi(p) \geq 0$ can be reduced to the statement of part (1), by using functional calculus for the element p.]

EXERCISE 3.22. Consider the framework of Example 3.10 ($\mathcal{A} = C(\Omega)$, where Ω is a compact Hausdorff space).

(1) Show that for every $a \in \mathcal{A}$ we have that $\mathrm{Sp}(a) = \{a(\omega) \mid \omega \in \Omega\}$ (i.e. it is the range of a when a is a viewed as a function from Ω to $\mathbb{C}$).

(2) Let a be an element in $\mathcal{A}$, and let f be a function in $C(\,\mathrm{Sp}(a)\,)$. Note that, from part (1) of this exercise, it makes sense to define the composition $f \circ a : \Omega \to \mathbb{C}$, by $(f \circ a)(\omega) = f(a(\omega))$, $\omega \in \Omega$. Prove that the functional calculus with continuous functions for $a \in \mathcal{A}$ gives the equality $f(a) = f \circ a$.

EXERCISE 3.23. Let $\mathcal{A}$ and $\mathcal{B}$ be unital C^*-algebras, and let $\Psi : \mathcal{A} \to \mathcal{B}$ be a unital $*$-homomorphism. Let a be a normal element of $\mathcal{A}$, and denote $\Psi(a) =: b$ (so b is a normal element of $\mathcal{B}$).

(1) Observe that $\mathrm{Sp}(b) \subset \mathrm{Sp}(a)$.

(2) Let f be a function in $C(\mathrm{Sp}(a))$, and denote the restriction of f to $\mathrm{Sp}(b)$ by f_o. Prove that $\Psi(f(a)) = f_o(b)$. [In other words, prove the "commutation relation" $\Psi(f(a)) = f(\Psi(a))$, for $f \in C(\mathrm{Sp}(a)$.]

EXERCISE 3.24. Consider the framework of Example 3.11 (where λ is the left regular representation of a discrete group G).

(1) Let $g_1, \ldots, g_n$ be some distinct elements of G, let $\alpha_1, \ldots, \alpha_n$ be in $\mathbb{C}$, and consider the operator $T = \sum_{i=1}^n \alpha_i \lambda(g_i) \in B(\,l^2(G)\,)$. Verify the equality $||T\xi_e||^2 = \sum_{i=1}^n |\alpha_i|^2$.

(2) By using part (1) of the exercise, prove that the family of operators $(\lambda(g))_{g \in G}$ is linearly independent in $B(\,l^2(G)\,)$.

LECTURE 4

Non-commutative joint distributions

The discussion of the concept of *joint* distribution is a point where things really start to have a different flavor in non-commutative probability, compared to their classical counterparts. To exemplify this, let us look for instance at the situation of selfadjoint elements in $*$-probability spaces. During the discussion in Lecture 1 the reader probably sensed the fact that, when taken in isolation, such an element is more or less the same thing as a classical real random variable – it is only that we allow this real random variable to live in a fancier (non-commutative) environment. Thus studying the distribution of one selfadjoint element in a $*$-probability space is not much of a departure from what one does in classical probability. In this lecture we will observe that the situation really becomes different when we want to study at the same time two or more selfadjoint elements which do not commute, and we look at the joint distribution of these elements.

Besides introducing the relevant definitions and some examples, the present lecture brings up only one (simple, but important) fact: the class of isomorphism of a $*$-algebra/C^*-algebra $\mathcal{A}$ is determined by knowledge of the joint $*$-distribution of a family of generators, with respect to a faithful expectation functional $\varphi : \mathcal{A} \to \mathbb{C}$. This is significant because it opens the way, at least in principle, to studying isomorphisms of C^*-algebras by starting from combinatorial data on $*$-moments of generators.

Joint distributions

NOTATIONS 4.1. Let s be a positive integer.

(1) We denote by $\mathbb{C}\langle X_1, \ldots, X_s\rangle$ the unital algebra freely generated by s non-commuting indeterminates $X_1, \ldots, X_s$. More concretely, $\mathbb{C}\langle X_1, \ldots, X_s\rangle$ can be described as follows. The monomials of the form $X_{r_1} X_{r_2} \cdots X_{r_n}$ where $n \geq 0$ and $1 \leq r_1, \ldots, r_n \leq s$ give a linear basis for $\mathbb{C}\langle X_1, \ldots, X_s\rangle$, and the multiplication of two such monomials is done by juxtaposition.

(2) Let $\mathcal{A}$ be a unital algebra, and let $a_1, \ldots, a_s$ be elements of $\mathcal{A}$. For every $P \in \mathbb{C}\langle X_1, \ldots, X_s\rangle$ we will denote by $P(a_1, \ldots, a_s)$ the

element of $\mathcal{A}$ which is obtained by replacing $X_1, \ldots, X_s$ with $a_1, \ldots, a_s$, respectively, in the explicit writing of P. Equivalently,

$$\mathbb{C}\langle X_1, \ldots, X_s\rangle \ni P \mapsto P(a_1, \ldots, a_s) \in \mathcal{A} \tag{4.1}$$

is the homomorphism of unital algebras uniquely determined by the fact that it maps X_r to a_r, for $1 \leq r \leq s$.

DEFINITION 4.2. Let $(\mathcal{A}, \varphi)$ be a non-commutative probability space, and let $a_1, \ldots, a_s$ be elements of $\mathcal{A}$.

(1) The family

$$\{\varphi(a_{r_1} \cdots a_{r_n}) \mid n \geq 1,\ 1 \leq r_1, \ldots, r_n \leq s\} \tag{4.2}$$

is called the **family of joint moments** of $a_1, \ldots, a_s$.

(2) The linear functional $\mu : \mathbb{C}\langle X_1, \ldots, X_s\rangle \to \mathbb{C}$ defined by

$$\mu(P) := \varphi(\,P(a_1, \ldots, a_s)\,),\ \ P \in \mathbb{C}\langle X_1, \ldots, X_s\rangle \tag{4.3}$$

is called the **joint distribution** of $a_1, \ldots, a_s$ in $(\mathcal{A}, \varphi)$.

The joint distribution of $a_1, \ldots, a_s$ is thus determined by the fact that it maps every monomial $X_{r_1} \cdots X_{r_n}$ into the corresponding joint moment, $\varphi(a_{r_1} \cdots a_{r_n})$, of $a_1, \ldots, a_s$.

REMARK 4.3. It is clear that the above definitions can, without any problems, be extended to the case of an arbitrary family $(a_i)_{i\in I}$ of random variables. (I is here an index set which might be infinite, even uncountable.) The joint distribution of $(a_i)_{i\in I}$ is then a linear functional on the unital algebra $\mathbb{C}\langle X_i \mid i \in I\rangle$, which is freely generated by non-commuting indeterminates X_i ($i \in I$). We leave it to the reader to write down the exact wording of Definition 4.2 for this case.

EXAMPLES 4.4. (1) Let $(\Omega, \mathcal{Q}, P)$ be a probability space, and let $f_1, \ldots, f_s : \Omega \to \mathbb{R}$ be bounded random variables. Then $f_1, \ldots, f_s$ are at the same time elements of the non-commutative probability space $L^\infty(\Omega, P)$ appearing in Example 1.4.1 (with $\varphi(a) = \int_\Omega a(\omega)\, dP(\omega)$ for $a \in L^\infty(\Omega, P)$). The joint distribution μ of $f_1, \ldots, f_s$ in $L^\infty(\Omega, P)$ is determined by the formula:

$$\mu(X_{r_1} \cdots X_{r_n}) = \int_\Omega f_{r_1}(\omega) \cdots f_{r_n}(\omega)\, dP(\omega), \tag{4.4}$$

holding for every $n \geq 1$ and $1 \leq r_1, \ldots, r_n \leq s$.

In this particular example, there exists a parallel concept of joint distribution of $f_1, \ldots, f_s$ coming from classical probability: this is the probability measure ν on the Borel σ-algebra of $\mathbb{R}^s$ which has, for every Borel set $E \subset \mathbb{R}^s$,

$$\nu(E) = P(\,\{\omega \in \Omega \mid (f_1(\omega), \ldots, f_s(\omega)) \in E\}\,). \tag{4.5}$$

(Note that the assumption that $f_1, \ldots, f_s$ are bounded implies that ν has compact support.) The functional μ of Equation (4.4) is closely related to this probability measure. Indeed, an argument very similar to the one shown in Example 1.11.1 gives us that for every $k_1, \ldots, k_s \geq 0$ we have:

$$\int_{\mathbb{R}^s} t_1^{k_1} \cdots t_s^{k_s} \, d\nu(t_1, \ldots, t_s) = \int_\Omega f_1(\omega)^{k_1} \cdots f_s(\omega)^{k_s} dP(\omega);$$

this implies that the above Equation (4.4) can be written as

$$\mu(X_{r_1} \cdots X_{r_n}) = \int_{\mathbb{R}^s} t_{r_1} \cdots t_{r_n} \, d\nu(t_1, \ldots, t_s) \tag{4.6}$$

(for $n \geq 1$ and $1 \leq r_1, \ldots, r_n \leq s$).

It is clear that the probability measure ν is better suited for studying the s-tuple $(f_1, \ldots, f_s)$ than the functional μ on $\mathbb{C}\langle X_1, \ldots, X_s\rangle$; this is not surprising, since the concept of non-commutative joint distribution is not meant to be particularly useful in commutative situations. However, what one should keep in mind in this example is that the non-commutative joint distribution for $f_1, \ldots, f_s$ is an algebraic (albeit clumsy) incarnation of its classical counterpart.

(2) Let d be a positive integer, and consider the $*$-probability space $(M_d(\mathbb{C}), \mathrm{tr})$ from Example 1.4.2 (the normalized trace on complex $d \times d$ matrices). Let $A_1, A_2 \in M_d(\mathbb{C})$ be Hermitian matrices. Their joint distribution $\mu : \mathbb{C}\langle X_1, X_2\rangle \to \mathbb{C}$ is determined by the formula

$$\mu(X_{r_1} \cdots X_{r_n}) = \mathrm{tr}(A_{r_1} \cdots A_{r_n}), \ \ \forall\, n \geq 1, \ \forall\, 1 \leq r_1, \ldots, r_n \leq 2.$$

Unless A_1 and A_2 happen to commute, the functional μ cannot be replaced by a simpler object (like a probability measure on $\mathbb{R}^2$) which records the same information.

EXAMPLE 4.5. Let $(\mathcal{A}, \varphi)$ be a $*$-probability space, and let x, y be selfadjoint elements of $\mathcal{A}$. For every $n \geq 1$ one can expand $(x+y)^n$ as a sum of 2^n non-commutative monomials in x and y (even though, of course, the usual binomial formula does not generally apply). As a consequence, the moments $\varphi(\,(x+y)^n\,)$, $n \geq 1$ (and hence the distribution of $x+y$) are determined by knowledge of the joint distribution of x and y.

On the other hand it is quite clear that, for x and y as above, just knowing what are the individual distributions of x and of y will not generally suffice in order to determine the distribution of $x+y$. In the remainder of this example we point out how this can be nicely illustrated in the situation of the group algebra (cf. Example 1.4.4).

Let G be a group and let $g, h \in G$ be two elements of infinite order. Consider the $*$-probability space $(\mathbb{C}G, \tau_G)$, as in Example 1.4.4. Recall that $\mathbb{C}G$ has a canonical linear basis indexed by G; the elements of this basis are denoted by the same letters as the group elements themselves, and they are unitaries in $\mathbb{C}G$. Thus we have in particular that $g, h \in \mathbb{C}G$, and that $g^* = g^{-1}$, $h^* = h^{-1}$.

As observed in Lecture 1 (cf. Remark 1.13) each of g and h becomes a Haar unitary in $(\mathbb{C}G, \tau_G)$; as a consequence, each of the selfadjoint elements $x := g + g^{-1}$ and $y := h + h^{-1}$ has an arcsine distribution (cf. Lecture 1, Example 1.14).

So, if in the framework of the preceding paragraph, we look at the element

$$\Delta := x + y = g + g^{-1} + h + h^{-1} \in \mathbb{C}G, \tag{4.7}$$

then Δ will always be a sum of two selfadjoint elements with arcsine distributions. Nevertheless, the distribution of Δ is not uniquely determined, but will rather depend on what group G and what elements $g, h \in G$ we started with. A way of understanding how the distribution of Δ relates to the geometry of the group G goes by considering the subgroup of G generated by g and h, and by looking at closed walks in the corresponding Cayley graph – see Exercise 4.15 below (which also contains the relevant definitions). In order to try one's hand at how this works in concrete situations, the reader could consider for instance the situations when

(1) $G = \mathbb{Z}^2$, with $g = (1, 0)$ and $h = (0, 1)$, or

(2) G is the non-commutative free group on two generators, $G = \mathbb{F}_2$, and g, h are two free generators of $\mathbb{F}_2$.

In situation (1) the corresponding Cayley graph is the lattice $\mathbb{Z}^2$, and the counting of closed walks which yields the moments of Δ is quite straightforward (see Exercise 4.16 at the end of the lecture). The formula obtained is

$$\tau_{\mathbb{Z}^2}(\Delta^n) = \begin{cases} 0 & \text{if } n \text{ is odd} \\ \binom{2p}{p}^2 & \text{if } n \text{ is even, } n = 2p. \end{cases} \tag{4.8}$$

In situation (2), the Cayley graph which appears is a tree (i.e. a graph without circuits), and the counting of closed walks which gives the moments of Δ is a well-known result of Kesten. One obtains a recurrence relation between moments, which can be expressed concisely as a formula giving the moment generating series:

$$\sum_{n=0}^{\infty} \tau_{\mathbb{F}_2}(\Delta^n) z^n = \frac{2\sqrt{1 - 12z^2} - 1}{1 - 16z^2} = 1 + 4z^2 + 28z^4 + 232z^6 + \cdots \tag{4.9}$$

Among the several possible derivations of the formula (4.9), there is one which illustrates the methods of free probability – this is because in situation (2) the elements $x = u + u^*$ and $y = v + v^*$ of $\mathbb{C}\mathbb{F}_2$ will turn out to be *freely independent* (in a sense to be defined precisely in the next Lecture 5), and consequently one can use the technique for computing the distribution of a sum of two freely independent elements – see Example 12.8.2 in Lecture 12.

Joint $*$-distributions

REMARK 4.6. Let $(\mathcal{A}, \varphi)$ be a $*$-probability space and let a be an element of $\mathcal{A}$. By looking at what is the $*$-distribution of a in algebraic sense (Definition 1.17), we see that this really is the same thing as the joint distribution of a and a^*, with the only difference that we re-denoted the indeterminate X_2 of $\mathbb{C}\langle X_1, X_2\rangle$ by X_1^*, and we used this notation to introduce a $*$-operation on $\mathbb{C}\langle X_1, X_2\rangle$. It will be convenient to have this formalism set up for tuples of elements as well. We thus introduce the following notations.

NOTATIONS 4.7. Let s be a positive integer.

(1) We denote by $\mathbb{C}\langle X_1, X_1^*, \ldots, X_s, X_s^*\rangle$ the unital algebra freely generated by $2s$ non-commuting indeterminates $X_1, X_1^*, \ldots, X_s, X_s^*$ (this is the same thing as $\mathbb{C}\langle X_1, \ldots, X_{2s}\rangle$ but where we re-denoted $X_{s+1}, \ldots, X_{2s}$ as $X_1^*, \ldots, X_s^*$, respectively). $\mathbb{C}\langle X_1, X_1^*, \ldots, X_s, X_s^*\rangle$ has a natural $*$-operation, determined by the requirement that the $*$-operation applied to X_r gives X_r^*, for $1 \leq r \leq s$.

(2) Let $\mathcal{A}$ be a unital $*$-algebra and let $a_1, \ldots, a_s$ be elements of $\mathcal{A}$. For every $Q \in \mathbb{C}\langle X_1, X_1^*, \ldots, X_s, X_s^*\rangle$ we will denote by $Q(a_1, \ldots, a_s)$ the element of $\mathcal{A}$ which is obtained by replacing X_1 with a_1, X_1^* with $a_1^*, \ldots, X_s$ with a_s, X_s^* with a_s^* in the explicit writing of Q. Equivalently,

$$\mathbb{C}\langle X_1, X_1^*, \ldots, X_s, X_s^*\rangle \ni Q \mapsto Q(a_1, \ldots, a_s) \in \mathcal{A} \tag{4.10}$$

is the unital $*$-homomorphism uniquely determined by the fact that it maps X_r to a_r, for $1 \leq r \leq s$.

DEFINITION 4.8. Let $(\mathcal{A}, \varphi)$ be a $*$-probability space, and let $a_1, \ldots, a_s$ be elements of $\mathcal{A}$.

(1) The family

$$\left\{ \varphi(a_{r_1}^{\varepsilon_1} \cdots a_{r_n}^{\varepsilon_n}) \;\middle|\; \begin{array}{l} n \geq 1,\ 1 \leq r_1, \ldots, r_n \leq s \\ \varepsilon_1, \ldots, \varepsilon_n \in \{1, *\} \end{array} \right\} \tag{4.11}$$

is called the **family of joint $*$-moments** of $a_1, \ldots, a_s$.

(2) The linear functional $\mu : \mathbb{C}\langle X_1, X_1^*, \ldots, X_s, X_s^*\rangle \to \mathbb{C}$ defined by

$$\mu(Q) := \varphi(\,Q(a_1, \ldots, a_s)\,), \quad Q \in \mathbb{C}\langle X_1, X_1^*, \ldots, X_s, X_s^*\rangle \tag{4.12}$$

is called the **joint $*$-distribution** of $a_1, \ldots, a_s$ in $(\mathcal{A}, \varphi)$.

In a certain sense, the main goal of this monograph is to study joint $*$-distributions which appear in connection to the framework of free independence. This means in particular that many interesting examples will come into play once we start to discuss free independence (beginning in the next lecture, and continuing throughout the rest of the book). For the time being let us have a quick look at an example which (by adjusting the corresponding name from C^*-theory) could be called "the $*$-algebra of the rotation by θ."

EXAMPLE 4.9. Let θ be a number in $[0, 2\pi]$. Suppose that $(\mathcal{A}, \varphi)$ is a $*$-probability space where the $*$-algebra $\mathcal{A}$ is generated by two unitaries u_1, u_2 which satisfy

$$u_1 u_2 = e^{i\theta} u_2 u_1, \tag{4.13}$$

and where $\varphi : \mathcal{A} \to \mathbb{C}$ is a faithful positive functional such that

$$\varphi(u_1^m u_2^n) = \begin{cases} 1 & \text{if } m = n = 0 \\ 0 & \text{otherwise,} \end{cases} \qquad \text{for } m, n \in \mathbb{Z}. \tag{4.14}$$

We will discuss later in this lecture why such a $*$-probability space does indeed exist; right now let us assume it does, and let us make some straightforward remarks about it. Observe that from (4.13) we get

$$\begin{cases} (u_1^m u_2^n) \cdot (u_1^p u_2^q) &= e^{-inp\theta}(u_1^{m+p} u_2^{n+q}) \\ (u_1^m u_2^n)^* &= e^{-imn\theta}(u_1^{-m} u_2^{-n}), \end{cases} \qquad m, n \in \mathbb{Z}. \tag{4.15}$$

This in turn implies that

$$\mathcal{A} = \text{span}\{u_1^m u_2^n \mid m, n \in \mathbb{Z}\} \tag{4.16}$$

(since the right-hand side of (4.16) is, as a consequence of (4.15), a unital $*$-algebra which contains u_1 and u_2). In particular this shows that the linear functional φ is completely described by Equation (4.14). Another fact which quickly follows is that φ is a trace. Indeed, verifying this fact reduces to checking that for every $m, n, p, q \in \mathbb{Z}$ we have

$$\varphi(\,(u_1^m u_2^n) \cdot (u_1^p u_2^q)\,) = \varphi(\,(u_1^p u_2^q) \cdot (u_1^m u_2^n)\,);$$

but (from (4.14) and (4.15)) both sides of this equation are equal to $e^{-imn\theta}$ when $(p, q) = -(m, n)$, and are equal to 0 in all other cases.

Let $\mu : \mathbb{C}\langle X_1, X_1^*, X_2, X_2^*\rangle \to \mathbb{C}$ be the joint $*$-distribution of the unitaries u_1 and u_2. Then for every $n \geq 1$ and $r_1, \ldots, r_n \in \{1, 2\}$,

$\varepsilon_1, \ldots, \varepsilon_n \in \{1, *\}$, the value of μ on the monomial $X_{r_1}^{\varepsilon_1} \cdots X_{r_n}^{\varepsilon_n}$ is either 0 or of the form $e^{ik\theta}$ for some $k \in \mathbb{Z}$. More precisely: an immediate computation (left to the reader) shows that $\mu(X_{r_1}^{\varepsilon_1} \cdots X_{r_n}^{\varepsilon_n})$ is non-zero precisely when the number of X_1 appearing in the sequence $X_{r_1}^{\varepsilon_1}, \ldots, X_{r_n}^{\varepsilon_n}$ is equal to the number of X_1^* appearing in the sequence, and the same when counting the occurrences of X_2 and of X_2^*. In the case when the latter conditions are fulfilled, we get that

$$\mu(X_{r_1}^{\varepsilon_1} \cdots X_{r_n}^{\varepsilon_n}) = e^{ik\theta}, \tag{4.17}$$

where $k \in \mathbb{Z}$ can be interpreted as the oriented area enclosed by a suitably traced walk on the lattice $\mathbb{Z}^2$ – see Exercise 4.17 at the end of the lecture.

Joint $*$-distributions and isomorphism

THEOREM 4.10. *Let $(\mathcal{A}, \varphi)$ and $(\mathcal{B}, \psi)$ be $*$-probability spaces such that φ and ψ are faithful. We denote the units of $\mathcal{A}$ and of $\mathcal{B}$ by $1_{\mathcal{A}}$ and $1_{\mathcal{B}}$, respectively. Suppose that $a_1, \ldots, a_s \in \mathcal{A}$ and $b_1, \ldots, b_s \in \mathcal{B}$ are such that:*

(i) *$a_1, \ldots, a_s$ and $1_{\mathcal{A}}$ generate $\mathcal{A}$ as a $*$-algebra;*

(ii) *$b_1, \ldots, b_s$ and $1_{\mathcal{B}}$ generate $\mathcal{B}$ as a $*$-algebra;*

(iii) *the joint $*$-distribution of $a_1, \ldots, a_s$ in $(\mathcal{A}, \varphi)$ is equal to the joint $*$-distribution of $b_1, \ldots, b_s$ in $(\mathcal{B}, \psi)$.*

Then there exists a $*$-isomorphism $\Phi : \mathcal{A} \to \mathcal{B}$, uniquely determined, such that $\Phi(a_1) = b_1, \ldots, \Phi(a_s) = b_s$. This Φ is also an isomorphism between $(\mathcal{A}, \varphi)$ and $(\mathcal{B}, \psi)$, i.e. it has the property that $\psi \circ \Phi = \varphi$.

PROOF. Observe that the hypotheses (i) and (ii) amount to

$$\begin{cases} \mathcal{A} = \{P(a_1, \ldots, a_s) \mid P \in \mathbb{C}\langle X_1, X_1^*, \ldots, X_s, X_s^* \rangle\} \\ \\ \mathcal{B} = \{P(b_1, \ldots, b_s) \mid P \in \mathbb{C}\langle X_1, X_1^*, \ldots, X_s, X_s^* \rangle\} \end{cases} \tag{4.18}$$

(since on the right-hand sides of Equations (4.18) we have unital $*$-subalgebras of $\mathcal{A}$ and of $\mathcal{B}$ which contain $a_1, \ldots, a_s$ and $b_1, \ldots, b_s$, respectively).

Let $\mu : \mathbb{C}\langle X_1, X_1^*, \ldots, X_s, X_s^* \rangle \to \mathbb{C}$ be the common joint $*$-distribution of $a_1, \ldots, a_s$ and of $b_1, \ldots, b_s$. From the definition of μ and the fact that the functionals φ and ψ are faithful, it is immediate that for $P, Q \in \mathbb{C}\langle X_1, X_1^*, \ldots, X_s, X_s^* \rangle$ we have:

$$\begin{aligned} P(a_1, \ldots, a_s) = Q(a_1, \ldots, a_s) \quad &\Leftrightarrow \quad \mu(\,(P-Q)^*(P-Q)\,) = 0 \\ &\Leftrightarrow \quad P(b_1, \ldots, b_s) = Q(b_1, \ldots, b_s). \end{aligned}$$

As a consequence, it makes sense to define a function $\Phi : \mathcal{A} \to \mathcal{B}$ by the formula

$$\Phi(\ P(a_1, \ldots, a_s)\ ) = P(b_1, \ldots, b_s), \quad P \in \mathbb{C}\langle X_1, X_1^*, \ldots, X_s, X_s^* \rangle,$$

and moreover this function is bijective. Indeed, from the equivalences observed above it follows that the definition of Φ is coherent and that Φ is injective, whereas Equations (4.18) imply that Φ is defined on all of $\mathcal{A}$ and it is surjective.

The formula defining Φ clearly implies that Φ is a unital $*$-homomorphism and that $\Phi(a_r) = b_r$, $1 \leq r \leq s$. Moreover, we have that $\psi \circ \Phi = \varphi$; indeed, this amounts to the equality

$$\psi(\ P(b_1, \ldots, b_s)\ ) = \varphi(\ P(a_1, \ldots, a_s)\ ), \quad \forall\, P \in \mathbb{C}\langle X_1, X_1^*, \ldots, X_s, X_s^* \rangle,$$

which is true, since both its sides are equal to $\mu(P)$. The uniqueness of Φ with the above properties is clear. □

We now upgrade the preceding theorem to the framework of a C^*-probability space. What is different in this framework is that, if $a_1, \ldots, a_s$ generate $\mathcal{A}$ as a unital C^*-algebra, then the polynomials $P(a_1, \ldots, a_s)$ (with $P \in \mathbb{C}\langle X_1, X_1^*, \ldots, X_s, X_s^* \rangle$) do not necessarily exhaust $\mathcal{A}$, they will only give us a dense unital $*$-subalgebra of $\mathcal{A}$. But this issue can be easily handled by using a norm-preservation argument.

THEOREM 4.11. *Let $(\mathcal{A}, \varphi)$ and $(\mathcal{B}, \psi)$ be C^*-probability spaces such that φ and ψ are faithful. We denote the units of $\mathcal{A}$ and of $\mathcal{B}$ by $1_{\mathcal{A}}$ and $1_{\mathcal{B}}$, respectively. Suppose that $a_1, \ldots, a_s \in \mathcal{A}$ and $b_1, \ldots, b_s \in \mathcal{B}$ are such that:*

(i) *$a_1, \ldots, a_s$ and $1_{\mathcal{A}}$ generate $\mathcal{A}$ as a C^*-algebra;*

(ii) *$b_1, \ldots, b_s$ and $1_{\mathcal{B}}$ generate $\mathcal{B}$ as a C^*-algebra;*

(iii) *the joint $*$-distribution of $a_1, \ldots, a_s$ in $(\mathcal{A}, \varphi)$ is equal to the joint $*$-distribution of $b_1, \ldots, b_s$ in $(\mathcal{B}, \psi)$.*

Then there exists an isometric $*$-isomorphism $\Phi : \mathcal{A} \to \mathcal{B}$, uniquely determined, such that $\Phi(a_1) = b_1, \ldots, \Phi(a_s) = b_s$. This Φ is also an isomorphism between $(\mathcal{A}, \varphi)$ and $(\mathcal{B}, \psi)$, i.e. it has the property that $\psi \circ \Phi = \varphi$.

PROOF. Let us denote

$$\mathcal{A}_0 := \{P(a_1, \ldots, a_s) \mid P \in \mathbb{C}\langle X_1, X_1^*, \ldots, X_s, X_s^* \rangle\},$$

and

$$\mathcal{B}_0 := \{P(b_1, \ldots, b_s) \mid P \in \mathbb{C}\langle X_1, X_1^*, \ldots, X_s, X_s^* \rangle\}.$$

It is clear that $\mathcal{A}_0$ is a unital $*$-subalgebra of $\mathcal{A}$, and hypothesis (i) of the theorem gives us that $\mathcal{A}_0$ is dense in $\mathcal{A}$ in the norm-topology. (Indeed, it is immediate that the closure of $\mathcal{A}_0$ in the norm-topology is the smallest unital C^*-subalgebra of $\mathcal{A}$ which contains $a_1, \ldots, a_s$.) Likewise, we have that $\mathcal{B}_0$ is a dense unital $*$-subalgebra of $\mathcal{B}$.

The $*$-probability spaces $(\mathcal{A}_0, \varphi|\mathcal{A}_0)$ and $(\mathcal{B}_0, \psi|\mathcal{B}_0)$ satisfy the hypotheses of Theorem 4.10 (with respect to the given $a_1, \ldots, a_s$ and $b_1, \ldots, b_s$). So from that theorem and its proof we know that the map $\Phi_0 : \mathcal{A}_0 \to \mathcal{B}_0$ defined by

$$\Phi_0(\, P(a_1, \ldots, a_s)\,) = P(b_1, \ldots, b_s)$$

(where P runs in $\mathbb{C}\langle X_1, X_1^*, \ldots, X_s, X_s^*\rangle$) is an isomorphism of $*$-probability spaces between $(\mathcal{A}_0, \varphi|\mathcal{A}_0)$ and $(\mathcal{B}_0, \psi|\mathcal{B}_0)$.

The point of the proof is to observe that the map Φ_0 is isometric on $\mathcal{A}_0$, i.e. that for every $P \in \mathbb{C}\langle X_1, X_1^*, \ldots, X_s, X_s^*\rangle$ we have

$$||P(a_1, \ldots, a_s)||_{\mathcal{A}} = ||P(b_1, \ldots, b_s)||_{\mathcal{B}}. \tag{4.19}$$

Indeed, given a polynomial $P \in \mathbb{C}\langle X_1, X_1^*, \ldots, X_s, X_s^*\rangle$, we compute:

$$\begin{aligned} ||P(a_1, \ldots, a_s)||_{\mathcal{A}} &= \lim_{n\to\infty} \varphi(\, (P(a_1, \ldots, a_s)^* P(a_1, \ldots, a_s))^n\,)^{1/2n} \\ &= \lim_{n\to\infty} \varphi(\, (P^*P)^n(a_1, \ldots, a_s)\,)^{1/2n} \\ &= \lim_{n\to\infty} \mu(\, (P^*P)^n\,)^{1/2n}, \end{aligned}$$

where μ denotes the common joint $*$-distribution of $a_1, \ldots, a_s$ and of $b_1, \ldots, b_s$, and where at the first equality sign we used Proposition 3.17 from the preceding lecture. Clearly, the same kind of calculation can be done for the norm $||P(b_1, \ldots, b_s)||_{\mathcal{B}}$, and (4.19) follows.

Now, a standard argument of extension by continuity shows that there exists a unique continuous function $\Phi : \mathcal{A} \to \mathcal{B}$ such that $\Phi|\mathcal{A}_0 = \Phi_0$. The properties of Φ_0 of being a $*$-homomorphism and of being isometric are immediately passed on to Φ, by continuity. We have that Φ is one-to-one because it is isometric. The range of Φ is complete (being an isometric image of $\mathcal{A}$), hence closed in $\mathcal{B}$; since $\text{ran}(\Phi)$ contains the dense $*$-subalgebra $\mathcal{B}_0$ of $\mathcal{B}$, it follows that $\text{ran}(\Phi) = \mathcal{B}$, hence that Φ is onto. Thus Φ has all the properties appearing in the statement of the theorem. The uniqueness of Φ follows from the fact that, in general, a unital $*$-homomorphism defined on $\mathcal{A}$ is determined by its values on $a_1, \ldots, a_s$. □

REMARKS 4.12. (1) The kind of isomorphism which appeared in Theorem 4.11 is suitable for the category of unital C^*-algebras, i.e. it includes the appropriate metric property of being isometric ($||\Phi(a)||_{\mathcal{B}} = ||a||_{\mathcal{A}}$, for every $a \in \mathcal{A}$). It is worth mentioning here that in fact a bijective unital $*$-homomorphism between unital C^*-algebras is always isometric (the metric property is an automatic consequence of the algebraic ones). See Exercise 4.18 at the end of the lecture.

(2) Theorem 4.11 has a version where the families $a_1, \ldots, a_s$ and $b_1, \ldots, b_s$ consist of *selfadjoint elements* (of $\mathcal{A}$ and of $\mathcal{B}$, respectively), and where hypothesis (iii) in the theorem is adjusted to require that the joint *distribution* of $a_1, \ldots, a_s$ in $(\mathcal{A}, \varphi)$ is equal to the joint distribution of $b_1, \ldots, b_s$ in $(\mathcal{B}, \psi)$. In order to obtain this version of the theorem one can either repeat (with trivial adjustments) the proof shown above, or one can invoke the actual statement of Theorem 4.11 in conjunction with the (trivial) trick described in Exercise 4.19

(3) Another possible generalization of Theorem 4.11 is in the direction of allowing the families of generators considered for $\mathcal{A}$ and $\mathcal{B}$ to be infinite. The precise statement appears in Exercise 4.20 at the end of the lecture.

EXAMPLE 4.13. We look again at the situation of Example 4.9, but now considered in the C^*-framework. So let θ be a fixed number in $[0, 2\pi]$. Suppose that $(\mathcal{A}, \varphi)$ is a C^*-probability space where the C^*-algebra $\mathcal{A}$ is generated by two unitaries u_1, u_2 which satisfy Equation (4.13), and where $\varphi : \mathcal{A} \to \mathbb{C}$ is a faithful positive functional satisfying Equation (4.14). Then exactly as in Example 4.9 we see that the relations (4.15) hold, and imply that

(i) $\mathcal{A} = \text{cl span}\{u_1^m u_2^n \mid m, n \in \mathbb{Z}\}$ (where "cl span" stands for "norm-closure of linear span"), and

(ii) φ is a trace.

Now, Theorem 4.11 implies that a C^*-probability space $(\mathcal{A}, \varphi)$ as described in the preceding paragraph is uniquely determined up to isomorphism. In particular, the isomorphism class of the C^*-algebra $\mathcal{A}$ involved in the example is uniquely determined; it therefore makes sense (and it is customary) to refer to such an $\mathcal{A}$ by calling it *the* C^*-algebra of rotation by θ.

Of course, in order to talk about the C^*-algebra of rotation by θ one must also show that it exists – i.e. one must construct an example of C^*-probability space $(\mathcal{A}, \varphi)$ where φ is faithful and where Equations (4.13) and (4.14) are satisfied. In the remainder of this example we show how this can be done.

Consider the Hilbert space $l^2(\mathbb{Z}^2)$, and denote its canonical orthonormal basis by $\{\xi_{(m,n)} \mid m, n \in \mathbb{Z}\}$. It is immediate that one can define two unitary operators U_1, U_2 on $l^2(\mathbb{Z}^2)$ by prescribing their action on the canonical orthonormal basis to be as follows:

$$\begin{cases} U_1 \xi_{(m,n)} = \xi_{(m+1,n)} \\ U_2 \xi_{(m,n)} = e^{-im\theta} \xi_{(m,n+1)}, \end{cases} \quad m, n \in \mathbb{Z}. \tag{4.20}$$

Let $\mathcal{A}$ be the C^*-subalgebra of $B(l^2(\mathbb{Z}^2))$ which is generated by U_1 and U_2, and let $\varphi : \mathcal{A} \to \mathbb{C}$ be the vector-state defined by the vector $\xi_{(0,0)}$; that is,

$$\varphi(T) = \langle T\xi_{(0,0)}, \xi_{(0,0)} \rangle, \quad T \in \mathcal{A}. \tag{4.21}$$

From (4.20) it is immediate that $U_1U_2 = e^{i\theta}U_2U_1$ (indeed, both U_1U_2 and $e^{i\theta}U_2U_1$ send $\xi_{(m,n)}$ to $e^{-im\theta}\xi_{(m+1,n+1)}$, for every $m, n \in \mathbb{Z}$). So in order for the C^*-probability space $(\mathcal{A}, \varphi)$ to have the required properties, we are only left to check that φ is faithful.

Observe that even without knowing that φ is faithful, we can see that it is a trace. This is checked exactly as in Example 4.9, where Equation (4.16) is now replaced by the fact that $\mathcal{A}$ equals $\text{cl span}\{U_1^m U_2^n \mid m, n \in \mathbb{Z}\}$; the details of this are left to the reader.

Now suppose that $T \in \mathcal{A}$ is such that $\varphi(T^*T) = 0$. Since $\varphi(T^*T) = ||T\xi_{(0,0)}||^2$, we thus have that $T\xi_{(0,0)} = 0$. But then for every $m, n, p, q \in \mathbb{Z}$ we can write:

$$\begin{aligned}
\langle T\xi_{(m,n)}, \xi_{(p,q)} \rangle &= \langle T(U_1^m U_2^n)\xi_{(0,0)}, (U_1^p U_2^q)\xi_{(0,0)} \rangle \\
&= \langle (U_1^p U_2^q)^* T (U_1^m U_2^n)\xi_{(0,0)}, \xi_{(0,0)} \rangle \\
&= \varphi(\, (U_1^p U_2^q)^* T (U_1^m U_2^n)\,) \\
&= \varphi(\, (U_1^m U_2^n)(U_1^p U_2^q)^* T\,) \qquad \text{(since } \varphi \text{ is a trace)} \\
&= \langle (U_1^m U_2^n)(U_1^p U_2^q)^* T\xi_{(0,0)}, \xi_{(0,0)} \rangle \\
&= 0 \qquad \text{(because } T\xi_{(0,0)} = 0\text{)}.
\end{aligned}$$

Hence $\langle T\xi_{(m,n)}, \xi_{(p,q)} \rangle = 0$ for all $m, n, p, q \in \mathbb{Z}$, and this clearly implies that $T = 0$ (thus completing the verification of the faithfulness of φ).

Without going into any details, we mention here that the universality and uniqueness properties of the C^*-algebra $\mathcal{A}$ of rotation by θ can be obtained without taking the canonical trace $\varphi : \mathcal{A} \to \mathbb{C}$ as part of our initial data (but then the arguments are no longer as simple as shown above).

Exercises

Exercises 4.14–4.16 fill in some of the details remaining from the discussion in Example 4.5.

EXERCISE 4.14. Let G be a group which is generated by two elements $g, h \in G$, both of infinite order and such that none of them generates G by itself. Consider the $*$-probability space $(\mathbb{C}G, \tau_G)$ and the element $\Delta = g + g^{-1} + h + h^{-1} \in \mathbb{C}G$ (as in Example 4.5). Verify that $\tau_G(\Delta) = 0$, $\tau_G(\Delta^2) = 4$, $\tau_G(\Delta^3) = 0$, but that the value of $\tau_G(\Delta^4)$

is not uniquely determined. What are the minimal and maximal values which $\tau_G(\Delta^4)$ can have, under the given hypotheses?

EXERCISE 4.15. Consider the framework of Exercise 4.14, and consider the Cayley graph of G with respect to the set of generators $\{g, g^{-1}, h, h^{-1}\}$. (The vertices of this graph are the elements of G. Two vertices $g_1, g_2 \in G$ are connected by an edge of the graph precisely when $g_1^{-1}g_2 \in \{g, g^{-1}, h, h^{-1}\}$, or, equivalently, when $g_2^{-1}g_1 \in \{g, g^{-1}, h, h^{-1}\}$.) Prove that for every $n \geq 1$, the moment $\tau_G(\Delta^n)$ is equal to the number of closed paths of length n in the Cayley graph, which begin and end at the unit element e of G.

EXERCISE 4.16. (1) Consider the framework of Exercises 4.14 and 4.15, where we set $G = \mathbb{Z}^2$ and $g = (1, 0), h = (0, 1)$. Observe that in this case the Cayley graph of G with respect to the set of generators $\{g, g^{-1}, h, h^{-1}\}$ is precisely the square lattice $\mathbb{Z}^2$.

(2) Prove that the number of closed paths in the square lattice $\mathbb{Z}^2$ which have length n and which begin and end at $(0, 0)$ is equal to

$$\begin{cases} 0 & \text{if } n \text{ is odd} \\ \binom{2p}{p}^2 & \text{if } n \text{ is even, } n = 2p. \end{cases}$$

Observe that this implies formula (4.8) stated in Example 4.5.

EXERCISE 4.17. Refer to the notations in the last paragraph of Example 4.9. Given a positive integer n and some values $r_1, \ldots, r_n \in \{1, 2\}$, $\varepsilon_1, \ldots, \varepsilon_n \in \{1, *\}$, consider the n-step walk γ in the lattice $\mathbb{Z}^2$ which starts at $(0, 0)$ and has its jth step $(1 \leq j \leq n)$ described as follows:

- if $r_j = 1$ and $\varepsilon_j = 1$, then the jth step of γ is towards East;
- if $r_j = 1$ and $\varepsilon_j = -1$, then the jth step of γ is towards West;
- if $r_j = 2$ and $\varepsilon_j = 1$, then the jth step of γ is towards North;
- if $r_j = 2$ and $\varepsilon_j = -1$, then the jth step of γ is towards South.

(1) Prove that $\mu(X_{r_1}^{\varepsilon_1} \cdots X_{r_n}^{\varepsilon_n})$ is different from 0 if and only if the path γ is closed (that is, γ ends at $(0, 0)$).

(2) Suppose that γ is closed. Verify the formula stated in Equation (4.17) of Example 4.9, where $k \in \mathbb{Z}$ denotes the signed area enclosed by the path γ that is, k is given by the contour integral

$$k = \int_\gamma x\, dy = -\int_\gamma y\, dx.$$

EXERCISE 4.18. (1) *(Detail left from the Remark 4.12)* Let $\mathcal{A}$ and $\mathcal{B}$ be unital C^*-algebras, and let $\Phi : \mathcal{A} \to \mathcal{B}$ be a bijective unital $*$-homomorphism. Prove that $||\Phi(a)||_{\mathcal{B}} = ||a||_{\mathcal{A}}, \forall\, a \in \mathcal{A}$.

(2) *(A generalization of part (1), which is used in Lecture 7)* Let $\mathcal{A}$ and $\mathcal{B}$ be unital C^*-algebras, and let $\Phi : \mathcal{A} \to \mathcal{B}$ be a unital $*$-homomorphism which is one-to-one. Prove that $||\Phi(a)||_{\mathcal{B}} = ||a||_{\mathcal{A}}$, $\forall\, a \in \mathcal{A}$.
[Hint: It suffices to check that $||\Phi(p)||_{\mathcal{B}} = ||p||_{\mathcal{A}}$ for $p \in \mathcal{A}^+$. In part (1) this is because $\mathrm{Sp}(p) = \mathrm{Sp}(\Phi(p))$. In part (2), one only has $\mathrm{Sp}(\Phi(p)) \subset \mathrm{Sp}(p)$; but if it happened that $||\Phi(p)|| < ||p||$, then one could use the functional calculus of p and Exercise 3.23 to obtain a contradiction.]

EXERCISE 4.19. (1) Let $(\mathcal{A}, \varphi)$ be a $*$-probability space, and let $a_1, \ldots, a_s$ be selfadjoint elements of $\mathcal{A}$. Let $\mu : \mathbb{C}\langle X_1, \ldots, X_s\rangle \to \mathbb{C}$ be the joint distribution of $a_1, \ldots, a_s$ and let $\widetilde{\mu} : \mathbb{C}\langle X_1, X_1^* \ldots, X_s, X_s^*\rangle \to \mathbb{C}$ be the joint $*$-distribution of $a_1, \ldots, a_s$ (in $(\mathcal{A}, \varphi)$). Prove the relation $\widetilde{\mu} = \mu \circ \Pi$, where Π is the unital homomorphism from $\mathbb{C}\langle X_1, X_1^* \ldots, X_s, X_s^*\rangle$ to $\mathbb{C}\langle X_1, \ldots, X_s\rangle$ uniquely determined by the condition that $\Pi(X_r) = \Pi(X_r^*) = X_r$, for $1 \leq r \leq s$.

(2) By using the first part of this exercise, give a proof of the self-adjoint version of Theorem 4.11 which is described in Remark 4.12.

EXERCISE 4.20. *(Generalization of Theorem 4.11 to the case of infinite families of generators)* Let $(\mathcal{A}, \varphi)$ and $(\mathcal{B}, \psi)$ be C^*-probability spaces such that φ and ψ are faithful. We denote the units of $\mathcal{A}$ and of $\mathcal{B}$ by $1_{\mathcal{A}}$ and $1_{\mathcal{B}}$, respectively. Suppose that $(a_i)_{i\in I}$ and $(b_i)_{i\in I}$ are families of elements of $\mathcal{A}$ and respectively of $\mathcal{B}$, indexed by the same index set I, such that:

(j) $\{a_i \mid i \in I\} \cup \{1_{\mathcal{A}}\}$ generate $\mathcal{A}$ as a C^*-algebra;

(jj) $\{b_i \mid i \in I\} \cup \{1_{\mathcal{B}}\}$ generate $\mathcal{B}$ as a C^*-algebra;

(jjj) for every finite subset $\{i_1, \ldots, i_s\}$ of I, the joint $*$-distribution of $a_{i_1}, \ldots, a_{i_s}$ in $(\mathcal{A}, \varphi)$ is equal to the joint $*$-distribution of $b_{i_1}, \ldots, b_{i_s}$ in $(\mathcal{B}, \psi)$.

Prove that there exists an isometric $*$-isomorphism $\Phi : \mathcal{A} \to \mathcal{B}$, uniquely determined, such that $\Phi(a_i) = b_i$ for every $i \in I$. Prove moreover that this Φ is also an isomorphism between $(\mathcal{A}, \varphi)$ and $(\mathcal{B}, \psi)$, i.e. it has the property that $\psi \circ \Phi = \varphi$.
[Hint: Reduce these statements to those of Theorem 4.11, by considering the unital C^*-subalgebras of $\mathcal{A}$ and of $\mathcal{B}$ which are generated by finite subfamilies of the a_i and the b_i.]

LECTURE 5

Definition and basic properties of free independence

In this lecture we will introduce the basic concept which refines "non-commutative probability theory" to "free probability theory" – the notion of free independence. As the name indicates, this concept should be seen as an analog of the notion of independence from classical probability theory. Thus, before we define free independence we recall this classical notion. Since we are working with algebras which might be non-commutative, it is more appropriate to formulate the concept of classical independence on this more general level, where it corresponds to the notion of a tensor product.

We will also derive some very basic properties of free independence in this lecture. A more systematic theory, however, will be deferred to Part 2.

The classical situation: tensor independence

DEFINITION 5.1. Let $(\mathcal{A}, \varphi)$ be a non-commutative probability space and let I be a fixed index set.

(1) Unital subalgebras $(\mathcal{A}_i)_{i\in I}$ are called **tensor independent**, if the subalgebras $\mathcal{A}_i$ commute (i.e. $ab = ba$ for all $a \in \mathcal{A}_i$ and all $b \in \mathcal{A}_j$ and all $i, j \in I$ with $i \neq j$) and if φ factorizes in the following way:

$$\varphi\Big(\prod_{j\in J} a_j\Big) = \prod_{j\in J} \varphi(a_j) \tag{5.1}$$

for all finite subsets $J \subset I$ and all $a_j \in \mathcal{A}_j$ $(j \in J)$.

(2) Tensor (or classical) independence of random variables is defined by tensor independence of the generated unital algebras; hence "a and b tensor independent" means nothing but a and b commute and mixed moments factorize, i.e.

$$ab = ba \qquad \text{and} \qquad \varphi(a^n b^m) = \varphi(a^n)\varphi(b^m) \quad \forall n, m \geq 0. \tag{5.2}$$

From a combinatorial point of view one can consider tensor independence as a special rule, namely (5.2), for calculating mixed moments of independent random variables from the moments of the single variables. Free independence will just be another such specific rule.

REMARK 5.2. Note that in the non-commutative context we have to specify many more mixed moments than in the commutative case. If a and b commute then every mixed moment in a and b can be reduced to a moment of the form $\varphi(a^n b^m)$ and thus the factorization rule in (5.2) for these contains the full information about the joint distribution of a and b, provided we know the distribution of a and the distribution of b. If, on the other hand, a and b do not commute then $\varphi(a^n b^m)$ is only a very small part of the joint distribution of a and b, because we have to consider moments like $\varphi(a^{n_1} b^{m_1} a^{n_2} b^{m_2} \cdots a^{n_k} b^{m_k})$, and these cannot be reduced in general to just $\varphi(a^n b^m)$. As first guess for a factorization rule for non-commutative situations one might think of a direct extension of the classical rule, namely

$$\varphi(a^{n_1} b^{m_1} \cdots a^{n_k} b^{m_k}) = \varphi(a^{n_1}) \cdot \varphi(b^{m_2}) \cdots \varphi(a^{n_k}) \cdot \varphi(b^{m_k}). \qquad (5.3)$$

This, however, is *not* the rule of free independence. One sees easily that (5.3) is not consistent in general if one puts, for example, some of the m_i or some of the n_i equal to 0. If one is willing to accept this deficiency then the rule (5.3) can be used to define the so-called "Boolean independence." One can develop elements of a Boolean probability theory, however, its structure is quite trivial compared to the depth of free probability theory. We will not elaborate more on this Boolean factorization rule, but want to present now the more interesting rule for free independence. As the reader might have guessed from the preceding remarks, the rule for free independence is not as straightforward as the above factorization rules. Actually, the definition of free independence might look somewhat artificial at first, but we will see throughout the rest of the book that this is a very important concept and deserves special attention. In the last section of this lecture we will also comment on the way in which free independence is, despite the more complicated nature of its rule for calculating mixed moments, a very natural concept.

Definition of free independence

DEFINITION 5.3. Let $(\mathcal{A}, \varphi)$ be a non-commutative probability space and let I be a fixed index set.

(1) Let, for each $i \in I$, $\mathcal{A}_i \subset \mathcal{A}$ be a unital subalgebra. The subalgebras $(\mathcal{A}_i)_{i\in I}$ are called **freely independent**, if

$$\varphi(a_1 \cdots a_k) = 0$$

whenever we have the following.

- k is a positive integer;
- $a_j \in \mathcal{A}_{i(j)}$ $(i(j) \in I)$ for all $j = 1, \dots, k$;
- $\varphi(a_j) = 0$ for all $j = 1, \dots, k$;
- and neighboring elements are from different subalgebras, i.e. $i(1) \neq i(2)$, $i(2) \neq i(3)$, $\dots$, $i(k-1) \neq i(k)$.

(2) Let $\mathcal{X}_i \subset \mathcal{A}$ $(i \in I)$ be subsets of $\mathcal{A}$. Then $(\mathcal{X}_i)_{i\in I}$ are called freely independent if $(\mathcal{A}_i)_{i\in I}$ are freely independent, where, for $i \in I$, $\mathcal{A}_i := \text{alg}(1, \mathcal{X}_i)$ is the unital algebra generated by $\mathcal{X}_i$.

(3) In particular, if the unital algebras $\mathcal{A}_i := \text{alg}(1, a_i)$ generated by elements $a_i \in \mathcal{A}$ $(i \in I)$ are freely independent, then $(a_i)_{i\in I}$ are called **freely independent random variables**.

(4) If, in the context of a $*$-probability space, the unital $*$-algebras $\mathcal{A}_i := \text{alg}(1, a_i, a_i^*)$ generated by the random variables a_i $(i \in I)$ are freely independent, then we call $(a_i)_{i\in I}$ **$*$-freely independent**.

REMARKS 5.4. (1) Clearly, free independence is a concept with respect to a linear functional φ; random variables which are freely independent with respect to some functional φ are in general not freely independent with respect to some other functional ψ. So a more precise name would be "freely independent with respect to φ." However, usually it is clear to which φ we refer. In particular, it is understood that if we are working in a non-commutative probability space $(\mathcal{A}, \varphi)$, then our free independence is with respect to this φ.

(2) Note: the condition on the indices in the definition of free independence is only on consecutive indices; $i(1) = i(3)$, for example, is allowed. We also do not require that the first and the last element are from different subalgebras, thus $i(1) = i(k)$ is in general allowed.

(3) Let us state more explicitly the requirement for freely independent random variables: $(a_i)_{i\in I}$ are freely independent if we have $\varphi\big(P_1(a_{i(1)}) \dots P_k(a_{i(k)})\big) = 0$ for all polynomials $P_1, \dots, P_k \in \mathbb{C}\langle X\rangle$ in one indeterminate X and all $i(1) \neq i(2) \neq \cdots \neq i(k)$, such that $\varphi(P_j(a_{i(j)})) = 0$ for all $j = 1, \dots, k$.

(4) Note that the index set I might be infinite, even uncountable; but this is not really an issue. Free independence of $(\mathcal{A}_i)_{i\in I}$ is the same as free independence of $(\mathcal{A}_j)_{j\in J}$ for all finite subsets J of I.

(5) Free independence of random variables is defined in terms of the generated algebras, but one should note that it extends also to the generated C^*-algebras; see Exercise 5.23.

(6) Sometimes we will have free independence between sets of random variables where some of the sets consist only of one element. Usually, we will replace these sets just by the random variables. So, for example, free independence between $\{a_1, a_2\}$ and b (for some random

variables a_1, a_2, b in a non-commutative probability space) means free independence between $\{a_1, a_2\}$ and $\{b\}$, which is by definition the same as free independence between the unital algebra generated by a_1 and a_2 and the unital algebra generated by b.

NOTATION 5.5. Instead of saying that algebras, sets, or random variables are "freely independent" we will often just say that they are **free**. In the same way, **$*$-free** means the same as "$*$-freely independent." Instead of "free independence" one often also uses **freeness**.

The example of a free product of groups

Before we start to take a closer look at the structure of our definition, we want to present one basic model for freely independent random variables. Indeed, this example was the starting point of Voiculescu and motivated the above definition of free independence. This example takes place in the probability space $(\mathbb{C}G, \tau_G)$ of the group algebra of a group G, in the special situation where the group G is the free product of subgroups G_i. Let us first recall what it means that subgroups are free. Freeness in the context of groups is a purely algebraic condition (i.e. does not depend on any linear functional) and means that we do not have non-trivial relations between elements from different subgroups.

DEFINITION 5.6. Let G be a group and $(G_i)_{i \in I}$ subgroups of G. By e we will denote the common neutral element of all these groups. The subgroups $(G_i)_{i \in I}$ are **free** if for all $k \geq 1$, all $i(1), \ldots, i(k) \in I$ with $i(1) \neq i(2) \neq \cdots \neq i(k)$ and all $g_1 \in G_{i(1)} \backslash \{e\}, \ldots, g_k \in G_{i(k)} \backslash \{e\}$ we have that $g_1 \cdots g_k \neq e$.

EXAMPLE 5.7. Let $\mathbb{F}_n$ be the **free group with n generators**, i.e. $\mathbb{F}_n$ is generated by n elements $f_1, \ldots, f_n$, which fulfill no other relations apart from the group axioms. Then we have $\mathbb{F}_1 = \mathbb{Z}$ and within

$$\mathbb{F}_{n+m} = \text{group generated by } f_1, \ldots, f_{m+n}$$

the groups

$$\mathbb{F}_m = \text{group generated by } f_1, \ldots, f_m$$

and

$$\mathbb{F}_n = \text{group generated by } f_{m+1}, \ldots, f_{m+n}$$

are free.

If one has the notion that groups are free then a canonical question is whether for any given collection $(G_i)_{i\in I}$ of groups (not necessarily subgroups of a bigger group) one can construct some group G such that the G_i are (isomorphic to) subgroups of this G and that, in addition, they are free in G. The affirmative answer to this is given by the free product construction.

DEFINITION 5.8. Let G_i $(i \in I)$ be groups with identity elements e_i $(i \in I)$, respectively. The **free product** $G := *_{i\in I} G_i$ is the group which is generated by all elements from all G_i $(i \in I)$ subject to the following relations:

(1) the relations within each G_i $(i \in I)$;

(2) the identity element e_i of G_i, for each $i \in I$, is identified with the identity element e of G:

$$e = e_i \qquad \text{for all } i \in I.$$

EXAMPLE 5.9. With the notion of the free product we can rephrase the content of Example 5.7 also as

$$\mathbb{F}_{m+n} = \mathbb{F}_m * \mathbb{F}_n.$$

REMARKS 5.10. (1) An important property of the free product of groups is its universality property. Whenever we have a group F and group homomorphisms $\eta_i : G_i \to F$ for all $i \in I$, then there exists a unique homomorphism $\eta : *_{i\in I} G_i \to F$, which extends the given η_i, i.e. $\eta|_{G_i} = \eta_i$, where we, of course, identify G_i with a subgroup of the free product. This universality property determines the free product uniquely (up to group isomorphism), the only non-trivial point is to see that such an object indeed exists. The above, more explicit definition, can be used to show this existence.

(2) Even more explicitly, we can describe the free product $G = *_{i\in I} G_i$ as follows:

$$G = \{e\} \cup \{g_1 \dots g_k \mid g_j \in G_{i(j)}, i(1) \neq i(2) \neq \dots \neq i(k), g_j \neq e_{i(j)}\},$$

and multiplication in G is given by juxtaposition and reduction to the above form by multiplication of neighboring terms from the same group.

(3) In particular, for $g_j \in G_{i(j)}$ such that $g_j \neq e$ $(j = 1, \dots, k)$ and $i(1) \neq \dots \neq i(k)$ we have $g_1 \cdots g_k \neq e$; i.e. the G_i $(i \in I)$ are indeed free in $*_{i\in I} G_i$.

The relation between the group algebra of the free product of groups and the concept of free independence is given in the following proposition. Of course, this relation is the reason for calling this concept "free" independence. Let us emphasize again, that whereas "free" in an algebraic context (as for groups or algebras) just means the absence of non-trivial algebraic relations, in our context of non-commutative probability spaces "free" is a very specific requirement for a fixed linear functional.

PROPOSITION 5.11. *Let G_i $(i \in I)$ be subgroups of a group G. Then the following are equivalent.*

(1) The groups $(G_i)_{i\in I}$ are free.

(2) The algebras $(\mathbb{C}G_i)_{i\in I}$ (considered as subalgebras of $\mathbb{C}G$) are freely independent in the non-commutative probability space $(\mathbb{C}G, \tau_G)$.

PROOF. First, we prove the implication (1) $\Rightarrow$ (2). Consider

$$a_j = \sum_{g\in G_{i(j)}} \alpha_g^{(j)} g \in \mathbb{C}G_{i(j)} \qquad (1 \le j \le k)$$

such that $i(1) \neq i(2) \neq \cdots \neq i(k)$ and $\tau_G(a_j) = 0$ (i.e. $\alpha_e^{(j)} = 0$) for all $1 \le j \le k$. Then we have

$$\begin{aligned}\tau_G(a_1 \cdots a_k) &= \tau_G\Big(\Big(\sum_{g_1\in G_{i(1)}} \alpha_{g_1}^{(1)} g_1\Big) \cdots \Big(\sum_{g_k\in G_{i(k)}} \alpha_{g_k}^{(k)} g_k\Big)\Big) \\ &= \sum_{g_1\in G_{i(1)},\dots,g_k\in G_{i(k)}} \alpha_{g_1}^{(1)} \cdots \alpha_{g_k}^{(k)} \tau_G(g_1 \cdots g_k).\end{aligned}$$

For all $g_1, \dots, g_k$ with $\alpha_{g_1}^{(1)} \dots \alpha_{g_k}^{(k)} \neq 0$ we have $g_j \neq e$ for all $j = 1, \dots, k$ and $i(1) \neq i(2) \neq \cdots \neq i(k)$, and thus, by Definition 5.6, that $g_1 \dots g_k \neq e$. This implies $\tau_G(a_1 \cdots a_k) = 0$, and thus the assertion.

Now let us prove (2) $\Rightarrow$ (1). Consider $k \in \mathbb{N}$, $i(1), \dots, i(k) \in I$ with $i(1) \neq i(2) \neq \cdots \neq i(k)$ and $g_1 \in G_{i(1)}\backslash\{e\}, \dots, g_k \in G_{i(k)}\backslash\{e\}$. The latter means that $\tau_G(g_j) = 0$ for all $j = 1, \dots, k$ and thus, by the definition of free independence, we also have $\tau_G(g_1 \cdots g_k) = 0$, which is exactly our assertion that $g_1 \cdots g_k \neq e$. □

REMARKS 5.12. (1) The group algebra $\mathbb{C}G$ can be extended in a canonical way to the so-called group von Neumann algebra $L(G)$. We will not address von Neumann algebras and the corresponding W^*-probability spaces in this book, but let us make at least some remarks about this on an informal level. In Example 3.11, we saw how one can extend the group algebra $\mathbb{C}G$ of a discrete group to a C^*-algebra $C^*_{\text{red}}(G)$ by taking the norm closure of $\mathbb{C}G$ in the left regular representation. If one takes instead a closure in a weaker topology, then one gets a bigger object, the group von Neumann algebra $L(G)$. The desire to understand the structure of such von Neumann algebras is a driving force for investigations in operator algebras. In particular, Voiculescu's motivation for starting free probability theory was to understand the structure of such $L(G)$ in the case where $G = G_1 * G_2$ is a free product group. Similarly, as $G_1 * G_2$ is built out of G_1 and G_2 and $\mathbb{C}(G_1 * G_2)$ is built out of $\mathbb{C}G_1$ and $\mathbb{C}G_2$, one would hope to understand $L(G_1 * G_2)$ by building it out of $L(G_1)$ and $L(G_2)$. However, this cannot be a purely algebraic operation. There is no useful way of saying that $L(G_1)$ and $L(G_2)$ are algebraically free in $L(G)$. (Note that by completing the group algebra in some topology we necessarily have to go over from finite sums over group elements to infinite sums.) What, however, can be extended from the level of group algebras to the level of von Neumann algebras is the characterization in terms of τ_G. τ_G extends to a faithful state (even trace) on $L(G)$ (in the same way as it extends to $C^*_{\text{red}}(G)$, see Example 3.11) and we still have that $L(G_1)$ and $L(G_2)$ are freely independent in $(L(G_1 * G_2), \tau_G)$. Thus understanding free independence may shed some light on the structure of $L(G_1 * G_2)$.

(2) In particular we have that $L(\mathbb{F}_n)$ and $L(\mathbb{F}_m)$ are freely independent in $(L(\mathbb{F}_{n+m}), \tau_{\mathbb{F}_{n+m}})$. This was the starting point of Voiculescu; in particular he wanted to attack the (still open) problem of the isomorphism of the free group factors, which asks the following: is it true that $L(\mathbb{F}_n)$ and $L(\mathbb{F}_m)$ are isomorphic as von Neumann algebras for all $n, m \geq 2$?

(3) Free independence has in the mean time provided a lot of information about the structure of $L(\mathbb{F}_n)$. The general philosophy is that these so-called free group factors are one of the most interesting class of von Neumann algebras after the well-understood hyperfinite ones and that free probability theory provides the right tools for studying this class.

Free independence and joint moments

Let us now start to examine the concept of free independence a bit more closely. Although not as obvious as in the case of tensor independence, free independence is from a combinatorial point of view nothing but a very special rule for calculating joint moments of freely independent variables out of the moments of the single variables. Or in other words, we have the following important fact.

If a family of random variables is freely independent, then the joint distribution of the family is completely determined by the knowledge of the individual distributions of the variables.

The proof of the following lemma shows how this calculation can be done in principle.

LEMMA 5.13. *Let $(\mathcal{A}, \varphi)$ be a non-commutative probability space and let the unital subalgebras $\mathcal{A}_i$ $(i \in I)$ be freely independent. Denote by $\mathcal{B}$ the algebra which is generated by all $\mathcal{A}_i$, $\mathcal{B} := \operatorname{alg}(\mathcal{A}_i \mid i \in I)$. Then $\varphi|_{\mathcal{B}}$ is uniquely determined by $\varphi|_{\mathcal{A}_i}$ for all $i \in I$ and by the free independence condition.*

PROOF. Each element of $\mathcal{B}$ can be written as a linear combination of elements of the form $a_1 \cdots a_k$ where $a_j \in \mathcal{A}_{i(j)}$ $(i(j) \in I)$. We can assume that $i(1) \neq i(2) \neq \cdots \neq i(k)$. (Otherwise, we just multiply some neighbors together to give a new element.) Let $a_1 \cdots a_k \in \mathcal{B}$ be such an element. We have to show that $\varphi(a_1 \dots a_k)$ is uniquely determined by the $\varphi|_{\mathcal{A}_i}$ $(i \in I)$.

We prove this by induction on k. The case $k = 1$ is clear because $a_1 \in \mathcal{A}_{i(1)}$. In the general case we put

$$a_j^o := a_j - \varphi(a_j)1 \in \mathcal{A}_{i(j)} \qquad (j = 1, \dots, k).$$

Then we have

$$\begin{aligned} \varphi(a_1 \cdots a_k) &= \varphi\big((a_1^o + \varphi(a_1)1) \cdots (a_k^o + \varphi(a_k)1)\big) \\ &= \varphi(a_1^o \cdots a_k^o) + \text{rest}, \end{aligned}$$

where

$$\text{rest} = \sum_{\cdots} \varphi(a_{p(1)}^o \cdots a_{p(s)}^o) \cdot \varphi(a_{q(1)}) \cdots \varphi(a_{q(k-s)}),$$

and the sum runs over all disjoint decompositions

$$\big((p(1) < \cdots < p(s)) \,\dot\cup\, (q(1) < \cdots < q(k-s)\big) = (1, \dots, k) \qquad (s < k).$$

Since $\varphi(a_j^o) = 0$ for all j it follows, by the definition of free independence, that $\varphi(a_1^o \cdots a_k^o) = 0$. On the other hand, all terms in rest are of length smaller than k, and thus are uniquely determined from the induction hypothesis. □

NOTATION 5.14. The operation of going over from some random variable a to

$$a^o := a - \varphi(a)1$$

is usually called the **centering** of a.

EXAMPLES 5.15. Let us look at some concrete examples. In the following we fix a non-commutative probability space $(\mathcal{A}, \varphi)$ and consider two free subalgebras $\tilde{\mathcal{A}}$ and $\tilde{\mathcal{B}}$. For elements $a, a_1, a_2 \in \tilde{\mathcal{A}}$ and $b, b_1, b_2 \in \tilde{\mathcal{B}}$ we want to calculate concretely some mixed moments of small length. The main trick is to reduce a general mixed moment to the special ones considered in the definition of free independence by centering the involved variables.

(1) According to the definition of free independence we have directly $\varphi(ab) = 0$ if $\varphi(a) = 0$ and $\varphi(b) = 0$. To calculate $\varphi(ab)$ in general we center our variables as in the proof of the lemma:

$$\begin{aligned} 0 &= \varphi\big((a - \varphi(a)1)(b - \varphi(b)1)\big) \\ &= \varphi(ab) - \varphi(a1)\varphi(b) - \varphi(a)\varphi(1b) + \varphi(a)\varphi(b)\varphi(1) \\ &= \varphi(ab) - \varphi(a)\varphi(b) \end{aligned}$$

which implies

$$\varphi(ab) = \varphi(a)\varphi(b) \qquad \text{if } a \text{ and } b \text{ are free.} \tag{5.4}$$

(2) In the same way we write

$$\varphi\big((a_1 - \varphi(a_1)1)(b - \varphi(b)1)(a_2 - \varphi(a_2)1)\big) = 0$$

implying

$$\varphi(a_1 b a_2) = \varphi(a_1 a_2)\varphi(b) \qquad \text{if } \{a_1, a_2\} \text{ and } b \text{ are free.} \tag{5.5}$$

(3) All the examples up to now yielded the same result as we would get for tensor independent random variables. To see the difference between "free independence" and "tensor independence" we consider now $\varphi(a_1 b_1 a_2 b_2)$. Starting from

$$\varphi\big((a_1 - \varphi(a_1)1)(b_1 - \varphi(b_1)1)(a_2 - \varphi(a_2)1)(b_2 - \varphi(b_2)1)\big) = 0$$

one arrives after some calculations at

$$\begin{aligned} \varphi(a_1 b_2 a_2 b_2) = \varphi(a_1 a_2)\varphi(b_1)\varphi(b_2) &+ \varphi(a_1)\varphi(a_2)\varphi(b_1 b_2) \\ &- \varphi(a_1)\varphi(b_1)\varphi(a_2)\varphi(b_2), \end{aligned} \tag{5.6}$$

if $\{a_1, a_2\}$ and $\{b_1, b_2\}$ are free.

Some basic properties of free independence

Although the above examples are only the tip of an iceberg they allow us to infer some general statements about freely independent random variables. In particular, one can see that the concept of free independence is a genuine non-commutative one and only trivial shadows of it can be seen in the commutative world.

REMARKS 5.16. (1) When can commuting random variables a and b be freely independent? We claim that this can only happen if at least one of them has vanishing variance, i.e. if

$$\varphi\big((a - \varphi(a)1)^2\big) = 0 \qquad \text{or} \qquad \varphi\big((b - \varphi(b)1)^2\big) = 0.$$

Indeed, let a and b be free and such that $ab = ba$. Then, by combining Equations (5.4) (for a^2 and b^2 instead of a and b) and (5.6) (for the case that $a_1 = a_2 = a$ and $b_1 = b_2 = b$), we have

$$\begin{aligned}\varphi(a^2)\varphi(b^2) &= \varphi(a^2b^2)\\ &= \varphi(abab)\\ &= \varphi(a^2)\varphi(b)^2 + \varphi(a)^2\varphi(b^2) - \varphi(a)^2\varphi(b)^2,\end{aligned}$$

and hence

$$\begin{aligned}0 &= \big(\varphi(a^2) - \varphi(a)^2\big)\big(\varphi(b^2) - \varphi(b)^2\big)\\ &= \varphi\big((a - \varphi(a)1)^2\big) \cdot \varphi\big((b - \varphi(b)1)^2\big),\end{aligned}$$

which implies that at least one of the two factors has to vanish.

(2) In particular, if a and b are classical random variables then they can only be freely independent if at least one of them is almost surely constant. This shows that free independence is really a non-commutative concept and cannot be considered as a special kind of dependence between classical random variables.

(3) A special case of the above is the following. If a is freely independent from itself then we have $\varphi(a^2) = \varphi(a)^2$. If we are in a $*$-probability space $(\mathcal{A}, \varphi)$ where φ is faithful, and if $a = a^*$, then this implies that a is a constant: $a = \varphi(a)1$. Another way of putting this is as follows. If the algebras $\mathcal{A}_1$ and $\mathcal{A}_2$ are $*$-free in the $*$-probability space $(\mathcal{A}, \varphi)$ and if φ is faithful then

$$\mathcal{A}_1 \cap \mathcal{A}_2 = \mathbb{C}1.$$

Another general statement about freely independent random variables which can be inferred directly from the definition is that constant random variables are freely independent from everything. Because of its importance we state this observation as a lemma.

LEMMA 5.17. *Let $(\mathcal{A}, \varphi)$ be a non-commutative probability space and $\mathcal{B} \subset \mathcal{A}$ a unital subalgebra. Then the subalgebras $\mathbb{C}1$ and $\mathcal{B}$ are freely independent.*

PROOF. Consider $a_1, \dots, a_k$ as in the definition of free independence, with $k \geq 2$ (the case $k = 1$ is clear). Then we have at least one $a_j \in \mathbb{C}1$ with $\varphi(a_j) = 0$. But this means $a_j = 0$, hence $a_1 \cdots a_k = 0$ and thus $\varphi(a_1 \cdots a_k) = 0$. □

In the next proposition we observe the fact that free independence behaves nicely with respect to the tracial property. To prove this we need a bit of information on how to calculate special mixed moments of freely independent random variables.

LEMMA 5.18. *Let $(\mathcal{A},\varphi)$ be a non-commutative probability space, and let $(\mathcal{A}_i)_{i\in I}$ be a freely independent family of unital subalgebras of $\mathcal{A}$. Let $a_1,\dots,a_k$ be elements of the algebras $\mathcal{A}_{i(1)},\dots,\mathcal{A}_{i(k)}$, respectively, where the indices $i(1),\dots,i(k)\in I$ are such that*

$$i(1)\neq i(2),\dots,i(k-1)\neq i(k),$$

and where we have $\varphi(a_1)=\cdots=\varphi(a_k)=0$. Likewise, let $b_1,\dots,b_l$ be elements of $\mathcal{A}_{j(1)},\dots,\mathcal{A}_{j(l)}$, respectively, such that

$$j(1)\neq j(2),\dots,j(l-1)\neq j(l),$$

and such that $\varphi(b_1)=\cdots=\varphi(b_l)=0$. Then we have

$$\varphi(a_1\cdots a_k b_l\cdots b_1) = \begin{cases} \varphi(a_1b_1)\cdots\varphi(a_kb_k) & \text{if } k=l,\ i(1)=j(1),\dots,i(k)=j(k) \\ 0 & \text{otherwise.} \end{cases} \tag{5.7}$$

PROOF. One has to iterate the following observation: either we have $i(k)\neq j(l)$, in which case

$$\varphi(a_1\cdots a_k b_l\cdots b_1)=0,$$

or we have $i(k)=j(l)$, which gives

$$\begin{aligned}\varphi(a_1\cdots a_kb_l\cdots b_1) &= \varphi\Big(a_1\dots a_{k-1}\cdot\big((a_kb_l)^o+\varphi(a_kb_l)1\big)\cdot b_{l-1}\dots b_1\Big)\\ &= 0+\varphi(a_kb_l)\cdot\varphi(a_1\dots a_{k-1}b_{l-1}\dots b_1).\end{aligned}$$

□

PROPOSITION 5.19. *Let $(\mathcal{A},\varphi)$ be a non-commutative probability space, let $(\mathcal{A}_i)_{i\in I}$ be a freely independent family of unital subalgebras of $\mathcal{A}$, and let $\mathcal{B}$ be the subalgebra of $\mathcal{A}$ generated by $\cup_{i\in I}\mathcal{A}_i$. If $\varphi|_{\mathcal{A}_i}$ is a trace for every $i\in I$, then $\varphi|_{\mathcal{B}}$ is a trace.*

PROOF. We have to prove that $\varphi(ab)=\varphi(ba)$ for all $a,b\in\mathcal{B}$. Since every element a from $\mathcal{B}$ can be written as a linear combination of 1 and elements of the form $a_1\cdots a_k$ (for $k\geq 1$, $a_p\in\mathcal{A}_{i(p)}$ such that $i(1)\neq i(2)\neq\ldots\neq i(k)$ and $\varphi(a_1)=\cdots=\varphi(a_k)=0$), it suffices to prove the assertion for a and b of the special form $a=a_1\cdots a_k$ and $b=b_l\cdots b_1$ with $a_p\in\mathcal{A}_{i(p)}$ and $b_q\in\mathcal{A}_{j(q)}$ where $i(1)\neq i(2)\neq\ldots\neq i(k)$ and $j(1)\neq j(2)\neq\ldots\neq j(l)$, and such that $\varphi(a_1)=\cdots=\varphi(a_k)=0$ and $\varphi(b_1)=\cdots=\varphi(b_l)=0$. But in this situation we can apply the previous Lemma 5.18 and get

$$\varphi(a_1\cdots a_kb_l\cdots b_1)=\delta_{kl}\cdot\delta_{i(k)j(k)}\cdots\delta_{i(1)j(1)}\cdot\varphi(a_1b_1)\cdots\varphi(a_kb_k)$$

and

$$\varphi(b_l\cdots b_1a_1\cdots a_k)=\delta_{lk}\cdot\delta_{j(1)i(1)}\cdots\delta_{j(l)i(l)}\cdot\varphi(b_la_l)\cdots\varphi(b_1a_1).$$

The assertion follows now from the assumption that φ is a trace on each $\mathcal{A}_i$, since this means that $\varphi(a_p b_p) = \varphi(b_p a_p)$ for all p. $\square$

For some more general observations which one can derive directly from the definition of free independence, see Exercises 5.22–5.25.

REMARK 5.20. One general observation which is worth spelling out explicitly is that free independence is (in the same way as classical independence) commutative and associative, in the sense that

$$\mathcal{A}_1, \mathcal{A}_2 \text{ free} \qquad \Longleftrightarrow \qquad \mathcal{A}_2, \mathcal{A}_1 \text{ free}$$

and

$$\left.\begin{array}{c} \mathcal{X}_1, \mathcal{X}_2 \cup \mathcal{X}_3 \text{ free} \\ \mathcal{X}_2, \mathcal{X}_3 \text{ free} \end{array}\right\} \Longleftrightarrow \left\{\begin{array}{c} \mathcal{X}_1 \cup \mathcal{X}_2, \mathcal{X}_3 \text{ free} \\ \mathcal{X}_1, \mathcal{X}_2 \text{ free} \end{array}\right\} \Longleftrightarrow \mathcal{X}_1, \mathcal{X}_2, \mathcal{X}_3 \text{ free}$$

The commutativity is obvious from the definition, associativity will be addressed in Exercise 5.25.

REMARK 5.21. The fact that the joint distribution of a free family is determined by the individual distributions can be combined with Theorem 4.11 of the preceding lecture – this will enable us to talk about C^*-algebras defined by a family of $*$-free generators with given $*$-distributions. (At least it will be clear that the class of isomorphism of such a C^*-algebra is uniquely determined. The issue of why the C^*-algebra in question does indeed exist will be discussed separately in the next two lectures.)

For instance one can talk about

$$\left\{\begin{array}{l} \text{“the unital } C^*\text{-algebra generated by 3 free} \\ \text{selfadjoint elements with arcsine distributions.”} \end{array}\right. \tag{5.8}$$

This means: a unital C^*-algebra $\mathcal{A}$ endowed with a faithful positive functional φ, and generated by 3 free selfadjoint elements x_1, x_2, x_3, where each of x_1, x_2, x_3 has arcsine distribution with respect to φ. (Recall that the arcsine distribution was discussed in Example 1.14.) The C^*-algebra $\mathcal{A}$ is, up to isomorphism, uniquely determined, by the fact that the above conditions determine the joint distribution of x_1, x_2, x_3, and by Theorem 4.11 (see also Remark 4.12.1).

Of course, when referring to a C^*-algebra introduced as in (5.8), one must also make sure that a C^*-algebra satisfying the required conditions does indeed exist. For the example at hand, this is (incidentally) very easy – we can just take $\mathcal{A}$ to be the unital C^*-subalgebra of $C^*_{\text{red}}(\mathbb{F}_3)$ which is generated by the real parts of the 3 canonical unitary generators of $C^*_{\text{red}}(\mathbb{F}_3)$. In general, this kind of approach does not necessarily take us to look at the reduced C^*-algebra of a free group, but rather to a more general type of free product construction, which is discussed in the next two lectures.

Are there other universal product constructions?

Before we continue our investigation of the structure of free independence we want to pause for a moment and consider the question how special free independence is. Has it some very special properties or is it just one of many examples of other forms of independence? We know that we have free independence and tensor independence. What are other examples?

Let us formalize a bit what we mean by a concept of independence. Independence of subalgebras $\mathcal{A}_i$ $(i \in I)$ should give us a prescription for calculating a linear functional on the algebra generated by all $\mathcal{A}_i$ if we know the value of the functional on each of the subalgebras (as Lemma 5.13 assures us for the case of free independence). This prescription should be universal in the sense that it does not depend on the actual choice of subalgebras, but works in the same way for all situations.

So what we are looking for are universal product constructions in the following sense: given any pair of non-commutative probability spaces $(\mathcal{A}_1, \varphi_1)$ and $(\mathcal{A}_2, \varphi_2)$, we would like to construct in a universal way a new non-commutative probability space $(\mathcal{A}, \varphi)$, where $\mathcal{A}$ consists of all linear combinations of possible words made of letters from $\mathcal{A}_1$ and from $\mathcal{A}_2$. (Thus, $\mathcal{A} := \mathcal{A}_1 * \mathcal{A}_2$ is the so-called algebraic free product of $\mathcal{A}_1$ and $\mathcal{A}_2$. We will say more about this in the next lecture.)

One can formulate this in an abstract way by using the language of category theory (one is looking for a construction which is natural, i.e. commutes with homomorphisms), but it can be shown that in the end this comes down to having formulas for mixed moments $\varphi(a_1b_1a_2b_2 \cdots a_nb_n)$ (where $a_i \in \mathcal{A}_1$ and $b_i \in \mathcal{A}_2$) which involve only products of moments of the a_i and moments of the b_i, such that in each such product all a_i and all b_i appear exactly once and in their original order.

Let us make the type of formulas a bit clearer by writing down some examples for small n. The case $n = 1$ yields only one possibility for such a product, thus

$$\varphi(a_1b_1) = \varepsilon\varphi_1(a_1)\varphi_2(b_1), \tag{5.9}$$

whereas $n = 2$ gives rise to four possible contributions:

$$\begin{aligned}\varphi(a_1b_1a_2b_2) &= \alpha\varphi_1(a_1a_2)\varphi_2(b_1b_2) + \beta\varphi_1(a_1)\varphi_1(a_2)\varphi_2(b_1b_2)\\ &+\gamma\varphi_1(a_1a_2)\varphi_2(b_1)\varphi_2(b_2) + \delta\varphi_1(a_1)\varphi_1(a_2)\varphi_2(b_1)\varphi_2(b_2).\end{aligned} \tag{5.10}$$

Universality of the construction means that the coefficients ε, α, β, γ, δ do not depend on the special choice of the non-commutative probability spaces $(\mathcal{A}_1, \varphi_1)$, $(\mathcal{A}_2, \varphi_2)$, nor the special choice of the elements a_i, b_i, but that these coefficients are just some fixed numbers.

The question is now: how much freedom do we have to choose the coefficients $\varepsilon, \alpha, \beta, \gamma, \delta$ in the above formulas? Of course, the coefficients cannot be totally arbitrary, because we clearly want to impose the following consistency requirements:

(1) the formulas must be consistent if we put some of the a_i or b_i equal to 1;

(2) the formulas must respect associativity, i.e. in the iterated case of three or more algebras the resulting formula must be independent of the order in which we iterate.

The first requirement, for example, gives us directly

$$1 = \varphi(1 \cdot 1) = \varepsilon\varphi_1(1)\varphi_2(1) = \varepsilon,$$

so for $n = 1$ we have no choice but

$$\varphi(ab) = \varphi_1(a)\varphi_2(b). \tag{5.11}$$

This means in particular that

$$\varphi(a) = \varphi_1(a) \qquad \forall\, a \in \mathcal{A}_1, \tag{5.12}$$

$$\varphi(b) = \varphi_2(b) \qquad \forall\, b \in \mathcal{A}_2, \tag{5.13}$$

which agrees with our expectation that such an universal product construction should be an extension of given states to the free product of the involved algebras.

For $n = 2$, we get, by putting $a_1 = a_2 = 1$ in (5.10):

$$\begin{aligned}\varphi_2(b_1b_2) &= \varphi(b_1b_2)\\ &= \alpha\varphi_1(1 \cdot 1)\varphi_2(b_1b_2) + \beta\varphi_1(1)\varphi_1(1)\varphi_2(b_1b_2)\\ &\quad + \gamma\varphi_1(1 \cdot 1)\varphi_2(b_1)\varphi_2(b_2) + \delta\varphi_1(1)\varphi_1(1)\varphi_2(b_1)\varphi_2(b_2)\\ &= (\alpha + \beta)\varphi_2(b_1b_2) + (\gamma + \delta)\varphi_2(b_1)\varphi_2(b_2).\end{aligned}$$

Since we can choose any probability space $(\mathcal{A}_2, \varphi_2)$ and arbitrary elements $b_1, b_2 \in \mathcal{A}_2$, the above equality has to be true for arbitrary $\varphi_2(b_1b_2)$ and arbitrary $\varphi_2(b_1)\varphi_2(b_2)$, which implies that

$$\alpha + \beta = 1 \qquad \text{and} \qquad \gamma + \delta = 0. \tag{5.14}$$

Similarly, by putting $b_1 = b_2 = 1$ in (5.10), we obtain

$$\alpha + \gamma = 1 \qquad \text{and} \qquad \beta + \gamma = 0. \tag{5.15}$$

Note that these relations imply in particular (by putting $b_2 = 1$ in (5.10)) that

$$\varphi(a_1 b_1 a_2) = \varphi(a_1 a_2)\varphi(b_1).$$

The relations (5.14) and (5.15) give us some substantial restrictions for our allowed coefficients in (5.10), but still it looks as if we could find a 1-parameter family of such formulas for $n = 2$.

However, associativity will produce more restrictions. Let us consider three probability spaces $(\mathcal{A}_1, \varphi_1)$, $(\mathcal{A}_2, \varphi_2)$, $(\mathcal{A}_3, \varphi_3)$ with elements $a_1, a_2 \in \mathcal{A}_1$, $b_1, b_2 \in \mathcal{A}_2$, $c_1, c_2 \in \mathcal{A}_3$ and let us calculate $\varphi(a_1 c_1 b_1 c_2 a_2 b_2)$. We can do this in two different ways. We can read it either as $\varphi(a_1(c_1 b_1 c_2) a_2 b_2)$ or as $\varphi((a_1 c_1) b_1 (c_2 a_2) b_2)$. In the first case we calculate

$$\begin{aligned}\varphi(a_1(c_1 b_1 c_2) a_2 b_2) &= \alpha\varphi(a_1 a_2)\varphi(c_1 b_1 c_2 b_2) + \beta\varphi(a_1)\varphi(a_2)\varphi(c_1 b_1 c_2 b_2) \\ &\quad + \gamma\varphi(a_1 a_2)\varphi(c_1 b_1 c_2)\varphi(b_2) + \delta\varphi(a_1)\varphi(a_2)\varphi(c_1 b_1 c_2)\varphi(b_2),\end{aligned}$$

and reformulate this further by using

$$\begin{aligned}\varphi(c_1 b_1 c_2 b_2) &= \alpha\varphi(c_1 c_2)\varphi(b_1 b_2) + \beta\varphi(c_1)\varphi(c_2)\varphi(b_1 b_2) \\ &\quad + \gamma\varphi(c_1 c_2)\varphi(b_1)\varphi(b_2) + \delta\varphi(c_1)\varphi(c_2)\varphi(b_1)\varphi(b_2)\end{aligned}$$

and

$$\varphi(c_1 b_1 c_2) = \varphi(c_1 c_2)\varphi(b_1).$$

This leads to a final expression where the term $\varphi(a_1 a_2)\varphi(b_1 b_2)\varphi(c_1 c_2)$ appears with coefficient α^2, i.e.,

$$\varphi(a_1(c_1 b_1 c_2) a_2 b_2) = \alpha^2\varphi(a_1 a_2)\varphi(b_1 b_2)\varphi(c_1 c_2) + \cdots .$$

On the other hand, in the second case we calculate

$$\begin{aligned}\varphi((a_1 c_1) b_1 (c_2 a_2) b_2) &= \alpha\varphi(a_1 c_1 c_2 a_2)\varphi(b_1 b_2) + \beta\varphi(a_1 c_1)\varphi(c_2 a_2)\varphi(b_1 b_2) \\ &\quad + \gamma\varphi(a_1 c_1 c_2 a_2)\varphi(b_1)\varphi(b_2) + \delta\varphi(a_1 c_1)\varphi(c_2 a_2)\varphi(b_1)\varphi(b_2).\end{aligned}$$

By using

$$\varphi(a_1 c_1 c_2 a_2) = \varphi(a_1(c_1 c_2) a_2) = \varphi(a_1 a_2)\varphi(c_1 c_2)$$

and

$$\varphi(a_1 c_1) = \varphi(a_1)\varphi(c_1), \qquad \varphi(c_2 a_2) = \varphi(c_2)\varphi(a_2),$$

we finally get an expression in which the term $\varphi(a_1 a_2)\varphi(b_1 b_2)\varphi(c_1 c_2)$ appears with coefficient α,

$$\varphi((a_1 c_1) b_1 (c_2 a_2) b_2) = \alpha\varphi(a_1 a_2)\varphi(b_1 b_2)\varphi(c_1 c_2) + \cdots .$$

Since the other appearing moments can be chosen independently from $\varphi(a_1 a_2)\varphi(b_1 b_2)\varphi(c_1 c_2)$, comparison of both calculations yields that $\alpha^2 = \alpha$. Thus we only remain with the two possibilities $\alpha = 1$ or $\alpha = 0$.

By (5.14) and (5.15), the value of α determines the other coefficients and finally we arrive at the conclusion that, for $n = 2$, we only have the two possibilities that either $\alpha = 1$ and $\beta = \gamma = \delta = 0$, which means

$$\varphi(a_1b_1a_2b_2) = \varphi(a_1a_2)\varphi(b_1b_2),$$

or that $\alpha = 0$, $\beta = \gamma = 1$, $\delta = -1$, which means

$$\varphi(a_1b_1a_1b_2) = \varphi(a_1a_2)\varphi(b_1)\varphi(b_2) + \varphi(a_1)\varphi(a_2)\varphi(b_1b_2) \\ - \varphi(a_1)\varphi(a_2)\varphi(b_1)\varphi(b_2).$$

But the first case is the formula which we get for tensor independent variables (see Equation (5.2)), whereas the second case, is exactly the formula (5.6), which describes freely independent random variables. Thus we see from these considerations that on the level of words of length 4 there are only two possibilities for having universal product constructions. It can be shown that this is also true for greater lengths: although the number of coefficients in universal formulas for expressions $\varphi(a_1b_1a_2b_2 \cdots a_nb_n)$ grows very fast with n, the consistency conditions (in particular, associativity) give such strong relations between the allowed coefficients that in the end only two possibilities survive – either one has tensor independence or one has free independence.

This shows that free independence, which might appear somewhat artificial on first look, is a very fundamental concept – it is the only other possibility for a universal product construction.

Exercises

EXERCISE 5.22. (1) Prove that functions of freely independent random variables are freely independent: if a and b are freely independent and f and g polynomials, then $f(a)$ and $g(b)$ are freely independent, too.

(2) Make the following statement precise and prove it: free independence is preserved via taking homomorphic images of algebras.

EXERCISE 5.23. Let $(\mathcal{A}, \varphi)$ be a C^*-probability space, and let $(\mathcal{A}_i)_{i\in I}$ be a freely independent family of unital $*$-subalgebras of $\mathcal{A}$. For every $i \in I$, let $\mathcal{B}_i$ be the closure of $\mathcal{A}_i$ in the norm-topology. Prove that the algebras $(\mathcal{B}_i)_{i\in I}$ are freely independent.

EXERCISE 5.24. Let $(\mathcal{A}, \varphi)$ be a $*$-probability space. Consider a unital subalgebra $\mathcal{B} \subset \mathcal{A}$ and a Haar unitary $u \in \mathcal{A}$ such that $\{u, u^*\}$ and $\mathcal{B}$ are free. Show that then also $\mathcal{B}$ and $u^*\mathcal{B}u$ are free. (The algebra $u^*\mathcal{B}u$ is of course $u^*\mathcal{B}u := \{u^*bu \mid b \in \mathcal{B}\} \subset \mathcal{A}$.)

EXERCISE 5.25. In this exercise we prove that free independence behaves well under successive decompositions and thus is associative. Let $(\mathcal{A}_i)_{i\in I}$ be unital subalgebras of $\mathcal{A}$ and, for each $i \in I$, $(\mathcal{B}_i^j)_{j\in J(i)}$ unital subalgebras of $\mathcal{A}_i$. Then we have the following.

(1) If $(\mathcal{A}_i)_{i\in I}$ are freely independent in $\mathcal{A}$ and, for each $i \in I$, $(\mathcal{B}_i^j)_{j\in J(i)}$ are freely independent in $\mathcal{A}_i$, then all $(\mathcal{B}_i^j)_{i\in I; j\in J(i)}$ are freely independent in $\mathcal{A}$.

(2) If all $(\mathcal{B}_i^j)_{i\in I; j\in J(i)}$ are freely independent in $\mathcal{A}$ and if, for each $i \in I$, $\mathcal{A}_i$ is the algebra generated by all $\mathcal{B}_i^j$ for $j \in J(i)$, then $(\mathcal{A}_i)_{i\in I}$ are freely independent in $\mathcal{A}$.

EXERCISE 5.26. If we consider, instead of unital algebras and unital linear functionals, just algebras and linear functionals, then we might also ask about the existence of universal product constructions in this frame. We have to give up the first consistency requirement about setting some of the random variables equal to 1, and we can only require associativity of the universal product construction. Of course, the tensor product and the free product are still examples of such products. Show that in such a frame there exists exactly one additional example of a universal product if we also require the natural extension properties (5.12), (5.13) and the factorization property (5.11) to hold. Describe this additional example.

LECTURE 6

Free product of $*$-probability spaces

In order to use free independence we have to be able to find sufficiently many situations where freely independent random variables arise. In particular, given a family of non-commutative probability spaces $(\mathcal{A}_i, \varphi_i)$, $i \in I$, we should be able to find "models" of the $(\mathcal{A}_i, \varphi_i)$ sitting inside some bigger non-commutative probability space $(\mathcal{A}, \varphi)$, such that the $\mathcal{A}_i$ are freely independent in $(\mathcal{A}, \varphi)$. To put it in other words: if free independence is to be a structure as powerful as classical independence, then it should allow us to make assumptions such as "let x_i be freely independent and identically distributed random variables" (with a given distribution). In classical probability theory it is of course the existence of product measures (or of tensor products in the more general algebraic frame) which ensures this. In this lecture we discuss the free counterpart of this construction – free products of non-commutative probability spaces.

Free product of unital algebras

Similar to the free product of groups discussed in the preceding lecture, the free product of a family $\{\mathcal{A}_i \mid i \in I\}$ of unital algebras will be a unital algebra $\mathcal{A}$ whose elements are words made with "letters" from the $\mathcal{A}_i$. Before describing how exactly we make words with letters from the $\mathcal{A}_i$, let us state the formal definition of the free product in terms of its universality property (this is analogous to the universality property stated for free products of groups in Remark 5.10.1 of the preceding lecture).

DEFINITION 6.1. Let $(\mathcal{A}_i)_{i\in I}$ be a family of unital algebras over $\mathbb{C}$. The (algebraic) **free product with identification of units** of the $\mathcal{A}_i$ is a unital algebra $\mathcal{A}$, given together with a family of unital homomorphisms $(V_i : \mathcal{A}_i \to \mathcal{A})_{i\in I}$, such that the following universality property holds: whenever $\mathcal{B}$ is a unital algebra over $\mathbb{C}$ and $(\Phi_i : \mathcal{A}_i \to \mathcal{B})_{i\in I}$ is a family of unital homomorphisms, there exists a unique unital homomorphism $\Phi : \mathcal{A} \to \mathcal{B}$ such that $\Phi \circ V_i = \Phi_i$, $\forall\, i \in I$.

NOTATIONS 6.2. We now make some clarifying comments related to the preceding definition (and also introduce at the same time a number of useful notations). So let us consider the setting of Definition 6.1. Quite clearly, the free product algebra $\mathcal{A}$ is determined up to isomorphism (in the obvious way, common to all situations when objects are defined by universality properties). On the other hand, the homomorphisms $V_i : \mathcal{A}_i \to \mathcal{A}$ turn out to be one-to-one (see the discussion in the next Remark 6.3); so by a slight notational abuse we can assume that $\mathcal{A}$ *contains* every $\mathcal{A}_i$ as a unital subalgebra. This makes the map V_i disappear (or rather, V_i simply becomes the inclusion of $\mathcal{A}_i$ into $\mathcal{A}$). This version of the free product algebra (which contains every $\mathcal{A}_i$ as a unital subalgebra) is somewhat more "canonical"; it is the one which is usually considered, and is denoted as

$$\mathcal{A} = *_{i \in I} \mathcal{A}_i. \tag{6.1}$$

We should warn the reader here that the simplified notation in Equation (6.1) comes together with the following convention: by relabeling the $\mathcal{A}_i$ if necessary, we assume that they all share the same unit, while on the other hand an intersection $\mathcal{A}_{i_1} \cap \mathcal{A}_{i_2}$ for $i_1 \neq i_2$ does not contain any element which is not a scalar multiple of the unit. (This is the case even if we are looking at a free product of the form, say, $\mathcal{A} = \mathcal{B} * \mathcal{B} * \mathcal{B}$, for some given unital algebra $\mathcal{B}$. Before being embedded inside $\mathcal{A}$, the 3 copies of $\mathcal{B}$ that we are dealing with have to be relabeled as $\mathcal{B}_1, \mathcal{B}_2, \mathcal{B}_3$, with $\mathcal{B}_i \cap \mathcal{B}_j = \mathbb{C}1_{\mathcal{A}}$ for $i \neq j$.)

The structure of the free product $\mathcal{A} = *_{i \in I} \mathcal{A}_i$ is better understood if one identifies $\mathcal{A}$ as being spanned by certain sets of "words" made with "letters" from the algebras $\mathcal{A}_i$. In order to describe this, let us choose inside every $\mathcal{A}_i$ a subspace $\mathcal{A}_i^o$ of codimension 1 which gives a complement for the scalar multiples of the unit of $\mathcal{A}_i$. (A way of finding such a subspace $\mathcal{A}_i^o$ which fits very well the spirit of these lectures is by setting $\mathcal{A}_i^o := \ker(\varphi_i)$, where $\varphi_i : \mathcal{A}_i \to \mathbb{C}$ is a linear functional such that $\varphi_i(1_{\mathcal{A}_i}) = 1$.) Once the subspaces $\mathcal{A}_i^o$ are chosen, we get a direct sum decomposition for the free product algebra $\mathcal{A} = *_{i \in I} \mathcal{A}_i$, as follows:

$$\mathcal{A} = \mathbb{C}1 \oplus \Big(\bigoplus_{n=1}^{\infty} \bigoplus_{\substack{i_1, \ldots, i_n \in I \\ i_1 \neq i_2, \ldots, i_{n-1} \neq i_n}} \mathcal{W}_{i_1, \ldots, i_n} \Big), \tag{6.2}$$

where for every $n \geq 1$ and every $i_1, \ldots, i_n \in I$ such that $i_1 \neq i_2, \ldots, i_{n-1} \neq i_n$ we set

$$\mathcal{W}_{i_1, \ldots, i_n} := \text{span}\{a_1 \cdots a_n \mid a_1 \in \mathcal{A}_{i_1}^o, \ldots, a_n \in \mathcal{A}_{i_n}^o\}. \tag{6.3}$$

Thus every $\mathcal{W}_{i_1,\ldots,i_n}$ is a set of "words" of a specified type, and the free product $*_{i\in I}\mathcal{A}_i$ can be understood in terms of such linear subspaces of words via Equation (6.2). It is also worth recording that for every $n \geq 1$ and every $i_1, \ldots, i_n \in I$ such that $i_1 \neq i_2, \ldots, i_{n-1} \neq i_n$, the space $\mathcal{W}_{i_1,\ldots,i_n}$ of Equation (6.3) is canonically isomorphic to the tensor product $\mathcal{A}_{i_1}^o \otimes \cdots \otimes \mathcal{A}_{i_n}^o$, via the linear map determined by

$$\mathcal{A}_{i_1}^o \otimes \cdots \otimes \mathcal{A}_{i_n}^o \ni a_1 \otimes \cdots \otimes a_n \mapsto a_1 \cdots a_n \in \mathcal{W}_{i_1,\ldots,i_n}, \tag{6.4}$$

for $a_1 \in \mathcal{A}_{i_1}^o, \ldots, a_n \in \mathcal{A}_{i_n}^o$.

REMARK 6.3. One might object at this point that our presentation of the free product $\mathcal{A} = *_{i\in I}\mathcal{A}_i$ lacks in the following respects: first we did not give a proof that an algebra $\mathcal{A}$ with the universality property stated in Definition 6.1 does indeed exist, and then in Notations 6.2 we presented some properties of this hypothetical algebra $\mathcal{A}$ which again we gave without proof. For the reader interested in filling in these gaps, let us make the observation that the two shortcomings mentioned above can be made to cancel each other, by reasoning in the following way. Let $(\mathcal{A}_i)_{i\in I}$ be a family of unital algebras for which we want to construct the free product. For every $i \in I$ consider a linear functional $\varphi_i : \mathcal{A}_i \to \mathbb{C}$ such that $\varphi_i(1_{\mathcal{A}_i}) = 1$, and the subspace $\mathcal{A}_i^o := \ker(\varphi_i) \subset \mathcal{A}_i$. Then consider the vector space

$$\mathcal{A} = \mathbb{C}1 \oplus \Big(\bigoplus_{n=1}^{\infty} \bigoplus_{\substack{i_1,\ldots,i_n \in I \\ i_1 \neq i_2,\ldots,i_{n-1}\neq i_n}} \mathcal{A}_{i_1}^o \otimes \cdots \otimes \mathcal{A}_{i_n}^o \Big). \tag{6.5}$$

The point to observe is that on this vector space $\mathcal{A}$ one can rigorously define a multiplication which reflects the intuitive idea of how "words" can be multiplied by concatenation. Thus $\mathcal{A}$ becomes a unital algebra over $\mathbb{C}$, and the algebras $\mathcal{A}_i$ are naturally embedded inside it (via $\mathcal{A}_i \simeq \mathbb{C}1 \oplus \mathcal{A}_i^o$, $i \in I$); finally, the universality properties known for tensor products and direct sums can be used in order to deduce that $\mathcal{A}$ indeed has the universality property required by Definition 6.1.

In this approach, the tedious details which have to be verified are then concentrated in the process of making sure that the "natural" definition of the multiplication on $\mathcal{A}$ indeed makes sense, and gives us an algebra. We will leave it as an exercise to the conscientious reader to work out the formula for how to multiply two general tensors $a_1' \otimes \cdots \otimes a_m'$ and $a_1'' \otimes \cdots \otimes a_n''$ – see Exercise 6.15. Here we will only discuss, for illustration, one simple example of such a multiplication.

Say for instance that we have chosen two distinct indices $i_1 \neq i_2$ in I and some elements $a_1, b_1 \in \mathcal{A}_{i_1}^o$, $a_2, b_2 \in \mathcal{A}_{i_2}^o$, and that we want to figure out the formula for multiplying the elements $a_1 \otimes a_2$ and $b_2 \otimes b_1$ of $\mathcal{A}$.

The first candidate which comes to mind as result of this multiplication is $a_1 \otimes (a_2 b_2) \otimes b_1$ (obtained by concatenating the two given tensors and by using the multiplication of $\mathcal{A}_{i_2}$). But $a_1 \otimes (a_2 b_2) \otimes b_1$ does not necessarily belong to any of the summands in the direct sum in (6.5), as $a_2 b_2$ may not belong to $\mathcal{A}_{i_2}^o$. In order to fix this, we thus consider (similar to the pattern of notation used in Lecture 5) the centering of $a_2 b_2$,

$$(a_2 b_2)^o := a_2 b_2 - \varphi_{i_2}(a_2 b_2) \cdot 1_{\mathcal{A}_{i_2}} \in \mathcal{A}_{i_2}^o.$$

Then the candidate for the product of $a_1 \otimes a_2$ and $b_2 \otimes b_1$ becomes:

$$a_1 \otimes (a_2 b_2)^o \otimes a_1 + \varphi_{i_2}(a_2 b_2) \cdot (a_1 b_1).$$

This is closer to what we need, but still requires the centering of $a_1 b_1$,

$$(a_1 b_1)^o := a_1 b_1 - \varphi_{i_1}(a_1 b_1) \cdot 1_{\mathcal{A}_{i_1}} \in \mathcal{A}_{i_1}^o.$$

By replacing $a_1 b_1$ by $(a_1 b_1)^o + \varphi_{i_1}(a_1 b_1) \cdot 1$ in the preceding form of the candidate for the product, we arrive at the correct definition:

$$\begin{aligned}(a_1 \otimes a_2) \cdot (b_2 \otimes b_1) = \big(\varphi_{i_1}(a_1 b_1)\varphi_{i_2}(a_2 b_2)\big) \cdot 1 \\ + \varphi_{i_2}(a_2 b_2) \cdot (a_1 b_1)^o + a_1 \otimes (a_2 b_2)^o \otimes b_1.\end{aligned}$$

(Thus $(a_1 \otimes a_2) \cdot (b_2 \otimes b_1)$ is an element of $\mathbb{C}1 \oplus \mathcal{A}_{i_1}^o \oplus (\mathcal{A}_{i_1}^o \otimes \mathcal{A}_{i_2}^o \otimes \mathcal{A}_{i_1}^o) \subset \mathcal{A}$.)

A final point: from the approach suggested in this remark, it would seem that the free product $*_{i \in I} \mathcal{A}_i$ actually depends on the choice of a family of linear functionals $\varphi_i : \mathcal{A}_i \to \mathbb{C}$, $i \in I$. It is indeed true that the decomposition of $\mathcal{A}$ appearing on the right-hand side of (6.5) depends on the choice of φ_i. But the class of isomorphism of $\mathcal{A}$ itself does not depend on the φ_i – this is immediate from the fact that $\mathcal{A}$ has the universality property required in Definition 6.1.

Free product of non-commutative probability spaces

DEFINITION 6.4. Let $(\mathcal{A}_i, \varphi_i)_{i \in I}$ be a family of non-commutative probability spaces. Consider the free product algebra $\mathcal{A} = *_{i \in I} \mathcal{A}_i$ and its direct sum decomposition as described in Equation (6.2) of Notations 6.2, where the subspaces $\mathcal{A}_i^o \subset \mathcal{A}_i$ are defined as $\mathcal{A}_i^o := \ker(\varphi_i)$, $i \in I$. The free product of the functionals $(\varphi_i)_{i \in I}$ is defined as the unique linear functional $\varphi : \mathcal{A} \to \mathbb{C}$ such that $\varphi(1_{\mathcal{A}}) = 1$ and such that $\varphi | \mathcal{W}_{i_1, \ldots, i_n} = 0$ for every $n \geq 1$ and every $i_1, \ldots, i_n \in I$ with $i_1 \neq i_2, i_2 \neq i_3, \ldots, i_{n-1} \neq i_n$ The notation used for this functional $\varphi : \mathcal{A} \to \mathbb{C}$ is $*_{i \in I} \varphi_i$. The corresponding non-commutative probability space $(\mathcal{A}, \varphi)$ is called the **free product of the non-commutative probability spaces** $(\mathcal{A}_i, \varphi_i)$ for $i \in I$, and one writes sometimes

$$(\mathcal{A}, \varphi) = *_{i \in I} (\mathcal{A}_i, \varphi_i)$$

(instead of $(\mathcal{A}, \varphi) = (*_{i\in I}\mathcal{A}_i, *_{i\in I}\varphi_i))$.

REMARK 6.5. In the situation of Definition 6.4, the restriction of the free product functional φ to an $\mathcal{A}_i$ is equal to the original $\varphi_i : \mathcal{A}_i \to \mathbb{C}$ which we started with. Indeed, φ is defined such that $\ker(\varphi) \supset \mathcal{A}_i^o = \ker(\varphi_i)$; hence the functionals $\varphi|_{\mathcal{A}_i}$ and φ_i coincide on $\mathcal{A}_i^o$ and on $\mathbb{C}1$, and must therefore be equal to each other. Thus if $(\mathcal{A}, \varphi) = *_{i\in I}(\mathcal{A}_i, \varphi_i)$, then every $(\mathcal{A}_i, \varphi_i)$ is indeed "a subspace" of $(\mathcal{A}, \varphi)$.

PROPOSITION 6.6. *Let $(\mathcal{A}_i, \varphi_i)_{i\in I}$ be a family of non-commutative probability spaces, and let $(\mathcal{A}, \varphi)$ be their free product. Then we have the following.*

(1) *The subalgebras $\mathcal{A}_i$, $i \in I$, are freely independent in $(\mathcal{A}, \varphi)$.*

(2) *$(\mathcal{A}, \varphi)$ has a universality property, described as follows. Let $(\mathcal{B}, \psi)$ be a non-commutative probability space, suppose that for every $i \in I$ we have a homomorphism Φ_i between $(\mathcal{A}_i, \varphi_i)$ and $(\mathcal{B}, \psi)$ (in the sense that $\Phi_i : \mathcal{A}_i \to \mathcal{B}$ is a unital homomorphism such that $\psi \circ \Phi_i = \varphi_i$), and suppose moreover that the images $(\Phi_i(\mathcal{A}_i))_{i\in I}$ are freely independent in $(\mathcal{B}, \psi)$. Then there exists a homomorphism Φ between $(\mathcal{A}, \varphi)$ and $(\mathcal{B}, \psi)$, uniquely determined, such that $\Phi|\mathcal{A}_i = \Phi_i$ for every $i \in I$.*

PROOF. (1) Let $i_1, \ldots, i_n \in I$ be such that $i_1 \neq i_2, \ldots, i_{n-1} \neq i_n$, and let $a_1 \in \mathcal{A}_{i_1}, \ldots, a_n \in \mathcal{A}_{i_n}$ be such that $\varphi(a_1) = \cdots = \varphi(a_n) = 0$. In the terminology used in Definition 6.4 we thus have $a_1 \in \mathcal{A}_{i_1}^o, \ldots, a_n \in \mathcal{A}_{i_n}^o$. But then $a_1 \cdots a_n \in \mathcal{W}_{i_1,\ldots,i_n} \subset \ker(\varphi)$, and we get that $\varphi(a_1 \cdots a_n) = 0$, as required by the definition of free independence.

(2) By the universality property of $\mathcal{A}$ (cf. Definition 6.1) we know that there exists a unique unital homomorphism $\Phi : \mathcal{A} \to \mathcal{B}$ such that $\Phi|\mathcal{A}_i = \Phi_i$, $\forall\, i \in I$. We have to show that Φ also has the property that $\psi \circ \Phi = \varphi$. In view of the definition of φ, it suffices to check that $\psi \circ \Phi$ vanishes on each of the linear subspaces $\mathcal{W}_{i_1,\ldots,i_n}$ of $\mathcal{A}$, for every $n \geq 1$ and every $i_1 \neq i_2, \ldots, i_{n-1} \neq i_n$ in I. So in other words it suffices to fix such n and $i_1, \ldots, i_n$, then to pick some elements $a_1 \in \mathcal{A}_{i_1}^o, \ldots, a_n \in \mathcal{A}_{i_n}^o$, and to prove that $(\psi \circ \Phi)(a_1 \cdots a_n) = 0$.

For the $a_1, \ldots, a_n$ picked above, let us denote

$$\Phi(a_1) = \Phi_{i_1}(a_1) =: b_1, \qquad \ldots, \qquad \Phi(a_n) = \Phi_{i_n}(a_n) =: b_n.$$

Then for every $1 \leq k \leq n$ we have that $b_k \in \Phi_{i_k}(\mathcal{A}_{i_k})$ and that

$$\begin{aligned} \psi(b_k) &= \psi(\Phi_{i_k}(a_k)) \\ &= \varphi_{i_k}(a_k) \qquad (\text{since } \psi \circ \Phi_{i_k} = \varphi_{i_k}) \\ &= 0 \qquad\qquad (\text{since } a_k \in \mathcal{A}_{i_k}^o). \end{aligned}$$

But now, since the algebras $(\,\Phi(\mathcal{A}_i)\,)_{i\in I}$ are free in $(\mathcal{B},\psi)$, it follows that $\psi(b_1\cdots b_k)=0$. Thus

$$(\psi\circ\Phi)(a_1\cdots a_n)=\psi(\Phi(a_1)\cdots\Phi(a_n))=\psi(b_1\cdots b_n)=0,$$

as desired. □

EXERCISE 6.7. Let $G_1,\ldots,G_m$ be groups and $G=G_1*\cdots*G_m$ the free product of these groups (as discussed in the part of Lecture 5 about free products of groups). Show that

$$(\mathbb{C}G_1,\tau_{G_1})*\cdots*(\mathbb{C}G_m,\tau_{G_m})=(\mathbb{C}G,\tau_G). \tag{6.6}$$

We conclude this section by noting that a free product of tracial non-commutative probability spaces is again tracial.

PROPOSITION 6.8. *Let $(\mathcal{A}_i,\varphi_i)_{i\in I}$, be a family of non-commutative probability spaces, and let $(\mathcal{A},\varphi)$ be their free product. If φ_i is a trace on $\mathcal{A}_i$ for every $i\in I$, then φ is a trace on $\mathcal{A}$.*

PROOF. This is an immediate consequence of Proposition 5.19: the subalgebras $(\mathcal{A}_i)_{i\in I}$ of $\mathcal{A}$ are freely independent and $\varphi|\mathcal{A}_i=\varphi_i$ is a trace for every $i\in I$, hence φ is a trace on the subalgebra generated by $\cup_{i\in I}\mathcal{A}_i$ (which is all of $\mathcal{A}$). □

Free product of $*$-probability spaces

REMARK 6.9. Let $(\mathcal{A}_i,\varphi_i)_{i\in I}$ be a family of $*$-probability spaces. One can of course view the $(\mathcal{A}_i,\varphi_i)$ as plain non-commutative probability spaces, and consider their free product $(\mathcal{A},\varphi)$ defined in the preceding section. It is moreover fairly easy to see that the algebra $\mathcal{A}=*_{i\in I}\mathcal{A}_i$ has in this case a natural $*$-operation on it, uniquely determined by the fact that it extends the $*$-operations existing on the algebras $\mathcal{A}_i$, $i\in I$. Referring to the direct sum decomposition

$$\mathcal{A}=\mathbb{C}1\oplus\Big(\bigoplus_{n=1}^{\infty}\bigoplus_{\substack{i_1,\ldots,i_n\in I\\ i_1\neq i_2,\ldots,i_{n-1}\neq i_n}}\mathcal{W}_{i_1,\ldots,i_n}\Big)$$

discussed in the preceding sections (cf. Equations (6.2) and (6.3) above), we have that the $*$-operation on $\mathcal{A}$ maps $\mathcal{W}_{i_1,\ldots,i_n}$ onto $\mathcal{W}_{i_n,\ldots,i_1}$, via the formula

$$(a_1\cdots a_n)^*=a_n^*\cdots a_1^*$$

(holding for $a_1\in\mathcal{A}_{i_1}^o,\ldots,a_n\in\mathcal{A}_{i_n}^o$, where $n\geq 1$ and where $i_1,\ldots,i_n\in I$ are such that $i_1\neq i_2,\ldots,i_{n-1}\neq i_n$). This immediately implies that the free product functional $\varphi = *_{i \in I} \varphi_i$ is selfadjoint on $\mathcal{A}$, in the sense that it satisfies the equation

$$\varphi(a^*) = \overline{\varphi(a)}, \quad \forall \, a \in \mathcal{A}.$$

Nevertheless, it is not clear from the outset that the free product $(\mathcal{A}, \varphi)$ of the $(\mathcal{A}_i, \varphi_i)$ is a $*$-probability space – indeed, it is not clear whether φ satisfies the positivity condition $\varphi(a^*a) \geq 0$, $a \in \mathcal{A}$. The main goal of the present section is to prove that the desired positivity of φ does actually take place. The proof will rely on some basic facts about positive matrices, which are reviewed next.

REMARK 6.10. Recall that a matrix $A \in M_n(\mathbb{C})$ is said to be **positive** when it satisfies one (hence all) of the following equivalent conditions:

(1) A is selfadjoint and all its eigenvalues are in $[0, \infty)$;

(2) A can be written in the form $A = X^*X$ for some $X \in M_n(\mathbb{C})$;

(3) one has $\langle A\xi, \xi \rangle \geq 0$ for every $\xi \in \mathbb{C}^n$, where $\langle \; , \; \rangle$ is the standard inner product on $\mathbb{C}^n$.

(The equivalence between (1) and (2) above is a particular case of Proposition 3.6, used for the C^*-algebra $M_n(\mathbb{C})$. But, of course, in this particular case we do not have to refer to Proposition 3.6, for example for (1) $\Rightarrow$ (2) one can simply find X by diagonalizing the matrix A.)

A fact about positive matrices which we want to use concerns the entry-wise product – also called Schur product – of matrices. Given $A = (a_{ij})_{i,j=1}^n$ and $B = (b_{ij})_{i,j=1}^n$ in $M_n(\mathbb{C})$, the **Schur product** of A and B is the matrix $S := (a_{ij}b_{ij})_{i,j=1}^n$.

LEMMA 6.11. *Let $A = (a_{ij})_{i,j=1}^n$ and $B = (b_{ij})_{i,j=1}^n$ be two positive matrices in $M_n(\mathbb{C})$. Then the Schur product $S = (a_{ij}b_{ij})_{i,j=1}^n$ is a positive matrix as well.*

PROOF. We will show that S satisfies condition (3) from Remark 6.10. For $\xi = (\xi_1, \ldots, \xi_n) \in \mathbb{C}^n$ we clearly have:

$$\langle S\xi, \xi \rangle = \sum_{i,j=1}^n a_{ij} b_{ij} \xi_j \overline{\xi_i}. \tag{6.7}$$

But then let us write $A = X^*X$, where $X = (x_{ij})_{i,j=1}^n \in M_n(\mathbb{C})$. For every $1 \leq i, j \leq n$ we have $a_{ij} = \sum_{k=1}^n \overline{x_{ki}} x_{kj}$, and substituting this in

(6.7) we get:

$$\begin{aligned}\langle S\xi, \xi\rangle &= \sum_{i,j,k=1}^{n} \overline{x_{ki}} x_{kj} b_{ij} \xi_j \overline{\xi_i} \\ &= \sum_{k=1}^{n} \Big(\sum_{i,j=1}^{n} b_{ij} (\xi_j x_{kj}) (\overline{\xi_i x_{ki}}) \Big) \\ &= \sum_{k=1}^{n} \langle B\eta_k, \eta_k\rangle \geq 0,\end{aligned}$$

where $\eta_k := (\xi_1 x_{k1}, \ldots, \xi_n x_{kn}) \in \mathbb{C}^n$ for $1 \leq k \leq n$. □

Positive matrices appear in the framework of a $*$-probability space in the following way.

LEMMA 6.12. *Consider a unital $*$-algebra $\mathcal{A}$ equipped with a linear functional $\varphi : \mathcal{A} \to \mathbb{C}$. Then the following statements are equivalent:*

(1) *φ is positive, i.e. we have $\varphi(a^*a) \geq 0$, $\forall\, a \in \mathcal{A}$;*

(2) *for all $n \geq 1$ and $a_1, \ldots, a_n \in \mathcal{A}$, the matrix $\big(\varphi(a_i^* a_j)\big)_{i,j=1}^n \in M_n(\mathbb{C})$ is positive.*

PROOF. (2) $\Rightarrow$ (1) is clear ((1) is the particular case "$n = 1$" of (2)).

(1) $\Rightarrow$ (2): Given $n \geq 1$ and $a_1, \ldots, a_n \in \mathcal{A}$, we verify that the matrix $A = \big(\varphi(a_i^* a_j)\big)_{i,j=1}^n$ satisfies condition (3) of Remark 6.10. Indeed, for every $\xi = (\xi_1, \ldots, \xi_n) \in \mathbb{C}^n$ we can write:

$$\begin{aligned}\langle A\xi, \xi\rangle = \sum_{i,j=1}^{n} \varphi(a_i^* a_j) \xi_j \overline{\xi_i} &= \varphi\Big(\sum_{i,j=1}^{n} \xi_j \overline{\xi_i} a_i^* a_j \Big) \\ &= \varphi\Big((\sum_{i=1}^{n} \xi_i a_i)^* \, (\sum_{i=1}^{n} \xi_i a_i) \Big) \geq 0.\end{aligned}$$

□

We can now give the positivity result announced at the beginning of this section.

THEOREM 6.13. *Let $(\mathcal{A}_i, \varphi_i)_{i \in I}$ be a family of $*$-probability spaces. Then the functional $\varphi := *_{i \in I} \mathcal{A}_i$ is positive, and hence the free product $(\mathcal{A}, \varphi) := *_{i \in I} (\mathcal{A}_i, \varphi_i)$ is a $*$-probability space.*

PROOF. In order to prove the positivity of φ, we will rely on the direct sum decomposition

$$\mathcal{A} = \bigoplus_{n=0}^{\infty} \bigoplus_{\substack{i_1,\dots,i_n \in I \\ i_1 \neq i_2,\dots,i_{n-1} \neq i_n}} \mathcal{W}_{i_1,\dots,i_n}.$$

This is the same as in Equation (6.2) above, with the additional convention that, for $n = 0$, the subspace $\mathcal{W}_{i_1,\dots,i_n}$ indexed by the empty 0-tuple is $\mathbb{C}1$. Observe that, as an immediate consequence of Lemma 5.18 from the preceding lecture we have that, for $i_1 \neq i_2, \dots, i_{n-1} \neq i_n$ and $j_1 \neq j_2, \dots, j_{m-1} \neq j_m$ in I:

$$(i_1, \dots, i_m) \neq (j_1, \dots, j_m) \Rightarrow \varphi(a^*b) = 0, \quad \begin{array}{l} \forall\, a \in \mathcal{W}_{i_1,\dots,i_n} \\ \forall\, b \in \mathcal{W}_{j_1,\dots,j_m}. \end{array} \tag{6.8}$$

Consider now an element $a \in \mathcal{A}$ and write it as

$$a = \sum_{n=0}^{N} \sum_{\substack{i_1,\dots,i_n \in I \\ i_1 \neq i_2 \neq \dots \neq i_n}} a_{i_1,\dots,i_n}$$

for some $N \geq 0$ and where $a_{i_1,\dots,i_n} \in \mathcal{W}_{i_1,\dots,i_n}$ for every $0 \leq n \leq N$ and every $i_1 \neq i_2 \neq \dots \neq i_n$ in I. Then we have

$$\begin{aligned} \varphi(a^*a) &= \sum_{m,n=0}^{N} \sum_{\substack{i_1,\dots,i_n \in I \\ i_1 \neq i_2 \neq \dots \neq i_n}} \sum_{\substack{j_1,\dots,j_m \in I \\ j_1 \neq j_2 \neq \dots \neq j_m}} \varphi(a^*_{i_1,\dots,i_n} a_{j_1,\dots,j_m}) \\ &= \sum_{n=0}^{N} \sum_{\substack{i_1,\dots,i_n \in I \\ i_1 \neq i_2 \neq \dots \neq i_n}} \varphi(a^*_{i_1,\dots,i_n} a_{i_1,\dots,i_n}), \end{aligned} \tag{6.9}$$

where at the last equality sign we made use of the implication (6.8).

In view of (6.9), we are clearly reduced to proving that $\varphi(b^*b) \geq 0$ when b belongs to a subspace $\mathcal{W}_{i_1,\dots,i_n}$. Fix such a b. We can write

$$b = \sum_{k=1}^{p} a_1^{(k)} a_2^{(k)} \dots a_n^{(k)},$$

where $a_m^{(k)} \in \mathcal{A}_{i_m}^o$ for $1 \leq m \leq n$, $1 \leq k \leq p$. We thus have:

$$\begin{aligned} \varphi(b^*b) &= \sum_{k,l=1}^{p} \varphi\Big((a_1^{(k)} \cdots a_n^{(k)})^* \cdot (a_1^{(l)} \cdots a_n^{(l)}) \Big) \\ &= \sum_{k,l=1}^{p} \varphi\Big(a_n^{(k)*} \cdots a_1^{(k)*} \cdot a_1^{(l)} \cdots a_n^{(l)} \Big) \end{aligned}$$

$$= \sum_{k,l=1}^{p} \varphi(a_1^{(k)*}a_1^{(l)}) \cdots \varphi(a_n^{(k)*}a_n^{(l)}) \qquad \text{(by Lemma 5.18)}.$$

Since $\varphi|\mathcal{A}_i = \varphi_i$ for all i, what we have obtained is that:

$$\varphi(b^*b) = \sum_{k,l=1}^{p} \varphi_{i_1}(a_1^{(k)*}a_1^{(l)}) \cdots \varphi_{i_n}(a_n^{(k)*}a_n^{(l)}). \tag{6.10}$$

Now for every $1 \leq m \leq n$ let us consider the matrix $B_m = (\varphi_{i_m}(a_m^{(k)*}a_m^{(l)}))_{k,l=1}^{p} \in M_p(\mathbb{C})$, and let S be the Schur product of the matrices $B_1, \ldots, B_n$. Lemma 6.12 gives us that each of $B_1, \ldots, B_n$ is positive, and a repeated application of Lemma 6.11 gives us that S is positive as well. Finally, we observe that Equation (6.10) amounts to the fact that $\varphi(b^*b)$ is the sum of all the entries of S; hence, by taking $\zeta = (1, 1, \ldots, 1) \in \mathbb{C}^p$, we have $\varphi(b^*b) = \langle S\zeta, \zeta\rangle \geq 0$. □

Finally, let us point out that the two basic properties of the expectation functional which were followed throughout the preceding lectures – traciality and faithfulness – are preserved when one forms free products of $*$-free probability spaces. The statement about traciality is a particular case of Proposition 6.8 from the preceding section, while the statement about faithfulness is treated in the next proposition.

PROPOSITION 6.14. *Let $(\mathcal{A}_i, \varphi_i)_{i\in I}$, be a family of $*$-probability spaces, and let $(\mathcal{A}, \varphi)$ be their free product. If φ_i is faithful on $\mathcal{A}_i$ for every $i \in I$, then φ is faithful on $\mathcal{A}$.*

PROOF. As in the proof of Theorem 6.13, we will use the direct sum decomposition of $\mathcal{A}$ into subspaces $\mathcal{W}_{i_1,\ldots,i_n}$ (for $n \geq 0$ and $i_1 \neq i_2 \neq \cdots \neq i_n$ in I). The very same calculation which led to Equation (6.9) in the proof of Theorem 6.13 shows that it suffices to prove the implication "$\varphi(b^*b) = 0 \Rightarrow b = 0$" for an element b which belongs to one of the subspaces $\mathcal{W}_{i_1,\ldots,i_n}$. We will prove this implication by induction on n.

The cases $n = 0$ and $n = 1$ of our proof by induction are clear. Indeed, in the case $n = 0$ we have that $b \in \mathbb{C}1$, hence the implication to be proved reduces to "$|\lambda|^2 = 0 \Rightarrow \lambda = 0$" (for some $\lambda \in \mathbb{C}$). In the case $n = 1$ we have that $b \in \mathcal{W}_i \subset \mathcal{A}_i$ for some $i \in I$, and the implication "$\varphi(b^*b) = 0 \Rightarrow b = 0$" follows from the hypothesis that $\varphi|\mathcal{A}_i$ (which is just φ_i) is faithful.

So it remains that we verify the induction step, $n - 1 \Rightarrow n$, for $n \geq 2$. Consider some indices $i_1 \neq i_2 \neq \cdots \neq i_n$ in I and an element $b \in \mathcal{W}_{i_1,\ldots,i_n}$ such that $\varphi(b^*b) = 0$. In view of how the space $\mathcal{W}_{i_1,\ldots,i_n}$ is defined, we can write

$$b = x_1 y_1 + \cdots + x_p y_p \tag{6.11}$$

for some $p \geq 1$, where $x_1, \ldots, x_p \in \mathcal{W}_{i_1}$ and $y_1, \ldots, y_p \in \mathcal{W}_{i_2,\ldots,i_n}$. Moreover, by appropriately regrouping the terms and by incorporating the necessary linear combinations, we can assume that in (6.11) the elements $x_1, \ldots, x_p$ are linearly independent. The fact that $\varphi(b^* b) = 0$ entails that:

$$0 = \varphi\Big((\sum_{k=1}^{p} x_k y_k)^* (\sum_{l=1}^{p} x_l y_l) \Big) = \sum_{k,l=1}^{p} \varphi(y_k^* x_k^* x_l y_l).$$

If we also make use of Lemma 5.18 from the preceding lecture, we thus see that we have obtained:

$$\sum_{k,l=1}^{p} \varphi(y_k^* y_l) \varphi(x_k^* x_l) = 0. \tag{6.12}$$

Now, the matrix $\big(\varphi(y_k^* y_l) \big)_{k,l=1}^{p}$ is positive (since φ is positive, and by Lemma 6.12), hence we can find a matrix $B = (\beta_{k,l})_{k,l=1}^{p}$ such that $\big(\varphi(y_k^* y_l) \big)_{k,l=1}^{p} = B^* B$. Written in terms of entries, this means that we have:

$$\varphi(y_k^* y_l) \;=\; \sum_{h=1}^{p} \overline{\beta_{hk}} \beta_{hl}, \;\; \forall\, 1 \leq k, l \leq p.$$

We substitute this in (6.12) and we get:

$$\begin{aligned} 0 &= \sum_{k,l=1}^{p} \Big(\sum_{h=1}^{p} \overline{\beta_{hk}} \beta_{hl} \Big) \varphi(x_k^* x_l) \\ &= \sum_{h=1}^{p} \varphi\Big(\sum_{k,l=1}^{p} \overline{\beta_{hk}} \beta_{hl} x_k^* x_l \Big) \\ &= \sum_{h=1}^{p} \varphi\Big((\sum_{k=1}^{p} \beta_{hk} x_k)^* \cdot (\sum_{k=1}^{p} \beta_{hk} x_k) \Big). \end{aligned}$$

By using the positivity of φ, we infer that:

$$\varphi\Big((\sum_{k=1}^{p} \beta_{hk} x_k)^* \cdot (\sum_{k=1}^{p} \beta_{hk} x_k) \Big) = 0, \;\forall\, 1 \leq h \leq p. \tag{6.13}$$

Moreover, since $\varphi|\mathcal{A}_{i_1}$ is φ_{i_1} and is thus faithful (by hypothesis), Equation (6.13) has as consequence that

$$\sum_{k=1}^{p} \beta_{hk} x_k = 0, \quad \forall\, 1 \leq h \leq p.$$

This in turn implies that $\beta_{hk} = 0$ for every $1 \leq h, k \leq p$, because the elements $x_1, \ldots, x_p \in \mathcal{W}_{i_1}$ are linearly independent. As a consequence, we obtain that

$$\varphi(y_k^* y_k) = \sum_{h=1}^{p} \overline{\beta_{hk}} \beta_{hk} = 0, \quad 1 \leq k \leq p.$$

But φ is faithful on $\mathcal{W}_{i_2,\ldots,i_n}$, by the induction hypothesis; so from the latter equalities we infer that $y_1 = \cdots = y_p = 0$, and we can conclude that $b = \sum_{k=1}^{p} x_k y_k = 0$. □

Exercises

EXERCISE 6.15. In the setting of Remark 6.3, describe precisely the multiplication operation on the vector space $\mathcal{A}$ introduced in Equation (6.5), and prove that in this way $\mathcal{A}$ becomes a unital algebra.
[Hint: In order to spell out the multiplication of two tensors $a_1' \otimes \cdots \otimes a_m'$ and $a_1'' \otimes \cdots \otimes a_n''$, one can proceed by induction on $m+n$. If $a_m' \in \mathcal{A}_{i_m}^o$ and $a_1'' \in \mathcal{A}_{j_1}^o$ with $i_m \neq j_1$, then the desired product is simply defined to be $a_1' \otimes \cdots \otimes a_m' \otimes a_1'' \otimes \cdots \otimes a_n''$. If $i_m = j_1 =: i$, then consider the element

$$b = (a_1' \otimes \cdots \otimes a_{m-1}') \cdot (a_2'' \otimes \cdots \otimes a_n'')$$

which is defined by the induction hypothesis, and define the product of $a_1' \otimes \cdots \otimes a_m'$ and $a_1'' \otimes \cdots \otimes a_n''$ to be

$$a_1' \otimes \cdots \otimes a_{m-1}' \otimes (a_m' a_1'')^o \otimes a_2'' \otimes \cdots \otimes a_n'' + \varphi_i(a_m' a_1'') \cdot b.]$$

EXERCISE 6.16. *(f.i.d. sequences)*
Let μ be a probability measure with compact support on $\mathbb{R}$. Show that one can find a $*$-probability space $(\mathcal{A}, \varphi)$ where φ is a faithful trace, and a sequence $(x_n)_{n\geq 1}$ of freely independent selfadjoint random variables in $\mathcal{A}$, such that each of the x_i has distribution μ.

EXERCISE 6.17. Let $(\mathcal{A}, \varphi)$ be a $*$-probability space and let a be an element of $\mathcal{A}$. Sometimes we need to make the following kind of assumption (see e.g. Lecture 15 below): "by enlarging $(\mathcal{A}, \varphi)$ (if necessary), we may assume that there exists a Haar unitary $u \in \mathcal{A}$ such that a and u are $*$-free." Explain why one can make such an assumption.

EXERCISE 6.18. State and prove an analog of Proposition 6.6, holding in the framework of $*$-probability spaces.

EXERCISE 6.19. Let $(\mathcal{A}, \varphi)$ be a $*$-probability space. Let $(x_i)_{i \in I}$ be a freely independent family of selfadjoint elements of $\mathcal{A}$, such that the unital $*$-algebra generated by $\{x_i \mid i \in I\}$ is all of $\mathcal{A}$. Suppose in addition that for every $i \in I$ the element x_i has distribution μ_i with respect to φ, where μ_i is a compactly supported probability measure on $\mathbb{R}$ (as in Remark 1.10 of Lecture 1), and such that the support of μ_i is an infinite set. Prove that φ is a faithful trace on $\mathcal{A}$.
[Hint: For faithfulness, it suffices to check that the restriction of φ to $\{P(x_i) \mid P \in \mathbb{C}[X]\}$ is faithful, for every $i \in I$. This happens because a non-zero polynomial in $\mathbb{C}[X]$ cannot vanish everywhere on the support of μ_i.]

LECTURE 7

Free product of C^*–probability spaces

After discussing free products for non-commutative probability spaces and for $*$-probability spaces in the preceding lecture, we will now look at the corresponding concept for C^*-probability spaces. We will restrict our attention to the technically simpler case of C^*-probability spaces $(\mathcal{A}, \varphi)$ where φ is a faithful trace. We will show how for such spaces the free product at the C^*-level can be obtained from the free product as $*$-probability spaces by using the basic concept of Gelfand–Naimark–Segal (or GNS for short) representation.

The GNS representation

In this section we consider the framework of $*$-probability spaces. Recall from Lecture 1 (Definition 1.6) that by a representation of a $*$-probability space $(\mathcal{A}, \varphi)$ we understand a triple $(\mathcal{H}, \pi, \xi)$ where $\mathcal{H}$ is a Hilbert space, $\pi : \mathcal{A} \to B(\mathcal{H})$ is a unital $*$-homomorphism and ξ is a vector in $\mathcal{H}$, such that the relation $\varphi(a) = \langle \pi(a)\xi, \xi \rangle$ holds for every $a \in \mathcal{A}$.

REMARK 7.1. *(The space $L^2(\mathcal{A}, \varphi)$)*

Let $(\mathcal{A}, \varphi)$ be a $*$-probability space. Consider the positive definite sesquilinear form on $\mathcal{A}$ defined by:

$$\langle a, b \rangle := \varphi(b^* a), \quad a, b \in \mathcal{A}. \tag{7.1}$$

By using the Cauchy–Schwarz inequality for φ (Lecture 1, Equation (1.1)), one sees immediately that $\mathcal{N} := \{a \in \mathcal{A} \mid \langle a, a \rangle = 0\}$ can also be described as $\{a \in \mathcal{A} \mid \langle a, b \rangle = 0 \text{ for all } b \in \mathcal{A}\}$, and is therefore a linear subspace of $\mathcal{A}$. It is a standard procedure to consider the quotient space $\mathcal{A}/\mathcal{N}$, endowed with the inner product inherited from the sesquilinear form (7.1), and then to take the completion of $\mathcal{A}/\mathcal{N}$ with respect to this inner product. The result is a Hilbert space which is customarily denoted as "$L^2(\mathcal{A}, \varphi)$."

Rather than remembering the (somewhat uncomfortable) procedure described above for constructing $L^2(\mathcal{A}, \varphi)$, it is easier to remember $L^2(\mathcal{A}, \varphi)$ in the following way: there exists a linear map

$$\mathcal{A} \ni a \mapsto \widehat{a} \in L^2(\mathcal{A}, \varphi) \tag{7.2}$$

such that:

$$\begin{aligned} &\text{(i)} \quad \{\widehat{a} \mid a \in \mathcal{A}\} \text{ is a dense subspace of } L^2(\mathcal{A}, \varphi), \text{ and} \\ &\text{(ii)} \quad \langle \widehat{a}, \widehat{b} \rangle = \varphi(b^*a), \quad \forall\, a, b \in \mathcal{A}. \end{aligned} \tag{7.3}$$

Referring to the notations of the preceding paragraph, the map from (7.2) sends an element $a \in \mathcal{A}$ to its coset in the quotient $\mathcal{A}/\mathcal{N} \subset L^2(\mathcal{A}, \varphi)$. But in the concrete manipulations of $L^2(\mathcal{A}, \varphi)$ this actually never appears, it is always the combination of properties (i) + (ii) from (7.3) that is used.

The GNS representation for $(\mathcal{A}, \varphi)$ is defined in the way described in the next proposition. Some comments around the condition (7.4) imposed on $\mathcal{A}$ in this proposition are made in Remark 7.4.

PROPOSITION 7.2. *Let $(\mathcal{A}, \varphi)$ be a $*$-probability space, and let us assume that*

$$\mathcal{A} = \text{span}\{u \mid u \in \mathcal{A}, u \textit{ is a unitary }\}. \tag{7.4}$$

Then for every $a \in \mathcal{A}$ there exists a unique bounded linear operator $\pi(a) \in B(L^2(\mathcal{A}, \varphi))$ such that

$$\pi(a)\widehat{b} = \widehat{ab}, \quad \forall\, b \in \mathcal{A}. \tag{7.5}$$

The map $\pi : \mathcal{A} \to B(L^2(\mathcal{A}, \varphi))$ so defined is a unital $*$-homomorphism. Moreover, the triple $(L^2(\mathcal{A}, \varphi), \pi, \widehat{1})$ is a representation of $(\mathcal{A}, \varphi)$, where $\widehat{1}$ is defined according to the conventions of notation in (7.3), with $1 = 1_{\mathcal{A}}$, the unit of $\mathcal{A}$.

DEFINITION 7.3. This special representation of $(\mathcal{A}, \varphi)$ described in the preceding proposition is called the **GNS representation**.

PROOF. Most of the verifications required in order to prove this proposition are trivial (and will be left to the reader). The only point that we will examine here is why the formula (7.5) defines a bounded linear operator on $L^2(\mathcal{A}, \varphi)$. It is immediate that (given $a \in \mathcal{A}$) it suffices to prove the existence of a constant $k(a) \geq 0$ such that

$$||\widehat{ab}||_{L^2(\mathcal{A},\varphi)} \leq k(a) \cdot ||\widehat{b}||_{L^2(\mathcal{A},\varphi)}, \quad \forall\, b \in \mathcal{A}; \tag{7.6}$$

indeed, once this is done, a standard continuity argument will extend the map $\widehat{b} \mapsto \widehat{ab}$ from the dense subspace $\{\widehat{b} \mid b \in \mathcal{A}\}$ to a bounded linear operator on $L^2(\mathcal{A}, \varphi)$.

Now, the set

$$\{a \in \mathcal{A} \mid \text{ there exists } k(a) \geq 0 \text{ such that (7.6) holds}\} \tag{7.7}$$

is a linear subspace of $\mathcal{A}$. The verification of this fact is immediate (the reader should have no difficulty in noticing that $k(a_1) + k(a_2)$ can serve as $k(a_1 + a_2)$ and that $|\alpha| k(a)$ can serve as $k(\alpha a)$). But on the other hand, let us observe that the set introduced in (7.7) contains all the unitaries of $\mathcal{A}$. Indeed, if $u \in \mathcal{A}$ is a unitary then (7.6) is satisfied for the constant $k(u) = 1$:

$$||\widehat{ub}||_{L^2(\mathcal{A},\varphi)} = \langle \widehat{ub}, \widehat{ub} \rangle^{1/2} = \varphi(b^* u^* ub)^{1/2} = \varphi(b^* b)^{1/2} = ||\widehat{b}||_{L^2(\mathcal{A},\varphi)},$$

for all $b \in \mathcal{A}$.

Consequently, the hypothesis that $\mathcal{A}$ is the linear span of its unitaries implies that the set appearing in (7.7) is all of $\mathcal{A}$ (as we wanted). □

REMARK 7.4. The hypothesis (7.4) that $\mathcal{A}$ is the span of its unitaries is for instance satisfied whenever $\mathcal{A}$ is a unital C^*-algebra – see Exercise 7.20 at the end of the lecture. It is fairly easy to relax this hypothesis without changing too much the argument presented above – see Exercise 7.22. On the other hand, one should be warned that this hypothesis cannot be simply removed (that is, the boundedness of the operators $\widehat{b} \mapsto \widehat{ab}$ on $L^2(\mathcal{A}, \varphi)$ cannot be obtained in the framework of an arbitrary $*$-probability space – see Exercise 7.23).

We next point out how GNS representations can be recognized (up to unitary equivalence) by using the concept of a cyclic vector.

DEFINITION 7.5. Let $\mathcal{A}$ be a unital $*$-algebra, let $\mathcal{H}$ be a Hilbert space, and let $\pi : \mathcal{A} \to B(\mathcal{H})$ be a unital $*$-homomorphism. A vector $\eta \in \mathcal{H}$ is said to be **cyclic** for π if it satisfies:

$$\text{cl}\{\pi(a)\eta \mid a \in \mathcal{A}\} = \mathcal{H}, \tag{7.8}$$

where "cl" denotes closure with respect to the norm-topology of $\mathcal{H}$.

It is obvious that, in the notations of Proposition 7.2, the vector $\widehat{1}$ is cyclic for the GNS representation $\pi : \mathcal{A} \to B(L^2(\mathcal{A}, \varphi))$ – indeed, the subspace $\{\pi(a)\widehat{1} \mid a \in \mathcal{A}\}$ is nothing but the dense subspace $\{\widehat{a} \mid a \in \mathcal{A}\}$ from (i) of (7.3). On the other hand, we have the following proposition, which says that in a certain sense the GNS representation is the "unique" representation $(\mathcal{H}, \pi, \xi)$ of $(\mathcal{A}, \varphi)$ such that ξ is cyclic for π.

PROPOSITION 7.6. *Let $(\mathcal{A}, \varphi)$ be a $*$-probability space, and assume that $(\mathcal{A}, \varphi)$ satisfies the hypothesis of Proposition 7.2 (hence that it has a GNS representation). Let $(\mathcal{H}, \rho, \xi)$ be a representation of $(\mathcal{A}, \varphi)$ such that ξ is cyclic for ρ. Then $(\mathcal{H}, \rho, \xi)$ is unitarily equivalent to the GNS representation $(L^2(\mathcal{A}, \varphi), \pi, \widehat{1})$, in the sense that there exists a linear operator $U : L^2(\mathcal{A}, \varphi) \to \mathcal{H}$ which is bijective and norm preserving, such that $U(\widehat{1}) = \xi$, and such that $U\pi(a)U^* = \rho(a)$ for every $a \in \mathcal{A}$.*

PROOF. Let us observe that

$$||\rho(a)\xi||_{\mathcal{H}} = ||\widehat{a}||_{L^2(\mathcal{A},\varphi)}, \quad \forall\, a \in \mathcal{A}. \tag{7.9}$$

Indeed, both sides of Equation (7.9) are equal to $\varphi(a^*a)^{1/2}$; for instance for the left-hand side we compute like this:

$$||\rho(a)\xi||^2_{\mathcal{H}} = \langle \rho(a)\xi, \rho(a)\xi\rangle_{\mathcal{H}} = \langle \rho(a^*a)\xi, \xi\rangle_{\mathcal{H}} = \varphi(a^*a).$$

Due to (7.9), it makes sense to define a function $U_0 : \{\widehat{a} \mid a \in \mathcal{A}\} \to \mathcal{H}$ by the formula

$$U_0(\widehat{a}) = \rho(a)\xi, \quad a \in \mathcal{A}. \tag{7.10}$$

Indeed, if a vector in the domain of U_0 can be written as both $\widehat{a}$ and $\widehat{b}$ for some $a, b \in \mathcal{A}$, then we get that $||\widehat{a-b}||_{L^2(\mathcal{A},\varphi)} = 0$, hence that $||\rho(a-b)\xi||_{\mathcal{H}} = 0$ (by (7.9)); and the latter fact implies that $\rho(a)\xi = \rho(b)\xi$.

It is immediate that the map U_0 defined by (7.10) is linear, and Equation (7.9) shows that U_0 is isometric. The usual argument of extension by continuity then shows that one can extend U_0 to a linear norm preserving operator $U : L^2(\mathcal{A}, \varphi) \to \mathcal{H}$. The range-space $\mathrm{ran}(U)$ is complete (since it is an isometric image of the complete space $L^2(\mathcal{A}, \varphi)$), hence it is closed in $\mathcal{H}$. But we also have that $\mathrm{ran}(U) \supset \mathrm{ran}(U_0) = \{\rho(a)\xi \mid a \in \mathcal{A}\}$, and the latter space is dense in $\mathcal{H}$, by the hypothesis that ξ is cyclic for ρ. In this way we obtain that U is surjective.

We have thus defined a linear operator $U : L^2(\mathcal{A}, \varphi) \to \mathcal{H}$ which is bijective and norm preserving, and has the property that $U(\widehat{a}) = \rho(a)\xi$ for every $a \in \mathcal{A}$. The latter property gives in particular that $U(\widehat{1}) = \rho(1_{\mathcal{A}})\xi = \xi$. From the same property we also infer that

$$U\pi(a)\widehat{b} = \rho(a)U\widehat{b}, \quad \forall\, a, b \in \mathcal{A} \tag{7.11}$$

(we leave it as an immediate exercise to the reader to check that both sides of (7.11) are equal to $\rho(ab)\xi$). Equation (7.11) implies in turn that $U\pi(a) = \rho(a)U$, $\forall\, a \in \mathcal{A}$, hence that $U\pi(a)U^* = \rho(a)$, $\forall\, a \in \mathcal{A}$. ☐

REMARK 7.7. We conclude this section with an observation concerning faithfulness. Let $\mathcal{A}$ be a unital $*$-algebra, let $\pi : \mathcal{A} \to B(\mathcal{H})$ be a unital $*$-homomorphism, and let η be a vector in the Hilbert space $\mathcal{H}$. It is customary to say that η is **separating** for π if the map $\mathcal{A} \ni a \mapsto \pi(a)\eta \in \mathcal{H}$ is one-to-one (equivalently, if for $a \in \mathcal{A}$ we have the implication $\pi(a)\eta = 0 \Rightarrow a = 0$).

Now let $(\mathcal{A}, \varphi)$ be a $*$-probability space which satisfies the hypothesis of Proposition 7.2, and consider the GNS representation

$(L^2(\mathcal{A},\varphi),\pi,\widehat{1})$. It is an immediate exercise, left to the reader, that we have the equivalence:

$$\varphi \text{ is faithful } \Leftrightarrow \widehat{1} \text{ is a separating vector for } \pi. \tag{7.12}$$

A consequence of (7.12) which is worth recording is that if φ is faithful, then $\pi : \mathcal{A} \to B(L^2(\mathcal{A},\varphi))$ is one-to-one (indeed, the injectivity of π is clearly implied by the existence of a separating vector).

Free product of C^*-probability spaces

We will restrict our attention to the main situation considered throughout these lectures, when the expectation functional is a faithful trace. The construction of a free product of C^*-probability spaces will be obtained from the corresponding construction at the level of $*$-probability spaces, by using the GNS representation. Before going into the precise description of this, it is useful to note the following fact.

LEMMA 7.8. *Let $(\mathcal{A}_o,\varphi_o)$ be a $*$-probability space such that φ_o is a faithful trace. Suppose that $\mathcal{A}_o$ satisfies the hypothesis of Proposition 7.2, and consider the GNS representation $(L^2(\mathcal{A}_o,\varphi_o),\pi,\widehat{1})$, as described in that proposition. Let us denote*

$$\mathcal{A} := \mathrm{cl}(\pi(\mathcal{A}_o)) \subset B(L^2(\mathcal{A}_o,\varphi_o)).$$

If $T \in \mathcal{A}$ is such that $T\widehat{1} = 0$, then $T = 0$.

PROOF. Observe that for every $a,b,c \in \mathcal{A}_o$ we have:

$$\langle \pi(c)\widehat{a},\widehat{b}\rangle = \langle \pi(c)\widehat{1},\widehat{ba^*}\rangle. \tag{7.13}$$

Indeed, the left-hand side of (7.13) is $\langle \widehat{ca},\widehat{b}\rangle = \varphi(b^*ca)$, while the right-hand side is $\langle \widehat{c},\widehat{ba^*}\rangle = \varphi(ab^*c)$. But $\varphi(b^*ca) = \varphi(ab^*c)$, due to the assumption that φ is a trace.

By approximating an arbitrary operator $T \in \mathcal{A}$ with operators of the form $\pi(c)$, $c \in \mathcal{A}_o$ (while $a,b \in \mathcal{A}_o$ are fixed), we immediately infer from (7.13) that we actually have

$$\langle T\widehat{a},\widehat{b}\rangle = \langle T\widehat{1},\widehat{ba^*}\rangle, \quad \forall\, T \in \mathcal{A}, \quad a,b \in \mathcal{A}_o. \tag{7.14}$$

Let now $T \in \mathcal{A}$ be such that $T\widehat{1} = 0$. From (7.14) we then obtain that $\langle T\widehat{a},\widehat{b}\rangle = 0$, $\forall\, a,b \in \mathcal{A}_o$. Since $\{\widehat{a} \mid a \in \mathcal{A}_o\}$ is a dense subspace of $L^2(\mathcal{A}_o,\varphi_o)$, and T is a bounded linear operator, this in turn gives us that $T = 0$. □

THEOREM 7.9. *Let $(\mathcal{A}_i,\varphi_i)_{i\in I}$ be a family of C^*-probability spaces such that the functionals $\varphi_i : \mathcal{A}_i \to \mathbb{C}$, $i \in I$, are faithful traces. Then there exists a C^*-probability space $(\mathcal{A},\varphi)$ with φ a faithful trace, and* *a family of norm-preserving unital $*$-homomorphisms $W_i : \mathcal{A}_i \to \mathcal{A}$, $i \in I$, such that:*

(i) $\varphi \circ W_i = \varphi_i, \quad \forall\, i \in I$;

(ii) *the unital C^*-subalgebras $(W_i(\mathcal{A}_i))_{i\in I}$ form a free family in $(\mathcal{A}, \varphi)$;*

(iii) $\cup_{i\in I} W_i(\mathcal{A}_i)$ *generates $\mathcal{A}$ as a C^*-algebra.*

Moreover, $(\mathcal{A}, \varphi)$ and $(W_i : \mathcal{A}_i \to \mathcal{A})_{i\in I}$ are uniquely determined up to isomorphism, in the sense that if $(\mathcal{A}', \varphi')$ and $(W_i' : \mathcal{A}_i \to \mathcal{A}')_{i\in I}$ have the same properties, then there exists a C^*-algebra isomorphism $\Phi : \mathcal{A} \to \mathcal{A}'$ such that $\varphi' \circ \Phi = \varphi$ and such that $\Phi \circ W_i = W_i'$, $\forall\, i \in I$.

PROOF. In order to construct $(\mathcal{A}, \varphi)$, let us first consider the free product of $*$-probability spaces $(\mathcal{A}_o, \varphi_o) = *_{i\in I}(\mathcal{A}_i, \varphi_i)$, as discussed in Lecture 6. Recall that in particular we have $\mathcal{A}_o = *_{i\in I}\mathcal{A}_i$ as in Equation (6.1) of Notations 6.2, and that every $\mathcal{A}_i$ is hence viewed as a unital $*$-subalgebra of $\mathcal{A}_o$. Let us also record here the fact that φ_o is a faithful trace on $\mathcal{A}_o$ (by Propositions 6.8 and 6.14).

We claim that the linear span $\mathcal{W} := \text{span}\{u \in \mathcal{A}_o \mid u \text{ is unitary}\}$ is all of $\mathcal{A}_o$. Indeed, for every $i \in I$ we have that

$$\mathcal{W} \supset \text{span}\{u \in \mathcal{A}_i \mid u \text{ is unitary}\} = \mathcal{A}_i$$

(with the latter equality holding because $\mathcal{A}_i$ is a C^*-algebra, and by Exercise 7.20). Hence $\mathcal{W} \supset \cup_{i\in I}\mathcal{A}_i$. But $\mathcal{W}$ is a unital $*$-subalgebra of $\mathcal{A}_o$ (immediate verification); so it follows that $\mathcal{W}$ contains the unital $*$-subalgebra of $\mathcal{A}_o$ generated by $\cup_{i\in I}\mathcal{A}_i$, which is all of $\mathcal{A}_o$.

We thus see that $(\mathcal{A}_o, \varphi_o)$ satisfies the hypothesis of Proposition 7.2, and we can therefore consider the GNS representation $(L^2(\mathcal{A}_o, \varphi_o), \pi, \widehat{1})$ for $(\mathcal{A}_o, \varphi_o)$. Since φ_o is faithful, we have that $\pi : \mathcal{A}_o \to B(L^2(\mathcal{A}_o, \varphi_o))$ is one-to-one (cf. Remark 7.7).

Let us consider the unital C^*-subalgebra

$$\mathcal{A} := \text{cl}(\pi(\mathcal{A}_o)) \subset B(L^2(\mathcal{A}_o, \varphi_o)).$$

Moreover, for every $i \in I$ let us denote by $W_i : \mathcal{A}_i \to \mathcal{A}$ the unital $*$-homomorphism which is obtained by suitably restricting π (this makes sense, since $\mathcal{A}_i$ is contained in the domain of π, while $\mathcal{A}$ contains its range). We have that W_i is one-to-one (because π was like that); in view of the fact that $\mathcal{A}_i$ and $\mathcal{A}$ are unital C^*-algebras, we can thus infer that W_i is norm preserving (cf. Exercise 4.18).

Let $\varphi : \mathcal{A} \to \mathbb{C}$ be the positive functional defined by the vector $\widehat{1} \in L^2(\mathcal{A}_o, \varphi_o)$; that is,

$$\varphi(T) := \langle T\widehat{1}, \widehat{1}\rangle, \quad T \in \mathcal{A}.$$

Then, clearly, $(\mathcal{A}, \varphi)$ is a C^*-probability space. Note that Lemma 7.8 gives us the faithfulness of φ on $\mathcal{A}$. Indeed, for $T \in \mathcal{A}$ we have that $\varphi(T^*T) = ||T\widehat{1}||^2$; so we get that $\varphi(T^*T) = 0 \Rightarrow T\widehat{1} = 0 \Rightarrow T = 0$, with the last implication given by Lemma 7.8.

By taking into account how the GNS representation is defined, it is immediate that we have

$$\varphi(\pi(a)) = \varphi_o(a), \quad \forall\, a \in \mathcal{A}_o. \tag{7.15}$$

Since the W_i considered above are obtained by restricting π, Equation (7.15) says in particular that $\varphi \circ W_i = \varphi_i$, for every $i \in I$.

In order to complete the required list of properties for $(\mathcal{A}, \varphi)$ and for $(W_i)_{i\in I}$, one is left to make the following three remarks.

(a) From (7.15) and the fact that $(\mathcal{A}_i)_{i\in I}$ are freely independent in $(\mathcal{A}_o, \varphi_o)$, it follows that the family $(W_i(\mathcal{A}_i))_{i\in I}$ is freely independent in $(\mathcal{A}, \varphi)$.

(b) From (7.15) and the fact that φ_o is a trace on $\mathcal{A}_o$ it follows (by using the density of $\pi(\mathcal{A}_o)$ in $\mathcal{A}$) that φ is a trace on $\mathcal{A}$.

(c) From the fact that $(\mathcal{A}_i)_{i\in I}$ generate $\mathcal{A}_o$ as a $*$-algebra and the density of $\pi(\mathcal{A}_o)$ in $\mathcal{A}$ it follows that $(W_i(\mathcal{A}_i))_{i\in I}$ generate $\mathcal{A}$ as a C^*-algebra.

The easy verifications required in the three remarks (a), (b), (c) listed above are left as an exercise to the reader.

Finally, the uniqueness part of the theorem is a consequence of Theorem 4.11 (in the version described in Exercise 4.20, which allows infinite families of generators). □

DEFINITION 7.10. Let $(\mathcal{A}_i, \varphi_i)_{i\in I}$ be a family of C^*-probability spaces such that the functionals $\varphi_i : \mathcal{A}_i \to \mathbb{C}$, $i \in I$, are faithful traces. A C^*-probability space $(\mathcal{A}, \varphi)$ together with a family of homomorphisms $(W_i : \mathcal{A}_i \to \mathcal{A})_{i\in I}$ as appearing in Theorem 7.9 will be called a **free product of the C^*-probability spaces** $(\mathcal{A}_i, \varphi_i)_{i\in I}$.

It was observed in Theorem 7.9 that, up to isomorphism, there actually exists only one free product $(\mathcal{A}, \varphi)$ of the C^*-probability spaces $(\mathcal{A}_i, \varphi_i)_{i\in I}$, and the corresponding homomorphisms $(W_i : \mathcal{A}_i \to \mathcal{A})_{i\in I}$ are one-to-one and norm preserving. Same as in the algebraic case (cf. Notations 6.2), we will make a slight notational abuse and assume that $\mathcal{A}$ *contains* every $\mathcal{A}_i$ as a unital C^*-subalgebra. This will make the W_i disappear out of the notations (they become the inclusion maps of the $\mathcal{A}_i$ into $\mathcal{A}$), and will give us a more "canonical" incarnation of $(\mathcal{A}, \varphi)$ which we will call *the* free product of the C^*-probability spaces $(\mathcal{A}_i, \varphi_i)_{i\in I}$. The customary notation for this canonical free product is:

$$(\mathcal{A}, \varphi) = *_{i\in I}(\mathcal{A}_i, \varphi_i). \tag{7.16}$$

(This happens to be the same notation as used for the free product of $*$-probability spaces in Lecture 6 – we will make sure in what follows to state explicitly which of the two is meant, whenever there can be some ambiguity about this.)

When we deal with the canonical C^*-free product of Equation (7.16), the C^*-algebra $\mathcal{A}$ will be a completion of the algebraic free product $*_{i\in I}\mathcal{A}_i$ which was described in Notations 6.2. In the operator algebra literature it is customary (for reasons that we do not go into here) to say that $\mathcal{A}$ is the **reduced free product** of the C^*-algebras $(\mathcal{A}_i)_{i\in I}$, with respect to the family of functionals $(\varphi_i : \mathcal{A}_i \to \mathbb{C})_{i\in I}$.

Example: semicircular systems and the full Fock space

In this section we present an important situation involving a C^*-algebra which appears as reduced free product – the C^*-algebra generated by a semicircular system.

DEFINITION 7.11. Let $(\mathcal{A}, \varphi)$ be a $*$-probability space. A **semicircular system** in $(\mathcal{A}, \varphi)$ is a family $x_1, \dots, x_k$ of selfadjoint elements of $\mathcal{A}$ such that:

(i) each of $x_1, \dots, x_k$ is a standard semicircular element in $(\mathcal{A}, \varphi)$ (in the sense of Definition 2.16 and Remarks 2.17);

(ii) $x_1, \dots, x_k$ are free with respect to φ.

REMARK 7.12. *(The C^*-algebra of a semicircular system)*

Let $(\mathcal{A}, \varphi)$ and $(\mathcal{B}, \psi)$ be C^*-probability spaces such that φ and ψ are faithful. Let $x_1, \dots, x_k$ be a semicircular system in $(\mathcal{A}, \varphi)$, and let $y_1, \dots, y_k$ be a semicircular system in $(\mathcal{B}, \psi)$. Let $\mathcal{M} \subset \mathcal{A}$ and $\mathcal{N} \subset \mathcal{B}$ be the unital C^*-subalgebras generated by $\{x_1, \dots, x_k\}$ and by $\{y_1, \dots, y_k\}$, respectively. Then the C^*-probability spaces $(\mathcal{M}, \varphi|\mathcal{M})$ and $(\mathcal{N}, \psi|\mathcal{N})$ satisfy the hypotheses of Theorem 4.11, with respect to their systems of generators $x_1, \dots, x_k$ and $y_1, \dots, y_k$. Hence, by Theorem 4.11, there exists a C^*-algebra isomorphism $\Phi : \mathcal{M} \to \mathcal{N}$ such that $\Phi(x_i) = y_i$, $1 \le i \le k$. (This is analogous to the discussion in Remark 5.21.)

Thus we see that all the semicircular systems of the kind appearing in the preceding paragraph generate, up to isomorphism, the same C^*-algebra $\mathcal{M}_k$. This $\mathcal{M}_k$ is called the **C^*-algebra of a semicircular system with k elements**.

Of course, in order to introduce the C^*-algebra $\mathcal{M}_k$, one must also make sure that semicircular systems with k elements can indeed be constructed in the C^*-framework. This is a direct consequence of the fact that one can form free products of C^*-probability spaces, as explained in the preceding section. In fact, it is immediate that $\mathcal{M}_k$ is nothing but the reduced free product of k copies of $C[-2,2]$, where the expectation functional $\varphi : C[-2,2] \to \mathbb{C}$ is integration against the semicircular density, $\varphi(f) = \frac{1}{2\pi}\int_{-2}^{2} f(t)\sqrt{4-t^2}\,dt$.

On the other hand it is noteworthy that semicircular systems may arise naturally without requiring us to manifestly perform a free product construction. The remaining part of this lecture will be devoted to showing how this happens in the framework of the so-called creation and annihilation operators on the full Fock space.

DEFINITIONS 7.13. Let $\mathcal{H}$ be a Hilbert space.

(1) The **full Fock space** over $\mathcal{H}$ is defined as

$$\mathcal{F}(\mathcal{H}) := \bigoplus_{n=0}^{\infty} \mathcal{H}^{\otimes n}. \tag{7.17}$$

The summand $\mathcal{H}^{\otimes 0}$ on the right-hand side of the above equation is a one-dimensional Hilbert space. It is customary to write it in the form $\mathbb{C}\Omega$ for a distinguished vector of norm one, which is called the **vacuum vector**.

(2) The vector state $\tau_{\mathcal{H}}$ on $B(\mathcal{F}(\mathcal{H}))$ given by the vacuum vector,

$$\tau_{\mathcal{H}}(T) := \langle T\Omega, \Omega\rangle, \qquad T \in B(\mathcal{F}(\mathcal{H})), \tag{7.18}$$

is called the **vacuum expectation state**.

(3) For each $\xi \in \mathcal{H}$, the operator $l(\xi) \in B(\mathcal{F}(\mathcal{H}))$ determined by the formula

$$\begin{cases} l(\xi)\Omega &= \xi \\ l(\xi)\xi_1 \otimes \cdots \otimes \xi_n &= \xi \otimes \xi_1 \otimes \cdots \otimes \xi_n, \\ & \quad \forall\, n \geq 1,\ \forall\, \xi_1, \dots, \xi_n \in \mathcal{H} \end{cases} \tag{7.19}$$

is called the **(left) creation operator** given by the vector ξ.

(4) As one can easily verify, the adjoint of $l(\xi)$ is described by the formula:

$$\begin{cases} l(\xi)^*\Omega &= 0 \\ l(\xi)^*\xi_1 &= \langle \xi_1, \xi\rangle\Omega, \quad \xi_1 \in \mathcal{H} \\ l(\xi)^*\xi_1 \otimes \cdots \otimes \xi_n &= \langle \xi_1, \xi\rangle \xi_2 \otimes \cdots \otimes \xi_n, \\ & \quad \forall\, n \geq 2,\ \forall\, \xi_1, \dots, \xi_n \in \mathcal{H}, \end{cases} \tag{7.20}$$

and is called the **(left) annihilation operator** given by the vector ξ.

REMARKS 7.14. Consider the framework of the preceding definitions.

(1) Instead of Equation (7.17), one could describe the full Fock space $\mathcal{F}(\mathcal{H})$ by using an orthonormal basis. More precisely: if an

orthonormal basis $\{\xi_i \mid i \in I\}$ of $\mathcal{H}$ is given, then (just from how tensor products and direct sums of Hilbert spaces are formed) we get an orthonormal basis of $\mathcal{F}(\mathcal{H})$ described as:

$$\{\Omega\} \cup \{\xi_{i_1} \otimes \cdots \otimes \xi_{i_n} \mid n \geq 1,\ i_1, \ldots, i_n \in I\}. \tag{7.21}$$

(2) The third part of the preceding definitions states implicitly that the formulas (7.19) do indeed define a bounded linear operator $l(\xi)$ on $\mathcal{F}(\mathcal{H})$, the adjoint of which acts by the formulas (7.20). A quick proof of the first of these two facts is obtained by considering an orthonormal basis $\{\xi_i \mid i \in I\}$ of $\mathcal{H}$ such that one of the ξ_i is a scalar multiple of ξ, and by examining how $l(\xi)$ acts on the corresponding basis (7.21) of $\mathcal{F}(\mathcal{H})$. What one gets is that, more than just being a bounded linear operator on $\mathcal{F}(\mathcal{H})$, $l(\xi)$ is actually a scalar multiple of an isometry. The verification that $l(\xi)^*$ acts indeed as stated in (7.20) is immediate, and is left to the reader.

(3) From (7.19) it is clear that the map $\mathcal{H} \ni \xi \mapsto l(\xi) \in B(\mathcal{F}(\mathcal{H}))$ is linear.

(4) Another important formula (also immediate to verify, and left as an exercise) is that

$$l(\xi)^* l(\eta) = \langle \eta, \xi \rangle 1_{\mathcal{F}(\mathcal{H})}, \quad \forall\, \xi, \eta \in \mathcal{H}. \tag{7.22}$$

As a consequence of this formula, note that a finite product of operators from $\{l(\xi) \mid \xi \in \mathcal{H}\} \cup \{l(\xi)^* \mid \xi \in \mathcal{H}\}$ can always be put in the form

$$\alpha \cdot l(\xi_1) \cdots l(\xi_m) l(\eta_1)^* \cdots l(\eta_n)^* \tag{7.23}$$

for some $\alpha \in \mathbb{C}$, $n, m \geq 0$, and $\xi_1, \ldots, \xi_m, \eta_1, \ldots, \eta_n \in \mathcal{H}$. (We carry the scalar α for convenience of notation – but clearly, α can be absorbed into $l(\xi_1)$ whenever $m \geq 1$, and α can be absorbed into $l(\eta_1)^*$ whenever $n \geq 1$.) Indeed, suppose we are starting with a product

$$\alpha \cdot l(\zeta_1)^{\varepsilon(1)} \cdots l(\zeta_p)^{\varepsilon(p)} \tag{7.24}$$

with $\alpha \in \mathbb{C}$, $\zeta_1, \ldots, \zeta_p \in \mathcal{H}$ and $\varepsilon(1), \ldots, \varepsilon(p) \in \{1, *\}$. If there exists $1 \leq k \leq p-1$ such that $\varepsilon(k) = *$ and $\varepsilon(k+1) = 1$, then:

$$l(\zeta_k)^{\varepsilon(k)} l(\zeta_{k+1})^{\varepsilon(k+1)} = l(\zeta_k)^* l(\zeta_{k+1}) = \langle \zeta_{k+1}, \zeta_k \rangle 1_{\mathcal{F}(\mathcal{H})};$$

thus (at the cost of adjusting the scalar α by a factor of $\langle \zeta_{k+1}, \zeta_k \rangle$) we can remove the kth and the $(k+1)$th factors in the product (7.24), and replace the monomial appearing there by one of a shorter length. By repeating this process of shortening the length as many times as possible, we will bring the monomial (7.24) to a stage where there is no $1 \leq k \leq p-1$ such that $\varepsilon(k) = *$ and $\varepsilon(k+1) = 1$; and when this is done, the monomial (7.24) will have to look as in (7.23).

The connection between free probability and the framework introduced in Definitions 7.13 comes from the fact that orthogonality of vectors translates into free independence of the corresponding creation and annihilation operators.

PROPOSITION 7.15. *Let $\mathcal{H}$ be a Hilbert space and consider the C^*-probability space $(B(\mathcal{F}(\mathcal{H})), \tau_{\mathcal{H}})$. Let $\mathcal{H}_1, \ldots, \mathcal{H}_k$ be a family of linear subspaces of $\mathcal{H}$, such that $\mathcal{H}_i \perp \mathcal{H}_j$ for $i \neq j$ $(1 \leq i, j \leq k)$. For every $1 \leq i \leq k$ let $\mathcal{A}_i$ be the unital C^*-subalgebra of $B(\mathcal{F}(\mathcal{H}))$ generated by $\{l(\xi) \mid \xi \in \mathcal{H}_i\}$. Then $\mathcal{A}_1, \ldots, \mathcal{A}_k$ are freely independent in $(B(\mathcal{F}(\mathcal{H})), \tau_{\mathcal{H}})$.*

PROOF. For $1 \leq i \leq k$ let $\mathcal{B}_i \subset \mathcal{A}_i$ be the unital $*$-algebra generated by $\{l(\xi) \mid \xi \in \mathcal{H}_i\}$. It will suffice to prove that $\mathcal{B}_1, \ldots, \mathcal{B}_k$ are freely independent (cf. Lecture 5, Exercise 5.23).

For $1 \leq i \leq k$, the elements of $\mathcal{B}_i$ are obtained as linear combinations of finite products of operators from $\{l(\xi) \mid \xi \in \mathcal{H}_i\} \cup \{l(\xi)^* \mid \xi \in \mathcal{H}_i\}$. By taking into account Remark 7.14.4, it then follows that every $T \in \mathcal{B}_i$ can be put in the form:

$$T = \alpha 1_{\mathcal{F}(\mathcal{H})} + \sum_{j=1}^{p} l(\xi_{j,1}) \cdots l(\xi_{j,m(j)}) l(\eta_{j,1})^* \cdots l(\eta_{j,n(j)})^*, \qquad (7.25)$$

where for $1 \leq j \leq p$ we have $(m(j), n(j)) \neq (0,0)$ and $\xi_{j,1}, \ldots, \xi_{j,m(j)}, \eta_{j,1}, \ldots, \eta_{j,n(j)} \in \mathcal{H}_i$.

Note also that for T as in (7.25) we have $\tau_{\mathcal{H}}(T) = \alpha$. This is because for every $1 \leq j \leq p$ we have:

$$\begin{aligned} &\tau_{\mathcal{H}}(l(\xi_{j,1}) \cdots l(\xi_{j,m(j)}) l(\eta_{j,1})^* \cdots l(\eta_{j,n(j)})^*) \\ &\qquad = \langle l(\eta_{j,1})^* \cdots l(\eta_{j,n(j)})^* \Omega, l(\xi_{j,m(j)})^* \cdots l(\xi_{j,1})^* \Omega \rangle = 0, \quad (7.26) \end{aligned}$$

where the last equality occurs because

$$m(j) \neq 0 \Rightarrow l(\xi_{j,m(j)})^* \cdots l(\xi_{j,1})^* \Omega = 0,$$

while

$$n(j) \neq 0 \Rightarrow l(\eta_{j,1})^* \cdots l(\eta_{j,n(j)})^* \Omega = 0.$$

A moment's thought shows that the discussion in the preceding paragraph has the following consequence: if for $1 \leq i \leq k$ we denote

$$\mathcal{B}_i^o := \{T \in \mathcal{B}_i \mid \tau_{\mathcal{H}}(T) = 0\},$$

then $\mathcal{B}_i^o$ can also be described as

$$\mathcal{B}_i^o = \text{span}\left\{ l(\xi_1) \cdots l(\xi_m) l(\eta_1)^* \cdots l(\eta_n)^* \;\middle|\; \begin{array}{c} (m,n) \neq (0,0) \\ \xi_1, \ldots, \eta_n \in \mathcal{H}_i \end{array} \right\}. \quad (7.27)$$

Now let us go ahead and prove the required free independence of $\mathcal{B}_1, \ldots, \mathcal{B}_k$. To this end we fix some indices $i_1, \ldots, i_p \in \{1, \ldots, k\}$ such that $i_1 \neq i_2, \ldots, i_{p-1} \neq i_p$ and some elements $T_1 \in \mathcal{B}_{i_1}^o, \ldots, T_p \in \mathcal{B}_{i_p}^o$. Our goal is to show that $\tau_{\mathcal{H}}(T_1 \cdots T_p) = 0$.

By taking (7.27) into account, we can assume without loss of generality that for every $1 \leq j \leq p$ the operator T_j is of the form

$$T_j = l(\xi_{j,1}) \cdots l(\xi_{j,m(j)}) l(\eta_{j,1})^* \cdots l(\eta_{j,n(j)})^* \tag{7.28}$$

for some $(m(j), n(j)) \neq (0,0)$ and some vectors $\xi_{j,1}, \ldots, \xi_{j,m(j)}$, $\eta_{j,1}, \ldots, \eta_{j,n(j)} \in \mathcal{H}_{i_j}$. We distinguish two possible cases.

Case 1. There exists $j \in \{1, \ldots, p-1\}$ such that $n(j) \neq 0$ and $m(j+1) \neq 0$.

In this case, when we replace T_j and T_{j+1} from (7.28) we get a product containing two neighboring factors $l(\eta_{j,n(j)})^*$ and $l(\xi_{j+1,1})$. But the product of these two factors is $\langle \xi_{j+1,1}, \eta_{j,n(j)} \rangle 1_{\mathcal{F}(\mathcal{H})}$, and is hence equal to 0, due to the hypothesis that $\mathcal{H}_{i_j} \perp \mathcal{H}_{i_{j+1}}$. So in this case we get that $T_j T_{j+1} = 0$, and the vanishing of $\tau_{\mathcal{H}}(T_1 \cdots T_p)$ follows.

Case 2. The situation of Case 1 does not hold. That is, for every $j \in \{1, \ldots, p-1\}$ we have that either $n(j) = 0$ or $m(j+1) = 0$.

In this case it is immediate that when we replace each of $T_1, \ldots, T_p$ from (7.28) we get a product of the form $l(\xi_1) \cdots l(\xi_m) l(\eta_1)^* \cdots l(\eta_n)^*$ with $m + n = \sum_{j=1}^p (m(j) + n(j)) > 0$. The vacuum expectation of this product is 0, by exactly the same argument as in (7.26). So we obtain that $\tau_{\mathcal{H}}(T_1 \cdots T_p)$ is equal to 0 in this case as well. □

Let us note, moreover, that semicircular elements also appear naturally in the framework of creation and annihilation operators on the full Fock space.

PROPOSITION 7.16. *Let $\mathcal{H}$ be a Hilbert space and consider the C^*-probability space $(B(\mathcal{F}(\mathcal{H})), \tau_{\mathcal{H}})$. Then for every $\xi \neq 0$ in $\mathcal{H}$, the element $l(\xi) + l(\xi)^*$ is semicircular of radius $2\|\xi\|$ (in the sense of Lecture 2, Definition 2.16).*

PROOF. Due to the linearity of $\xi \mapsto l(\xi)$ we may assume that $\|\xi\| = 1$. Then we have $l(\xi)^* l(\xi) = 1_{\mathcal{F}(\mathcal{H})}$ (by (7.22)), while $l(\xi) l(\xi)^* \neq 1_{\mathcal{F}(\mathcal{H})}$ (as implied for instance by the fact that $l(\xi)^* \Omega = 0$). Also, by exactly the same argument as in (7.26) we see that we have

$$\tau_{\mathcal{H}}\Big(l(\xi)^m (l(\xi)^*)^n \Big) = \begin{cases} 1 & \text{if } m = n = 0 \\ 0 & \text{otherwise.} \end{cases}$$

Now, let us consider again the $*$-probability space $(\mathcal{A}, \varphi)$ and the special non-unitary isometry $a \in \mathcal{A}$ which were considered (and fixed) throughout Lecture 2. Based on the properties of $l(\xi)$ which were put into evidence in the preceding paragraph, we can proceed exactly as in the discussion of Remark 2.5 in order to define a unital $*$-homomorphism $\Phi : \mathcal{A} \to B(\mathcal{F}(\mathcal{H}))$ such that $\Phi(a) = l(\xi)$ and such that $\tau_{\mathcal{H}} \circ \Phi = \varphi$. Then we have that $\Phi(a + a^*) = l(\xi) + l(\xi)^*$, and it follows that the distribution of $l(\xi) + l(\xi)^*$ in $(B(\mathcal{F}(\mathcal{H})), \tau_{\mathcal{H}})$ coincides with the distribution of $a + a^*$ in $(\mathcal{A}, \varphi)$. But the latter distribution is indeed the semicircular one of radius 2, as verified in Proposition 2.15. □

As a consequence of the preceding two propositions, we see that semicircular systems do indeed arise in the framework of the full Fock space.

COROLLARY 7.17. *Let $\mathcal{H}$ be a Hilbert space and consider the C^*-probability space $(B(\mathcal{F}(\mathcal{H})), \tau_{\mathcal{H}})$. Let $\xi_1, \ldots, \xi_k$ be an orthonormal system of vectors in $\mathcal{H}$. Then the elements*

$$l(\xi_1) + l(\xi_1)^*, \ldots, l(\xi_k) + l(\xi_k)^*$$

form a semicircular system in $(B(\mathcal{F}(\mathcal{H})), \tau_{\mathcal{H}})$.

PROOF. The free independence of $l(\xi_1) + l(\xi_1)^*, \ldots, l(\xi_k) + l(\xi_k)^*$ follows from Proposition 7.15, and the fact that every $l(\xi_j) + l(\xi_j)^*$ is standard semicircular follows from Proposition 7.16. □

We will conclude this discussion by pointing out that the above considerations on the full Fock space really give us a concrete realization of the C^*-algebra $\mathcal{M}_k$ introduced in Remark 7.12. The only thing which prevents us from plainly applying Remark 7.12 to the operators $l(\xi_1) + l(\xi_1)^*, \ldots, l(\xi_k) + l(\xi_k)^*$ is that, obviously, the vacuum state $\tau_{\mathcal{H}}$ is not faithful on $B(\mathcal{F}(\mathcal{H}))$. A way of circumventing this problem is indicated by the next proposition.

PROPOSITION 7.18. *Suppose that $\mathcal{H}$ is a Hilbert space of dimension k, and that $\xi_1, \ldots, \xi_k$ is an orthonormal basis of $\mathcal{H}$. Consider the C^*-probability space $(B(\mathcal{F}(\mathcal{H})), \tau_{\mathcal{H}})$, and let $\mathcal{M}$ denote the unital C^*-subalgebra of $B(\mathcal{F}(\mathcal{H}))$ generated by $l(\xi_1)+l(\xi_1)^*, \ldots, l(\xi_k)+l(\xi_k)^*$. Let us also make the notation $\tau_{\mathcal{H}}|\mathcal{M} =: \varphi$. Then*

(1) *the vacuum vector Ω is cyclic for $\mathcal{M}$ – that is, $\{T\Omega \mid T \in \mathcal{M}\}$ is a dense subspace of $\mathcal{F}(\mathcal{H})$,*

(2) *φ is a trace on $\mathcal{M}$,*

(3) *φ is faithful on $\mathcal{M}$.*

PROOF. Throughout the proof we will denote by $\mathcal{M}_o$ the unital $*$-algebra generated by $l(\xi_1) + l(\xi_1)^*, \ldots, l(\xi_k) + l(\xi_k)^*$ (thus $\mathcal{M}_o$ is a dense unital $*$-subalgebra of $\mathcal{M}$).

(1) Let us denote $\{T\Omega \mid T \in \mathcal{M}_o\} =: \mathcal{F}_o$ (linear subspace of $\mathcal{F}(\mathcal{H})$). Observe that $\Omega \in \mathcal{F}_o$ (since $\Omega = 1_{B(\mathcal{F}(\mathcal{H}))}\Omega$), and that $\xi_i \in \mathcal{F}_o$ for every $1 \leq i \leq k$ (since $\xi_i = (l(\xi_i) + l(\xi_i^*))\Omega$). Going one step further, we see that $\xi_{i_1} \otimes \xi_{i_2} \in \mathcal{F}_o$, $\forall\, 1 \leq i_1, i_2 \leq k$ – indeed, we can write

$$\xi_{i_1} \otimes \xi_{i_2} = (l(\xi_{i_1}) + l(\xi_{i_1})^*)\xi_{i_2} - \delta_{i_1,i_2}\Omega \in \mathcal{F}_o.$$

In general, it is easy to prove by induction on n that

$$\xi_{i_1} \otimes \cdots \otimes \xi_{i_n} \in \mathcal{F}_o, \quad \forall\, n \geq 1, \ \forall\, 1 \leq i_1, \ldots, i_n \leq k. \tag{7.29}$$

The induction step "$n-1 \Rightarrow n$" (for $n \geq 3$) follows immediately by using the identity, for all $1 \leq i_1, \ldots, i_n \leq k$,

$$\xi_{i_1} \otimes \cdots \otimes \xi_{i_n} = (l(\xi_{i_1} + l(\xi_{i_1})^*)(\xi_{i_2} \otimes \cdots \otimes \xi_{i_n}) - \delta_{i_1,i_2}(\xi_{i_3} \otimes \cdots \otimes \xi_{i_n}).$$

From (7.29) (and the fact that $\mathcal{F}_o \ni \Omega$) we infer that $\mathcal{F}_o$ contains an orthonormal basis of $\mathcal{F}(\mathcal{H})$. This implies that $\mathcal{F}_o$ is a dense subspace of $\mathcal{F}(\mathcal{H})$, and the same must then be true for $\{T\Omega \mid T \in \mathcal{M}\} \supset \mathcal{F}_o$.

(2) Proposition 5.19 gives us that φ is a trace on $\mathcal{M}_o$; then a straightforward approximation argument shows that φ must also be a trace on $\mathcal{M} = \mathrm{cl}(\mathcal{M}_o)$.

(3) This is a repetition of the argument presented in Lemma 7.8. We start by observing that

$$\langle T(A\Omega), B\Omega\rangle = \langle T\Omega, BA^*\Omega\rangle, \quad \forall\, A, B, T \in \mathcal{M}. \tag{7.30}$$

Indeed, the left-hand side of (7.30) is $\varphi(B^*TA)$, while the right-hand side is $\varphi(AB^*T)$, and these quantities are equal to each other due to the traciality of φ.

Now let $T \in \mathcal{M}$ be such that $\varphi(T^*T) = 0$; this means that $T\Omega = 0$ (since $\varphi(T^*T) = ||T\Omega||^2$). But then from (7.30) we get that $\langle T(A\Omega), B\Omega\rangle = 0$, for all $A, B \in \mathcal{M}$. Since (by part (1) of the proposition) $A\Omega$ and $B\Omega$ are covering a dense subspace of $\mathcal{F}(\mathcal{H})$, we conclude that $T = 0$. □

COROLLARY 7.19. *In the notations of the preceding proposition we have that $\mathcal{M} \cong \mathcal{M}_k$, the C^*-algebra of a semicircular system with k elements.*

PROOF. One only has to apply the considerations from Remark 7.12 to $(\mathcal{M}, \varphi)$ where $\varphi = \tau_{\mathcal{H}}|\mathcal{M}$. □

Exercises

EXERCISE 7.20. Let $\mathcal{A}$ be a unital C^*-algebra. Prove that $\mathcal{A} = \text{span}\{u \in \mathcal{A} \mid u \text{ is unitary}\}$.
[Hint: It suffices to take an element $x \in \mathcal{A}$ such that $x = x^*$ and $||x|| \leq 1$, and write it as $x = (u+v)/2$ with u, v unitaries. Find such u and v by using the functional calculus of x.]

In the next exercise we will use the following definition.

DEFINITION 7.21. Let $\mathcal{A}$ be a unital $*$-algebra.

(1) An element $p \in \mathcal{A}$ is said to be a **projection** if it satisfies $p = p^* = p^2$.

(2) An element $v \in \mathcal{A}$ is said to be an **isometry** if it satisfies $v^*v = 1_{\mathcal{A}}$. (In particular every unitary is an isometry.)

(3) An element $w \in \mathcal{A}$ is said to be a **partial isometry** if both w^*w and ww^* are projections. (In particular every isometry is a partial isometry, and every projection is a partial isometry.)

EXERCISE 7.22. Prove that the conclusion of Proposition 7.2 still holds if the hypothesis (7.4) is replaced by the weaker condition that $\mathcal{A}$ is generated (as a $*$-algebra) by the set

$$\mathcal{W} = \{w \in \mathcal{A} \mid w \text{ is a partial isometry }\}.$$

[Hint: Prove that the set appearing in (7.7) during the proof of Proposition 7.2 is a subalgebra of $\mathcal{A}$, which contains $\mathcal{W}$.]

EXERCISE 7.23. Let γ be the standard normal distribution on $\mathbb{R}$, that is,

$$d\gamma(t) = \frac{1}{\sqrt{2\pi}} e^{-t^2/2}\, dt.$$

Consider the $*$-probability space $(\mathcal{A}, \varphi)$ where

$$\mathcal{A} := L^{\infty-}(\mathbb{R}, \gamma) = \cap_{1 \leq p < \infty} L^p(\mathbb{R}, \gamma),$$

and where $\varphi : \mathcal{A} \to \mathbb{C}$ is defined by

$$\varphi(f) := \int_{-\infty}^{\infty} f(t)\, d\gamma(t), \quad f \in \mathcal{A}.$$

Show that $L^2(\mathcal{A}, \varphi) = L^2(\mathbb{R}, \gamma)$ and that there exists $f \in \mathcal{A}$ such that $\widehat{g} \mapsto \widehat{fg}$ is not a bounded operator on $L^2(\mathcal{A}, \varphi)$.

The next two exercises take place in the framework of the full Fock space $\mathcal{F}(\mathcal{H})$ over a Hilbert space $\mathcal{H}$. In addition to creation and annihilation operators, one can also consider operators on $\mathcal{F}(\mathcal{H})$ defined as follows.

DEFINITION 7.24. Let $\mathcal{H}$ be a Hilbert space, and let $\mathcal{F}(\mathcal{H})$ be the full Fock space over $\mathcal{H}$ (as in Definitions 7.13). For every $T \in B(\mathcal{H})$, the operator $\Lambda(T) \in B(\mathcal{F}(\mathcal{H}))$ defined by the formula:

$$\begin{cases} \Lambda(T)\Omega &= 0 \\ \Lambda(T)\xi_1 \otimes \cdots \otimes \xi_n &= (T\xi_1) \otimes \xi_2 \otimes \cdots \otimes \xi_n, \\ & \quad \forall\, n \geq 1,\ \forall\, \xi_1, \ldots, \xi_n \in \mathcal{H} \end{cases} \tag{7.31}$$

is called the **gauge operator** associated to T.

EXERCISE 7.25. In the framework of the preceding definition, check the following properties of the gauge operators $\Lambda(T)$.

(1) For every $T \in B(\mathcal{H})$, Equation (7.31) does indeed define a bounded linear operator $\Lambda(T)$ on $\mathcal{F}(\mathcal{H})$, and we have $\|\Lambda(T)\| = \|T\|$.

(2) The map $T \mapsto \Lambda(T)$ is a unital $*$-homomorphism from $B(\mathcal{H})$ to $B(\mathcal{F}(\mathcal{H}))$.

(3) For all $\xi, \eta \in \mathcal{H}$ and all $T \in B(\mathcal{H})$ we have that

$$l(\xi)^*\Lambda(T)l(\eta) = \langle T\eta, \xi\rangle 1_{\mathcal{F}(\mathcal{H})}. \tag{7.32}$$

EXERCISE 7.26. Let $\mathcal{H}$ be a Hilbert space and consider the C^*-probability space $(B(\mathcal{F}(\mathcal{H})), \tau_{\mathcal{H}})$. Let $\mathcal{H}_1, \ldots, \mathcal{H}_k$ be a family of linear subspaces of $\mathcal{H}$, such that $\mathcal{H}_i \perp \mathcal{H}_j$ for $i \neq j$ ($1 \leq i, j \leq k$). For every $1 \leq i \leq k$ let $\mathcal{A}_i$ be the unital C^*-subalgebra of $B(\mathcal{F}(\mathcal{H}))$ generated by $\{l(\xi) \mid \xi \in \mathcal{H}_i\} \cup \{\Lambda(T) \mid T \in B(\mathcal{H}), T(\mathcal{H}_i) \subset \mathcal{H}_i$ and T vanishes on $\mathcal{H} \ominus \mathcal{H}_i\}$. Prove that $\mathcal{A}_1, \ldots, \mathcal{A}_k$ are freely independent in $(B(\mathcal{F}(\mathcal{H})), \tau_{\mathcal{H}})$.

EXERCISE 7.27. By using the framework of the full Fock space, prove the following statement. Let $(\mathcal{A}, \varphi)$ be a $*$-probability space and let $a_1, a_2 \in \mathcal{A}$ be semicircular elements of radii r_1 and respectively r_2, such that a_1 is free from a_2. Then $a_1 + a_2$ is a semicircular element of radius $\sqrt{r_1^2 + r_2^2}$.

EXERCISE 7.28. Let $(\mathcal{A}, \varphi)$ be a C^*-probability space such that φ is faithful, and let x be a selfadjoint element of $\mathcal{A}$. Suppose that the distribution of x is of the form $\rho(t)\,dt$ on an interval $[a, b] \subset \mathbb{R}$, where $\rho : [a, b] \to \mathbb{R}$ is a continuous function such that $\int_a^b \rho(t)\,dt = 1$ and such that ρ is not identically equal to zero on any subinterval $(c, d) \subset [a, b]$ ($a \leq c < d \leq b$). Prove that there exists an element $y = y^* \in \mathcal{A}$ such that:

(i) the unital C^*-subalgebra of $\mathcal{A}$ generated by y is equal to the unital C^*-subalgebra generated by x, and

(ii) the distribution of y is precisely the uniform distribution on the interval $[0, 1]$.

[Hint: Try $y = f(x)$ where $f : [a, b] \to [0, 1]$ is defined by $f(t) = \int_a^t \rho(s)\,ds$, $a \leq t \leq b$.]

EXERCISE 7.29. (1) Consider the unital C^*-algebra $\mathcal{A}$ described in expression (5.8), in Remark 5.21. Prove that $\mathcal{A}$ is isomorphic to the C^*-algebra (denoted in the above Remark 7.12 by $\mathcal{M}_3$) of a semicircular system with 3 elements.

(2) Generalize part (1) of the exercise to the unital C^*-algebra generated by k free selfadjoint elements $x_1, \dots, x_k$ such that the distribution of each of $x_1, \dots, x_k$ satisfies the hypotheses of Exercise 7.28.

Part 2

Cumulants

LECTURE 8

Motivation: free central limit theorem

One of the main ideas in free probability theory is to consider the notion of free independence in analogy with the notion of classical or tensor independence. In this spirit, the first investigations of Voiculescu in free probability theory focused on free analogs of some of the most fundamental statements from classical probability theory. In particular, he proved a free analog of a central limit theorem and introduced and described a free analog of "convolution." His investigations were quite analytical and centered around the concept of the "R-transform," an analytic function which plays the same role in free probability theory as the logarithm of the Fourier transform in classical probability theory. However, in this analytic approach it is not so obvious why the R-transform and the logarithm of the Fourier transform should be analogous.

Our approach to free probability theory is much more combinatorial in nature and will reveal in a clearer way the parallelism between classical and free probability theory.

In order to see what kind of combinatorial objects are relevant for free probability theory, we will begin by giving an algebraic proof of the free central limit theorem. This approach will show the similar nature of classical and free probability theory very clearly, because the same kind of proof can be given for the classical central limit theorem. Most of the arguments will be the same, only in the very end one has to distinguish whether one is in the classical or in the free situation. For convenience, we will restrict the discussion to the simplest case where we have identically distributed variables.

Convergence in distribution

Let $(\mathcal{A}, \varphi)$ be a $*$-probability space and $a_1, a_2, \ldots \in \mathcal{A}$ a sequence of identically distributed selfadjoint random variables which are either tensor independent or freely independent. Furthermore, assume that the variables are centered, $\varphi(a_r) = 0$ $(r \in \mathbb{N})$, and denote by $\sigma^2 := \varphi(a_r^2)$ the common variance of the variables. (Note that $\varphi(a_r^2) \geq 0$ because φ is positive and a_r selfadjoint.) A central limit theorem asks about the limit behavior of

$$\frac{a_1 + \cdots + a_N}{\sqrt{N}}$$

if N tends to infinity.

Of course, one has to specify the kind of convergence, and the only meaningful way for this is convergence of all moments or "convergence in distribution." So let us first define this concept.

DEFINITION 8.1. Let $(\mathcal{A}_N, \varphi_N)$ $(N \in \mathbb{N})$ and $(\mathcal{A}, \varphi)$ be non-commutative probability spaces and consider random variables $a_N \in \mathcal{A}_N$ for each $N \in \mathbb{N}$, and $a \in \mathcal{A}$. We say that a_N **converges in distribution** towards a for $N \to \infty$, and denote this by

$$a_N \xrightarrow{\text{distr}} a,$$

if we have

$$\lim_{N\to\infty} \varphi_N(a_N^n) = \varphi(a^n) \qquad \forall n \in \mathbb{N}.$$

REMARKS 8.2. (1) This form of convergence seems to be weaker than the usual form of convergence appearing in the classical central limit theorem. There statements are usually in terms of "weak convergence." If a_N and a have distributions in analytical sense μ_N and μ, respectively – which are, by our Definition 1.8, compactly supported probability measures on $\mathbb{R}$ – then the classical notion of "convergence in distribution" (or "convergence in law") of the random variables a_N to the random variable a means by definition the weak convergence of μ_N towards μ, i.e.

$$\lim_{N\to\infty} \int f(t)\, d\mu_N(t) = \int f(t)\, d\mu(t) \qquad \text{for all bounded continuous } f.$$

Clearly, by an application of Stone–Weierstrass, the convergence of all moments is enough to ensure the convergence of all continuous functions f on the compact support of μ, and thus our notion of convergence in distribution coincides in this situation with the corresponding classical notion.

(2) Note that the above remark applies only to situations where the limit element a has a compactly supported distribution (as it is required in our Definition 1.8 of "distribution in analytical sense"). Thus this remark does not seem to be relevant for the classical central limit theorem. Since the normal density does not have compact support, a classical normal random variable does not have a distribution in our analytical sense and Stone–Weierstrass is not enough to ensure that the convergence of moments in the classical central limit theorem implies weak convergence. However, the normal distribution is still

"nice" enough to allow this conclusion, namely it is determined by its moments.

DEFINITION 8.3. Let μ be a probability measure on $\mathbb{R}$ with moments

$$m_n := \int_{\mathbb{R}} t^n \, d\mu(t).$$

We say that μ is **determined by its moments**, if μ is the only probability measure on $\mathbb{R}$ with these moments, i.e. if for any probability measure ν on $\mathbb{R}$ we have

$$\int_{\mathbb{R}} t^n \, d\nu(t) = m_n \quad \forall n \in \mathbb{N} \qquad \Longrightarrow \qquad \nu = \mu.$$

REMARKS 8.4. (1) It makes sense to push our definition of distribution in analytical sense a bit further and allow probability measures which are determined by their moments as candidates for such a distribution, even if they do not have compact support. This gives us more flexibility in connecting our combinatorial considerations with classical analytical considerations. We will point out explicitly if we want to consider distributions in analytical sense in this more general frame.

(2) The relevance for us of probability measures determined by their moments comes from the following two well-known facts from classical probability theory.

(i) The normal distribution is determined by its moments.
(ii) Let probability measures μ and μ_N ($N = 1, 2, \dots$) on $\mathbb{R}$ be given such that μ is determined by its moments and that the μ_N have moments of all orders. If we have

$$\lim_{N \to \infty} \int_{\mathbb{R}} t^n \, d\mu_N(t) = \int_{\mathbb{R}} t^n \, d\mu(t) \qquad \forall n = 1, 2, \dots,$$

then μ_N converges weakly to μ.

These two facts imply that for the weak convergence of classical random variables to a normal distribution it is enough to check the convergence of all moments. Thus, in order to prove the classical central limit theorem (in the case where all involved random variables possess moments of all orders) it is enough to prove the convergence of all moments – which is exactly what our notion of convergence in distribution asks for.

General central limit theorem

In order to see that we have convergence in distribution of $(a_1 + \cdots + a_N)/\sqrt{N}$ we should calculate the limit $N \to \infty$ of all moments of $(a_1 + \cdots + a_N)/\sqrt{N}$. Let us first see how much we can say about such moments for finite N. In the following we fix a positive integer n. Then we have

$$\varphi\big((a_1 + \cdots + a_N)^n\big) = \sum_{1 \le r(1),\ldots,r(n) \le N} \varphi(a_{r(1)} \ldots a_{r(n)}).$$

Since all a_r have the same distribution we have

$$\varphi(a_{r(1)} \ldots a_{r(n)}) = \varphi(a_{p(1)} \ldots a_{p(n)})$$

whenever

$$r(i) = r(j) \quad \Longleftrightarrow \quad p(i) = p(j) \qquad \forall \quad 1 \le i, j \le n.$$

(This is a consequence of the fact that both tensor independence and free independence give a rule for calculating mixed moments from the values of the moments of the variables.) Thus the value of $\varphi(a_{r(1)} \ldots a_{r(n)})$ depends on the tuple $(r(1), \ldots, r(n))$ only through the information on which of the indices are the same and which are different. We will encode this information by a partition (i.e. a decomposition into disjoint subsets) $\pi = \{V_1, \ldots, V_s\}$ of the set $\{1, \ldots, n\}$. This partition π is determined as follows. Two numbers p and q belong to the same block V_m of π (for some $m = 1, \ldots, s$) if and only if $r(p) = r(q)$. We will write $(r(1), \ldots, r(n)) \hat{=} \pi$ in this case,

$$\big[(r(1), \ldots, r(n)) \hat{=} \pi\big] \Longleftrightarrow \big[r(p) = r(q) \text{ if and only if } p \sim_\pi q\big]. \quad (8.1)$$

Furthermore we denote the common value of $\varphi(a_{r(1)} \ldots a_{r(n)})$ for all tuples $(r(1), \ldots, r(n))$ with $(r(1), \ldots, r(n)) \hat{=} \pi$ by κ_π.

For illustration, consider the following example. Since a_1 has the same moments as a_7, since a_2 has the same moments as a_5, and since a_3 has the same moments as a_8, the free/tensor independence of a_1, a_2, a_3 produces for $\varphi(a_1 a_2 a_1 a_1 a_2 a_3)$ the same result as the free/tensor independence of a_7, a_5, a_8 for $\varphi(a_7 a_5 a_7 a_7 a_5 a_8)$, and we denote the common value of both expressions by

$$\kappa_{\{(1,3,4),(2,5),(6)\}} = \varphi(a_1 a_2 a_1 a_1 a_2 a_3) = \varphi(a_7 a_5 a_7 a_7 a_5 a_8).$$

With these notations we can continue the above calculation with

$$\varphi\big((a_1 + \cdots + a_N)^n\big) = \sum_{\pi \text{ partition of } \{1,\ldots,n\}} \kappa_\pi \cdot A_\pi^N,$$

where A_π^N is the number of tuples corresponding to π, i.e.

$$A_\pi^N := \#\{(r(1), \ldots, r(n)) \hat{=} \pi \mid 1 \le r(1), \ldots, r(n) \le N\}.$$

Note that the number of terms in the above sum does not depend on N, the only dependence on N is via the numbers A_π^N. It remains to examine the contribution of the different partitions. We will see that most of them will give no contribution in the normalized limit, only very special ones survive.

First, we will argue that partitions with singletons do not contribute: Consider a partition $\pi = \{V_1, \dots, V_s\}$ with a singleton, i.e. we have $V_m = \{r\}$ for some m and some r. Then we have

$$\kappa_\pi = \varphi(a_{r(1)} \dots a_r \dots a_{r(n)}) = \varphi(a_r) \cdot \varphi(a_{r(1)} \dots \check{a}_r \dots a_{r(n)}),$$

because $\{a_{r(1)}, \dots, \check{a}_r, \dots, a_{r(n)}\}$ is either tensor independent or freely independent from a_r. (This factorization follows from Equation (5.2) in the tensor case, and from Equation (5.5) in the free case.) However, since our variables are centered, we get $\kappa_\pi = 0$. Thus only such partitions π contribute which have no singletons, i.e. only $\pi = \{V_1, \dots, V_s\}$ where each block V_m $(m = 1, \dots, s)$ has at least two elements. Note that this implies in particular that in our sum we can restrict to $\pi = \{V_1, \dots, V_s\}$ for which $s \leq n/2$.

Consider now a $\pi = \{V_1, \dots, V_s\}$; then we have N possibilities for the common index corresponding to the first block V_1, $N-1$ possibilities for the common index corresponding to the second block V_2 (since this index has to be different from the one of the first block), and so on. Thus, if we denote by $|\pi|$ the number of blocks of π, we have that

$$A_\pi^N = N(N-1) \cdots (N - |\pi| + 1),$$

which grows asymptotically like $N^{|\pi|}$ for large N. Thus

$$\begin{aligned} \lim_{N \to \infty} \varphi\left(\left(\frac{a_1 + \cdots + a_N}{\sqrt{N}}\right)^n\right) &= \lim_{N \to \infty} \sum_\pi \frac{A_\pi^N}{N^{n/2}} \kappa_\pi \\ &= \lim_{N \to \infty} \sum_\pi N^{|\pi| - (n/2)} \kappa_\pi. \end{aligned}$$

Now note that for each appearing π the factor $N^{|\pi|-(n/2)}$ has a limit (because only $|\pi| \leq n/2$ appear in our sum), and that this limit is either 1 or 0, depending on whether $|\pi| = n/2$ or $|\pi| < n/2$. This means, in the limit $N \to \infty$ all partitions with $|\pi| < n/2$ are suppressed and we get exactly a contribution κ_π for each π which has the property that it has no singleton and that its number of blocks is equal to $n/2$. This means of course that π has to be a **pair partition** or **pairing**, i.e. a partition where each block V_m consists of exactly two elements.

Thus, we have now arrived at the following result:

$$\lim_{N \to \infty} \varphi\left(\left(\frac{a_1 + \cdots + a_N}{\sqrt{N}}\right)^n\right) = \sum_{\substack{\pi \text{ pair partition} \\ \text{of } \{1, \dots, n\}}} \kappa_\pi \tag{8.2}$$

Up to now, there has been no difference between the case of tensor independence and the case of free independence. The structure of the limiting moments is in both cases the same, namely they are calculated by summing over pair partitions. However, we still have not determined the weighting factors κ_π for these pair partitions. That is the point where we have to distinguish the two cases.

However, before we do this, let us note that the general formula (8.2) is enough to conclude that odd moments vanish in both cases. This conclusion comes from the simple observation that there are no pair partitions of a set with an odd number of elements. Thus:

$$\lim_{N\to\infty} \varphi\left(\left(\frac{a_1+\cdots+a_N}{\sqrt{N}}\right)^n\right) = 0 \qquad \text{for } n \text{ odd.} \tag{8.3}$$

Classical central limit theorem

The actual calculation of the limit distribution will now depend on whether we have classically independent or freely independent variables. Let us first consider the classical case. The factorization rule (5.2) for tensor independent random variables gives directly that for any pair partition π, the corresponding κ_π factorizes into a product of second moments, thus we have $\kappa_\pi = \sigma^n$. So we get in this case

$$\lim_{N\to\infty} \varphi\left(\left(\frac{a_1+\cdots+a_N}{\sqrt{N}}\right)^n\right) = \sigma^n \cdot \Big(\#\text{pair partitions of } \{1,\dots,n\}\Big). \tag{8.4}$$

It is easy to see that the number of pair partitions of the set $\{1,\dots,n\}$ is, in the case n even, given by $(n-1)\cdot(n-3)\cdots 5\cdot 3\cdot 1$.

On the other hand, one can also check quite easily that these numbers are exactly the moments of a centered normal distribution. We leave this as an exercise to the reader, see Exercise 8.22.

Putting all this together, we have thus proved the following version of the classical central limit theorem.

THEOREM 8.5. **(Classical central limit theorem)**
Let $(\mathcal{A},\varphi)$ be a $*$-probability space and $a_1, a_2, \dots \in \mathcal{A}$ a sequence of independent and identically distributed selfadjoint random variables. Furthermore, assume that all variables are centered, $\varphi(a_r) = 0$ $(r \in \mathbb{N})$, and denote by $\sigma^2 := \varphi(a_r^2)$ the common variance of the variables. Then we have

$$\frac{a_1+\cdots+a_N}{\sqrt{N}} \xrightarrow{\text{distr}} x,$$

where x is a normally distributed random variable of variance σ^2.

REMARKS 8.6. (1) Let us recall that this statement means explicitly

$$\lim_{N\to\infty} \varphi\left(\left(\frac{a_1+\cdots+a_N}{\sqrt{N}}\right)^n\right) = \frac{1}{\sqrt{2\pi\sigma^2}}\int_{\mathbb{R}} t^n e^{-t^2/2\sigma^2}dt \qquad \forall n\in\mathbb{N}.$$

(2) According to our Remarks 8.4, the normal distribution is determined by its moments and our algebraic form of the classical central limit theorem is equivalent to the usual formulation in terms of weak convergence.

(3) Note also that it is implicit in the definition of a $*$-probability space $(\mathcal{A},\varphi)$ that all variables have moments of all orders. In our algebraic frame we are not able to deal with situations where some moments do not exist.

Free central limit theorem

Now we want to switch to the free case. So we start off again with the general formula (8.2) and it remains to specify what the weighting factors κ_π are in the case of freely independent random variables.

Since we know that the odd moments vanish in this case, too, it suffices to consider even moments. So let $n = 2k$ be even and consider a pair partition $\pi = \{V_1,\ldots,V_k\}$. Let $(r(1),\ldots,r(n))$ be an index-tuple corresponding to this π, $(r(1),\ldots,r(n)) \hat{=} \pi$. Then there exist the following two possibilities.

(1) All consecutive indices are different:

$$r(1)\neq r(2)\neq\cdots\neq r(n).$$

Since $\varphi(a_{r(m)}) = 0$ for all $m = 1,\ldots,n$, we have by the definition of free independence

$$\kappa_\pi = \varphi(a_{r(1)}\cdots a_{r(n)}) = 0.$$

(2) Two consecutive indices coincide, i.e.

$$r(m) = r(m+1) = r \qquad \text{for some } m = 1,\ldots,n-1.$$

Because the index r does not appear any more among the other indices we have that $\{a_{r(1)},\ldots,a_{r(m-1)},a_{r(m+2)},\ldots,a_{r(n)}\}$ is free from $a_{r(m)}a_{r(m+1)} = a_r a_r$ and we get by the factorization property (5.5) that

$$\begin{aligned}\kappa_\pi &= \varphi(a_{r(1)}\cdots a_r a_r\cdots a_{r(n)})\\ &= \varphi(a_{r(1)}\cdots a_{r(m-1)}a_{r(m+2)}\cdots a_{r(n)})\cdot\varphi(a_r a_r)\\ &= \varphi(a_{r(1)}\cdots a_{r(m-1)}a_{r(m+2)}\cdots a_{r(n)})\cdot\sigma^2.\end{aligned}$$

It is clear that we can repeat the above argument in the second case and either get zero for κ_π or reduce the length of the considered moment further. We repeat this iteration until either we get zero in one of the steps or until we arrive at the moment $\varphi(1)$. In the latter case, the corresponding pairing will give a contribution σ^n. Thus we see that in the free case only special pairings will make a contribution. These special pairings are exactly those for which in each iteration step we are in the second case, i.e. we successively can find consecutive indices which coincide.

Let us consider a pairing π which does not have this property (i.e. which will contribute $\kappa_\pi = 0$). We want to see that we can characterize such a pairing in a geometrical way. The fact $\kappa_\pi = 0$ means that eventually our iterative procedure produces a pairing τ of $m > 0$ elements to which case (1) applies. Thus τ does not pair any neighbors. Take any pair $a_1 < a_2$ of τ. This does not consist of neighbors, thus there must be some elements between a_1 and a_2. If we find another pair $a_1' < a_2'$ between a_1 and a_2 (i.e. $a_1 < a_1' < a_2' < a_2$) then we rename this pair (a_1', a_2') to (a_1, a_2). We continue in this way until we find no other pair between $a_1 < a_2$. But there must still be at least one other element b with $a_1 < b < a_2$ (otherwise the pair a_1, a_2 would consist of neighbors), and this b must be paired with a c with either $c < a_1$ or $c > a_2$. Thus we see that τ must be "crossing" in the sense that there exist $p_1 < q_1 < p_2 < q_2$ such that p_1 is paired with p_2 and q_1 is paired with q_2. Clearly, the original π must exhibit the same crossing property. Thus we have seen that $\kappa_\pi = 0$ implies that π must be crossing in the above sense. On the other hand, if $\kappa_\pi = \sigma^n$, which means that we can reduce π by iterated application of case (1) to the empty pairing, then π cannot have this crossing property.

So we have arrived at the conclusion that in the free case exactly those pairings contribute which are not crossing in the above sense. This "non-crossing" feature is the basic property on which our description of free probability theory will rest.

NOTATION 8.7. A pairing of $\{1, \dots, n\}$ is called **non-crossing** if there does not exist $1 \leq p_1 < q_1 < p_2 < q_2 \leq n$ such that p_1 is paired with p_2 and q_1 is paired with q_2. The set of non-crossing pairings of $\{1, \dots, n\}$ is denoted by $NC_2(n)$.

Thus we have shown

$$\lim_{N\to\infty} \varphi\left(\left(\frac{a_1 + \cdots + a_N}{\sqrt{N}}\right)^n\right) = D_n \sigma^n, \tag{8.5}$$

where

$$D_n := \#\{\pi \mid \pi \text{ non-crossing pair partition of } \{1, \dots, n\}\}.$$

EXAMPLES 8.8. It is quite easy to calculate the number of non-crossing pair partitions explicitly. Let us first count them for small n. Of course, for odd n we have $D_n = 0$; thus we only have to consider even n. In the pictures below, the geometrical meaning of the property "non-crossing" will become obvious; namely, a pairing π of $\{1, \dots, n\}$ is non-crossing if we can draw the connections for the pairs of π in the half-plane below the numbers $1, \dots, n$ in such a way that these connections do not cross.

- $D_2 = 1$; there is only one pairing of 2 elements, and this is also non-crossing:

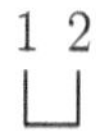

- $D_4 = 2$; here are the two non-crossing pairings of 4 elements:

Note that there is one additional, crossing, pairing for $n = 4$, namely

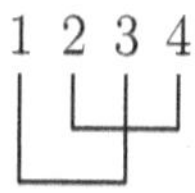

- $D_6 = 5$; here are the five non-crossing pairings of 6 elements:

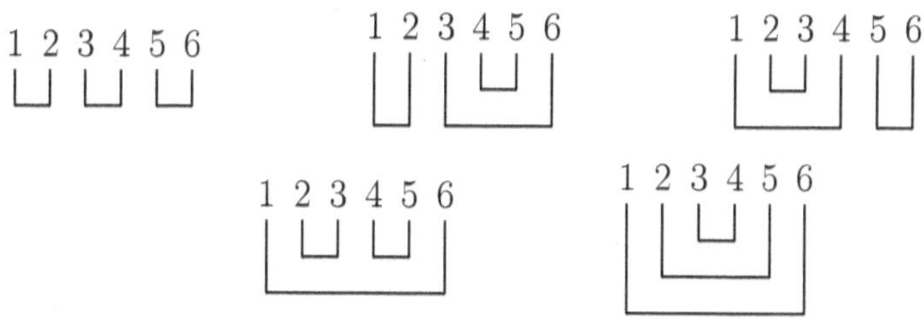

The other 10 of the 15 pairings of 6 elements are crossing.

We see that the Catalan numbers $1, 2, 5$ show up here. It is quite easy to see that this is true in general

LEMMA 8.9. *The number D_{2k} of non-crossing pair partitions of the set $\{1, \dots, 2k\}$ is given by the Catalan number C_k.*

PROOF. Since clearly $D_2 = 1 = C_1$, it suffices to check that the D_{2k} fulfill the recurrence relation of the Catalan numbers. Let $\pi = \{V_1, \dots, V_k\}$ be a non-crossing pair partition. We denote by V_1 that block of π which contains the element 1, i.e. it has to be of the form $V_1 = (1, m)$. Then the property "non-crossing" enforces that, for each

V_j $(j \neq 1)$, we cannot have a crossing between V_1 and V_j, i.e. we have either $1 < V_j < m$ or $1 < m < V_j$. (In particular, this implies that m has to be even, $m = 2l$.) This means that π restricted to $\{2, \ldots, m-1\}$ is a non-crossing pair partition of $\{2, \ldots, m-1\}$ and π restricted to $\{m+1, \ldots, n\}$ is a non-crossing pair partition of $\{m+1, \ldots, n\}$. There exist D_{m-2} many non-crossing pair partitions of $\{2, \ldots, m-1\}$ and D_{n-m} many non-crossing pair partitions of $\{m+1, \ldots, n\}$, where we put consistently $D_0 := 1$. Both these possibilities can appear independently from each other and $m = 2l$ can run through all even numbers from 2 to n. Hence we get

$$D_{2k} = \sum_{l=1}^{k} D_{2(l-1)} D_{2(k-l)}.$$

But this is the recurrence relation for the Catalan numbers, so the assertion follows. □

Another possibility for proving $D_{2k} = C_k$ is addressed in Exercise 8.23.

Since we know from Lecture 2 that the Catalan numbers are also the moments of a semicircular variable, we have thus proved the following version of the free central limit theorem.

THEOREM 8.10. **(Free central limit theorem)** *Let $(\mathcal{A}, \varphi)$ be a $*$-probability space and $a_1, a_2, \ldots \in \mathcal{A}$ a sequence of freely independent and identically distributed selfadjoint random variables. Assume furthermore $\varphi(a_r) = 0$ $(r \in \mathbb{N})$ and denote by $\sigma^2 := \varphi(a_r^2)$ the common variance of the variables. Then we have*

$$\frac{a_1 + \cdots + a_N}{\sqrt{N}} \xrightarrow{\text{distr}} s,$$

where s is a semicircular element of variance σ^2.

REMARKS 8.11. (1) According to the free central limit theorem the semicircular distribution has to be considered as the free analog of the normal distribution and is thus one of the basic distributions in free probability theory.

(2) As in the classical case, the assumptions in the free central limit theorem can be weakened considerably. For example, the assumption "identically distributed" is essentially chosen to simplify the argument; the same proof works if one replaces this by

$$\sup_{i \in \mathbb{N}} |\varphi(a_i^n)| < \infty \qquad \forall n \in \mathbb{N}$$

and

$$\sigma^2 := \lim_{N\to\infty} \frac{1}{N} \sum_{i=1}^{N} \varphi(a_i^2).$$

Our parallel treatment of the classical and the free central limit theorem shows the similarity between these two theorems very clearly. In particular, we have learned the simplest manifestation (on the level of pairings) of the following basic observation: the transition from classical to free probability theory is equivalent, on a combinatorial level, to the transition from all partitions to non-crossing partitions.

The multi-dimensional case

One of the main advantages of our combinatorial approach to free probability theory is the fact that, in contrast to an analytical treatment, a lot of arguments can be extended from one variable to several variables without any problems. In the following we want to demonstrate this for the free central limit theorem.

EXAMPLE 8.12. To motivate the problem, let us consider the case of two variables. So we now have two sequences $a_1, a_2, \dots$ and $b_1, b_2, \dots$ of selfadjoint variables such that the sets $\{a_1, b_1\}, \{a_2, b_2\}, \dots$ are free and have the same joint distribution. We do not necessarily assume that the a are free from the b. Then, under the assumption that all our variables are centered, we get from our one-dimensional free central limit theorem 8.10 that

$$\frac{a_1 + \cdots + a_N}{\sqrt{N}} \xrightarrow{\text{distr}} s_1$$

for a semicircular element s_1 and that

$$\frac{b_1 + \cdots + b_N}{\sqrt{N}} \xrightarrow{\text{distr}} s_2$$

for another semicircular element s_2. However, what we want to know in addition is the relation between s_1 and s_2. We will see that the joint distribution of the pair (s_1, s_2) is determined by knowledge of the covariance matrix

$$\begin{pmatrix} \varphi(a_r a_r) & \varphi(a_r b_r) \\ \varphi(b_r a_r) & \varphi(b_r b_r) \end{pmatrix}$$

of a_r and b_r (which is independent of r by our assumption on identical joint distribution of the families $\{a_r, b_r\}$). Furthermore, calculation of the joint distribution of s_1 and s_2 from this covariance matrix is very similar to calculation of the moments of a semicircular element from its variance σ^2. This will be the content of our multi-dimensional free central limit theorem 8.17. We will present and prove this in the following in full generality for arbitrarily many sequences; however, it might be illuminating for the reader to restrict its statement to the case of two sequences $a_1, a_2, \dots$ and $b_1, b_2, \dots$, as considered in this example.

We will now treat the general multi-dimensional case by looking at arbitrarily many sequences, which will be indexed by a fixed index set I (which might be infinite). So, we replace each a_r from the one-dimensional case by a family of selfadjoint random variables $(a_r^{(i)})_{i\in I}$ and assume that all these families are free and each of them has the same joint distribution and that all appearing random variables are centered. We want to investigate the convergence of the joint distribution of the random variables $\big((a_1^{(i)} + \cdots + a_N^{(i)})/\sqrt{N}\big)_{i\in I}$ when N tends to infinity. Let us first define the obvious generalization of our notion of convergence to this multi-dimensional setting.

DEFINITIONS 8.13. (1) Let $(\mathcal{A}_N, \varphi_N)$ $(N \in \mathbb{N})$ and $(\mathcal{A}, \varphi)$ be non-commutative probability spaces. Let I be an index set and consider for each $i \in I$ random variables $a_N^{(i)} \in \mathcal{A}_N$ and $a_i \in \mathcal{A}$. We say that $(a_N^{(i)})_{i\in I}$ **converges in distribution** towards $(a_i)_{i\in I}$ and denote this by

$$(a_N^{(i)})_{i\in I} \xrightarrow{\text{distr}} (a_i)_{i\in I},$$

if we have that each joint moment of $(a_N^{(i)})_{i\in I}$ converges towards the corresponding joint moment of $(a_i)_{i\in I}$, i.e. if we have for all $n \in \mathbb{N}$ and all $i(1), \dots, i(n) \in I$

$$\lim_{N\to\infty} \varphi_N(a_N^{(i(1))} \cdots a_N^{(i(n))}) = \varphi(a_{i(1)} \cdots a_{i(n)}). \tag{8.6}$$

(2) In the context of $*$-probability spaces we will say that $(a_N^{(i)})_{i\in I}$ **converges in $*$-distribution** towards $(a_i)_{i\in I}$ and denote this by

$$(a_N^{(i)})_{i\in I} \xrightarrow{*-\text{distr}} (a_i)_{i\in I},$$

if we have that each joint $*$-moment of $(a_N^{(i)})_{i\in I}$ converges towards the corresponding joint $*$-moment of $(a_i)_{i\in I}$, i.e. if

$$\big(a_N^{(i)}, (a_N^{(i)})^*\big)_{i\in I} \xrightarrow{\text{distr}} (a_i, a_i^*)_{i\in I}.$$

REMARK 8.14. Since free independence is equivalent to the validity of a collection of equations between moments, it is an easy but important observation that free independence goes over to the limit under convergence in distribution. Exercise 8.25 will ask for a proof of that statement. An application of this idea will appear later in the proof of Proposition 8.19.

Let us now look at our multi-dimensional version of the free central limit theorem. The calculation of the joint distribution of our normalized sums $\big((a_1^{(i)} + \cdots + a_N^{(i)})/\sqrt{N}\big)_{i\in I}$ works in the same way as in the one-dimensional case. Namely, we now have to consider for all $n \in \mathbb{N}$ and all $i(1), \ldots, i(n) \in I$

$$\varphi\Big(\big(a_1^{(i(1))} + \cdots + a_N^{(i(1))}\big) \cdots \big(a_1^{(i(n))} + \cdots + a_N^{(i(n))}\big)\Big)$$
$$= \sum_{1\le r(1),\ldots,r(n)\le N} \varphi\big(a_{r(1)}^{(i(1))} \cdots a_{r(n)}^{(i(n))}\big). \quad (8.7)$$

Again, we have that the value of $\varphi(a_{r(1)}^{(i(1))} \cdots a_{r(n)}^{(i(n))})$ depends on the tuple $(r(1), \ldots, r(n))$ only through the information on which of the indices are the same and which are different, which we will encode as before by a partition π of $\{1, \ldots, n\}$. The common value of $\varphi(a_{r(1)}^{(i(1))} \cdots a_{r(n)}^{(i(n))})$ for all tuples $(r(1), \ldots, r(n)) \hat{=} \pi$ will now, in addition, also depend on the tuple $(i(1), \ldots, i(n))$ and we will denote it by $\kappa_\pi[i(1), \ldots, i(n)]$. The next steps are the same as before. Singletons do not contribute because of the centeredness assumption and only pair partitions give the leading order in N and survive in the limit. Thus we arrive at

$$\lim_{N\to\infty} \varphi\Big(\frac{a_1^{(i(1))} + \cdots + a_N^{(i(1))}}{\sqrt{N}} \cdots \frac{a_1^{(i(n))} + \cdots + a_N^{(i(n))}}{\sqrt{N}}\Big)$$
$$= \sum_{\substack{\pi \text{ pair partition} \\ \text{of } \{1,\ldots,n\}}} \kappa_\pi[i(1), \ldots, i(n)].$$

It only remains to identify the contribution $\kappa_\pi[i(1), \ldots, i(n)]$ for a pair partition π. As before, the free independence assumption implies that $\kappa_\pi[i(1), \ldots, i(n)] = 0$ for crossing π. So consider finally a non-crossing π. Remember that in this case we can find two consecutive indices which coincide, i.e. $r(m) = r(m+1) = r$ for some m. Then we have

$$\begin{aligned}\kappa_\pi[i(1), \ldots, i(n)] &= \varphi\big(a_{r(1)}^{(i(1))} \cdots a_r^{(i(m))} a_r^{(i(m+1))} \cdots a_{r(n)}^{(i(n))}\big) \\ &= \varphi\big(a_{r(1)}^{(i(1))} \cdots a_{r(m-1)}^{(i(m-1))} a_{r(m+2)}^{(i(m+2))} \cdots a_{r(n)}^{(i(n))}\big) \cdot \varphi\big(a_r^{(i(m))} a_r^{(i(m+1))}\big) \\ &= \varphi\big(a_{r(1)}^{(i(1))} \cdots a_{r(m-1)}^{(i(m-1))} a_{r(m+2)}^{(i(m+2))} \cdots a_{r(n)}^{(i(n))}\big) \cdot c_{i(m)i(m+1)},\end{aligned}$$

where $(c_{ij})_{i,j\in I}$ with $c_{ij} := \varphi(a_r^{(i)} a_r^{(j)})$ is the covariance matrix of $(a_r^{(i)})_{i\in I}$.

Iterating this will lead to the final result that $\kappa_\pi[i(1),\dots,i(n)]$ is, for a non-crossing pairing π, given by the product of covariances $\prod_{(p,q)\in\pi} c_{i(p)i(q)}$ (one factor for each block (p,q) of π).

This form of the limiting moments motivates the following generalization of the notion of a semicircular element to the multi-dimensional case.

DEFINITION 8.15. Let $(c_{ij})_{i,j\in I}$ be a positive definite matrix. A family $(s_i)_{i\in I}$ of selfadjoint random variables in some $*$-probability space is called a **semicircular family of covariance** $(c_{ij})_{i,j\in I}$, if its distribution is of the following form: for all $n\in\mathbb{N}$ and all $i(1),\dots,i(n)\in I$ we have

$$\varphi(s_{i(1)}\cdots s_{i(n)}) = \sum_{\substack{\pi \text{ non-crossing pair partition}\\ \text{of } \{1,\dots,n\}}} \kappa_\pi[s_{i(1)},\dots,s_{i(n)}], \tag{8.8}$$

where

$$\kappa_\pi[s_{i(1)},\dots,s_{i(n)}] = \prod_{(p,q)\in\pi} c_{i(p)i(q)}. \tag{8.9}$$

EXAMPLES 8.16. (1) For illustration, let us write down the formulas (8.8) explicitly for small moments:

$$\varphi(s_a s_b) = c_{ab}, \qquad \varphi(s_a s_b s_c s_d) = c_{ab}c_{cd} + c_{ad}c_{bc},$$

$$\begin{aligned}\varphi(s_a s_b s_c s_d s_e s_f) &= c_{ab}c_{cd}c_{ef} + c_{ab}c_{cf}c_{de}\\ &\quad + c_{ad}c_{bc}c_{ef} + c_{af}c_{bc}c_{de} + c_{af}c_{be}c_{cd}.\end{aligned}$$

(2) If I consists just of one element then the above definition reduces, of course, to the definition of a semicircular element. More generally, each element s_j from a semicircular family is a semicircular element. Note, however, that in general the s_i are not free. In Corollary 8.20 we will see that the free independence of the s_i is equivalent to the diagonality of the covariance matrix. So in the case that the covariance matrix is just the identity matrix, our semicircular family reduces to a "semicircular system" in the sense of our Definition 7.11. We have to warn the reader that many authors mean by "semicircular family" the more restricted notion of a "semicircular system."

With our notion of a semicircular family we can summarize our calculations in the following multi-dimensional version of the free central limit theorem.

THEOREM 8.17. **(Free CLT, multi-dimensional version)** *Let $(\mathcal{A},\varphi)$ be a $*$-probability space and $\{a_1^{(i)}\}_{i\in I}, \{a_2^{(i)}\}_{i\in I}, \dots \subset \mathcal{A}$ a sequence of freely independent sets of selfadjoint random variables with* *the same joint distribution of* $(a_r^{(i)})_{i\in I}$ *for all* $r \in \mathbb{N}$ *– the latter meaning that, for any choice of* $n \in \mathbb{N}$ *and* $i(1), \dots, i(n) \in I$*, the moment* $\varphi(a_r^{(i(1))} \cdots a_r^{(i(n))})$ *does not depend on* r*. Assume furthermore that all variables are centered*

$$\varphi(a_r^{(i)}) = 0 \qquad (r \in \mathbb{N}, i \in I)$$

and denote the covariance of the variables by

$$c_{ij} := \varphi(a_r^{(i)} a_r^{(j)}) \qquad (i, j \in I).$$

Then we have

$$\left(\frac{a_1^{(i)} + \cdots + a_N^{(i)}}{\sqrt{N}}\right)_{i\in I} \xrightarrow{\text{distr}} (s_i)_{i\in I}, \tag{8.10}$$

where $(s_i)_{i\in I}$ *is a semicircular family of covariance* $(c_{ij})_{i,j\in I}$.

REMARKS 8.18. (1) Clearly, we can also prove a multi-dimensional version of the classical central limit theorem in the same way. Then the limit is a "Gaussian family" (multivariate normal distribution), whose joint moments are given by a similar formula as for semicircular families, the only difference is again that the summation runs over all pairings instead of non-crossing pairings. So for a Gaussian family $(x_i)_{i\in I}$ of covariance $(c_{ij})_{i,j\in I}$ we have

$$\varphi(x_1x_2x_3x_4) = c_{12}c_{34} + c_{14}c_{23} + c_{13}c_{24}$$

and the moment $\varphi(x_1x_2x_3x_4x_5x_6)$ is given as a sum over the 15 pairings of 6 elements. This combinatorial description of the joint moments of Gaussian families usually goes under the name of "Wick formula" (in particular, in the physical community).

(2) According to the previous remark, a semicircular family is to be considered as the free analog of a multivariate normal distribution.

A simple special case of a semicircular family is given if the covariance is a diagonal matrix. We can use our free central limit theorem to conclude quite easily that this is equivalent to having freely independent semicircular elements. This is a special case of the following proposition.

PROPOSITION 8.19. *Let* $(s_i)_{i\in I}$ *be a semicircular family of covariance* $(c_{ij})_{i,j\in I}$ *and consider a disjoint decomposition* $I = \dot{\bigcup}_{p=1}^d I_p$*. Then the following two statements are equivalent.*

(1) *The sets* $\{s_i \mid i \in I_1\}, \dots, \{s_i \mid i \in I_d\}$ *are freely independent.*

(2) *We have* $c_{ij} = 0$ *whenever* $i \in I_p$ *and* $j \in I_q$ *with* $p \neq q$.

PROOF. Assume first that the families $\big(\{s_i \mid i \in I_p\}\big)_{p=1,\dots,d}$ are free and consider $i \in I_p$ and $j \in I_q$ with $p \neq q$. Then the free independence of s_i and s_j implies in particular

$$c_{ij} = \varphi(s_i s_j) = \varphi(s_i)\varphi(s_j) = 0.$$

If however we have $c_{ij} = 0$ whenever $i \in I_p$ and $j \in I_q$ with $p \neq q$, then we can use our free central limit theorem in the following way. Choose in some $*$-probability space $(\mathcal{B}, \psi)$ a freely independent sequence of sets $\{a_1^{(i)}\}_{i\in I}, \{a_2^{(i)}\}_{i\in I}, \dots$ of random variables such that $(a_r^{(i)})_{i\in I}$ has for each $r = 1, 2, \dots$ the same joint distribution, which is prescribed in the following way:

- for each $p = 1, \dots, d$, the family $(a_r^{(i)})_{i\in I_p}$ has the same joint distribution as the family $(s_i)_{i\in I_p}$;
- the sets $\{a_r^{(i)} \mid i \in I_1\}, \dots, \{a_r^{(i)} \mid i \in I_d\}$ are free.

Note that the free product construction for $*$-probability spaces from Lecture 6 ensures that we can find such elements $a_r^{(i)}$. Furthermore, by the free independence between elements corresponding to different sets I_p, we have for $i \in I_p$ and $j \in I_q$ with $p \neq q$ that

$$\psi(a_r^{(i)} a_r^{(j)}) = \psi(a_r^{(i)}) \cdot \psi(a_r^{(j)}) = 0 = \varphi(s_i s_j).$$

Thus the covariance of the family $(a_r^{(i)})_{i\in I}$ is the same as the covariance of our given semicircular family $(s_i)_{i\in I}$. But now our free central limit theorem tells us that

$$\left(\frac{a_1^{(i)} + \cdots + a_N^{(i)}}{\sqrt{N}}\right)_{i\in I} \overset{\text{distr}}{\longrightarrow} (s_i)_{i\in I}, \tag{8.11}$$

where the limit is given exactly by the semicircular family from which we started (because this has the right covariance). But by our construction of the $a_r^{(i)}$ we have now in addition that the sets $\{(a_1^{(i)} + \cdots + a_N^{(i)})/\sqrt{N}\}_{i\in I_p}$ $(p = 1, \dots, d)$ are freely independent. As we observed in Remark 8.14 (see also Exercise 8.25), free independence passes over to the limit, and so we get the wanted result that the sets $\{s_i\}_{i\in I_p}$ $(p = 1, \dots, d)$ are freely independent. □

Note that this proposition implies that $\kappa_\pi[s_{i(1)}, \dots, s_{i(n)}]$ vanishes if the blocks of π couple elements which are free.

COROLLARY 8.20. *Consider a semicircular family* $(s_i)_{i\in I}$ *of covariance* $(c_{ij})_{i,j\in I}$. *Then the following are equivalent.*

(1) *The covariance matrix* $(c_{ij})_{i,j\in I}$ *is diagonal.*
(2) *The random variables* $(s_i)_{i\in I}$ *are free.*

EXAMPLE 8.21. Assume s_1 and s_2 are two semicircular elements which are free. Let us also assume that both have variance 1. Then the above tells us that their mixed moments are given by counting the non-crossing pairings which connect s_1 with s_1 and s_2 with s_2 (no blocks connecting s_1 with s_2 are allowed). For example, we have in such a situation

$$\varphi(s_1 s_1 s_2 s_2 s_1 s_2 s_2 s_1) = 2,$$

because there are two contributing non-crossing pairings, namely

and

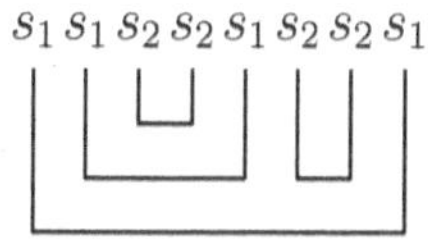

Conclusion and outlook

The general conclusion which we draw from this lecture is that non-crossing partitions appear quite naturally in free probability. From a combinatorial point of view, the transition from classical probability theory to free probability theory consists of replacing all partitions by non-crossing partitions.

But there are also more specific features shown by our treatment of the free central limit theorem. In the next lectures we will generalize to arbitrary distributions what we have learned from the case of semicircular families, namely:

(1) it seems to be canonical to write moments as sums over non-crossing partitions;

(2) the summands κ are multiplicative in the sense that they factorize in a product according to the block structure of π;

(3) the summands κ_π reflect free independence quite clearly, since $\kappa_\pi[s_{i(1)}, \dots, s_{i(n)}]$ vanishes if the blocks of π couple elements which are freely independent.

More concretely, we will write moments of random variables as

$$\varphi(a_1 \cdots a_n) = \sum_{\pi \text{ non-crossing partition}} \kappa_\pi[a_1, \dots, a_n], \tag{8.12}$$

where the κ_π ("free cumulants") factorize in a product according to the block structure of π. The difference from the present case is that we do not only have to consider non-crossing pairings, but we have to sum over all non-crossing partitions. Before we introduce free cumulants in full generality, we have to talk about the definition and basic properties of non-crossing partitions. In particular, we should also understand how to invert the relation (8.12) by so-called "Möbius inversion." This will be the content of the next two lectures.

Exercises

EXERCISE 8.22. Show that

$$\frac{1}{\sqrt{2\pi\sigma^2}}\int_{\mathbb{R}} t^n e^{-t^2/2\sigma^2}dt = \begin{cases} 0 & \text{if } n \text{ odd} \\ \sigma^n \cdot (n-1)\cdot(n-3)\cdots 5\cdot 3\cdot 1 & \text{if } n \text{ even.} \end{cases}$$

EXERCISE 8.23. Another possibility for proving $D_{2k} = C_k$ is to present a bijection between non-crossing pair partitions and Dyck paths. Here is one: we map a non-crossing pair partition π to a Dyck path $(i_1, \dots, i_n)$ by

$$i_m = +1 \quad \Longleftrightarrow \quad m \text{ is the first element in some } V_j \in \pi$$
$$i_m = -1 \quad \Longleftrightarrow \quad m \text{ is the second element in some } V_j \in \pi$$

Here are some examples for this mapping:

- $n = 2$

- $n = 4$

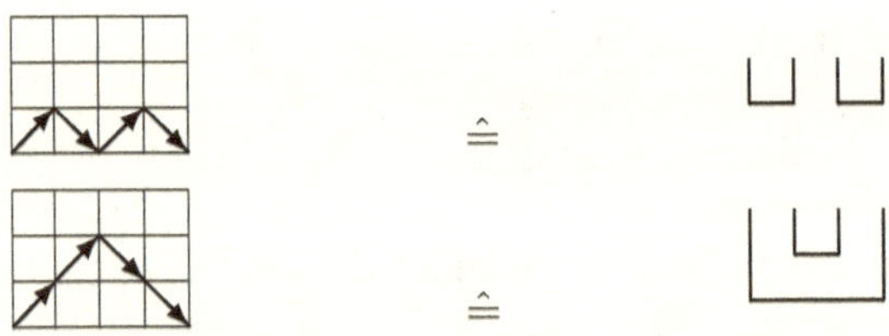

- $n = 6$

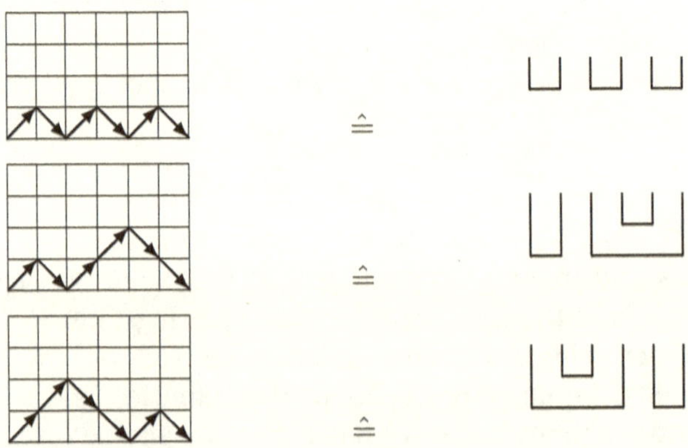

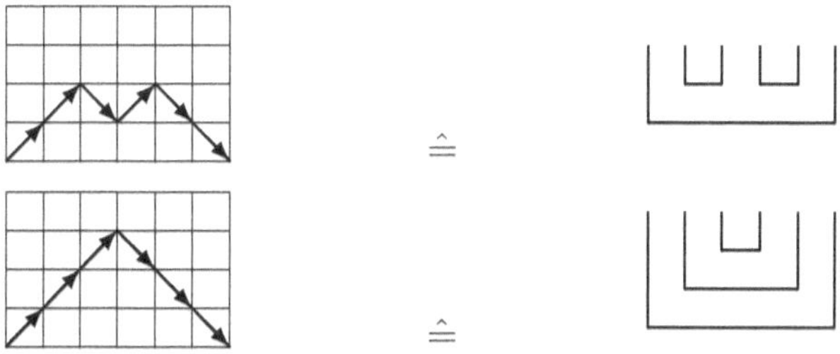

Prove that this mapping gives a bijection between Dyck paths and non-crossing pair partitions.

EXERCISE 8.24. (1) Prove that for every positive definite matrix $(c_{ij})_{i,j\in I}$ one can find a semicircular family of covariance $(c_{ij})_{i,j\in I}$ in some $*$-probability space.
[Hint: one possibility is to use the free central limit theorem and the fact that positivity is preserved under limit in distribution; another possibility is to use the next part of this problem.]

(2) Show that each semicircular family can be written as a linear combination of free semicircular elements.

EXERCISE 8.25. Let $(\mathcal{A},\varphi)$ and $(\mathcal{A}_N,\varphi_N)$ $(N\in\mathbb{N})$ be non-commutative probability spaces, and consider random variables $a,b\in\mathcal{A}$ and $a_N,b_N\in\mathcal{A}_N$ $(N\in\mathbb{N})$ such that $(a_N,b_N)\stackrel{\text{distr}}{\longrightarrow}(a,b)$. Assume that for each $N\in\mathbb{N}$ the random variables a_N and b_N are free (with respect to φ_N). Show that then also a and b are free (with respect to φ).

EXERCISE 8.26. Fill in the details in the following use of the free central limit theorem to infer that the distribution of the sum of creation and annihilation operators on a full Fock space has a semicircular distribution.

Consider in the non-commutative probability space $(B(\mathcal{F}(\mathcal{H})),\tau_{\mathcal{H}})$ for a fixed $f\in\mathcal{H}$ the variable $l(f)+l^*(f)$. Show that, for each natural N, this has the same distribution as the random variable

$$l\Big(\frac{f\oplus\cdots\oplus f}{\sqrt{N}}\Big)+l^*\Big(\frac{f\oplus\cdots\oplus f}{\sqrt{N}}\Big) \tag{8.13}$$

in the non-commutative probability space $(B(\mathcal{F}(\mathcal{H}_N)),\tau_{\mathcal{H}_N})$ with

$$\mathcal{H}_N:=\underbrace{\mathcal{H}\oplus\cdots\oplus\mathcal{H}}_{N\text{ times}}.$$

Show that the random variable (8.13) is the sum of N free random variables and apply the free central limit theorem to infer that the random variable $l(f)+l^*(f)$ is a semicircular element.

LECTURE 9

Basic combinatorics I: non-crossing partitions

In the preceding lecture we saw that a special type of partitions seems to lie underneath the structure of free probability. These are the so-called "non-crossing" partitions. The study of the lattices of non-crossing partitions was started by combinatorialists quite some time before the development of free probability. In this and the next lecture we will introduce these objects in full generality and present their main combinatorial properties which are of relevance for us.

The preceding lecture has also told us that, from a combinatorial point of view, classical probability and free probability should behave as all partitions versus non-crossing partitions. Thus, we will also keep an eye on similarities and differences between these two cases.

Non-crossing partitions of an ordered set

DEFINITIONS 9.1. Let S be a finite totally ordered set.

(1) We call $\pi = \{V_1, \dots, V_r\}$ a **partition** of the set S if and only if the V_i $(1 \le i \le r)$ are pairwise disjoint, non-void subsets of S such that $V_1 \cup \dots \cup V_r = S$. We call $V_1, \dots, V_r$ the **blocks** of π. The number of blocks of π is denoted by $|\pi|$. Given two elements $p, q \in S$, we write $p \sim_\pi q$ if p and q belong to the same block of π.

(2) The set of all partitions of S is denoted by $\mathcal{P}(S)$. In the special case $S = \{1, \dots, n\}$, we denote this by $\mathcal{P}(n)$.

(3) A partition π of the set S is called **crossing** if there exist $p_1 < q_1 < p_2 < q_2$ in S such that $p_1 \sim_\pi p_2 \not\sim_\pi q_1 \sim_\pi q_2$:

$$\cdots \; p_1 \; \cdots \; q_1 \; \cdots \; p_2 \; \cdots \; q_2 \; \cdots$$

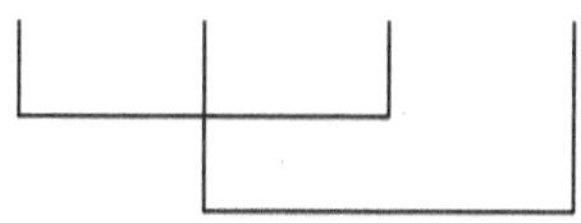

If π is not crossing, then it is called **non-crossing**.

(4) The set of all non-crossing partitions of S is denoted by $NC(S)$. In the special case $S = \{1, \dots, n\}$, we denote this by $NC(n)$.

We get a linear graphical representation of a partition π by writing all elements from S in a linear way, supplying each with a vertical line under it and joining the vertical lines of the elements in the same block with a horizontal line. For example, consider the partition $\{\{1,4,5,7\},\{2,3\},\{6\}\}$ of the set $\{1,2,3,4,5,6,7\}$. Graphically this looks as follows:

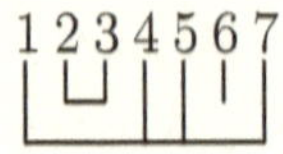

The name "non-crossing" becomes evident in such a representation. An example of a crossing partition is $\pi = \{\{1,3,5\},\{2,4\}\}$ which looks like this:

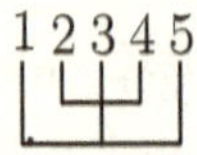

REMARKS 9.2. (1) Of course, $NC(S)$ depends only on the number of elements in S. In the following we will use the natural identification $NC(S_1) \cong NC(S_2)$ for $\#S_1 = \#S_2$ without further comment.

(2) In many cases the following recursive description of non-crossing partitions is of great use: a partition π of $\{1,\dots,n\}$ is non-crossing if and only if at least one block $V \in \pi$ is an interval and $\pi \backslash V$ is non-crossing, i.e. $V \in \pi$ has the form $V = \{k, k+1, \dots, k+p\}$ for some $1 \le k \le n$ and $p \ge 0$, $k+p \le n$ and we have

$$\pi \backslash V \in NC(\{1,\dots,k-1,k+p+1,\dots,n\}) \cong NC(n-(p+1)).$$

As an example consider the partition

$$\pi = \{\{1,10\},\{2,5,9\},\{3,4\},\{6\},\{7,8\}\}$$

of $\{1,\dots,10\}$:

Let us verify that $\pi \in NC(10)$ by doing successive "interval-stripping" operations. We first remove the intervals $\{3,4\}$, $\{6\}$, and $\{7,8\}$, which reduces us to:

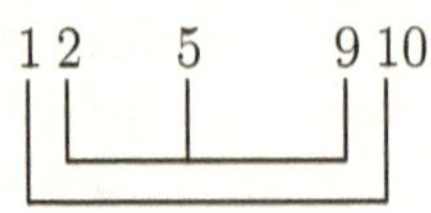

Now $\{2,5,9\}$ is an interval and can be removed, so that we are left with the interval $\{1,10\}$:

NOTATIONS 9.3. (1) If S (totally ordered) is the union of two disjoint sets S_1 and S_2 then, for $\pi_1 \in NC(S_1)$ and $\pi_2 \in NC(S_2)$, we let $\pi_1 \cup \pi_2$ be the partition of S which has as blocks the blocks of π_1 and the blocks of π_2. Note that $\pi_1 \cup \pi_2$ is not automatically non-crossing.

(2) Let S be a totally ordered set. Let W be a non-empty subset of S, on which we consider the order induced from S. For $\pi \in NC(S)$ we will denote by $\pi|_W$ the restriction of π to W, i.e.

$$\pi|_W := \{V \cap W \mid V \text{ block of } \pi\}. \tag{9.1}$$

It is immediately verified that $\pi|W \in NC(W)$. Note that in the particular case when W is a union of some of the blocks of π, the above equation reduces to just

$$\pi|_W = \{V \in \pi \mid V \subset W\} \in NC(W).$$

Note that for $n \leq 3$ all partitions are non-crossing and accordingly we have $\#NC(1) = 1$, $\#NC(2) = 2$, $\#NC(3) = 5$. Out of the 15 partitions of 4 elements exactly one is crossing, thus we have $\#NC(4) = 14$. We recognize here the occurrence of the first few Catalan numbers 1, 2, 5, 14 (cf. Notation 2.9). The next proposition shows this is not an accident.

PROPOSITION 9.4. *The number of elements in $NC(n)$ is equal to the Catalan number C_n.*

PROOF. For $n \geq 1$ let us denote $\#NC(n) =: D_n$, and let us also set $D_0 := 1$. We will verify that the numbers D_n satisfy

$$D_n = \sum_{i=1}^{n} D_{i-1} D_{n-i}, \qquad n \geq 1. \tag{9.2}$$

Since this recursion characterizes the Catalan numbers (as discussed e.g. in Lecture 2, Remark 2.12), the verification of (9.2) will give us the assertion.

For $n \geq 1$ and $1 \leq i \leq n$ let us denote by $NC^{(i)}(n)$ the set of non-crossing partitions $\pi \in NC(n)$ for which the block containing 1 contains i as its largest element. Because of the non-crossing condition, a partition $\pi \in NC^{(i)}(n)$ decomposes canonically into $\pi = \pi_1 \cup \pi_2$, where $\pi_1 \in NC^{(i)}(i)$ and $\pi_2 \in NC(\{i+1, \ldots, n\})$; thus we have

$$NC^{(i)}(n) \cong NC^{(i)}(i) \times NC(n-i).$$

However, by restricting π_1 to $\{1, \ldots, i-1\}$ we see that $NC^{(i)}(i)$ is in bijection with $NC(i-1)$. It follows that

$$NC^{(i)}(n) \cong NC(i-1) \times NC(n-i),$$

and, by taking cardinalities, that:

$$\#NC^{(i)}(n) = D_{i-1}D_{n-i}, \qquad \text{for } 1 \leq i \leq n. \tag{9.3}$$

(The cases $i = 1$ and $i = n$ of (9.3) involve the appropriate use of a set $NC(0)$ with $\#NC(0) = 1$, or can simply be checked directly.)

Since $NC(n) = \cup_{i=1}^{n} NC^{(i)}(n)$ and this is a disjoint union, we get that (9.2) follows from (9.3). □

REMARK 9.5. For $n \geq 1$, let $NC_2(2n)$ denote the set of non-crossing pair-partitions of $\{1, \ldots, 2n\}$, as discussed in the preceding lecture. Comparing the preceding proposition with Lemma 8.9 we see that we have

$$\#NC(n) = \#NC_2(2n), \qquad n \geq 1, \tag{9.4}$$

One can in fact check this equality by a direct bijective argument – see Exercise 9.42 at the end of the lecture.

In the remaining part of this section we will show how paths on $\mathbb{Z}^2$ can be used to obtain a more refined enumeration of the set $NC(n)$. The paths being used are the so-called *Lukasiewicz paths*; they provide a generalization of the Catalan paths from Lecture 2.

DEFINITIONS 9.6. (1) We will use the term **almost-rising path** for a path in $\mathbb{Z}^2$ which starts at $(0, 0)$ and makes steps of the form $(1, i)$ where $i \in \mathbb{N} \cup \{-1, 0\}$. (Thus an almost-rising path has "rising" steps, except for some possible "flat" steps of the form $(1, 0)$ and some "falling" steps of the form $(1, -1)$.)

(2) A **Lukasiewicz path** is an almost-rising path γ which ends on the x-axis, and never goes strictly below the x-axis. That is, all the lattice points visited by γ are of the form (i, j) with $j \geq 0$, and the last of them is $(n, 0)$, where n is the number of steps of γ.

The set of all Lukasiewicz paths with n steps will be denoted as $Luk(n)$.

REMARKS 9.7. (1) Let γ be an almost-rising path with n steps, and let the steps of γ be denoted (in the order they are made) as $(1, \lambda_1), \ldots, (1, \lambda_n)$. Information about the path is then completely recorded by the n-tuple

$$(\lambda_1, \ldots, \lambda_n) \in \Big(\mathbb{N} \cup \{-1, 0\}\Big)^n,$$

which will be referred to as the **rise-vector** of the path γ. Indeed, if the rise-vector $(\lambda_1, \ldots, \lambda_n)$ of γ is given, then γ can be described as the path which starts at $(0, 0)$ and visits successively the lattice points $(1, \lambda_1)$, $(2, \lambda_1 + \lambda_2), \ldots, (n, \lambda_1 + \cdots + \lambda_n)$.

Concrete example: here is the almost-rising path of length 6 which has rise-vector $(2, -1, 0, -1, -1, 1)$.

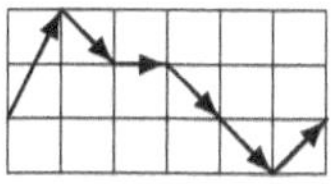

This path ends on the x-axis, but is not a Lukasiewicz path (as it goes under the x-axis after 5 steps).

(2) Let γ be an almost-rising path with rise-vector $(\lambda_1, \ldots, \lambda_n)$. It is clear that the condition for γ to be a Lukasiewicz path is expressed in terms of the rise-vector as follows:

$$\begin{cases} \lambda_1 + \cdots + \lambda_j \geq 0, \quad \forall\, 1 \leq j < n, \\ \\ \lambda_1 + \cdots + \lambda_n = 0. \end{cases} \tag{9.5}$$

PROPOSITION 9.8. *Let n be a positive integer.*

(1) *Let $\pi = \{V_1, \ldots, V_r\}$ be a partition in $NC(n)$. For $1 \leq i \leq r$ let us denote the minimal element of V_i by a_i. Consider the numbers $\lambda_1, \ldots, \lambda_n \in \mathbb{N} \cup \{-1, 0\}$ defined as follows:*

$$\lambda_m = \begin{cases} |V_i| - 1 & \text{if } m = a_i, \text{ for some } 1 \leq i \leq r \\ -1 & \text{otherwise.} \end{cases} \tag{9.6}$$

Then $(\lambda_1, \ldots, \lambda_n)$ is the rise-vector of a unique Lukasiewicz path with n steps.

(2) *For every $\pi \in NC(n)$, let us denote by $\Lambda(\pi)$ the Lukasiewicz path obtained from π in the way described in part (1) of the proposition. Then $\pi \mapsto \Lambda(\pi)$ is a bijection between $NC(n)$ and the set $Luk(n)$ of Lukasiewicz paths with n steps.*

PROOF. (1) For $1 \leq m \leq n$ we have

$$\lambda_1 + \cdots + \lambda_m = \sum_{\substack{1 \leq p \leq m \\ p \in V_1}} \lambda_p + \cdots + \sum_{\substack{1 \leq p \leq m \\ p \in V_r}} \lambda_p, \tag{9.7}$$

where (for $1 \leq j \leq r$) we make the convention that the sum $\sum_{\substack{1 \leq p \leq m \\ p \in V_j}} \lambda_p$ is equal to 0 if $\{1 \leq p \leq m \mid p \in V_j\} = \emptyset$. From the definition of $\lambda_1, \ldots, \lambda_n$ it is clear that each of the r sums on the right-hand side of (9.7) is non-negative, and it is equal to 0 in the case when $m = n$. This implies that the n-tuple $(\lambda_1, \ldots, \lambda_n)$ satisfies (9.5), and is thus the rise-vector of a Lukasiewicz path.

(2) We will prove that Λ is bijective by explicitly describing its inverse function, $\Pi : Luk(n) \to NC(n)$.

So given a path $\gamma \in Luk(n)$, we have to indicate the recipe for how to construct the partition $\pi = \Pi(\gamma)$ of $\{1, \ldots, n\}$. Let r be the number of non-falling steps of γ, and let us denote by $1 = a_1 < a_2 < \cdots < a_r \leq n$ the positions of these r steps among the n steps of γ. For every $1 \leq i \leq r$ the two lattice points connected by the a_ith step of γ must be of the form

$$(a_i - 1, h_i) \text{ and } (a_i, h_i + \xi_i), \quad \text{with } h_i, \xi_i \geq 0. \tag{9.8}$$

(That is, the a_ith step of the path γ is $(1, \xi_i)$, and makes the path go from $(a_i - 1, h_i)$ to $(a_i - 1, h_i) + (1, \xi_i)$.) In order to describe the partition $\pi = \Pi(\gamma)$, we first stipulate that π will have r blocks $V_1, \ldots, V_r$, and that $V_i \ni a_i$ for every $1 \leq i \leq r$. Then what is left (if we want to determine π completely) is to take the elements from $\{1, \ldots, n\} \setminus \{a_1, \ldots, a_r\}$, and assign each of them to one of the blocks $V_1, \ldots, V_r$. The recipe for doing this is described as follows. Let b be an element of $\{1, \ldots, n\} \setminus \{a_1, \ldots, a_r\}$. The bth step of γ is a falling one, hence the two lattice points connected by it are of the form $(b-1, l)$ and $(b, l-1)$ for some $l \geq 1$. There have to exist values of i, $1 \leq i \leq r$, such that

$$a_i < b \text{ and } [l-1, l] \subset [h_i, h_i + \xi_i] \tag{9.9}$$

(with h_i, ξ_i as in (9.8)); this is clearly seen when one draws the path γ, and observes that the first $b-1$ steps of γ give a piecewise linear graph connecting $(0,0)$ with $(b-1, l)$. We assign b to the block V_i, where i is the *largest* value in $\{1, \ldots, r\}$ for which (9.9) holds.

[Concrete example: suppose that $n = 8$ and that $\gamma \in Luk(8)$ has rise-vector $(2, 1, -1, -1, 0, -1, 1, -1)$:

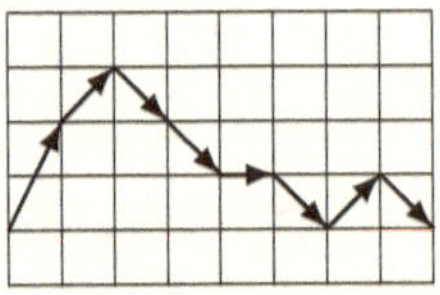

This path has four non-falling steps, the 1st, the 2nd, the 5th and the 7th step. Thus the partition $\pi = \Pi(\gamma)$ of $\{1, \ldots, 8\}$ must have four blocks $V_1, \ldots, V_4$, such that $V_1 \ni 1$, $V_2 \ni 2$, $V_3 \ni 5$, $V_4 \ni 7$. In order to complete the description of π, we have to consider the remaining elements $3, 4, 6, 8$ of $\{1, \ldots, 8\}$, and assign every one of them to one of the blocks $V_1, \ldots, V_4$. By using the recipe described above we get:

$$3 \in V_2, \ 4 \in V_1, \ 6 \in V_1, \ 8 \in V_4, \tag{9.10}$$

thus arriving to $\pi = \{\ \{1,4,6\}, \{2,3\}, \{5\}, \{7,8\}\ \}$. Referring to the picture of γ, one can describe the assignments in (9.10) as "projecting towards left." For instance, the 6th step of γ connects the lattice points $(5,1)$ and $(6,0)$; if one shoots an arrow (so to speak) horizontally towards left from this segment, then the arrow will land on the rising step which goes from $(0,0)$ to $(1,2)$ – this leads to the assignment "$6 \in V_1$".]

We leave it as an exercise to the reader to check that the construction described above always produces a partition π of $\{1,\ldots,n\}$ which is non-crossing, and where (referring to the notations in (9.8)) the blocks $V_1,\ldots,V_r$ of π satisfy:

$$\min(V_i) = a_i \qquad \text{and} \qquad |V_i| = 1+\xi_i, \;\; 1 \le i \le r. \tag{9.11}$$

Finally, it is easily verified that the map $\Pi : Luk(n) \to NC(n)$ obtained from the above construction is indeed an inverse for Λ. □

REMARK 9.9. The bijection $NC(n) \ni \pi \leftrightarrow \gamma \in Luk(n)$ found in Proposition 9.8 has good properties when one compares the block structure of π versus the "step-structure" of γ. Indeed, it is clear that when π and γ correspond to each other, the number of blocks of π is equal to the number of non-falling steps of γ. Even more precisely, for any given $k \in \{1,\ldots,n\}$, the number of blocks with k elements in π can be retrieved as the number of steps of the form $(1,k-1)$ in γ.

REMARK 9.10. An important benefit of the direct bijection between $NC(n)$ and $Luk(n)$ comes from the fact that Lukasiewicz paths can be nicely enumerated via a *"cyclic permutation"* trick (which is also known as *Raney's lemma*). The idea when performing this trick goes as follows. Take a Lukasiewicz path with n steps, add to it a falling step $(1,-1)$, and do a cyclic permutation of the total $n+1$ steps. This results in an almost-rising path with $n+1$ steps, going from $(0,0)$ to $(n+1,-1)$.

In order to clarify what is being done, let us look at a concrete example. Suppose that we start with the example of $\gamma \in Luk(8)$ which appeared in the proof of Proposition 9.8. We add to it a 9th falling step $(1,-1)$, and then we decide to read the 9 steps by starting with the 5th step, and by going cyclically. The result is the following almost-rising path with 9 steps:

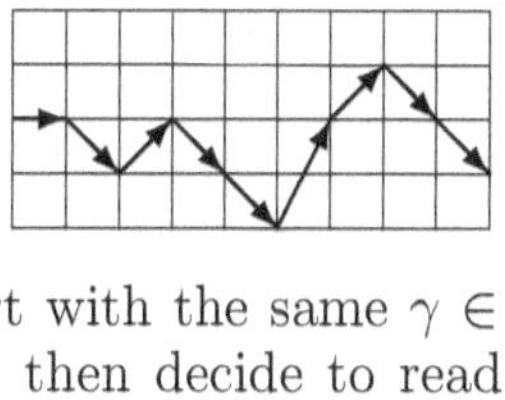

Or suppose that we start with the same $\gamma \in Luk(8)$, add to it a 9th falling step $(1,-1)$, and then decide to read the 9 steps by starting with the 9th step (the one we have just added). The result is the path in the following picture:

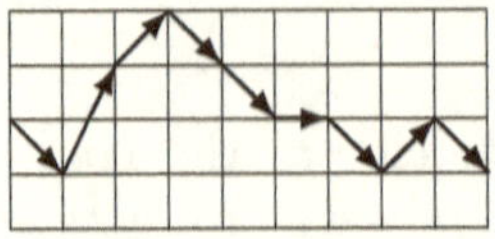

So to summarize, the input for performing the cyclic permutation trick consists of a path $\gamma \in Luk(n)$ and a number $m \in \{1, \ldots, n+1\}$ (the number m indicates where to start the cyclically permuted reading of the steps). The output is an almost-rising path with $n+1$ steps $\widetilde{\gamma}$, which goes from $(0,0)$ to $(n+1,-1)$. If the rise-vector of γ is $(\lambda_1, \ldots, \lambda_n)$, then $\widetilde{\gamma}$ can be defined formally by saying that its rise-vector $(\widetilde{\lambda}_1, \ldots, \widetilde{\lambda}_{n+1})$ is:

$$(\widetilde{\lambda}_1, \ldots, \widetilde{\lambda}_{n+1}) = (\lambda_m, \ldots, \lambda_n, -1, \lambda_1, \ldots, \lambda_{m-1}), \tag{9.12}$$

where the right-hand side of (9.12) is to be read as "$(\lambda_1, \ldots, \lambda_n, -1)$" if $m = 1$ and as "$(-1, \lambda_1, \ldots, \lambda_n)$" if $m = n+1$.

What makes the cyclic permutation trick useful is that it is bijective.

PROPOSITION 9.11. *Let n be a positive integer. The construction formalized by Equation (9.12) gives a bijection between $Luk(n) \times \{1, \ldots, n+1\}$ and the set of all almost-rising paths going from $(0,0)$ to $(n+1,-1)$.*

PROOF. We show that the map defined by the construction in (9.12) is one-to-one on $Luk(n) \times \{1, \ldots, n+1\}$. So suppose that $\widetilde{\gamma}$ is associated by this construction to $(\gamma, m) \in Luk(n) \times \{1, \ldots, n+1\}$; we want to prove that γ and m can be retrieved from $\widetilde{\gamma}$. Quite clearly, we will be done if we can find m; indeed, knowing m will tell us what component of the rise-vector $(\widetilde{\lambda}_1, \ldots, \widetilde{\lambda}_{n+1})$ of $\widetilde{\gamma}$ we have to delete, and then how to cyclically permute what is left, in order to obtain the rise-vector $(\lambda_1, \ldots, \lambda_n)$ of γ.

Now, here is the observation which tells us how to determine m from the knowledge of $\widetilde{\gamma}$. Look at the heights (i.e. second components) of the points in $\mathbb{Z}^2$ which are visited by $\widetilde{\gamma}$. Let h be the smallest such height, and suppose that the first time that $\widetilde{\gamma}$ visits a point of height h is after q steps, $1 \leq q \leq n+1$. (Note that $h \leq -1$, since $\widetilde{\gamma}$ ends at $(n+1,-1)$. This explains why we cannot have $q = 0$ – it is because $\widetilde{\gamma}$ starts at height $0 > h$.) So h and q are found by only looking at $\widetilde{\gamma}$; but on the other hand it is easily seen that they are related to the rise-vector $(\lambda_1, \ldots, \lambda_n)$ of γ and to m via the formulas:

$$h = \lambda_m + \cdots + \lambda_n - 1, \quad q = (n+2) - m.$$

In particular it follows that m can be retrieved from q by the formula

$$m = (n+2) - q \in \{1, \ldots, n+1\}; \tag{9.13}$$

this completes the proof of the injectivity of the map considered in the proposition.

For proving surjectivity, one essentially repeats the same arguments as above: find an $m \in \{1, \ldots, n+1\}$ by the formula (9.13), and then find a path $\gamma \in Luk(n)$, all by starting from $\widetilde{\gamma}$, only now $\widetilde{\gamma}$ is allowed to be an *arbitrary* almost-rising path going from $(0,0)$ to $(n+1,-1)$. We leave it as an exercise to the reader to adjust the arguments for how m and γ are obtained, and to verify the fact that (γ, m) is indeed mapped onto the given $\widetilde{\gamma}$. □

Now let us show some concrete applications of the Lukasiewicz paths to the enumeration of non-crossing partitions.

COROLLARY 9.12. *Let n be a positive integer, and let $r_1, \ldots, r_n \in \mathbb{N} \cup \{0\}$ be such that $r_1 + 2r_2 + \cdots + nr_n = n$. The number of partitions $\pi \in NC(n)$ which have r_1 blocks with 1 element, r_2 blocks with 2 elements, ..., r_n blocks with n elements is equal to*

$$\frac{n!}{r_1! r_2! \cdots r_n! \big(n+1-(r_1+r_2+\cdots+r_n)\big)!}. \tag{9.14}$$

PROOF. The bijection observed in the part (2) of Proposition 9.8 puts the given set of non-crossing partitions in one-to-one correspondence with the set of paths

$$\mathcal{L}_{r_1,\ldots,r_n} := \left\{ \gamma \in Luk(n) \;\middle|\; \begin{array}{l} \gamma \text{ has } r_k \text{ steps } (1, k-1), \text{ for } 1 \le k \le n, \\ \text{and } n - (r_1 + \cdots + r_n) \text{ steps } (1,-1) \end{array} \right\}.$$

The bijection from Proposition 9.11 puts $\mathcal{L}_{r_1,\ldots,r_n} \times \{1, \ldots, n+1\}$ into one-to-one correspondence with

$$\widetilde{\mathcal{L}}_{r_1,\ldots,r_n} := \left\{ \widetilde{\gamma} \;\middle|\; \begin{array}{l} \widetilde{\gamma} \text{ almost-rising path with } r_k \text{ steps} \\ \text{of the form } (1, k-1), \text{ for } 1 \le k \le n, \\ \text{and with } (n+1) - (r_1 + \cdots + r_n) \text{ steps of } (1,-1) \end{array} \right\}$$

So the number of non-crossing partitions counted in this corollary is equal to $(\#\widetilde{\mathcal{L}}_{r_1,\ldots,r_n})/(n+1)$. But on the other hand it is clear that $\#\widetilde{\mathcal{L}}_{r_1,\ldots,r_n}$ is equal to the multinomial coefficient

$$\frac{(n+1)!}{r_1! \cdots r_n! \Big((n+1) - (r_1 + \cdots + r_n)\Big)!},$$

and the result follows. □

By the same method, one can prove the following.

COROLLARY 9.13. *For $1 \le k \le n$ we have that*

$$\#\{\pi \in NC(n) \mid \pi \text{ has } k \text{ blocks}\} = \frac{1}{n}\binom{n}{k}\binom{n}{k-1}. \tag{9.15}$$

The proof of this corollary is left as an exercise to the reader (cf. Exercise 9.39 at the end of the lecture). Let us mention here that the numbers of the form appearing on the right-hand side of (9.15) are usually called **Narayana numbers**.

The lattice structure of $NC(n)$

$NC(n)$ is not just a collection of partitions, but is a quite structured set. Namely, $NC(n)$ is a *poset* (short for "partially ordered set"), where the partial order is defined as follows.

DEFINITIONS 9.14. (1) Let $\pi, \sigma \in NC(n)$ be two non-crossing partitions. We write $\pi \le \sigma$ if each block of π is completely contained in one of the blocks of σ (that is, if π can be obtained out of σ by refining the block structure). The partial order obtained in this way on $NC(n)$ is called the **reversed refinement order**.

(2) The maximal element of $NC(n)$ with respect to the reversed refinement order is the partition consisting of only one block and is denoted by 1_n. The partition consisting of n blocks, each of which has one element, is the minimal element of $NC(n)$ and is denoted by 0_n.

An important feature of the reversed refinement order on $NC(n)$ is that it makes $NC(n)$ into a *lattice*.

DEFINITIONS 9.15. Let P be a finite partially ordered set.

(1) Let π, σ be in P. If the set $U = \{\tau \in P \mid \tau \ge \pi \text{ and } \tau \ge \sigma\}$ is non-empty and has a minimum τ_0 (that is, an element $\tau_0 \in U$ which is smaller than all the other elements of U) then τ_0 is called the **join** of π and σ, and is denoted as $\pi \vee \sigma$.

(2) Let π, σ be in P. If the set $L = \{\rho \in P \mid \rho \le \pi \text{ and } \rho \le \sigma\}$ is non-empty and has a maximum ρ_0 (that is, an element $\rho_0 \in L$ which is larger than all the other elements of L) then ρ_0 is called the **meet** of π and σ, and is denoted as $\pi \wedge \sigma$.

(3) The poset P is said to be a **lattice** if every two elements $\pi, \sigma \in P$ have a join $\pi \vee \sigma$ and a meet $\pi \wedge \sigma$.

REMARKS 9.16. (1) Let P be a finite lattice. An immediate induction argument shows that every finite family of elements $\pi_1, \dots, \pi_k \in P$ have a join (i.e. the smallest common upper bound) $\pi_1 \vee \cdots \vee \pi_k$ and a meet (i.e. the largest common lower bound) $\pi_1 \wedge \cdots \wedge \pi_k$.

In particular, by taking $\pi_1, \ldots, \pi_k$ to be a list of all the elements of P we see that P must have a maximum element, usually denoted by 1_P, and a minimum element, usually denoted by 0_P. (Thus 0_P and 1_P are such that $0_P \leq \pi \leq 1_P$, $\forall\, \pi \in P$.)

(2) Let P be a finite poset. It is useful to note that if P has a maximum element 1_P and if every two elements $\pi, \sigma \in P$ have a meet, then P is a lattice. Indeed, if every two elements have a meet then it follows by induction on k that $\rho_1 \wedge \cdots \wedge \rho_k$ exists for every finite family $\rho_1, \ldots, \rho_k \in P$. Now let $\pi, \sigma \in P$ be arbitrary, and consider the set $U = \{\tau \in P \mid \tau \geq \pi,\ \tau \geq \sigma\}$. This set is non-empty (it contains e.g. the maximum element 1_P of P), so we can list it as $U = \{\rho_1, \ldots, \rho_k\}$ and we can consider the element $\rho_0 = \rho_1 \wedge \cdots \wedge \rho_k$. It is immediately verified that $\pi \vee \sigma$ exists, and is equal to ρ_0.

Of course, one could also dualize the above argument, and obtain that if P has a minimum element 0_P and if every two elements $\pi, \sigma \in P$ have a join, then P is a lattice.

PROPOSITION 9.17. *The partial order by reversed refinement induces a lattice structure on $NC(n)$.*

PROOF. In view of the preceding remark, and since $NC(n)$ has a maximum element 1_n, it will suffice to show that any two partitions $\pi, \sigma \in NC(n)$ have a meet $\pi \wedge \sigma$. And indeed, for $\pi = \{V_1, \ldots, V_r\}$ and $\sigma = \{W_1, \ldots, W_s\}$, it is immediate that the formula

$$\{V_i \cap W_j \mid 1 \leq i \leq r,\ 1 \leq j \leq s,\ V_i \cap W_j \neq \emptyset\} \tag{9.16}$$

defines a partition in $NC(n)$ which is smaller (in the reversed refinement order) than π and than σ, and is on the other hand the largest partition in $NC(n)$ having these properties. $\square$

EXAMPLE 9.18. An example for the partial order on $NC(8)$ is

$$\{(1,3),(2),(4,5),(6,8),(7)\} \leq \{(1,3,6,7,8),(2),(4,5)\},$$

or graphically

Examples for join and meet are

 and 

REMARK 9.19. The partial order by reversed refinement can also be considered on the set $\mathcal{P}(n)$ of all the partitions of $\{1, \ldots, n\}$, and turns $\mathcal{P}(n)$ into a lattice as well.

The formula describing the meet $\pi \wedge \sigma$ of two partitions $\pi, \sigma \in \mathcal{P}(n)$ is exactly the same as in Equation (9.16) in the proof of Proposition 9.17. As a consequence, it follows that for $\pi, \sigma \in NC(n)$ the partition "$\pi \wedge \sigma$" is the same, no matter whether the meet is considered in $NC(n)$ or in $\mathcal{P}(n)$.

The join $\pi \vee \sigma$ of two partitions $\pi, \sigma \in \mathcal{P}(n)$ can be described as follows: two elements $a, b \in \{1, \ldots, n\}$ belong to the same block of $\pi \vee \sigma$ if and only if there exist $k \geq 1$ and elements $a_0, a_1, \ldots, a_{2k} \in \{1, \ldots, n\}$ such that $a_0 = a$, $a_{2k} = b$, and we have

$$a_0 \sim_\pi a_1 \sim_\sigma a_2 \sim_\pi \cdots \sim_\pi a_{2k-1} \sim_\sigma a_{2k}. \tag{9.17}$$

(We can require the above sequence of equivalences to begin with $\sim_\pi$ and end with $\sim_\sigma$ due to the fact that we are allowing for example $a_0 = a_1$, or $a_{2k-1} = a_{2k}$.)

Unlike the situation with the meet, the join of two partitions $\pi, \sigma \in NC(n)$ may not be the same in $\mathcal{P}(n)$ as it is in $NC(n)$. For example in $NC(4)$ we have

$$\{\{1,3\},\{2\},\{4\}\} \vee \{\{1\},\{2,4\},\{3\}\} = \{\{1,2,3,4\}\} = 1_4;$$

if we calculated the same join in the lattice of all partitions $\mathcal{P}(4)$ the result would be the crossing partition $\{\{1,3\},\{2,4\}\}$. In what follows, our joins will be considered (unless specified otherwise) in $NC(n)$ rather than in $\mathcal{P}(n)$.

REMARK 9.20. The following picture shows the partitions in the lattice $NC(4)$, arranged according to the number of blocks. This arrangement corresponds to the poset structure, at least in the rough sense that 1_4 is at the top, 0_4 is at the bottom, and in general the larger partitions in $NC(4)$ "tend to occupy higher positions" in the picture. (See also Exercise 10.30 in Lecture 10.)

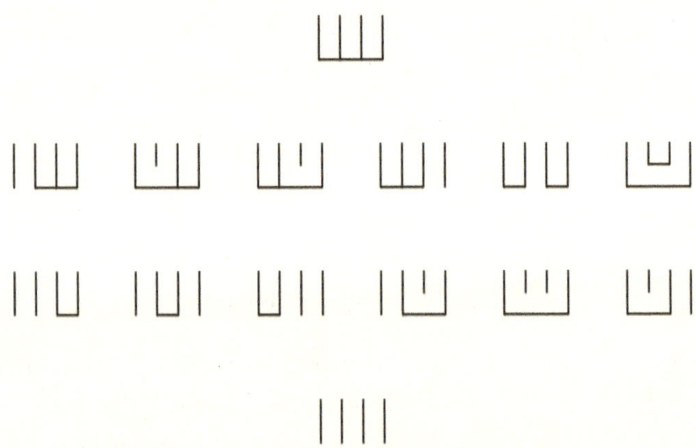

Note the up-down symmetry for the numbers of partitions sitting at each level in the above picture. This reflects a property of $NC(n)$ called *self-duality*. Moreover, it turns out that there exists an important anti-isomorphism $K : NC(n) \to NC(n)$, called the *Kreweras complementation map*, which implements this self-duality.

We should point out immediately that here the analogy with the lattice of all partitions breaks down. The lattice $\mathcal{P}(n)$ is not self-dual (as one can see, for example, by looking at the picture of $\mathcal{P}(4)$) and thus cannot have a complementation map. This combinatorial difference between all and non-crossing partitions will result in properties of free probability theory for which there is no classical analog (see, in particular, lecture 14 on products of freely independent random variables).

DEFINITION 9.21. The complementation map $K : NC(n) \to NC(n)$ is defined as follows. We consider additional numbers $\bar{1}, \dots, \bar{n}$ and interlace them with $1, \dots, n$ in the following alternating way:

$$1\,\bar{1}\,2\,\bar{2} \dots n\,\bar{n}\,.$$

Let π be a non-crossing partition of $\{1, \dots, n\}$. Then its **Kreweras complement** $K(\pi) \in NC(\bar{1}, \dots, \bar{n}) \cong NC(n)$ is defined to be the biggest element among those $\sigma \in NC(\bar{1}, \dots, \bar{n})$ which have the property that

$$\pi \cup \sigma \in NC(1, \bar{1}, \dots, n, \bar{n})\,.$$

EXAMPLE 9.22. Consider the partition

$$\pi := \{\{1,2,7\}, \{3\}, \{4,6\}, \{5\}, \{8\}\} \in NC(8).$$

For the complement $K(\pi)$ we get

$$K(\pi) = \{\{1\}, \{2,3,6\}, \{4,5\}, \{7,8\}\}\,,$$

as can be seen from the graphical representation:

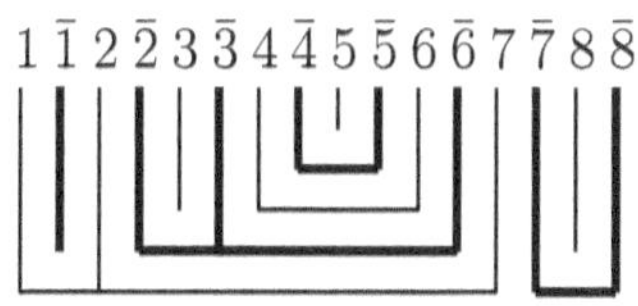

The following exercise contains some fundamental properties of the complementation map K, which follow directly from its definition.

EXERCISE 9.23. Let $K : NC(n) \to NC(n)$ be the Kreweras complementation map.

(1) Give a precise formulation and a proof for the following statement. "For every $\pi \in NC(n)$, the partition $K^2(\pi)$ is obtained by a

cyclic permutation of π." Observe in particular that $K^2(\pi)$ always has the same block structure as π.

(2) Observe that K^{2n} is the identity map of $NC(n)$. As a consequence show that K is a bijection. Describe the inverse K^{-1}.

(3) Show that K is a lattice anti-isomorphism, i.e. that $\pi \leq \sigma$ implies that $K(\sigma) \leq K(\pi)$. Note in particular that $K(0_n) = 1_n$ and $K(1_n) = 0_n$.

(4) Show that for any $\pi \in NC(n)$ we have

$$|\pi| + |K(\pi)| = n + 1. \tag{9.18}$$

REMARK 9.24. Once the Kreweras complementation map is introduced, one can look at the following enumeration problem, which is an analog of Corollary 9.12. Given $r_1, \ldots, r_n, q_1, \ldots, q_n \in \mathbb{N} \cup \{0\}$, how many partitions $\pi \in NC(n)$ are there such that π has r_i blocks with i elements and $K(\pi)$ has q_i blocks with i elements, for every $1 \leq i \leq n$? Of course, in order for any such π to exist, the numbers $r_1, \ldots, r_n, q_1, \ldots, q_n$ must fulfill the conditions

$$r_1 + 2r_2 + \cdots + nr_n = n = q_1 + 2q_2 + \cdots + nq_n$$

and (in view of the above relation (9.18))

$$r_1 + \cdots + r_n + q_1 + \cdots + q_n = n + 1.$$

If these conditions are satisfied, then it turns out that the enumeration problem stated above has a nice answer, the required number of partitions is equal to

$$n \cdot \frac{(r_1 + \cdots + r_n - 1)!(q_1 + \cdots + q_n - 1)!}{r_1! \cdots r_n! \cdot q_1! \cdots q_n!} \tag{9.19}$$

The factorization of intervals in NC

An important property of non-crossing partitions is that intervals in $NC(n)$ factorize into products of other $NC(k)$.

Let us first recall what we mean by an interval in a poset.

NOTATION 9.25. For a poset P and $\pi, \sigma \in P$ with $\pi \leq \sigma$ we denote by $[\pi, \sigma]$ the interval

$$[\pi, \sigma] := \{\tau \in P \mid \pi \leq \tau \leq \sigma\}.$$

REMARK 9.26. Clearly, $[\pi, \sigma]$ inherits the poset structure from P. Note moreover that if P is a lattice then for any $\pi \leq \tau_1, \tau_2 \leq \sigma$ we have $\pi \leq \tau_1 \vee \tau_2, \tau_1 \wedge \tau_2 \leq \sigma$ and thus $[\pi, \sigma]$ is itself a lattice.

Next we recall the notion of direct product for posets.

DEFINITION 9.27. Let $P_1, \dots, P_n$ be posets. The **direct product of the partial orders** on $P_1, \dots, P_n$ is the partial order on $P_1 \times \cdots \times P_n$ defined by

$$(\pi_1, \dots, \pi_n) \le (\sigma_1, \dots, \sigma_n) \iff \pi_i \le \sigma_i \qquad \forall\, i = 1, \dots, n.$$

REMARK 9.28. If all P_i are lattices then $P_1 \times \cdots \times P_n$ is also a lattice, and the meet and the join on P are given by the componentwise meet and join on P_i, respectively:

$$(\pi_1, \dots, \pi_n) \vee (\sigma_1, \dots, \sigma_n) = (\pi_1 \vee \sigma_1, \dots, \pi_n \vee \sigma_n),$$
$$(\pi_1, \dots, \pi_n) \wedge (\sigma_1, \dots, \sigma_n) = (\pi_1 \wedge \sigma_1, \dots, \pi_n \wedge \sigma_n).$$

Now we can state the factorization property for intervals in NC.

THEOREM 9.29. *For any $\pi, \sigma \in NC(n)$ with $\pi \le \sigma$ there exists a canonical sequence $(k_1, \dots, k_n)$ of non-negative integers such that we have the lattice-isomorphism*

$$[\pi, \sigma] \cong NC(1)^{k_1} \times NC(2)^{k_2} \times \cdots \times NC(n)^{k_n}. \tag{9.20}$$

PROOF. We clearly have

$$[\pi, \sigma] \cong \prod_{V \in \sigma} [\pi|_V, \sigma|_V].$$

But it is immediate that for every $V \in \sigma$, the order-preserving bijection from V onto $\{1, \dots, |V|\}$ will identify $[\pi|_V, \sigma|_V]$ to an interval of the form $[\tau, 1_{|V|}]$ for some $\tau \in NC(|V|)$. Thus it remains to perform the factorization for intervals of the form $[\tau, 1_k]$, $k \ge 1$.

For an interval of the form $[\tau, 1_k]$, we proceed as follows. By applying the complementation map K on $NC(k)$, we see that $[\tau, 1_k]$ is anti-isomorphic to $[K(1_k), K(\tau)] = [0_k, K(\tau)]$. But for the latter we know again that

$$[0_k, K(\tau)] \cong \prod_{W \in K(\tau)} [0_k|_W, K(\tau)|_W].$$

Since each $[0_k|_W, K(\tau)|_W]$ is just $NC(W)$ ($\cong NC(|W|)$), we thus obtain that $[\tau, 1_k]$ is anti-isomorphic to the product $\prod_{W \in K(\tau)} NC(|W|)$. Finally the latter product is anti-isomorphic to itself (by applying the product of complementation maps), and hence it gives the desired factorization of $[\tau, 1_k]$. $\square$

DEFINITION 9.30. The product decomposition observed in the above proof will be called the **canonical factorization** of $[\pi, \sigma]$.

EXAMPLE 9.31. To illustrate the above let us present the canonical factorization for $[\pi, \sigma] \subset NC(12)$, where

$$\pi = \{(1,9),(2,5),(3),(4),(6),(7,8),(10),(11),(12)\} \text{ and}$$
$$\sigma = \{(1,6,9,12),(2,4,5),(3),(7,8),(10,11)\}.$$

First we get

$$\begin{aligned}
[\pi, \sigma] &\cong [\{(1,9),(6),(12)\}, 1_{\{1,6,9,12\}}] \times [\{(2,5),(4)\}, 1_{\{2,4,5\}}] \\
&\quad \times [\{(3)\}, 1_{\{3\}}] \times [\{(7,8)\}, 1_{\{7,8\}}] \times [\{(10),(11)\}, 1_{\{10,11\}}] \\
&\cong [\{(1,3),(2),(4)\}, 1_4] \times [\{(1,3),(2)\}, 1_3] \\
&\qquad \times [1_1, 1_1] \times [1_2, 1_2] \times [0_2, 1_2].
\end{aligned}$$

By invoking the complement we get

$$\begin{aligned}
[\{(1,3),(2),(4)\}, 1_4] &\cong NC(2)^2 \\
[\{(1,3),(2)\}, 1_3] &\cong NC(1) \times NC(2) \\
[1_1, 1_1] &\cong NC(1) \\
[1_2, 1_2] &\cong NC(1)^2,
\end{aligned}$$

which yields finally that

$$[\pi, \sigma] \cong NC(1)^4 \times NC(2)^4.$$

The specifics of working with non-crossing partitions can be seen well in the above decomposition of $[\{(1,3),(2),(4)\}, 1_4]$. Since we are not allowed to start our chain from $\{(1,3),(2),(4)\}$ to 1_4 by putting together the blocks (2) and (4) (otherwise we would get a crossing partition), this decomposition is $NC(2)^2$, and not $NC(3)$, as one might expect on first glance (and as it is in the lattice of all partitions).

REMARK 9.32. In Theorem 9.29, the emphasis was on the fact that the factorization of the interval $[\pi, \sigma]$ is *canonical*, i.e. there is a precise recipe for how to obtain the exponents $k_1, \dots, k_n$ on the right-hand side of Equation (9.20). The canonical nature of this factorization will be discussed further in Lecture 18, in connection to the concept of relative Kreweras complement $K_\sigma(\pi)$ (cf. Lemma 18.6).

On the other hand, it is natural to ask: could there also exist some "non-canonical" factorization for the interval $[\pi, \sigma] \subset NC(n)$? In other words, is it the case that the exponents $k_1, \dots, k_n$ on the right-hand side of (9.20) are in fact uniquely determined? Phrased like this, the question clearly has a negative answer, as the exponent k_1 is not determined at all (which in turn happens because $\#NC(1) = 1$). But it is nevertheless interesting to observe that the other exponents $k_2, \dots, k_n$ *are* uniquely determined. In the remaining part of this section we will outline an argument which proves this via the enumeration of *multi-chains* in the lattices $NC(n)$.

NOTATION 9.33. Let P be a poset. For every $k \geq 1$ we denote

$$P^{(k)} := \{(\pi_1, \ldots, \pi_k) \in P^k \mid \pi_1 \leq \pi_2 \leq \cdots \leq \pi_k\}. \tag{9.21}$$

(So in particular $P^{(1)} = P$, while $P^{(2)}$ is essentially the set of all intervals of P.) The k-tuples in $P^{(k)}$ are called **multi-chains** of length $k-1$ in the poset P.

The enumeration of multi-chains in the lattices $NC(n)$ involves a generalization of the Catalan numbers, described as follows.

NOTATION 9.34. For every $n, k \geq 1$ we will denote

$$C_n^{(k)} := \frac{1}{nk+1} \binom{n(k+1)}{n}. \tag{9.22}$$

The numbers of the form $C_n^{(k)}$ are called **Fuss–Catalan numbers**. Note that in the particular case $k = 1$ we get

$$C_n^{(1)} = \frac{1}{n+1} \binom{2n}{n} = C_n,$$

the Catalan numbers which have repeatedly appeared in the preceding lectures.

PROPOSITION 9.35. *The multi-chains of length $k-1$ in $NC(n)$ are counted by the Fuss–Catalan number $C_n^{(k)}$:*

$$\#NC(n)^{(k)} = C_n^{(k)}, \qquad \forall\, n, k \geq 1. \tag{9.23}$$

Note that the case $k = 1$ of Equation (9.23) corresponds to the fact, proved earlier in this lecture, that $\#NC(n)$ is equal to the Catalan number C_n. We will not elaborate here on how that argument could be extended in order to prove the general case of Equation (9.23), but an alternative way of obtaining Proposition 9.35 will be outlined in Example 10.24 of the next lecture. (Let us mention here that the only place where we use Proposition 9.35 is in the proof of Proposition 9.38, which in turn is not used anywhere else in the remainder of the book.)

LEMMA 9.36. *Suppose that $r, s \geq 1$ and $m_1, \ldots, m_r, n_1, \ldots, n_s \geq 2$ are such that*

$$C_{m_1}^{(k)} \cdots C_{m_r}^{(k)} = C_{n_1}^{(k)} \cdots C_{n_s}^{(k)}, \tag{9.24}$$

for every $k \geq 1$. Then

$$\max(m_1, \ldots, m_r) = \max(n_1, \ldots, n_s).$$

PROOF. Assume by contradiction that the two maxima are not equal to each other, e.g. that

$$m := \max(m_1, \ldots, m_r) > \max(n_1, \ldots, n_s).$$

Note that $m \geq 3$ (since $m > n_1 \geq 2$). Let k be a positive integer such that $p := mk + (m-1)$ is a prime. (The existence of such a k is ensured by a well-known theorem of Dirichlet, which guarantees that if a and b are relatively prime, then the arithmetic progression $\{ak + b | k \geq 1\}$ contains infinitely many prime numbers.) For this choice of k and p observe that the Fuss–Catalan number $C_m^{(k)}$ is divisible by p. This is because one can write

$$C_m^{(k)} = \frac{(mk+2)\cdots(mk+(m-1))(mk+m)}{m!},$$

where p divides the numerator of the fraction but does not divide the denominator. In particular it follows that p divides the product appearing on the left-hand side of Equation (9.24).

On the other hand let us observe that (for the k and p found above) the prime number p does not divide any Fuss–Catalan number $C_n^{(k)}$ with $n < m$. This is because when we write

$$C_n^{(k)} = \frac{(nk+2)\cdots(nk+(n-1))(nk+n)}{n!}$$

we have that p is larger than any of the factors appearing in the numerator of the latter fraction (indeed, it is clear that $p = mk + (m-1) > nk + n$). But then p does not divide the right-hand side of Equation (9.24) – contradiction! □

LEMMA 9.37. *Suppose that $r, s \geq 1$ and $m_1, \ldots, m_r, n_1, \ldots, n_s \geq 2$ are such that (9.24) holds for every $k \geq 1$. Then $r = s$, and the s-tuple $n_1, \ldots, n_s$ is obtained by a permutation of $m_1, \ldots, m_r$.*

PROOF. This is immediate, by induction on $r + s$ and by using the preceding lemma. □

We finally arrive to the following proposition, which clearly implies that the exponents $k_2, \ldots, k_n$ in the canonical factorization of Theorem 9.29 are uniquely determined.

PROPOSITION 9.38. *Suppose that $r, s \geq 1$ and that $m_1, \ldots, m_r$, $n_1, \ldots, n_s \geq 2$ are such that*

$$NC(m_1) \times \cdots \times NC(m_r) \cong NC(n_1) \times \cdots \times NC(n_s). \qquad (9.25)$$

Then $r = s$, and the s-tuple $n_1, \ldots, n_s$ is obtained by a permutation of $m_1, \ldots, m_r$.

PROOF. It is immediate that if $P_1, \ldots, P_r$ are finite posets and if $P = P_1 \times \cdots \times P_r$, then we have canonical identifications

$$P^{(k)} \cong P_1^{(k)} \times \cdots \times P_r^{(k)}, \quad k \geq 1,$$

where the superscripts "(k)" are in reference to multi-chains of length $k-1$, as in Notation 9.33. By applying this to the specific case when $P = NC(m_1) \times \cdots \times NC(m_r)$ and by using Proposition 9.35, we thus see that the number of multi-chains of length $k-1$ in this lattice is equal to $C_{m_1}^{(k)} \cdots C_{m_r}^{(k)}$. If we do the same for the lattice $Q = NC(n_1) \times \cdots \times NC(n_s)$, and if we take into account that P and Q must have the same number of multi-chains of length $k-1$, we obtain that (9.24) holds for every $k \geq 1$. The assertion then follows from Lemma 9.37. □

Exercises

EXERCISE 9.39. Supply a proof of the Corollary 9.13.
[Hint: use Propositions 9.8 and 9.11, then do a direct enumeration for the set of almost-rising paths from $(0,0)$ to $(n+1,-1)$ which have exactly $n+1-k$ falling steps.]

NOTATION 9.40. Let n be a positive integer, and let S_n denote the group of all permutations of $\{1, \ldots, n\}$. For $\alpha \in S_n$ and $\pi = \{V_1, \ldots, V_r\} \in \mathcal{P}(n)$ one can form a new partition $\{\alpha(V_1), \ldots, \alpha(V_r)\} \in \mathcal{P}(n)$, which will be denoted as $\alpha \cdot \pi$.

EXERCISE 9.41. Let n be a positive integer.

(1) Let Φ be an automorphism of the poset $NC(n)$ (which means that $\Phi : NC(n) \to NC(n)$ is bijective, and has the property that $\pi \leq \sigma \Leftrightarrow \Phi(\pi) \leq \Phi(\sigma)$, for $\pi, \sigma \in NC(n)$). Prove that there exists $\alpha \in S_n$ such that $\Phi(\pi) = \alpha \cdot \pi$ for every $\pi \in NC(n)$.

(2) Prove that the group of automorphisms of the poset $NC(n)$ is isomorphic to the dihedral group with $2n$ elements.
[Hint: let $\gamma \in S_n$ be the cyclic permutation which has $\gamma(i) = i+1$ for $1 \leq i \leq n-1$, and $\gamma(n) = 1$. Let $\beta \in S_n$ be the order-reversing permutation which has $\beta(i) = n+1-i$ for $1 \leq i \leq n$. Prove that a permutation $\alpha \in S_n$ has the property that $\alpha \cdot \pi \in NC(n)$ for every $\pi \in NC(n)$ if and only if α belongs to the subgroup of S_n generated by β and γ.]

EXERCISE 9.42. Let n be a positive integer, and let $NCE(2n)$ be the set of partitions $\pi \in NC(2n)$ with the property that every block of π has even cardinality. Let K_{2n} denote the Kreweras complementation map on $NC(2n)$.

(1) Prove that

$$K_{2n}(NCE(2n)) = \left\{ \pi \in NC(2n) \;\middle|\; \begin{array}{l} \text{every block of } \pi \text{ is contained} \\ \text{either in } \{1,3,\ldots,2n-1\} \\ \text{or in } \{2,4,\ldots,2n\} \end{array} \right\}.$$

(2) Recall that $NC_2(2n)$ denotes the set of non-crossing pair-partitions of $\{1,\ldots,2n\}$. Prove that the map

$$NC_2(2n) \ni \pi \mapsto (K_{2n}(\pi))|\{1,3,\ldots,2n-1\}$$

is a bijection between $NC_2(2n)$ and $NC(\{1,3,\ldots,2n-1\}) \cong NC(n)$. [Hint: $NC_2(2n)$ is the set of minimal elements of $NCE(2n)$, with respect to the partial order induced from $NC(2n)$. Hence $K_{2n}(NC_2(2n))$ must be the set of maximal elements of $K_{2n}(NCE(2n))$.]

EXERCISE 9.43. Let $\sigma \in NC(n)$ be an interval partition, which means that all its blocks consist of consecutive numbers, i.e. all $V \in \sigma$ are of the form $V = \{p, p+1, p+2, \ldots, p+r\}$ for some $1 \le p \le n$ and $r \ge 0$ such that $p+r \le n$. Prove that in such a case the join of σ with any element in $NC(n)$ is the same in $\mathcal{P}(n)$ and in $NC(n)$: we have for any $\pi \in NC(n)$ that $\sigma \vee_{NC(n)} \pi = \sigma \vee_{\mathcal{P}(n)} \pi$.

LECTURE 10

Basic combinatorics II: Möbius inversion

Motivated by our combinatorial description of the free central limit theorem, in the following we will use the non-crossing partitions to write moments $\varphi(a_1 \cdots a_n)$ of non-commutative random variables in the form

$$\varphi(a_1 \cdots a_n) = \sum_{\pi \in NC(n)} \kappa_\pi[a_1, \ldots, a_n],$$

where the κ_π are some new quantities called free cumulants. Of course, we should be able to invert this equation in order to define the free cumulants in terms of the moments. This is a special case of the general theory of Möbius inversion and Möbius function – a unifying concept in modern combinatorics which provides a common frame for a variety of situations.

We will use the framework of a finite poset P. Suppose we are given two functions $f, g : P \to \mathbb{C}$ which are connected as follows:

$$f(\pi) = \sum_{\substack{\sigma \in P \\ \sigma \leq \pi}} g(\sigma), \quad \pi \in P.$$

This is a quite common situation and it is often useful to invert the above relation, i.e. to write an equation which expresses the values of g in terms of those of f. This is indeed possible, by using a two-variable function μ (which depends only on the poset P, but not on the particular choice of the functions f and g). The function μ is called the *Möbius function* of the poset P, and the formula retrieving the values of g from those of f is called the *Möbius inversion formula.* In order to present it, we will introduce a suitable concept of convolution in the poset framework.

Convolution in the framework of a poset

DEFINITION 10.1. Let P be a finite poset, and let us denote (same as in Lecture 9, Notation 9.33)

$$P^{(2)} := \{(\pi, \sigma) \mid \pi, \sigma \in P, \ \pi \leq \sigma\}. \tag{10.1}$$

For $F, G : P^{(2)} \to \mathbb{C}$, their **convolution** $F * G$ is the function from $P^{(2)}$ to $\mathbb{C}$ defined by:

$$(F * G)(\pi, \sigma) := \sum_{\substack{\rho \in P \\ \pi \leq \rho \leq \sigma}} F(\pi, \rho) G(\rho, \sigma). \tag{10.2}$$

Moreover, we will also consider convolutions $f * G$ for $f : P \to \mathbb{C}$ and $G : P^{(2)} \to \mathbb{C}$; the function $f * G$ is defined from P to $\mathbb{C}$, and is described by the formula

$$(f * G)(\sigma) := \sum_{\substack{\rho \in P \\ \rho \leq \sigma}} f(\rho) G(\rho, \sigma). \tag{10.3}$$

REMARK 10.2. On occasion it is useful to keep in mind that the convolution operations defined above can be regarded as matrix multiplications. In order to regard them in this way, let us denote $\#P =: n$ and let us consider a way of listing

$$P = \{\pi_1, \ldots, \pi_n\} \tag{10.4}$$

which has the following property: for every $1 \leq i < j \leq n$, either the elements π_i and π_j are incomparable or they are such that $\pi_i < \pi_j$. (That is, the listing in (10.4) is made such that it never happens that $i < j$ and $\pi_i > \pi_j$. Such listings can always be found – cf. Exercise 10.25 at the end of the lecture.) Then to every function $F : P^{(2)} \to \mathbb{C}$ let us associate an upper triangular $n \times n$ matrix $T_F = (t_{ij})_{i,j=1}^n$, where

$$t_{ij} = \begin{cases} 0 & \text{if } i > j \\ F(\pi_i, \pi_i) & \text{if } i = j \\ F(\pi_i, \pi_j) & \text{if } i < j \text{ and } \pi_i < \pi_j \\ 0 & \text{if } i < j \text{ and } \pi_i, \pi_j \text{ not comparable.} \end{cases}$$

It is immediately verified that the convolution of two functions $F, G : P^{(2)} \to \mathbb{C}$ amounts to the multiplication of the corresponding upper triangular matrices:

$$T_{F*G} = T_F T_G. \tag{10.5}$$

Moreover, by using the same listing (10.4) let us also associate to every function $f : P \to \mathbb{C}$ a $1 \times n$ matrix (or row-vector) v_f which has the ith component equal to $f(\pi_i)$, for $1 \leq i \leq n$. Then the second kind of convolution introduced in Definition 10.1 immediately reduces to matrix multiplication via the formula

$$v_{f*G} = v_f T_G, \tag{10.6}$$

holding for any $f : P \to \mathbb{C}$ and any $G : P^{(2)} \to \mathbb{C}$.

REMARK 10.3. The convolution operations defined above have the natural properties one would expect: they are associative and they are distributive with respect to the operation of taking linear combinations of functions on $P^{(2)}$ or on P. The associativity property, for instance, amounts to the fact that we have

$$(F * G) * H = F * (G * H), \quad (f * G) * H = f * (G * H)$$

for $F, G, H : P^{(2)} \to \mathbb{C}$ and $f : P \to \mathbb{C}$. All these properties can be easily verified directly from the definitions, or by using the interpretation via matrix multiplication described in Remark 10.2.

Another immediate observation is that the function $\delta : P^{(2)} \to \mathbb{C}$ defined by

$$\delta(\pi, \sigma) = \begin{cases} 1 & \text{if} \quad \pi = \sigma \\ 0 & \text{if} \quad \pi < \sigma \end{cases} \tag{10.7}$$

is the unit for our convolution operations. This is because, in the interpretation of Remark 10.2, the corresponding upper triangular matrix T_δ is precisely the unit $n \times n$ matrix.

Moving one step further, let us also record the situation with inverses under convolution.

PROPOSITION 10.4. *Let P be a finite poset, and consider the convolution operation for functions on $P^{(2)}$ as in Definition 10.1. A function $F : P^{(2)} \to \mathbb{C}$ is invertible with respect to convolution if and only if it has $F(\pi, \pi) \neq 0$ for every $\pi \in P$.*

PROOF. "$\Rightarrow$" If F has an inverse under convolution G, then for every $\pi \in P$ we have:

$$1 = \delta(\pi, \pi) = (F * G)(\pi, \pi) = F(\pi, \pi)G(\pi, \pi),$$

and this implies that $F(\pi, \pi) \neq 0$.

"$\Leftarrow$" Let us consider a listing of P as in Remark 10.2, and the relation between convolution and matrix multiplication described there. Let us denote

$$\mathcal{T} := \{T \in M_n(\mathbb{C}) \mid \text{there exists } F : P^{(2)} \to \mathbb{C} \text{ such that } T = T_F\}.$$

From the observations made in Remark 10.2 it is clear that $\mathcal{T}$ is closed under addition, multiplication, and scalar multiplication. It is also immediate that $\mathcal{T}$ contains all the $n \times n$ diagonal matrices. Based on these facts, it is easy to verify that if $T \in \mathcal{T}$ is invertible, then we must have $T^{-1} \in \mathcal{T}$ as well (cf. Exercise 10.26 at the end of the lecture).

But then let $F : P^{(2)} \to \mathbb{C}$ be such that $F(\pi, \pi) \neq 0$ for every $\pi \in P$. Then the matrix $T_F \in \mathcal{T}$ is invertible (because it is upper triangular with non-zero diagonal entries). As observed above, it follows that

$T_F^{-1} \in \mathcal{T}$, i.e. there exists a function $G : P^{(2)} \to \mathbb{C}$ such that $T_F^{-1} = T_G$. For this G we have that $T_{F*G} = T_F T_G = T_\delta$, which implies that $F * G = \delta$ (and the relation $G * F = \delta$ is obtained in exactly the same way). □

It is now easy to formalize the idea about Möbius inversion which was discussed in the introduction to this lecture.

DEFINITION 10.5. Let P be a finite poset. The **zeta function** of P is $\zeta : P^{(2)} \to \mathbb{C}$ defined by

$$\zeta(\pi, \sigma) = 1, \ \ \forall\ (\pi, \sigma) \in P^{(2)}.$$

The inverse of ζ under convolution is called the **Möbius function** of P, and is denoted by μ.

Note that the definition of the Möbius function as an inverse makes sense due to Proposition 10.4 and to the fact that $\zeta(\pi, \pi) = 1 \neq 0$, for all $\pi \in P$.

If there is a possibility of ambiguity on which poset we are referring to, we will write ζ_P and respectively μ_P instead of just ζ and μ.

PROPOSITION 10.6. *Let P be a finite poset, and let μ be the Möbius function of P. For two functions $f, g : P \to \mathbb{C}$ the statement that*

$$f(\pi) = \sum_{\substack{\sigma \in P \\ \sigma \leq \pi}} g(\sigma) \qquad \forall \pi \in P \tag{10.8}$$

is equivalent to

$$g(\pi) = \sum_{\substack{\sigma \in P \\ \sigma \leq \pi}} f(\sigma)\mu(\sigma, \pi) \qquad \forall \pi \in P. \tag{10.9}$$

PROOF. Equation (10.8) amounts to saying that $f = g * \zeta$, while (10.9) amounts to $g = f * \mu$. But it is clear that the latter two facts are equivalent to each other (by the associativity of convolution, and since ζ and μ are inverse to each other). □

REMARK 10.7. Let P be a finite poset, and let μ be the Möbius function of P. We have that μ is uniquely determined by the relation $\mu * \zeta = \delta$, which amounts to the following system of equations in the values of μ:

$$\sum_{\substack{\tau \in P \\ \pi \leq \tau \leq \sigma}} \mu(\pi, \tau) = \begin{cases} 1 & \text{if } \pi = \sigma \\ 0 & \text{if } \pi < \sigma. \end{cases} \tag{10.10}$$

Equivalently, μ is determined by the relation $\zeta * \mu = \delta$, which gives the system of equations

$$\sum_{\substack{\tau \in P \\ \pi \leq \tau \leq \sigma}} \mu(\tau, \sigma) = \begin{cases} 1 & \text{if } \pi = \sigma \\ 0 & \text{if } \pi < \sigma. \end{cases} \tag{10.11}$$

When computing μ in specific examples, the systems of equations in (10.10) and/or (10.11) can be used to determine the values $\mu(\pi, \sigma)$ by induction on the length of the interval $[\pi, \sigma]$. Let us briefly state here the poset terminology needed in order to continue the discussion.

DEFINITIONS 10.8. Let P be a finite poset.

(1) For $\pi < \sigma$ in P, the **length** of the interval $[\pi, \sigma]$ is the largest integer l having the following property: one can find elements $\rho_0, \rho_1, \ldots, \rho_l \in P$ such that $\pi = \rho_0 < \rho_1 < \cdots < \rho_l = \sigma$. (By convention, if $\pi = \sigma$, then the length of $[\pi, \sigma]$ is taken to be 0.)

(2) In the particular case when the length of the interval $[\pi, \sigma]$ is equal to 1, we will say that σ **covers** π. This is clearly equivalent to the fact that $\pi < \sigma$ and there is no element $\rho \in P$ such that $\pi < \rho < \sigma$.

REMARK 10.9. Let us return to the framework of Remark 10.7. Let us pick one of the systems of equations (those in (10.10), say) which appeared there, and let us explain how they can be used to compute inductively the values of μ. We can think of these equations like this: first we have $\mu(\pi, \pi) = 1$ for every $\pi \in P$, after which we have $\mu(\pi, \sigma) = -1$ whenever σ covers π. (Indeed, if σ covers π then the corresponding Equation (10.10) becomes $\mu(\pi, \pi) + \mu(\pi, \sigma) = 0$, giving $\mu(\pi, \sigma) = -\mu(\pi, \pi) = -1$.) The determination of the values of μ can continue in this manner, by induction on the length of the interval $[\pi, \sigma]$: if $[\pi, \sigma]$ has length l and if we assume the values of μ to be known for all the intervals of length smaller than l, then Equation (10.10) can be rewritten as

$$\mu(\pi, \sigma) = - \sum_{\substack{\tau \in P \\ \pi \leq \tau < \sigma}} \mu(\pi, \tau); \tag{10.12}$$

this will determine the value of $\mu(\pi, \sigma)$, since all the intervals appearing on the right-hand side of (10.12) have length strictly smaller than l.

EXAMPLE 10.10. Let us compute the Möbius function for the lattice $NC(3)$ of non-crossing partitions of $\{1, 2, 3\}$. One can write explicitly $NC(3) = \{0_3, \tau_1, \tau_2, \tau_3, 1_3\}$, where 0_3 and 1_3 are the minimal and maximal partitions of $\{1, 2, 3\}$ and where $\tau_1 = \{\ \{1\}, \{2, 3\}\ \}$, $\tau_2 = \{\ \{1, 3\}, \{2\}\ \}$, $\tau_3 = \{\ \{1, 2\}, \{3\}\ \}$. We have that every τ_i covers

0_3 and that 1_3 covers every τ_i, hence

$$\mu(0_3, \tau_i) = \mu(\tau_i, 1_3) = -1, \qquad 1 \le i \le 3.$$

Besides this, the only value of μ which is left to be computed is $\mu(0_3, 1_3)$. We obtain this value by using Equation (10.12):

$$\begin{aligned}\mu(0_3, 1_3) &= -\Big(\mu(0_3, 0_3) + \mu(0_3, \tau_1) + \mu(0_3, \tau_2) + \mu(0_3, \tau_3)\Big) \\ &= -(1 - 1 - 1 - 1) \\ &= 2.\end{aligned}$$

Möbius inversion in a lattice

The following proposition shows that, in the context of a lattice, one can also write down "partial versions" of the Möbius inversion formula. This will be of prominent importance in the next lecture where it will yield directly an important property of free cumulants.

PROPOSITION 10.11. *Let P be a finite lattice and let μ be the Möbius function of P. Consider two functions $f, g : P \to \mathbb{C}$ which are related by*

$$f(\tau) = \sum_{\substack{\pi \in P \\ \pi \le \tau}} g(\pi) \qquad \forall \tau \in P.$$

Then, for all $\omega, \tau \in P$ with $\omega \le \tau$, we have the relation:

$$\sum_{\substack{\sigma \in P \\ \omega \le \sigma \le \tau}} f(\sigma)\mu(\sigma, \tau) = \sum_{\substack{\pi \in P \\ \pi \vee \omega = \tau}} g(\pi). \tag{10.13}$$

PROOF. We have

$$\sum_{\substack{\sigma \in P \\ \omega \le \sigma \le \tau}} f(\sigma)\mu(\sigma, \tau) = \sum_{\substack{\sigma \in P \\ \omega \le \sigma \le \tau}} \sum_{\substack{\pi \in P \\ \pi \le \sigma}} g(\pi)\mu(\sigma, \tau) = \sum_{\substack{\pi \in P \\ \pi \le \tau}} \sum_{\substack{\sigma \in P \\ \pi \vee \omega \le \sigma \le \tau}} \mu(\sigma, \tau) g(\pi).$$

Consider now $\pi \in P$ with $\pi \le \tau$. Since also $\omega \le \tau$, we have $\pi \vee \omega \le \tau$. We will distinguish the two cases either $\pi \vee \omega = \tau$ or $\pi \vee \omega < \tau$. In the first case, the corresponding sum over σ reduces to one term for $\sigma = \pi \vee \omega = \tau$ and gives the contribution $\mu(\sigma, \tau)g(\pi) = \mu(\tau, \tau)g(\pi) = g(\pi)$. In the second case, the corresponding sum over σ vanishes because

$$\sum_{\substack{\sigma \in P \\ \pi \vee \omega \le \sigma \le \tau}} \mu(\sigma, \tau) = 0 \qquad \text{if } \pi \vee \omega < \tau$$

by the recursion formula (10.11) for the Möbius function. Thus our above calculation leads finally to

$$\sum_{\substack{\sigma \in P \\ \omega \le \sigma \le \tau}} f(\sigma)\mu(\sigma, \tau) = \sum_{\substack{\pi \in P \\ \pi \le \tau,\, \pi \vee \omega = \tau}} g(\pi).$$

Since the requirement $\pi \vee \omega = \tau$ includes $\pi \leq \tau$, this is exactly the assertion. $\square$

REMARK 10.12. Note that in the case when $\omega = 0_P$ (the minimal element of P) the formula (10.13) reduces to the Möbius inversion formula of Proposition 10.6,

$$\sum_{\substack{\sigma \in P \\ \sigma \leq \tau}} f(\sigma)\mu(\sigma, \tau) = g(\tau)$$

(because $\pi \vee 0_P = \pi$ for all $\pi \in P$). On the other hand, if in Proposition 10.11 we consider the case when $\omega = \tau$, then we just get

$$f(\tau) = \sum_{\substack{\pi \in P \\ \pi \leq \tau}} g(\pi)$$

(because $\pi \vee \tau = \tau$ is equivalent to $\pi \leq \tau$). In general, the formula (10.13) can be viewed as a kind of a partial Möbius inversion, standing "in between" the two equations which relate f and g in Proposition 10.6.

We now present an immediate consequence of Proposition 10.11, which is helpful in concrete computations of Möbius functions (in particular it will help us find the Möbius function of $NC(n)$ in the next section). One can think of the next corollary as a version of Equation (10.12) from Remark 10.9, where the following improvement is made: if in Corollary 10.13 the element $\omega \in P$ is picked to be "close to 0_P," then the summation in (10.14) will not have too many terms, and will give us a shot at a tractable formula for $\mu(0_P, 1_P)$.

COROLLARY 10.13. *Let P be a finite lattice and let μ be the Möbius function of P. Then, for every $\omega \neq 0_P$ we have that*

$$\sum_{\substack{\pi \in P \\ \pi \vee \omega = 1_P}} \mu(0_P, \pi) = 0. \tag{10.14}$$

PROOF. Consider the function $g : P \to \mathbb{C}$ defined by

$$g(\pi) := \mu(0_P, \pi), \quad \pi \in P.$$

Let us look at the function $f := g * \zeta$ obtained from g by partial summations: for every $\tau \in P$ we have

$$\begin{aligned} f(\sigma) = \sum_{\substack{\pi \in P \\ \pi \leq \sigma}} g(\pi) &= \sum_{\substack{\pi \in P \\ 0_P \leq \pi \leq \sigma}} \mu(0_P, \pi)\zeta(\pi, \sigma) \\ &= (\mu * \zeta)(0_P, \sigma) = \begin{cases} 1 & \text{if } \sigma = 0_P \\ 0 & \text{otherwise.} \end{cases} \end{aligned}$$

We apply the preceding proposition to these functions f and g, where in Equation (10.13) we pick the element ω given in the statement of the corollary and we make $\tau = 1_P$. We obtain:

$$\sum_{\substack{\sigma \in P \\ \omega \le \sigma \le 1_P}} f(\sigma)\mu(\sigma, 1_P) = \sum_{\substack{\pi \in P \\ \pi \vee \omega = 1_P}} g(\pi). \qquad (10.15)$$

The left-hand side of Equation (10.15) is equal to 0, because we have $f(\sigma) = 0$ for all the elements σ involved in that summation. Thus the right-hand side of (10.15) must vanish as well. □

The Möbius function of NC

The example which gave the name to the Möbius inversion is, naturally, due to Möbius and occurred in number theory (where P is the set of positive integers equipped with the partial order given by divisibility). For the present lectures, our main interest is however in the Möbius function of the poset $P = NC(n)$ of non-crossing partitions of $\{1, \ldots, n\}$.

Computation of the Möbius functions of the $NC(n)$ is largely simplified by the canonical factorization of intervals observed in Lecture 9, combined with the following general fact.

PROPOSITION 10.14. (1) *Let P and Q be finite posets, and suppose that $\Phi : P \to Q$ is a poset isomorphism. Then $\mu_Q(\Phi(\pi), \Phi(\sigma)) = \mu_P(\pi, \sigma)$ for every $\pi, \sigma \in P$ such that $\pi \le \sigma$ (and where μ_P and μ_Q are the Möbius functions of the lattices P and Q, respectively).*

(2) *Let $P_1, P_2, \ldots, P_k$ be finite posets, and consider their direct product $P = P_1 \times P_2 \times \cdots \times P_k$ (with partial order as described in Definition 9.27). Then for $\pi_1 \le \sigma_1$ in $P_1, \ldots, \pi_k \le \sigma_k$ in P_k we have*

$$\mu_P((\pi_1, \ldots, \pi_k), (\sigma_1, \ldots, \sigma_k)) = \mu_{P_1}(\pi_1, \sigma_1) \cdots \mu_{P_k}(\pi_k, \sigma_k). \qquad (10.16)$$

The statements in Proposition 10.14 have routine verifications, which are left to the reader (cf. Exercise 10.27 at the end of the lecture).

Now let us turn to the lattices of non-crossing partitions $NC(n)$. For every $n \ge 1$ we will denote the Möbius function of $NC(n)$ by μ_n. Moreover, let us use the notation

$$s_n := \mu_n(0_n, 1_n), \quad n \ge 1, \qquad (10.17)$$

where 0_n and 1_n are the minimum and maximum elements of $NC(n)$, respectively. Due to the factorization result observed in the last section of Lecture 9, the values of the Möbius functions of the lattices $NC(n)$ will be completely (and explicitly) determined as soon as we figure out what are the numbers s_n. Indeed, let $\pi \leq \sigma$ be in $NC(n)$ and suppose that the interval $[\pi, \sigma]$ has the canonical factorization

$$[\pi, \sigma] \cong NC(1)^{k_1} \times NC(2)^{k_2} \times \cdots \times NC(n)^{k_n};$$

then Proposition 10.14 clearly gives us that

$$\mu_n(\pi, \sigma) = s_1^{k_1} s_2^{k_2} \cdots s_n^{k_n}. \tag{10.18}$$

The first few of the numbers s_n are $s_1 = 1$ (since in $NC(1)$ we have $0_1 = 1_1$), $s_2 = -1$ (because in $NC(2)$ we have that 1_2 covers 0_2 – cf. Remark 10.9), and $s_3 = 2$ (see Example 10.10). The general formula for the s_n turns out again to be related to the Catalan numbers.

PROPOSITION 10.15. *For every $n \geq 1$, $\mu_n(0_n, 1_n)$ is a signed Catalan number,*

$$\mu_n(0_n, 1_n) = (-1)^{n-1} C_{n-1}. \tag{10.19}$$

PROOF. We will use the short-hand notation $\mu_n(0_n, 1_n) =: s_n$ introduced in Equation (10.17).

Let us fix for the moment an $n \geq 4$. We will invoke Corollary 10.13 in the situation when the poset P (appearing in the corollary) is $P = NC(n)$, and when the special element $\omega \in P$ is the non-crossing partition

$$\omega := \{\{1\}, \{2\}, \ldots, \{n-2\}, \{n-1, n\}\}.$$

In order to apply Corollary 10.13 to the above P and ω more efficiently, let us observe that the set $\{\pi \in NC(n) \mid \pi \vee \omega = 1_n\}$ which appears in the corollary can be listed explicitly: it is $\{1_n\} \cup \{\pi_1, \ldots, \pi_{n-1}\}$, where $\pi_1 := \{\{1, \ldots, n-1\}, \{n\}\}$ and where for every $2 \leq i \leq n-1$ we set $\pi_i := \{\{i, \ldots, n-1\}, \{1, \ldots, i-1, n\}\}$. Indeed, let $\pi \in NC(n)$ be such that $\pi \vee \omega = 1_n$. Then π cannot have a block which is completely contained in $\{1, \ldots, n-2\}$. (To see why, suppose such a block exists, and denote its minimal and maximal elements by a and by b, respectively. Denote $\{a, \ldots, b\} =: V$ and $\{1, \ldots, n\} \setminus V =: W$. Then the partition with two blocks $\rho = \{V, W\} \in NC(n)$ is a common upper bound for π and for ω, contradicting the fact that $\pi \vee \omega = 1_n$.) So every block of π either contains $n-1$ or it contains n. If $n-1$ and n lie in the same block of π then it follows that $\pi = 1_n$, while in the opposite case we get that π has exactly two blocks. In the latter case we denote by i the minimal element of the block of π which contains $n-1$, and an immediate non-crossing argument proves that we must have $\pi = \pi_i$ (where π_i was defined above).

Now let us apply Corollary 10.13 to the situation at hand. We obtain that

$$\mu_n(0_n, 1_n) + \sum_{i=1}^{n-1} \mu_n(0_n, \pi_i) = 0.$$

For each $1 \leq i \leq n-1$ it is immediate that the canonical factorization of the interval $[0_n, \pi_i]$ is

$$[0_n, \pi_i] \cong NC(i) \times NC(n-i),$$

hence the formula (10.18) gives us that $\mu_n(0_n, \pi_i) = s_i s_{n-i}$. The equation obtained by applying Corollary 10.13 is thus:

$$s_n + \sum_{i=1}^{n-1} s_i s_{n-i} = 0. \tag{10.20}$$

The above equation was derived for $n \geq 4$, but it is immediate (by plugging in the values of s_1, s_2, s_3 observed preceding this proposition) that it actually holds for $n \geq 2$.

If we now set $c_n := (-1)^n s_{n+1}$, for $n \geq 0$, and we rewrite Equation (10.20) in terms of the c_k, we get

$$c_{n-1} - \sum_{i=1}^{n-1} c_{i-1} c_{n-i-1} = 0, \quad n \geq 2.$$

This means that we have encountered again the recurrence relation

$$c_n = \sum_{i=1}^{n} c_{i-1} c_{n-i}, \quad n \geq 1,$$

which determines the sequence of Catalan numbers (cf. Lecture 2, Remark 2.12). Thus c_n is equal to the nth Catalan number C_n, and the result follows. □

Multiplicative functions on NC

The discussion of the preceding section suggests that the Möbius functions of the lattices $NC(n)$ should be looked at together (for all values of n at the same time). Moreover Equation (10.18) shows that the Möbius functions on the $NC(n)$ form a multiplicative family, in the sense of the following definition.

DEFINITION 10.16. Let $(\alpha_n)_{n\geq 1}$ be a sequence of complex numbers. Define a family of functions $F_n : NC(n)^{(2)} \to \mathbb{C}$, $n \geq 1$, by the following formula: if $\pi \leq \sigma$ in $NC(n)$ and if the canonical factorization of the interval $[\pi, \sigma]$ is

$$[\pi, \sigma] \cong NC(1)^{k_1} \times NC(2)^{k_2} \times \cdots \times NC(n)^{k_n},$$

then we have

$$F_n(\pi, \sigma) := \alpha_1^{k_1} \alpha_2^{k_2} \cdots \alpha_n^{k_n}. \tag{10.21}$$

Then $(F_n)_{n\geq 1}$ is called the **multiplicative family of functions on $\boldsymbol{NC^{(2)}}$** determined by the sequence $(\alpha_n)_{n\geq 1}$.

In general, a family of functions $(F_n : NC(n)^{(2)} \to \mathbb{C})_{n\geq 1}$ will be said to be **multiplicative** if it arises from some sequence of α_n in the way described above. (The α_n in this case will be uniquely determined, $\alpha_n = F_n(0_n, 1_n)$ for $n \geq 1$.)

REMARK 10.17. The algorithm for determining the canonical factorization of an interval $[\pi, \sigma] \subset NC(n)$ is fairly straightforward, and, as a consequence, so is the algorithm for computing a specific value for a multiplicative family on $NC(n)^{(2)}$. See for instance the concrete Example 9.31; for the interval $[\pi, \sigma] \subset NC(12)$ considered there, Equation (10.21) simply reduces to

$$F_{12}(\pi, \sigma) = \alpha_1^4 \alpha_2^4.$$

A way to capture what is going on here is to describe explicitly the following properties of a multiplicative family $(F_n)_{n=1}^{\infty}$.

(i) Let $\pi \leq \sigma$ be in $NC(n)$, where $\sigma = \{V_1, \dots, V_r\}$. For every $1 \leq k \leq r$ consider the unique order-preserving bijection from V_k to $\{1, \dots, |V_k|\}$, and let $\pi_k \in NC(|V_k|)$ be the image of $\pi|V_k$ by this bijection. Then

$$F_n(\pi, \sigma) = F_{|V_1|}(\pi_1, 1_{|V_1|}) \cdots F_{|V_r|}(\pi_r, 1_{|V_r|}). \tag{10.22}$$

(ii) For every $n \geq 1$ and every $\pi \in NC(n)$ we have that

$$F_n(\pi, 1_n) = F_n(0_n, K(\pi)), \tag{10.23}$$

where $K(\pi)$ is the Kreweras complement of π.

(iii) For $\pi = \{V_1, \dots, V_r\} \in NC(n)$ we have that

$$F_n(0_n, \pi) = \alpha_{|V_1|} \cdots \alpha_{|V_r|}, \tag{10.24}$$

where the α_k's are as in Definition 10.16, that is, $\alpha_k = F_k(0_k, 1_k)$ for $k \geq 1$.

EXERCISE 10.18. (1) Verify the properties of a multiplicative family on $NC^{(2)}$ which are stated in (i), (ii), (iii) of the preceding remark.

(2) Conversely, let $(\alpha_n)_{n\geq 1}$ be a sequence of complex numbers and let $(F_n : NC(n)^{(2)} \to \mathbb{C})_{n\geq 1}$ be a family of functions such that (i), (ii) and (iii) of Remark 10.17 hold. Prove that $(F_n)_{n\geq 1}$ is the multiplicative family of functions on $NC^{(2)}$ determined by the sequence $(\alpha_n)_{n\geq 1}$.

The solution to Exercise 10.18 is obtained by essentially copying the proof of Theorem 9.29, and is left to the reader.

Let us now also consider the one-variable version of the concept of a multiplicative family.

DEFINITION 10.19. Given a sequence $(\alpha_n)_{n\geq 1}$ of complex numbers, define a family of functions $f_n : NC(n) \to \mathbb{C}$, $n \geq 1$, by the following formula: if $\pi = \{V_1, \ldots, V_r\}$ in $NC(n)$, then

$$f_n(\pi) := \alpha_{|V_1|} \cdots \alpha_{|V_r|}. \tag{10.25}$$

Then $(f_n)_{n\geq 1}$ is called the **multiplicative family of functions on *NC*** determined by the sequence $(\alpha_n)_{n\geq 1}$.

In general, a family of functions $(f_n : NC(n) \to \mathbb{C})_{n\geq 1}$ will be said to be **multiplicative** if it arises from some sequence of α_n in the way described above. (The α_n in this case will be uniquely determined, $\alpha_n = f_n(1_n)$ for $n \geq 1$.)

So the multiplicativity of a family $(f_n)_{n\geq 1}$ on NC means that one has a factorization of the f_n according to the block structure of the non-crossing partitions that the f_n are applied to. For example, if $(f_n)_{n\geq 1}$ and $(\alpha_n)_{n\geq 1}$ are as above and if we look at $\pi = \{(1,10), (2,5,9), (3,4), (6), (7,8)\} \in NC(10)$, then we have $f_{10}(\pi) = \alpha_1\alpha_2^3\alpha_3$ (which is simply because π has 1 block with 1 element, has 3 blocks with 2 elements, and has 1 block with 3 elements).

In connection to Definition 10.19, we will also use the following notation.

NOTATION 10.20. Let $(\alpha_n)_{n\geq 1}$ be a sequence of complex numbers, and let $(f_n)_{n\geq 1}$ be the multiplicative family of functions on NC which is determined by $(\alpha_n)_{n\geq 1}$. Then we will use the notation

$$\alpha_\pi := f_n(\pi) \qquad \text{for } \pi \in NC(n),$$

and we will refer to the family of numbers $(\alpha_\pi)_{n\in\mathbb{N},\pi\in NC(n)}$ as the **multiplicative extension** of $(\alpha_n)_{n\geq 1}$. When using the multiplicative extension $(\alpha_\pi)_{n\in\mathbb{N},\pi\in NC(n)}$ we will occasionally just say that "$\pi \mapsto \alpha_\pi$ is multiplicative."

PROPOSITION 10.21. *Let $(f_n)_{n\geq 1}$ be a multiplicative family on NC, and let $(F_n)_{n\geq 1}$ be a multiplicative family on $NC^{(2)}$. Then the family $(f_n * F_n)_{n\geq 1}$ is multiplicative on NC as well.*

PROOF. For every $n \geq 1$ we denote $f_n * F_n =: g_n$, and we also make the notation

$$\beta_n := g_n(1_n) = \sum_{\tau \in NC(n)} f_n(\tau) F_n(\tau, 1_n).$$

Let us fix for the remaining of the proof an $n \geq 1$ and a partition $\pi = \{V_1, \ldots, V_r\} \in NC(n)$. Our goal for the proof is to verify that

$$g_n(\pi) = \beta_{|V_1|} \cdots \beta_{|V_r|}.$$

We consider the canonical lattice isomorphism between the interval $[0_n, \pi] \subset NC(n)$ and the direct product $NC(|V_1|) \times \cdots \times NC(|V_r|)$, as discussed in Lecture 9 (cf. the proof of Theorem 9.29). This isomorphism is explicitly described as

$$[0_n, \pi] \ni \tau \mapsto (\tau_1, \ldots, \tau_r) \in NC(|V_1|) \times \cdots \times NC(|V_r|),$$

where τ_k is the image of $\tau|V_k$ under the order-preserving bijection between V_k and $\{1, \ldots, |V_k|\}$, $1 \leq k \leq r$. Note that for τ and $(\tau_1, \ldots, \tau_r)$ as above, property (i) of Remark 10.17 gives us that

$$F_n(\tau, \pi) = F_{|V_1|}(\tau_1, 1_{|V_1|}) \cdots F_{|V_r|}(\tau_r, 1_{|V_r|}). \tag{10.26}$$

On the other hand, for the same τ and $(\tau_1, \ldots, \tau_r)$ the multiplicativity of f_n immediately gives us that

$$f_n(\tau) = f_{|V_1|}(\tau_1) \cdots f_{|V_r|}(\tau_r). \tag{10.27}$$

But then we can compute:

$$\begin{aligned} g_n(\pi) &= \sum_{\tau \in [0_n, \pi]} f_n(\tau) F_n(\tau, \pi) \\ &= \sum_{\tau_1, \ldots, \tau_r} \Big(f_{|V_1|}(\tau_1) \cdots f_{|V_r|}(\tau_r) \Big) \cdot \Big(F_{|V_1|}(\tau_1, 1_{|V_1|}) \cdots F_{|V_r|}(\tau_r, 1_{|V_r|}) \Big). \end{aligned}$$

In the latter sum we have that $\tau_1, \ldots, \tau_r$ run in $NC(|V_1|), \ldots, NC(|V_r|)$, respectively. This sum is obtained by performing the "change of variable" $\tau \leftrightarrow (\tau_1, \ldots, \tau_r)$, and then by using Equations (10.26) and (10.27). Finally, it is clear that the last expression obtained for $g_n(\pi)$ can be factored as

$$\prod_{k=1}^{r} \Big(\sum_{\tau_k \in NC(|V_k|)} f_{|V_k|}(\tau_k) F_{|V_k|}(\tau_k, 1_{|V_k|}) \Big),$$

which is $\prod_{k=1}^{r} \beta_{|V_k|}$, as desired. □

REMARK 10.22. With a bit more effort, one can prove the analogous proposition involving two multiplicative families on $NC^{(2)}$. More precisely, it turns out that if $(F_n)_{n \geq 1}$ and $(G_n)_{n \geq 1}$ are multiplicative families of functions on $NC^{(2)}$ then we always have that

1) $F_n * G_n = G_n * F_n$, $\forall\, n \geq 1$, and

2) $(F_n * G_n)_{n \geq 1}$ is a multiplicative family as well.

We will discuss this in more detail in Lecture 18.

Functional equation for convolution with μ_n

We conclude the lecture by returning to its main theme, and by making the F_n of Proposition 10.21 become the Möbius multiplicative family $(\mu_n)_{n\geq 1}$. So in this case we are dealing with two multiplicative families $(f_n)_{n\geq 1}$ and $(g_n)_{n\geq 1}$ on NC, related by

$$g_n = f_n * \mu_n \text{ (or equivalently: } f_n = g_n * \zeta_n), \quad n \geq 1. \tag{10.28}$$

For the theory of free cumulants which will be developed in the next lectures it will be important to have an alternative description of Equation (10.28), expressed in terms of power series. This goes as follows.

THEOREM 10.23. *Let $(f_n)_{n\geq 1}$ and $(g_n)_{n\geq 1}$ be two multiplicative families on NC, which are related as in Equation (10.28). Let $(\alpha_n)_{n\geq 1}$ and $(\beta_n)_{n\geq 1}$ be the sequences of numbers which determine these two multiplicative families; that is, we denote $f_n(1_n) =: \alpha_n$ and $g_n(1_n) =: \beta_n$, $n \geq 1$. Consider moreover the power series:*

$$u(z) = 1 + \sum_{n=1}^{\infty} \alpha_n z^n \qquad \text{and} \qquad v(z) = 1 + \sum_{n=1}^{\infty} \beta_n z^n. \tag{10.29}$$

Then u and v satisfy the functional equations

$$v\big(zu(z)\big) = u(z) \qquad \text{and} \qquad u\Big(\frac{z}{v(z)}\Big) = v(z). \tag{10.30}$$

PROOF. From $f_n = g_n * \zeta_n$ we have that $\alpha_n = f_n(1_n) = \sum_{\pi\in NC(n)} g_n(\pi)$. We rewrite the latter sum in the way that we fix the first block V_1 of π (i.e. that block which contains the element 1) and sum over all possibilities for the other blocks. We get:

$$\alpha_n = \sum_{s=1}^{n} \sum_{V_1 \text{ with } |V_1|=s} \sum_{\substack{\pi\in NC(n) \\ \text{where } \pi=\{V_1,\dots\}}} g_n(\pi).$$

If we write explicitly $V_1 = \{v_1, v_2, \dots, v_s\}$ with $1 = v_1 < v_2 < \cdots < v_s$, then from the non-crossing condition it is immediate that the π appearing in the above formula have to be of the form

$$\pi = V_1 \cup \tilde{\pi}_1 \cup \cdots \cup \tilde{\pi}_s,$$

where $\tilde{\pi}_j$ is a non-crossing partition of $\{v_j + 1, v_j + 2, \dots, v_{j+1} - 1\}$ (and where we make the convention that $v_{s+1} := n$). Putting

$$i_j := v_{j+1} - v_j - 1$$

we identify $\tilde{\pi}_j$ with an element in $NC(i_j)$ (where the appropriate convention is made in the case when $i_j = 0$). The multiplicativity of g

gives us that

$$g_n(\pi) = \beta_s g_{i_1}(\tilde{\pi}_1) \cdots g_{i_s}(\tilde{\pi}_s)$$

(where $g_{i_j}(\tilde{\pi}_j)$ is simply taken to be 1 in the cases when $i_j = 0$). We thus obtain:

$$\begin{aligned}
\alpha_n &= \sum_{s=1}^{n} \sum_{\substack{i_1,\dots,i_s \in \{0,1,\dots,n-s\} \\ i_1+\cdots+i_s+s=n}} \sum_{\substack{\pi = V_1 \cup \tilde{\pi}_1 \cup \cdots \cup \tilde{\pi}_s \\ \tilde{\pi}_j \in NC(i_j)}} \beta_s g_{i_1}(\tilde{\pi}_1) \cdots g_{i_s}(\tilde{\pi}_s) \\
&= \sum_{s=1}^{n} \beta_s \sum_{\substack{i_1,\dots,i_s \in \{0,1,\dots,n-s\} \\ i_1+\cdots+i_s+s=n}} \Big(\sum_{\tilde{\pi}_1 \in NC(i_1)} g_{i_1}(\tilde{\pi}_1) \Big) \cdots \Big(\sum_{\tilde{\pi}_s \in NC(i_s)} g_{i_s}(\tilde{\pi}_s) \Big) \\
&= \sum_{s=1}^{n} \sum_{\substack{i_1,\dots,i_s \in \{0,1,\dots,n-s\} \\ i_1+\cdots+i_s+s=n}} \beta_s \alpha_{i_1} \cdots \alpha_{i_s}.
\end{aligned}$$

This can now be used to rewrite the corresponding formal power series in the following way:

$$\begin{aligned}
u(z) &= 1 + \sum_{n=1}^{\infty} \alpha_n z^n \\
&= 1 + \sum_{n=1}^{\infty} \sum_{s=1}^{n} \sum_{\substack{i_1,\dots,i_s \in \{0,1,\dots,n-s\} \\ i_1+\cdots+i_s=n-s}} (\beta_s z^s)(\alpha_{i_1} z^{i_1}) \cdots (\alpha_{i_s} z^{i_s}) \\
&= 1 + \sum_{s=1}^{\infty} \beta_s z^s \Big(\sum_{i=0}^{\infty} \alpha_i z^i \Big)^s \\
&= v\big(zu(z)\big).
\end{aligned}$$

To get the second version of the functional equation (10.30) we put $x = zu(z)$, which yields

$$z = \frac{x}{u(z)} = \frac{x}{v(x)}.$$

But then we have

$$v(x) = v(zu(z)) = u(z) = u\Big(\frac{x}{v(x)}\Big).$$

$\square$

EXAMPLE 10.24. Recall the statement made in Proposition 9.35 of the preceding lecture, that the multi-chains of length $k-1$ in the lattice $NC(n)$ are counted by the Fuss–Catalan number $C_n^{(k)}$. We outline here a possible way of deriving this fact, relying on the functional equation obtained in the preceding theorem.

The Fuss–Catalan numbers $C_n^{(k)}$ $(n,k \geq 1)$ were introduced via Equation (9.22) of Notation 9.34, which generalizes the formula defining Catalan numbers. We will accept here that, for every fixed $k \geq 1$, the sequence $(C_n^{(k)})_{n=1}^{\infty}$ could alternatively be defined via the recurrence relation

$$C_n^{(k)} = \sum_{\substack{i_1,\dots,i_{k+1} \geq 0 \\ i_1+\cdots+i_{k+1}=n-1}} C_{i_1}^{(k)} C_{i_2}^{(k)} \cdots C_{i_{k+1}}^{(k)}, \quad n \geq 1, \tag{10.31}$$

where we make the convention that $C_0^{(k)} := 1$. (Note that in the particular case when $k = 1$, this is precisely the Catalan recurrence relation (2.8) from Remark 2.10, which has repeatedly appeared throughout these lectures.) It is immediately seen that the recurrence (10.31) can be concisely rewritten as an equation for the corresponding power series; this equation is

$$u_k(z) = 1 + z u_k(z)^{k+1}, \quad k \geq 1, \tag{10.32}$$

where for every $k \geq 1$ we put

$$u_k(z) := \sum_{n=0}^{\infty} C_n^{(k)} z^n. \tag{10.33}$$

Let us now pretend we do not know the number of multi-chains of length $k-1$ in $NC(n)$, and let us denote this number by $Z_n^{(k)}$ $(n,k \geq 1)$. For every $k \geq 1$ let $(f_n^{(k)})_{n \geq 1}$ be the multiplicative family of functions on NC which is determined by the sequence $(Z_n^{(k)})_{n \geq 1}$ (in the sense of Definition 10.19). It is easily verified by induction on k that

$$f_n^{(k)} = \underbrace{\zeta_n * \zeta_n * \cdots * \zeta_n}_{k+1}, \quad \forall\, n,k \geq 1.$$

But then, for every $k \geq 1$, the power series

$$w_{k+1}(z) := 1 + \sum_{n=1}^{\infty} Z_n^{(k+1)} z^n \qquad \text{and} \qquad w_k(z) := 1 + \sum_{n=1}^{\infty} Z_n^{(k)} z^n$$

can play the roles of u and respectively v in Theorem 10.23, and must therefore satisfy the functional equation of that theorem,

$$w_k(z w_{k+1}(z)) = w_{k+1}(z). \tag{10.34}$$

The functional equation (10.34) can in turn be used to show (via an easy induction on k) that w_k satisfies Equation (10.32). It follows that w_k must coincide with the series u_k defined in (10.33), and the equality $Z_n^{(k)} = C_n^{(k)}$ follows for all $n,k \geq 1$.

Exercises

EXERCISE 10.25. Let P be a poset with n elements. Prove that one can find a way of listing $P = \{\pi_1, \ldots, \pi_n\}$ which has the following property: for every $1 \leq i < j \leq n$, either the elements π_i and π_j are incomparable or they are such that $\pi_i < \pi_j$.
[Hint: there has to exist an $\omega \in P$ with no majorants – i.e. such that there exists no $\pi \in P$ with $\omega < \pi$. Set $\pi_n := \omega$ and proceed by induction.]

EXERCISE 10.26. Let $\mathcal{T}$ be a set of $n \times n$ upper triangular matrices which has the following properties:

- $\mathcal{T}$ is closed under addition, multiplication, and scalar multiplication;
- every diagonal matrix is in $\mathcal{T}$.

Suppose that $T \in \mathcal{T}$ is an invertible matrix. Prove that $T^{-1} \in \mathcal{T}$.
[Hint: write $T = D - N$ where D is diagonal and N is strictly upper triangular. Observe that D is invertible and that $D^{-1}, N \in \mathcal{T}$. Then write $T^{-1} = (I_n - D^{-1}N)^{-1} \cdot D^{-1} = \big(\sum_{k=0}^{n-1}(D^{-1}N)^k\big)D^{-1}$.]

EXERCISE 10.27. Prove the statements made in Proposition 10.14 of this lecture.
[Hint: for statement (2), verify that the function on $P^{(2)}$ defined by the right-hand side of Equation (10.16) satisfies the relations which were discussed in Remark 10.7, and which determine μ_P uniquely.]

The next exercise will use the following definition.

DEFINITION 10.28. Let P be a finite poset which has a minimum element 0_P.

(1) A family $\pi_0 < \pi_1 < \cdots < \pi_k$ of elements of P is called a **chain** in P; the non-negative integer k is called the **length** of the chain.

(2) A **saturated chain** in P is a chain $\pi_0 < \pi_1 < \cdots < \pi_k$ in P which has the property that π_i covers π_{i-1} for every $1 \leq i \leq k$.

(3) Consider the following condition on P: whenever $\pi_0 < \pi_1 < \cdots < \pi_k$ and $\sigma_0 < \sigma_1 < \cdots < \sigma_l$ are saturated chains in P such that $\pi_0 = \sigma_0$ and $\pi_k = \sigma_l$, it follows that $k = l$. If this condition is fulfilled, then we say that P is a **graded** poset.

(4) Suppose that P is a graded poset. Then for every $\pi \in P$ we define the **rank** of π to be the length of an arbitrary saturated chain $\pi_0 < \pi_1 < \cdots < \pi_k$ such that $\pi_0 = 0_P$ and $\pi_k = \pi$.

EXERCISE 10.29. Let n be a positive integer.

(1) Let $\pi, \sigma \in NC(n)$ be such that $\pi \leq \sigma$. Describe what it means for σ to cover π, and observe that this is equivalent to the equality $|\pi| = |\sigma| + 1$.

(2) Show that $NC(n)$ is a ranked poset, where the rank of $\pi \in NC(n)$ is equal to $n - |\pi|$.

EXERCISE 10.30. Give an alternative derivation for the Möbius functions on the lattices of non-crossing partitions by using the concept of multiplicative family on NC, and the functional equations provided by Theorem 10.23.

Exercises 10.31–10.33 are about the Möbius function for the lattice $\mathcal{P}(n)$ of all partitions of $\{1, \dots, n\}$. We will denote this Möbius function by $\mu_{\mathcal{P}(n)}$. The explicit formula for $\mu_{\mathcal{P}(n)}$ is derived on the same line as shown above for the Möbius function on $NC(n)$, and thus starts with a factorization result (which is even more straightforward than the one from the non-crossing framework).

EXERCISE 10.31. Let π and σ be partitions in $\mathcal{P}(n)$ such that $\pi \leq \sigma$. Let us write explicitly $\sigma = \{V_1, \dots, V_r\}$ and

$$\pi = \{W_{1,1}, \dots, W_{1,k_1}, \dots, W_{r,1}, \dots, W_{r,k_r}\},$$

where $W_{i,1} \cup \cdots \cup W_{i,k_i} = V_i$, $1 \leq i \leq r$. Prove that the interval $[\pi, \sigma]$ of $\mathcal{P}(n)$ is isomorphic to the direct product $\mathcal{P}(k_1) \times \cdots \times \mathcal{P}(k_r)$.

EXERCISE 10.32. For every $n \geq 1$, let us denote

$$\mu_{\mathcal{P}(n)}(\,[0_n, 1_n]\,) \;=:\; a_n,$$

where 0_n and 1_n are the minimum and respectively the maximum elements of $\mathcal{P}(n)$. By using an argument which parallels that in the proof of Proposition 10.15, show that

$$a_n \;=\; (-1)^{n-1}(n-1)!, \quad \forall\, n \geq 1.$$

[Hint: one obtains a recurrence for the a_n, in exactly the same way as Equation (10.20) was obtained in the proof of Proposition 10.15. The recurrence reads as follows:

$$a_n + \sum_{A \subset \{1,\dots,n-2\}} a_{1+|A|} \cdot a_{n-1-|A|} = 0, \quad n \geq 2.\,]$$

EXERCISE 10.33. Consider again the partitions $\pi \leq \sigma$ in $\mathcal{P}(n)$ which appeared in Exercise 10.31 (and where the blocks of π and of σ are listed explicitly in the same way as in Exercise 10.31). Prove that

$$\mu_{\mathcal{P}(n)}(\,[\pi, \sigma]\,) \;=\; \prod_{i=1}^{r} (-1)^{k_i - 1}(k_i - 1)!$$

LECTURE 11

Free cumulants: definition and basic properties

We will now introduce our main combinatorial tool for dealing with free independence, the "free cumulants." Motivated by our treatment of the free central limit theorem we expect these free cumulants κ_π to be determined by the "moment-cumulant formula"

$$\varphi(a_1 \cdots a_n) = \sum_{\pi \in NC(n)} \kappa_\pi[a_1, \dots, a_n]$$

and by the fact that κ_π factorizes according to the block structure of π. This fits in the frame of multiplicative functions on NC and the Möbius inversion of the last two lectures, with the small difference that our multiplicative functions are now not determined by a sequence of numbers but by a sequence of multilinear functionals on an algebra $\mathcal{A}$. We will thus first extend our notion of multiplicative functions to this setting.

Given all these preparations the definition of free cumulants will then be quite straightforward. That this is indeed a useful definition in the context of free probability theory will become clear from the main result of this lecture: *free independence can be characterized by the vanishing of mixed cumulants.* An important technical tool for deriving this characterization is a formula for free cumulants where the arguments are products of random variables. This formula is actually at the basis of many of our forthcoming results in later lectures and allows elegant proofs of many statements.

Multiplicative functionals on NC

DEFINITION 11.1. Let $\mathcal{A}$ be a unital algebra. Given a sequence $(\rho_n)_{n \geq 1}$ of multilinear functionals on $\mathcal{A}$,

$$\begin{aligned} \rho_n : \qquad & \mathcal{A}^n \to \mathbb{C} \\ & (a_1, \dots, a_n) \mapsto \rho_n(a_1, \dots, a_n), \end{aligned}$$

we extend this to a family of multilinear functionals ρ_π ($n \geq 1$, $\pi \in NC(n)$),

$$\begin{aligned} \rho_\pi : \qquad & \mathcal{A}^n \to \mathbb{C} \\ & (a_1, \dots, a_n) \mapsto \rho_\pi[a_1, \dots, a_n], \end{aligned}$$

by the following formula. If $\pi = \{V_1, \dots, V_r\} \in NC(n)$, then:

$$\rho_\pi[a_1, \dots, a_n] := \rho(V_1)[a_1, \dots, a_n] \cdots \rho(V_r)[a_1, \dots, a_n], \tag{11.1}$$

where we used the notation

$$\rho(V)[a_1, \dots, a_n] := \rho_s(a_{i_1}, \dots, a_{i_s}) \qquad \text{for } V = (i_1 < \cdots < i_s). \tag{11.2}$$

Then $(\rho_\pi)_{n \geq 1, \pi \in NC(n)}$ is called the **multiplicative family of functionals on *NC*** determined by the sequence $(\rho_n)_{n \geq 1}$. Note the distinction between our use of round brackets for the ρ_n and square brackets for the ρ_π. The ρ_π are indeed an extension of the ρ_n, because we have $\rho_n \hat{=} \rho_{1_n}$, i.e.

$$\rho_n(a_1, \dots, a_n) = \rho_{1_n}[a_1, \dots, a_n]$$

for all $n \geq 1$ and all $a_1, \dots, a_n$.

In general, a family of multilinear functionals

$$(\rho_\pi : \mathcal{A}^n \to \mathbb{C})_{n \geq 1, \pi \in NC(n)}$$

will be said to be **multiplicative** if it arises from some sequence of multilinear functionals $\rho_n : \mathcal{A}^n \to \mathbb{C}$ in the way described above. (The ρ_n will be in this case uniquely determined, $\rho_n \hat{=} \rho_{1_n}$ for $n \geq 1$.)

So the multiplicativity of a family $(\rho_\pi)_{n \geq 1, \pi \in NC(n)}$ means that one has a factorization of the ρ_π according to the block structure of the non-crossing partitions π. In addition to the case of multiplicative families of functions from the last lecture we must now also distribute our arguments $a_1, \dots, a_n$ according to the blocks of π. For example, for

$$\pi = \{(1, 10), (2, 5, 9), (3, 4), (6), (7, 8)\} \in NC(10),$$

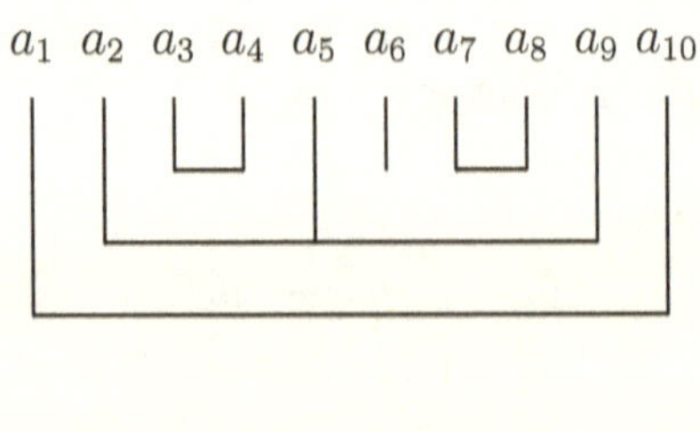

we have

$$\rho_\pi[a_1, \dots, a_{10}] = \rho_2(a_1, a_{10}) \cdot \rho_3(a_2, a_5, a_9) \cdot \rho_2(a_3, a_4) \cdot \rho_1(a_6) \cdot \rho_2(a_7, a_8).$$

Definition of free cumulants

First, we have to make out of our functional φ a sequence of multilinear functions φ_n and then extend this to the corresponding moment functionals φ_π.

NOTATION 11.2. Let $\mathcal{A}$ be a unital algebra and $\varphi : \mathcal{A} \to \mathbb{C}$ a unital linear functional. This gives rise to a sequence of multilinear functionals $(\varphi_n)_{n\in\mathbb{N}}$ on $\mathcal{A}$ via

$$\varphi_n(a_1, \dots, a_n) := \varphi(a_1 \cdots a_n). \tag{11.3}$$

We extend these to the corresponding multiplicative functionals on non-crossing partitions by $(a_1, \dots, a_n \in \mathcal{A})$

$$\varphi_\pi[a_1, \dots, a_n] := \prod_{V\in\pi} \varphi(V)[a_1, \dots, a_n], \tag{11.4}$$

where $\varphi(V)[a_1, \dots, a_n]$ is defined as in (11.2).

Now we can define the free cumulants by Möbius inversion.

DEFINITION 11.3. Let $(\mathcal{A}, \varphi)$ be a non-commutative probability space. The corresponding **free cumulants** $(\kappa_\pi)_{\pi\in NC}$ are, for each $n \in \mathbb{N}$, $\pi \in NC(n)$, multilinear functionals

$$\begin{aligned} \kappa_\pi : \qquad & \mathcal{A}^n \to \mathbb{C}, \\ & (a_1, \dots, a_n) \mapsto \kappa_\pi[a_1, \dots, a_n] \end{aligned}$$

which are defined as follows:

$$\kappa_\pi[a_1, \dots, a_n] := \sum_{\substack{\sigma\in NC(n)\\ \sigma\leq\pi}} \varphi_\sigma[a_1, \dots, a_n]\mu(\sigma, \pi), \tag{11.5}$$

where μ is the Möbius function on $NC(n)$.

For each $n \geq 1$, we put $\kappa_n := \kappa_{1_n}$.

Note that all our arguments from the last lecture about multiplicative functions and Möbius inversion remain valid in the more general context of multiplicative families of functionals (provided we take care to distribute the arguments $a_1, \dots, a_n$ at the right positions). Thus our results from the last lecture yield the following basic statements about free cumulants.

PROPOSITION 11.4. (1) *The free cumulant function $\pi \mapsto \kappa_\pi$ is a multiplicative family of functionals, i.e. we have*

$$\kappa_\pi[a_1, \dots, a_n] := \prod_{V\in\pi} \kappa(V)[a_1, \dots, a_n]. \tag{11.6}$$

(2) *In particular, all information about the free cumulants is contained in the sequence of cumulants* $(\kappa_n)_{n\in\mathbb{N}}$ *where, for* $n \in \mathbb{N}$, $\kappa_n := \kappa_{1_n}$. *Definition 11.3 of free cumulants is equivalent to the statement that* $\pi \mapsto \kappa_\pi$ *is a multiplicative family of functionals and that for all* $n \in \mathbb{N}$ *and all* $a_1, \dots, a_n \in \mathcal{A}$ *we have*

$$\kappa_n(a_1, \dots, a_n) = \sum_{\sigma\in NC(n)} \varphi_\sigma[a_1, \dots, a_n]\mu(\sigma, 1_n). \tag{11.7}$$

(3) *Definition 11.3 of free cumulants is equivalent to the statements that* $\pi \mapsto \kappa_\pi$ *is a multiplicative family of functionals and that for all* $n \in \mathbb{N}$ *and all* $a_1, \dots, a_n \in \mathcal{A}$ *we have*

$$\varphi(a_1 \cdots a_n) = \sum_{\sigma\in NC(n)} \kappa_\pi[a_1, \dots, a_n]. \tag{11.8}$$

PROOF. The fact that multiplicativity of φ and multiplicativity of κ are equivalent follows from Proposition 10.21. The equivalence between the relations (11.7) and (11.8) is just an instance of general Möbius inversion, Proposition 10.6. $\square$

NOTATION 11.5. We will call Equations (11.7) and (11.8) the **moment-cumulant formulas.**

EXAMPLES 11.6. We want to determine the concrete form of $\kappa_n(a_1, \dots, a_n)$ for small values of n.

(1) $n = 1$: Clearly, here we have

$$\kappa_1(a_1) = \varphi(a_1).$$

(2) $n = 2$: There are only two partitions in $NC(2)$ and the values of the Möbius function are

$$\mu(\sqcup, \sqcup) = 1, \qquad \mu(\mid\mid, \sqcup) = -1.$$

Thus we have

$$\begin{aligned}\kappa_2(a_1, a_2) &= \varphi_{\sqcup}[a_1, a_2] - \varphi_{\mid\mid}[a_1, a_2] \\ &= \varphi(a_1 a_2) - \varphi(a_1)\varphi(a_2).\end{aligned}$$

(3) $n = 3$: We have five partitions and the relevant values of the Möbius function are

$$\mu(\sqcup\!\sqcup, \sqcup\!\sqcup) = 1, \qquad \mu(\mid\sqcup, \sqcup\!\sqcup) = -1$$

$$\mu(\sqcup\mid, \sqcup\!\sqcup) = -1, \qquad \mu(\sqcup\!\!\!\sqcup, \sqcup\!\sqcup) = -1, \qquad \mu(\mid\mid\mid, \sqcup\!\sqcup) = 2.$$

With this we obtain

$$\begin{aligned}\kappa_3(a_1,a_2,a_3) &= \varphi_{\sqcup\!\sqcup}[a_1,a_2,a_3] - \varphi_{|\sqcup}[a_1,a_2,a_3] - \varphi_{\sqcup|}[a_1,a_2,a_3] \\ &\quad - \varphi_{\sqcup\!\!\!\!\!\;|\;}[a_1,a_2,a_3] + 2\varphi_{|||}[a_1,a_2,a_3] \\ &= \varphi(a_1a_2a_3) - \varphi(a_1)\varphi(a_2a_3) - \varphi(a_1a_2)\varphi(a_3) \\ &\quad - \varphi(a_1a_3)\varphi(a_2) + 2\varphi(a_1)\varphi(a_2)\varphi(a_3).\end{aligned}$$

(4) $n = 4$: In this case we consider only the special situation where all $\varphi(a_i) = 0$. Then we have

$$\kappa_4(a_1,a_2,a_3,a_4) = \varphi(a_1a_2a_3a_4) - \varphi(a_1a_2)\varphi(a_3a_4) - \varphi(a_1a_4)\varphi(a_2a_3).$$

Another way to look at the cumulants κ_n for $n \geq 2$ is that they organize in a special way the information about how much φ ceases to be a homomorphism.

PROPOSITION 11.7. *Let $(\kappa_n)_{n\geq 1}$ be the cumulants corresponding to φ. Then φ is a homomorphism if and only if κ_n vanishes for all $n \geq 2$.*

PROOF. Let φ be a homomorphism. Note that, for any $\sigma \in NC(n)$, this means

$$\varphi_\sigma[a_1,\dots,a_n] = \varphi(a_1)\varphi(a_2)\cdots\varphi(a_n) = \varphi_{0_n}[a_1,\dots,a_n]$$

for all $a_1,\dots,a_n \in \mathcal{A}$, thus we have $\varphi_\sigma = \varphi_{0_n}$ for all $\sigma \in NC(n)$. Thus we get

$$\kappa_n = \sum_{\sigma\leq 1_n} \varphi_\sigma \mu(\sigma,1_n) = \varphi_{0_n} \sum_{0_n\leq\sigma\leq 1_n} \mu(\sigma,1_n),$$

which is, by the recurrence relation (10.11) for the Möbius function, equal to zero if $0_n \neq 1_n$, i.e. for $n \geq 2$.

To see the other direction, one only has to observe the following: if κ_2 vanishes, then we have for all $a_1,a_2 \in \mathcal{A}$

$$0 = \kappa_2(a_1,a_2) = \varphi(a_1a_2) - \varphi(a_1)\varphi(a_2),$$

i.e. $\varphi(a_1a_2) = \varphi(a_1)\varphi(a_2)$ for all $a_1,a_2 \in \mathcal{A}$, and thus φ is a homomorphism. □

REMARK 11.8. In particular, this means that on constants only the first order cumulants are different from zero:

$$\kappa_n(1,\dots,1) = \delta_{n1}. \tag{11.9}$$

Products as arguments

In this section, we want to focus on properties of our free cumulants with respect to the algebraic structure of our underlying algebra $\mathcal{A}$. Of course, the behavior of free cumulants with respect to the linear structure of $\mathcal{A}$ is clear, because our cumulants are multilinear functionals. Thus it remains to see whether there is anything to say about the relation of the free cumulants with the multiplicative structure of the algebra.

The crucial property in a multiplicative context is associativity. On the level of moments this just means that we can put brackets arbitrarily and for our moment functionals φ_n we do not have to bother where to place commas; for example, we have

$$\varphi_2(a_1a_2, a_3) = \varphi((a_1a_2)a_3) = \varphi(a_1(a_2a_3)) = \varphi_2(a_1, a_2a_3).$$

The corresponding statement on the level of cumulants is, of course, not true, i.e. $\kappa_2(a_1a_2, a_3) \neq \kappa_2(a_1, a_2a_3)$ in general. However, there is a treatable and nice replacement for associativity, which allows us to deal with free cumulants whose entries are products of random variables. This formula will be fundamental for our forthcoming investigations on free cumulants.

Consider random variables $a_1, \ldots, a_n \in \mathcal{A}$, multiply some of the "neighboring" ones together, and look on a free cumulant with these products as entries, i.e. we are interested in

$$\kappa_\tau[a_1 \cdots a_{i(1)}, a_{i(1)+1} \cdots a_{i(2)}, \ldots, a_{i(m-1)+1} \cdots a_{i(m)}]$$

for some fixed increasing sequence of integers $1 \leq i(1) < i(2) < \cdots < i(m) := n$. Thus our cumulant has m arguments and τ is some partition in $NC(m)$. Our aim is to express this cumulant in terms of cumulants of the original random variables, i.e. in terms of $\kappa_\pi[a_1, a_2, \ldots, a_n]$, where π are now some cumulants in $NC(n)$. Since in such a formula τ and π must be somehow related, we need a way to put them both in the same lattice of non-crossing partitions. It will turn out that the following embedding of $NC(m)$ into $NC(n)$ will do this.

NOTATION 11.9. For fixed natural numbers $m, n \in \mathbb{N}$ with $m < n$ and a fixed sequence of integers

$$i(0) := 0 < i(1) < i(2) < \cdots < i(m) := n$$

we define an embedding from $NC(m)$ into $NC(n)$, $\tau \mapsto \hat{\tau}$, as follows: $\hat{\tau}$ is that partition which we get from τ by replacing each $j \in \{1, \ldots, m\}$ by $(i(j-1)+1, \ldots, i(j)) \subset (1, \ldots, n)$, i.e.

$$i(j-1)+1 \sim_{\hat{\tau}} i(j-1)+2 \sim_{\hat{\tau}} \cdots \sim_{\hat{\tau}} i(j)$$

and $i(k) \sim_{\hat{\tau}} i(l)$ if and only if $k \sim_\tau l$. It is easily checked that $\hat{\tau}$ is really non-crossing.

Another way of stating the definition of $\hat{\tau}$ is to say that it is given as the pullback $\hat{\tau} = f^{-1} \circ \tau$ of the function $f : \{1, \dots, n\} \to \{1, \dots, m\}$ defined by $f(l) = k$ for $i(k-1) < l \leq i(k)$.

Example 11.10. As an example, consider $n = 6$, $m = 3$, and $i(1) = 1 < i(2) = 4 < i(3) = 6$. It is most illustrative to index the partitions with the random variables instead of integers. So let dummy random variables $a_1, \dots, a_6$ be given and we are interested in cumulants in the new variables $A_1 := a_1$, $A_2 := a_2a_3a_4$, $A_3 := a_5a_6$.

Consider the partition $\tau = \{(1,2),(3)\} \in NC(3)$, which we draw as

$$A_1 A_2 A_3$$

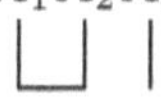

Then $\hat{\tau}$ is obtained by replacing A_1 with a_1, A_2 with $a_2a_3a_4$, and A_3 with a_5a_6 which leads to

$$a_1\ a_2\ a_3\ a_4\ a_5\ a_6$$

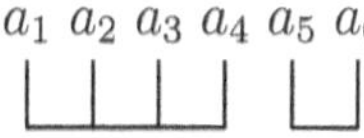

thus $\hat{\tau} = \{(1,2,3,4),(5,6)\} \in NC(6)$.

As another example, consider $\sigma = \{(1,3),(2)\} \in NC(3)$, i.e.

$$A_1 A_2 A_3$$

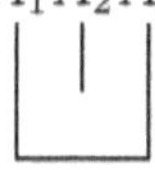

The corresponding partition in terms of the a_i looks like

$$a_1\ a_2\ a_3\ a_4\ a_5\ a_6$$

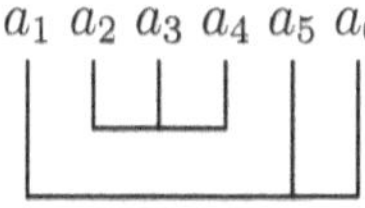

thus $\hat{\sigma} = \{(1,5,6),(2,3,4)\} \in NC(6)$.

Remarks 11.11. We want here to collect some basic properties of the mapping $\tau \mapsto \hat{\tau}$, which follow directly from the definition.

(1) The mapping $\tau \mapsto \hat{\tau}$ is injective, we have that $\hat{1}_m = 1_n$ and

$$\hat{0}_m = \{(1, \dots, i(1)), (i(1)+1, \dots, i(2)), \dots, (i(m-1)+1, \dots, i(m))\}.$$

Furthermore, the mapping $\tau \mapsto \hat{\tau}$ preserves the partial order, i.e. $\sigma \leq \pi$ for $\sigma, \pi \in NC(m)$ implies $\hat{\sigma} \leq \hat{\pi}$, and the image of $NC(m)$ is

$$\widehat{NC(m)} = [\hat{0}_m, \hat{1}_m] = [\hat{0}_m, 1_n] \subset NC(n).$$

We can summarize all these facts by saying that the mapping $\tau \mapsto \hat{\tau}$ is a lattice isomorphism between $NC(m)$ and $[\hat{0}_m, 1_n] \subset NC(n)$.

(2) Since the value $\mu(\sigma, \pi)$ of the Möbius function depends only on the interval $[\sigma, \pi]$, the lattice isomorphism between $[\sigma, \pi] \subset NC(m)$ and $[\hat{\sigma}, \hat{\pi}] \subset NC(n)$ implies, in particular, that for all $\sigma, \pi \in NC(m)$ we have $\mu(\sigma, \pi) = \mu(\hat{\sigma}, \hat{\pi})$.

THEOREM 11.12. *Consider a non-commutative probability space $(\mathcal{A}, \varphi)$ and let $(\kappa_\pi)_{\pi \in NC}$ be the corresponding free cumulants. Let $m,n \in \mathbb{N}$ and $1 \leq i(1) < i(2) < \cdots < i(m) = n$ be given and consider the corresponding embedding $\tau \mapsto \hat{\tau}$, defined in Notation 11.9. Recall that*

$$\hat{0}_m = \{(1, \ldots, i(1)), \ldots, (i(m-1)+1, \ldots, i(m))\} \in NC(m).$$

Consider now random variables $a_1, \ldots, a_n \in \mathcal{A}$.

(1) *For a non-crossing partition $\tau \in NC(m)$ the following equation holds:*

$$\kappa_\tau[a_1 \cdots a_{i(1)}, \ldots, a_{i(m-1)+1} \cdots a_{i(m)}] = \sum_{\substack{\pi \in NC(n) \\ \pi \vee \hat{0}_m = \hat{\tau}}} \kappa_\pi[a_1, \ldots, a_n] . \quad (11.10)$$

(2) *In particular, for $\tau = 1_m$ we have*

$$\kappa_m(a_1 \cdots a_{i(1)}, \ldots, a_{i(m-1)+1} \cdots a_{i(m)}) = \sum_{\substack{\pi \in NC(n) \\ \pi \vee \hat{0}_m = 1_n}} \kappa_\pi[a_1, \ldots, a_n]. \quad (11.11)$$

PROOF. Let us denote $A_j := a_{i(j-1)+1} \cdots a_{i(j)}$ $(1 \leq j \leq m)$. Then, by using the basic properties of the embedding $\tau \mapsto \hat{\tau}$ from Remarks 11.11, we can calculate as follows:

$$\begin{aligned} k_\tau[A_1, \ldots, A_m] &= \sum_{\substack{\pi \in NC(m) \\ \pi \leq \tau}} \varphi_\pi[A_1, \ldots, A_m] \mu(\pi, \tau) \\ &= \sum_{\substack{\pi \in NC(m) \\ \pi \leq \tau}} \varphi_{\hat{\pi}}[a_1, \ldots, a_n] \mu(\hat{\pi}, \hat{\tau}) \\ &= \sum_{\substack{\sigma \in NC(n) \\ \hat{0}_m \leq \sigma \leq \hat{\tau}}} \varphi_\sigma[a_1, \ldots, a_n] \mu(\sigma, \hat{\tau}). \end{aligned}$$

But the last formula is now exactly a partial Möbius inversion as we considered in the last lecture, and Proposition 10.11 yields directly our assertion. □

EXAMPLE 11.13. As an illustrative example for the statement of the theorem let us consider the following example. We take $m = 2$, $n = 3$, $i(1) = 2 < i(2) = 3$, and $\tau = 1_2$. This means we are looking at the cumulant $\kappa_2(a_1a_2, a_3)$. Our theorem tells us that we have to sum over all $\pi \in NC(3)$ which have the property that $\pi \vee \{(1,2),(3)\} = 1_3$. This means that π has to connect the block $(1,2)$ to the block (3). The $\pi \in NC(3)$ which do so are given in the following picture:

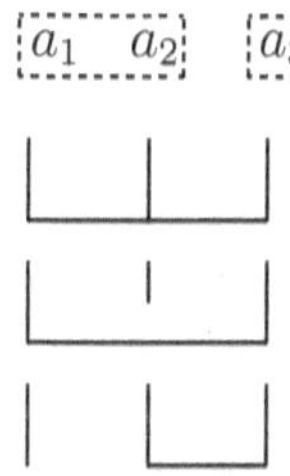

The other two elements in $NC(3)$ do not have this property:

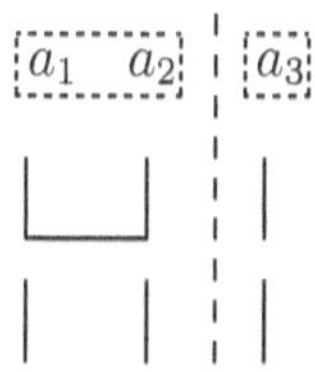

Thus our theorem claims that

$$\begin{aligned}\kappa_2(a_1a_2, a_3) &= \kappa_{\sqcup\!\sqcup}[a_1, a_2, a_3] + \kappa_{|\sqcup}[a_1, a_2, a_3] + \kappa_{\sqcup\!\!|\,}[a_1, a_2, a_3] \\ &= \kappa_3(a_1, a_2, a_3) + \kappa_1(a_1)\kappa_2(a_2, a_3) + \kappa_2(a_1, a_3)\kappa_1(a_2)\,.\end{aligned}$$

Of course, one can easily check directly that this true.

REMARK 11.14. If $\sigma \in NC(n)$ is an interval partition as above, i.e. if all blocks V of σ consist of consecutive numbers, then according to Exercise 9.43, the join $\sigma \vee \pi$ for any $\pi \in NC(n)$ is the same in $NC(n)$ as in $\mathcal{P}(n)$. In particular, the condition $\sigma \vee \pi = 1_n$ amounts to the fact that for any two blocks V and W of σ we can find a chain of points $p_1, \ldots, p_r$ such that $p_1 \in V$, $p_r \in W$, and such that alternatingly the points are in the same block of π or σ,

$$V \ni p_1 \sim_\pi p_2 \sim_\sigma p_3 \sim_\pi \cdots \sim_\sigma p_{r-1} \sim_\pi p_r \in W.$$

Thus we will also address, in the case that σ is an interval partition, the condition $\sigma \vee \pi = 1_n$ by saying that π **couples the blocks of** σ.

Free independence and free cumulants

Now we want to present the main reason why free cumulants are an important tool in free probability theory: free independence can be described very easily and effectively in terms of cumulants. Roughly speaking, random variables are freely independent if and only if their mixed cumulants vanish. Let us start with a special case of this. Since 1 is, by Lemma 5.17, free from everything we should have that free cumulants of lengths greater than one vanish whenever at least one of their arguments is 1. Note that for $n = 1$ we have, of course, $\kappa_1(1) = \varphi(1) = 1$.

PROPOSITION 11.15. *Let $(\mathcal{A}, \varphi)$ be a non-commutative probability space and let $(\kappa_n)_{n\in\mathbb{N}}$ be the corresponding free cumulants. Consider $n \geq 2$ and $a_1, \dots, a_n \in \mathcal{A}$. Then we have $\kappa_n(a_1, \dots, a_n) = 0$ if there exists at least one i, $1 \leq i \leq n$, such that $a_i = 1$.*

PROOF. To simplify notation we consider the case $a_n = 1$, i.e. we want to show that $\kappa_n(a_1, \dots, a_{n-1}, 1) = 0$. We will prove this by induction on n.

For $n = 2$, the assertion is true, since

$$\kappa_2(a, 1) = \varphi(a1) - \varphi(a)\varphi(1) = 0.$$

Now assume we have proved the assertion for all $k < n$ and let us show it for n. We have

$$\begin{aligned}\varphi(a_1 \cdots a_{n-1}1) &= \sum_{\pi\in NC(n)} \kappa_\pi[a_1, \dots, a_{n-1}, 1] \\ &= \kappa_n(a_1, \dots, a_{n-1}, 1) + \sum_{\substack{\pi\in NC(n)\\ \pi\neq 1_n}} \kappa_\pi[a_1, \dots, a_{n-1}, 1].\end{aligned}$$

According to our induction hypothesis, a partition $\pi \neq 1_n$ contributes to the above sum only if (n) is a one-element block of π, i.e. if $\pi = \sigma \cup (n)$ with $\sigma \in NC(n-1)$. For such a partition π we have

$$\kappa_\pi[a_1, \dots, a_{n-1}, 1] = \kappa_\sigma[a_1, \dots, a_{n-1}]\kappa_1(1) = \kappa_\sigma[a_1, \dots, a_{n-1}],$$

hence

$$\begin{aligned}\varphi(a_1 \cdots a_{n-1}1) &= \kappa_n(a_1, \dots, a_{n-1}, 1) + \sum_{\sigma\in NC(n-1)} k_\sigma[a_1, \dots, a_{n-1}] \\ &= \kappa_n(a_1, \dots, a_{n-1}, 1) + \varphi(a_1 \cdots a_{n-1}).\end{aligned}$$

Since $\varphi(a_1 \cdots a_{n-1}1) = \varphi(a_1 \cdots a_{n-1})$, we obtain finally our assertion $\kappa_n(a_1, \dots, a_{n-1}, 1) = 0$. □

Now we can prove our main theorem, which states that free independence is equivalent to the "vanishing of mixed cumulants."

THEOREM 11.16. **(Vanishing of mixed cumulants)**
Let $(\mathcal{A}, \varphi)$ be a non-commutative probability space and let $(\kappa_n)_{n\in\mathbb{N}}$ be the corresponding free cumulants. Consider unital subalgebras $(\mathcal{A}_i)_{i\in I}$ of $\mathcal{A}$. Then the following two statements are equivalent.

(i) *$(\mathcal{A}_i)_{i\in I}$ are freely independent.*

(ii) *We have for all $n \geq 2$ and for all $a_j \in \mathcal{A}_{i(j)}$ $(j = 1, \ldots, n)$ with $i(1), \ldots, i(n) \in I$ that $\kappa_n(a_1, \ldots, a_n) = 0$ whenever there exist $1 \leq l, k \leq n$ with $i(l) \neq i(k)$.*

REMARK 11.17. This characterization of freeness in terms of cumulants is the translation to cumulants of the definition of freeness (which was in terms of moments), by using the moment-cumulant formula (11.7). One should note that, in contrast to the characterization in terms of moments, we do not require that $i(1) \neq i(2) \neq \cdots \neq i(n)$ nor that $\varphi(a_j) = 0$. Hence the characterization of freeness in terms of cumulants is much easier to use than the characterization in terms of moments.

PROOF. (i) $\Longrightarrow$ (ii): If all a_j are centered, i.e. $\varphi(a_j) = 0$ for all $j = 1, \ldots, n$, and alternating, i.e. $i(1) \neq i(2) \neq \cdots \neq i(n)$, then the assertion follows directly by the moment-cumulant formula

$$\kappa_n(a_1, \ldots, a_n) = \sum_{\pi \in NC(n)} \varphi_\pi[a_1, \ldots, a_n]\mu(\pi, 1_n),$$

because at least one factor of φ_π is of the form $\varphi(a_l a_{l+1} \cdots a_{l+p})$, which vanishes by the definition of free independence.

The essential part of the proof consists in showing that on the level of cumulants the assumption "centered" is not needed and "alternating" can be weakened to "mixed."

Let us start by getting rid of the assumption "centered." Since $n \geq 2$, the above Proposition 11.15 implies that we have for arbitrary $a_1, \ldots, a_n \in \mathcal{A}$ the relation

$$\kappa_n(a_1, \ldots, a_n) = k_n\big(a_1 - \varphi(a_1)1, \ldots, a_n - \varphi(a_n)1\big), \qquad (11.12)$$

i.e. we can center the arguments of our cumulants κ_n $(n \geq 2)$ without changing the value of the cumulants.

Thus we have proved the following statement. Consider $n \geq 2$ and $a_j \in \mathcal{A}_{i(j)}$ $(j = 1, \ldots, n)$ with $i(1) \neq i(2) \neq \cdots \neq i(n)$. Then we have $\kappa_n(a_1, \ldots, a_n) = 0$.

To prove (ii) in full generality we will use induction on the length of our cumulants. For $n = 2$ and a_1, a_2 free we have $\varphi(a_1 a_2) = \varphi(a_1)\varphi(a_2)$

and thus

$$\kappa_2(a_1, a_2) = \varphi(a_1 a_2) - \varphi(a_1)\varphi(a_2) = 0.$$

Fix now $n \geq 3$ and assume we have proved (ii) for all κ_l with $l < n$. Consider $a_j \in \mathcal{A}_{i(j)}$ $(j = 1, \dots, n)$. Assume that there exist k, l with $i(k) \neq i(l)$. We have to show that $\kappa_n(a_1, \dots, a_n) = 0$. If $i(1) \neq i(2) \neq \dots \neq i(n)$, then the assertion is already proved. If the elements are not alternating then we multiply neighboring elements from the same algebra together, i.e. we write $a_1 \cdots a_n = A_1 \cdots A_m$ such that neighboring A come from different subalgebras. Note that $m \geq 2$ because of our assumption $i(k) \neq i(l)$. Then, by Theorem 11.12, we have

$$\begin{aligned}\kappa_m(A_1, \dots, A_m) &= \sum_{\substack{\pi \in NC(n), \\ \pi \vee \sigma = 1_n}} \kappa_\pi[a_1, \dots, a_n] \\ &= \kappa_n(a_1, \dots, a_n) + \sum_{\substack{\pi \in NC(n), \pi \neq 1_n \\ \pi \vee \sigma = 1_n}} \kappa_\pi[a_1, \dots, a_n]\end{aligned}$$

where $\sigma \in NC(n)$ is that partition whose blocks encode the information about which elements a_j we have to multiply in order to get the A_i; that is $\sigma = \hat{0}_m$ in Notation 11.9. Since the A are alternating we have $\kappa_m(A_1, \dots, A_m) = 0$. Furthermore, for $\pi \neq 1_n$, the term $\kappa_\pi[a_1, \dots, a_n]$ is a product of cumulants of lengths smaller than n. Thus our induction hypothesis applies to them and we see that $\kappa_\pi[a_1, \dots, a_n]$ can only be different from zero if each block of π contains only elements from the same subalgebra. So all blocks of σ that are coupled by π must correspond to the same subalgebra. (Note that by the definition of σ, each of its blocks contains only elements from the same subalgebra.) However, we are only looking at π with the additional property that $\pi \vee \sigma = 1_n$, which means that π has to couple all blocks of σ, compare Remark 11.14. Hence all appearing elements must be from the same subalgebra, which is in contradiction with $m \geq 2$. Thus there is no non-vanishing contribution in the above sum and we get $\kappa_n(a_1, \dots, a_n) = 0$.

(ii) $\Longrightarrow$ (i): Consider $a_j \in \mathcal{A}_{i(j)}$ $(j = 1, \dots, n)$ with $i(1) \neq i(2) \neq \dots \neq i(n)$ and $\varphi(a_j) = 0$ for all $j = 1, \dots, n$. Then we have to show that $\varphi(a_1 \cdots a_n) = 0$. But this is clear because we have $\varphi(a_1 \cdots a_n) = \sum_{\pi \in NC(n)} \kappa_\pi[a_1, \dots, a_n]$ and each product $\kappa_\pi[a_1, \dots, a_n] = \prod_{V \in \pi} \kappa(V)[a_1, \dots, a_n]$ contains at least one factor of the form $\kappa_{p+1}(a_l, a_{l+1}, \dots, a_{l+p})$ which vanishes in any case (for $p = 0$ because our variables are centered and for $p \geq 1$ because of our assumption on the vanishing of mixed cumulants). $\square$

Cumulants of random variables

When we are dealing with random variables $(a_i)_{i\in I}$ living in some non-commutative probability space $(\mathcal{A}, \varphi)$, then we are mainly interested in the collection of all their joint moments or their joint distribution, i.e. in φ restricted to the algebra $\mathcal{A}_0 := \text{alg}(a_i \mid i \in I\}$. In the same spirit, we will mainly consider free cumulants restricted to $\mathcal{A}_0$, more specifically to cumulants whose arguments are the random variables themselves.

NOTATION 11.18. Let $(a_i)_{i\in I}$ be random variables in some non-commutative probability space $(\mathcal{A}, \varphi)$ and let $(\kappa_n)_{n\in\mathbb{N}}$ be the corresponding free cumulant functionals.

(1) We will call **free cumulants of $(a_i)_{i\in I}$** all expressions of the form $\kappa_n(a_{i(1)}, a_{i(2)}, \dots, a_{i(n)})$ for $n \in \mathbb{N}$ and $i(1), \dots, i(n) \in I$.

(2) If $(\mathcal{A}, \varphi)$ is a $*$-probability space, then by the **free $*$-cumulants of $(a_i)_{i\in I}$** we will mean the free cumulants of $(a_i, a_i^*)_{i\in I}$.

(3) If we have only one random variable a, then we will also use the notation $\kappa_n^a := \kappa_n(a, \dots, a)$.

REMARKS 11.19. (1) It is clear that the knowledge of all cumulants of $(a_i)_{i\in I}$ contains the same information as the family of joint moments of $(a_i)_{i\in I}$.

(2) In order to recognize cumulants of given random variables $(a_i)_{i\in I}$ it is worth giving the following explicit reformulation of Proposition 11.4. Assume we are given some complex numbers $\tilde{\kappa}_\pi[a_{i(1)}, \dots, a_{i(n)}]$ for all $n \in \mathbb{N}$, $\pi \in NC(n)$, $1 \le i(1), \dots, i(n) \le m$ such that:

(i) the $\tilde{\kappa}_\pi$ are multiplicative in the sense

$$\tilde{\kappa}_\pi[a_{i(1)}, \dots, a_{i(n)}] = \prod_{V\in\pi} \tilde{\kappa}(V)[a_1, \dots, a_n],$$

where, for $V = (r_1 < \dots < r_s) \in \pi$, we use the notation (11.2),

$$\tilde{\kappa}(V)[a_{i(1)}, \dots, a_{i(n)}] := \tilde{\kappa}_{1_s}(a_{i(r_1)}, \dots, a_{i(r_s)});$$

(ii) we can write the moments of $(a_i)_{i\in I}$ as

$$\varphi(a_{i(1)} \cdots a_{i(n)}) = \sum_{\pi\in NC(n)} \tilde{\kappa}_\pi[a_{i(1)}, \dots, a_{i(n)}]$$

for all $n \in \mathbb{N}$ and all $i(1), \dots, i(n) \in I$.

Then these $\tilde{\kappa}$ are the cumulants of $(a_i)_{i\in I}$, i.e.

$$\kappa_\pi[a_{i(1)}, \dots, a_{i(n)}] = \tilde{\kappa}_\pi[a_{i(1)}, \dots, a_{i(n)}]$$

for all $n \in \mathbb{N}$ and $\pi \in NC(n)$.

Let us now consider the question of how to recognize free independence between random variables $(a_i)_{i\in I}$ by looking at their cumulants. Theorem 11.16 tells us that we can decide on this by checking the vanishing of mixed cumulants, however, there we have to examine all mixed cumulants with entries from the subalgebras generated by our random variables. In the spirit of the present section we might hope that it is enough to consider the free cumulants of the random variables themselves (without having to invoke the generated subalgebras). The next theorem shows that this is indeed the case.

THEOREM 11.20. *Consider a non-commutative probability space* $(\mathcal{A}, \varphi)$ *and let* $(\kappa_n)_{n\in\mathbb{N}}$ *be the corresponding free cumulant functionals. Consider random variables* $(a_i)_{i\in I}$ *in* $\mathcal{A}$. *Then the following two statements are equivalent.*

(i) $(a_i)_{i\in I}$ *are freely independent.*

(ii) *We have for all* $n \geq 2$ *and for all* $i(1), \dots, i(n) \in I$ *that* $\kappa_n(a_{i(1)}, \dots, a_{i(n)}) = 0$ *whenever there exist* $1 \leq l, k \leq n$ *with* $i(l) \neq i(k)$.

PROOF. That (i) implies (ii) follows of course directly from Theorem 11.16. But the other way around is not immediately clear, since we have to show that our present assumption (ii) also implies the apparently stronger assumption (ii) for the case of algebras. Thus let $\mathcal{A}_i$ be the unital algebra generated by the element a_i and consider elements $b_j \in \mathcal{A}_{r(j)}$ $(j = 1, \dots, n)$ with $r(1), \dots, r(n) \in I$ such that $r(l) \neq r(k)$ for some l, k. Then we have to show that $\kappa_n(b_1, \dots, b_n)$ vanishes. As each b_j is a polynomial in $a_{r(j)}$ and since cumulants with a 1 as entry vanish always for $n \geq 2$, it suffices, by the multilinearity of the cumulants, to consider the case where each b_j is some power of $a_{r(j)}$. If we write $b_1 \cdots b_n$ as $a_{i(1)} \cdots a_{i(m)}$ then we have

$$\kappa_n(b_1, \dots, b_n) = \sum_{\substack{\pi \in NC(m) \\ \pi \vee \sigma = 1_m}} \kappa_\pi[a_{i(1)}, \dots, a_{i(m)}],$$

where the blocks of σ denote the neighboring elements which have to be multiplied to give the b_i. In order that $\kappa_\pi[a_{i(1)}, \dots, a_{i(m)}]$ is different from zero, we must have, by our assumption (ii), that $i(p) = i(q)$ whenever $p \sim_\pi q$. So all blocks of σ which are coupled by π must correspond to the same a_i. However, we only consider π for which we have $\pi \vee \sigma = 1_m$, which means that all blocks of σ have to be coupled by π. Thus all a_i should be the same, in contradiction with the fact that we consider a mixed cumulant. Hence there is no non-vanishing contribution in the above sum and we finally get our assertion $\kappa_n(b_1, \dots, b_n) = 0$. □

Example: semicircular and circular elements

EXAMPLES 11.21. Let us record here the cumulants of semicircular elements and families.

(1) The second part of Remarks 11.19 allows us to extract directly the cumulants of semicircular variables from knowledge of their moments. Recall from Lectures 2 and 8 (in particular, Corollary 2.14 and Lemma 8.9) that the moments of a semicircular element s of variance σ^2 are given by the number of non-crossing pairings, i.e.

$$\varphi(s^{2k}) = \sigma^{2k} \cdot \#NC_2(2k) = \sigma^{2k} \sum_{\pi \in NC_2(2k)} 1 = \sum_{\pi \in NC(2k)} \prod_{V \in \pi} \kappa(V),$$

where

$$\kappa(V) = \begin{cases} \sigma^2 & \text{if } \#V = 2 \\ 0 & \text{otherwise.} \end{cases}$$

This tells us that the second order cumulant of s is equal to σ^2 and all other cumulants are zero,

$$\kappa_n^s = \delta_{n2}\sigma^2. \tag{11.13}$$

(2) More generally, Definition 8.15 of a semicircular family $(s_i)_{i\in I}$ of covariance $(c_{ij})_{i\in I}$ can be stated in the equivalent form that

$$\kappa_n(s_{i(1)}, \ldots, s_{i(n)}) = \delta_{n2} c_{i(1)i(2)}. \tag{11.14}$$

Another important random variable in free probability is the non-normal version of a semicircular element – the circular element. This is the replacement of a complex normal distribution in the world of free probability.

DEFINITION 11.22. An element c of the form $c = \frac{1}{\sqrt{2}}(s_1 + is_2)$, where s_1 and s_2 are two freely independent semicircular elements of variance 1, is called a **circular element**.

EXAMPLE 11.23. The vanishing of mixed cumulants in free variables gives directly the cumulants of a circular element. Since only second order cumulants of semicircular elements are different from zero, the only non-vanishing cumulants of a circular element are also of second order and for these we have

$$\kappa_2(c, c) = \kappa_2(c^*, c^*) = \frac{1}{2} - \frac{1}{2} = 0$$
$$\kappa_2(c, c^*) = \kappa_2(c^*, c) = \frac{1}{2} + \frac{1}{2} = 1.$$

Even elements

NOTATIONS 11.24. Let $(\mathcal{A}, \varphi)$ be a non-commutative probability space.

(1) We call an element $x \in \mathcal{A}$ **even** if all its odd moments vanish, i.e. if $\varphi(x^{2k+1}) = 0$ for all $k \geq 0$.

(2) Let x be an even element. We will call $(\alpha_n)_{n\geq 1}$ with $\alpha_n := \kappa_{2n}^x$ the **determining sequence** of the variable x.

It is immediately seen that the vanishing of all odd moments is equivalent to the vanishing of all odd cumulants and thus the determining sequence contains all information about the distribution of an even element. Another way of encoding the information about an even element x is by looking at the distribution of x^2. Actually, there is a very precise way of relating these two descriptions.

PROPOSITION 11.25. *Let x be an even element with determining sequence $(\alpha_n)_{n\geq 1}$. Then the cumulants of x^2 are given as follows:*

$$\kappa_n(x^2, \ldots, x^2) = \sum_{\pi \in NC(n)} \alpha_\pi, \tag{11.15}$$

where α_π is the multiplicative extension to NC of the determining sequence, i.e. for any $\pi \in NC(n)$

$$\alpha_\pi = \prod_{V \in \pi} \alpha_{|V|}.$$

The proof of this statement relies mainly on our formula for cumulants with products as arguments and a detailed study of what the condition on $\pi \vee \sigma$ means in this case. This proof is very typical for many of our investigations in free probability theory and various variants of these arguments will show up at different places in the rest of Part 2.

PROOF. Applying Theorem 11.12 yields

$$\kappa_n(xx, \ldots, xx) = \sum_{\substack{\pi \in NC(2n) \\ \pi \vee \sigma = 1_{2n}}} \kappa_\pi[x, x, \ldots, x, x]$$

with $\sigma = \{(1,2), (3,4), \ldots, (2n-1, 2n)\} \in NC(2n)$.

We claim now the following:

$$\{\pi \in NC(2n) \mid \pi \vee \sigma = 1_{2n}\} =$$
$$\{\pi \in NC(2n) \mid 1 \sim_\pi 2n,\ 2k \sim_\pi 2k+1\ \forall k = 1, \ldots, n-1\}. \tag{11.16}$$

Since the right-hand side of (11.16) is in canonical bijection with $NC(n)$ and since $k_\pi[x, x, \ldots, x, x]$ goes under this bijection to the product α_π, this gives the assertion directly.

So it remains to prove the claim. It is clear that a partition which has the claimed property does also fulfill $\pi \vee \sigma = 1_{2n}$. So we only have to prove the other direction.

Let V be the block of π which contains the element 1. Since x is even, the last element of this block has to be an even number. (Otherwise there would be an odd number of x which must be coupled among themselves by π.) If this even number were not $2n$, but $2k$ for $1 \leq k < n$, then we would have a situation as follows ...

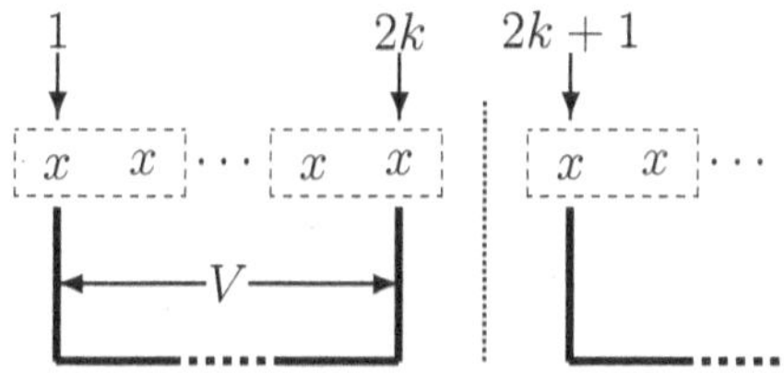

...and V would not be connected in $\pi \vee \sigma$ to the block containing $2k+1$. Hence $\pi \vee \sigma = 1_{2n}$ implies that the block containing the first element 1 contains also the last element $2n$.

Now fix a $k = 1, \ldots, n-1$ and let V be the block of π containing the element $2k$. Assume that V does not contain the element $2k+1$. Then there are two possibilities.

Either $2k$ is not the last element in V, i.e. there exists a next element in V, which is necessarily of the form $2l+1$ with $l > k$:

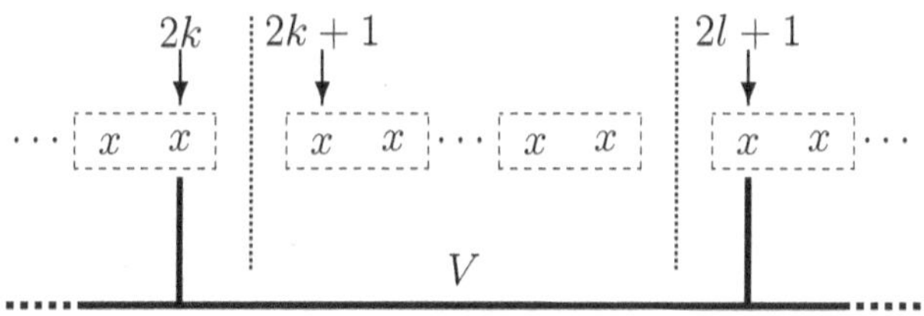

Or $2k$ is the last element in V, in which case the first element of V is of the form $2m+1$ with $0 \leq m \leq k-1$:

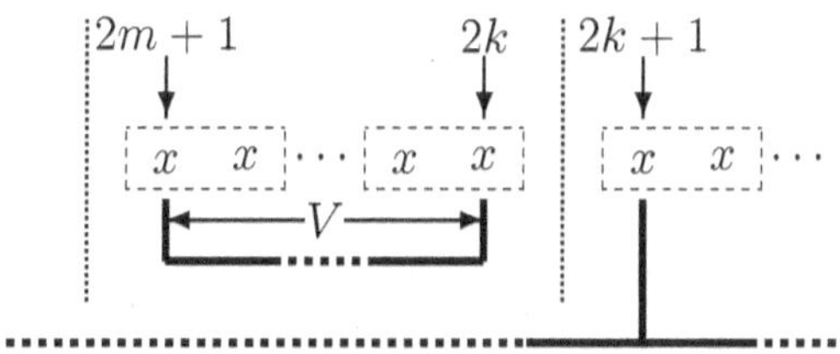

But in both cases the block V is not connected with the element $2k+1$ in $\pi \vee \sigma$, so that this cannot give 1_{2n}. Hence the condition $\pi \vee \sigma = 1_{2n}$ forces $2k$ and $2k+1$ to lie in the same block. This proves our claim and hence the assertion.

□

EXAMPLE 11.26. As an example for the application of the previous proposition, let us calculate the cumulants for the square of a semicircular variable s of variance σ^2. Then we have $\alpha_n = \delta_{n1}$ and thus

$$\alpha_\pi = \begin{cases} \sigma^{2n} & \text{if } \pi = 0_n \\ 0 & \text{otherwise.} \end{cases}$$

Hence (11.15) says in this case

$$\kappa_n(s^2, \dots, s^2) = \sigma^{2n}.$$

We will come back to the relevance of this result in the next lecture in the context of free Poisson distributions, see Proposition 12.13.

Appendix: classical cumulants

The theory of free cumulants is of course inspired by an analogous theory of classical cumulants. As we have pointed out repeatedly, from the combinatorial point of view the difference between classical probability theory and free probability theory consists in replacing the lattice of all partitions by the lattice of non-crossing partitions. In this section, we want to be a bit more explicit on this and provide, for comparison, the definition and some main properties of classical cumulants.

NOTATIONS 11.27. (1) Recall from Lecture 9, Definition 9.1 and Remark 9.19, that $\mathcal{P}(n)$ denotes the lattice of all partitions of the set $\{1, \dots, n\}$, equipped with the usual reversed refinement order as partial order. Furthermore, we use the notation

$$\mathcal{P} := \bigcup_{n=1}^{\infty} \mathcal{P}(n).$$

(2) We extend a linear functional on an algebra $\mathcal{A}$ to a corresponding multiplicative function on all partitions in the same way as we did in Notation 11.2 for non-crossing partitions, namely by ($\pi \in \mathcal{P}(n)$, $a_1, \dots, a_n \in \mathcal{A}$)

$$\varphi_\pi[a_1, \dots, a_n] := \prod_{V \in \pi} \varphi(V)[a_1, \dots, a_n], \tag{11.17}$$

where we use our usual notation $\varphi(V)[a_1, \dots, a_n] := \varphi_s(a_{i_1}, \dots, a_{i_s})$ for $V = (i_1 < \dots < i_s) \in \pi$.

DEFINITION 11.28. Let $(\mathcal{A}, \varphi)$ be a non-commutative probability space. Then, for $\pi \in \mathcal{P}(n)$, we define the **classical cumulants** c_π as a multilinear functional by

$$c_\pi[a_1, \ldots, a_n] = \sum_{\substack{\sigma \in \mathcal{P}(n) \\ \sigma \leq \pi}} \varphi_\sigma[a_1, \ldots, a_n] \cdot \mu_{\mathcal{P}(n)}(\sigma, \pi), \qquad (11.18)$$

where $\mu_{\mathcal{P}(n)}$ denotes the Möbius function on $\mathcal{P}(n)$. Sometimes, these classical cumulants are also called **semi-invariants**.

The above definition is, by Möbius inversion on $\mathcal{P}(n)$, equivalent to

$$\varphi(a_1 \cdots a_n) = \sum_{\pi \in \mathcal{P}(n)} c_\pi[a_1, \ldots, a_n].$$

As in the non-crossing case, one shows that $(c_\pi)_{n \in \mathbb{N}, \pi \in \mathcal{P}(n)}$ is a multiplicative family of functions on $\mathcal{P}$ and it is thus determined by the values of

$$c_n(a_1, \ldots, a_n) := c_{1_n}[a_1, \ldots, a_n].$$

EXAMPLE 11.29. Let us compare free and classical cumulants for small values of n. Since $NC(n)$ and $\mathcal{P}(n)$ agree up to $n = 3$, we have that $c_1 = \kappa_1$, $c_2 = \kappa_2$, and $c_3 = \kappa_3$. For $n = 4$, let us consider the special case of centered variables, $\varphi(a_i) = 0$ for $i = 1, \ldots, 4$. Then we have

$$\begin{aligned} c_4(a_1, a_2, a_3, a_4) &= \varphi(a_1 a_2 a_3 a_4) \\ &\quad - \varphi(a_1 a_2)\varphi(a_3 a_4) - \varphi(a_1 a_4)\varphi(a_2 a_3) - \varphi(a_1 a_3)\varphi(a_2 a_4), \end{aligned}$$

whereas

$$\kappa_4(a_1, a_2, a_3, a_4) = \varphi(a_1 a_2 a_3 a_4) - \varphi(a_1 a_2)\varphi(a_3 a_4) - \varphi(a_1 a_4)\varphi(a_2 a_3).$$

One has now the analogs of Theorems 11.12 and 11.16 for classical cumulants. For the first one, observe that our map $\tau \mapsto \hat{\tau}$ can be extended in a canonical way to an embedding $\mathcal{P}(m) \to \mathcal{P}(n)$.

THEOREM 11.30. *Consider a non-commutative probability space $(\mathcal{A}, \varphi)$ and let $(c_\pi)_{\pi \in \mathcal{P}}$ be the corresponding free cumulants. Let $m, n \in \mathbb{N}$ and $1 \leq i(1) < i(2) < \cdots < i(m) = n$ be given and denote by $\tau \mapsto \hat{\tau}$ the corresponding embedding $\mathcal{P}(m) \to \mathcal{P}(n)$. Recall that*

$$\hat{0}_m = \{(1, \ldots, i(1)), \ldots, (i(m-1)+1, \ldots, i(m))\}.$$

Consider now random variables $a_1, \ldots, a_n \in \mathcal{A}$. Then we have

$$c_m(a_1 \cdots a_{i(1)}, \ldots, a_{i(m-1)+1} \cdots a_{i(m)}) = \sum_{\substack{\pi \in \mathcal{P}(n) \\ \pi \vee \hat{0}_m = 1_n}} c_\pi[a_1, \ldots, a_n]. \quad (11.19)$$

EXAMPLE 11.31. Let us compare the statement of this theorem for $c_2(a_1a_2, a_3a_4)$ with the corresponding statement for $\kappa_2(a_1a_2, a_3a_4)$. In order to reduce the number of involved terms we consider the special case where $\varphi(a_i) = 0$ for all $i = 1, 2, 3, 4$. In the classical case there are three partitions $\pi \in \mathcal{P}(4)$ without singletons which satisfy

$$\pi \vee \{(1,2),(3,4)\} = 1_4,$$

namely

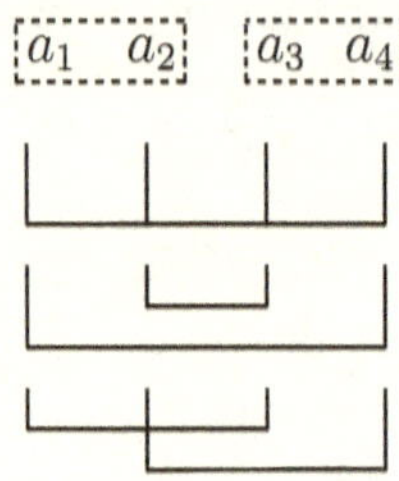

and thus Theorem 11.30 gives in this case

$$c_2(a_1a_2, a_3a_4) = c_4(a_1, a_2, a_3, a_4) \\ + c_2(a_1, a_4)c_2(a_2, a_3) + c_2(a_1, a_3)c_2(a_2, a_4).$$

In the free case, only the first two, non-crossing partitions contribute and the corresponding formula from Theorem 11.12 yields

$$\kappa_2(a_1a_2, a_3a_4) = \kappa_4(a_1, a_2, a_3, a_4) + \kappa_2(a_1, a_4)\kappa_2(a_2, a_3).$$

Classical cumulants have been considered in classical probability theory – usually in terms of Fourier transforms, see Exercise 11.37 – for a long time. Their relevance comes of course from the following characterization, which is the perfect analog of Theorem 11.16.

THEOREM 11.32. *Consider a non-commutative probability space* $(\mathcal{A}, \varphi)$ *and let* $(c_n)_{n\in\mathbb{N}}$ *be the corresponding classical cumulants. Consider unital subalgebras* $(\mathcal{A}_i)_{i\in I}$ *of* $\mathcal{A}$ *which commute. Then the following two statements are equivalent.*

(i) $(\mathcal{A}_i)_{i\in I}$ *are tensor independent*

(ii) *We have for all* $n \geq 2$ *and for all* $a_j \in \mathcal{A}_{i(j)}$ $(j = 1, \ldots, n)$ *with* $i(1), \ldots, i(n) \in I$ *that* $c_n(a_1, \ldots, a_n) = 0$ *whenever there exist* $1 \leq l, k \leq n$ *with* $i(l) \neq i(k)$.

Exercises

EXERCISE 11.33. Let $(\mathcal{A}, \varphi)$ be a probability space and $\mathcal{X}_1, \mathcal{X}_2 \subset \mathcal{A}$ two subsets of $\mathcal{A}$. Show that the following two statements are equivalent.

(i) We have for all $n \in \mathbb{N}$, $1 \leq k < n$ and all $a_1, \ldots, a_k \in \mathcal{X}_1$ and $a_{k+1}, \ldots, a_n \in \mathcal{X}_2$ that $\varphi(a_1 \cdots a_k a_{k+1} \cdots a_n) = \varphi(a_1 \cdots a_k) \cdot \varphi(a_{k+1} \cdots a_n)$.

(ii) We have for all $n \in \mathbb{N}$, $1 \leq k < n$ and all $a_1, \ldots, a_k \in \mathcal{X}_1$ and $a_{k+1}, \ldots, a_n \in \mathcal{X}_2$ that $\kappa_n(a_1, \ldots, a_k, a_{k+1}, \ldots, a_n) = 0$.

EXERCISE 11.34. We will use the following notations. A partition $\pi \in \mathcal{P}(n)$ is called decomposable if there exists an interval $I = \{k, k+1, \ldots, k+r\} \neq \{1, \ldots, n\}$ (for some $k \geq 1$, $0 \leq r \leq n-r$), such that π can be written in the form $\pi = \pi_1 \cup \pi_2$, where $\pi_1 \in \mathcal{P}(\{k, k+1, \ldots, k+r\})$ is a partition of I and $\pi_2 \in \mathcal{P}(1, \ldots, k-1, k+r+1, \ldots, n\})$ is a partition of $\{1, \ldots, n\} \backslash I$. If there does not exist such a decomposition of π, then we call π indecomposable. A function $t : \bigcup_{n \in \mathbb{N}} \mathcal{P}(n) \to \mathbb{C}$ is called NC-multiplicative, if we have for each decomposition $\pi = \pi_1 \cup \pi_2$ as above that $t(\pi_1 \cup \pi_2) = t(\pi_1) \cdot t(\pi_2)$ (π_1 and π_2 are here identified with partitions in $\mathcal{P}(r+1)$) and $\mathcal{P}(n-r-1)$, respectively, in the obvious way.)

Consider now a random variable a whose moments are given by the formula

$$\varphi(a^n) = \sum_{\pi \in \mathcal{P}(n)} t(\pi), \tag{11.20}$$

where t is a NC-multiplicative function on the set of all partitions. Show that the free cumulants of a are then given by

$$\kappa_n(a, \ldots, a) = \sum_{\substack{\pi \in \mathcal{P}(n) \\ \pi \text{ indecomposable}}} t(\pi). \tag{11.21}$$

EXERCISE 11.35. Let b be a symmetric Bernoulli variable, i.e. a self-adjoint random variable whose distribution is the probability measure $\frac{1}{2}(\delta_{-1} + \delta_1)$. In terms of moments this means:

$$\varphi(b^n) = \begin{cases} 1 & \text{if } n \text{ even} \\ 0 & \text{if } n \text{ odd.} \end{cases} \tag{11.22}$$

Show that the cumulants of b are given by the following formula:

$$\kappa_n(b, \ldots, b) = \begin{cases} (-1)^{k-1} C_{k-1} & \text{if } n = 2k \text{ even} \\ 0 & \text{if } n \text{ odd.} \end{cases} \tag{11.23}$$

EXERCISE 11.36. This exercise refers to classical cumulants

(1) Prove Theorem 11.30.

(2) Prove Theorem 11.32.

(3) Calculate the classical cumulants of a Gaussian distribution and of a classical Poisson distribution.

EXERCISE 11.37. In this exercise we want to establish the connection between the combinatorial Definition 11.28 of classical cumulants and the more common formulation in terms of Fourier transforms. We restrict the problem here to the case of one random variable.

(1) Let $(m_n)_{n\geq 1}$ be the moments of a random variable and $(c_n)_{n\geq 1}$ the corresponding classical cumulants. Consider the exponential generating power series

$$A(z) := 1 + \sum_{n=1}^{\infty} \frac{m_n}{n!} z^n$$

and

$$B(z) := \sum_{n=1}^{\infty} \frac{c_n}{n!} z^n.$$

Show that the combinatorial relation

$$m_n = \sum_{\pi \in \mathcal{P}(n)} c_\pi$$

between the coefficients of these power series is equivalent to the relation

$$B(z) = \log(A(z))$$

between the power series themselves.

(2) Use the previous part of this exercise to prove the following. Let ν be a compactly supported probability measure on $\mathbb{R}$ and $\mathcal{F}$ its Fourier transform, defined by

$$\mathcal{F}(t) := \int_{\mathbb{R}} e^{-itx}\, d\nu(x).$$

Then, with $(m_n)_{n\geq 1}$ and $(c_n)_{n\geq 1}$ denoting the moments and classical cumulants, respectively, of ν, we have the power series expansions

$$\mathcal{F}(t) = 1 + \sum_{n=1}^{\infty} \frac{(-it)^n}{n!} m_n$$

and

$$\log \mathcal{F}(t) = \sum_{n=1}^{\infty} \frac{(-it)^n}{n!} c_n.$$

LECTURE 12

Sums of free random variables

Our main concern in this lecture will be the understanding and effective description of the sum of freely independent random variables. How can we calculate the distribution of $a + b$ if a and b are free and if we know the distribution of a and the distribution of b. Of particular interest is the case of selfadjoint random variables x and y in a C^*-probability space. In this case their distributions can be identified with probability measures on $\mathbb{R}$ and thus taking the sum of free random variables gives rise to a binary operation on probability measures on $\mathbb{R}$. We will call this operation "free convolution," in analogy with the usual concept of convolution of probability measures which corresponds to taking the sum of classically independent random variables. Our combinatorial approach to free probability theory, resting on the notion of free cumulants, will give us very easy access to the main results of Voiculescu on this free convolution via the so-called "$\mathcal{R}$-transform."

Free convolution

DEFINITION 12.1. Let μ and ν be probability measures on $\mathbb{R}$ with compact support. Let x and y be selfadjoint random variables in some C^*-probability space such that x has distribution μ, y has distribution ν, and such that x and y are freely independent. Then the distribution of the sum $x + y$ is called the **free convolution** of μ and ν and is denoted by $\mu \boxplus \nu$.

REMARKS 12.2. (1) Note that, for given μ and ν as above, one can always find x and y as required. For example, we can realize x and y as multiplication operators with the identity function on the Hilbert spaces $L^2(\mu)$ and $L^2(\nu)$, respectively, and then take the reduced free product of these C^*-probability spaces to make x and y freely independent. Furthermore, by Lemma 5.13, the distribution of the sum only depends on the distribution μ of x and on the distribution ν of y and not on the concrete realizations of x and y. Thus $\mu \boxplus \nu$ is well-defined.

(2) Since $x + y$ is selfadjoint and bounded, its distribution is also a compactly supported probability measure on $\mathbb{R}$. Thus $\boxplus$ is a binary operation on all compactly supported probability measures on $\mathbb{R}$. Without pursuing this line, we would like to mention that by adequate truncations one can extend the definition of (and the main results on) $\boxplus$ also to arbitrary probability measures on $\mathbb{R}$.

(3) We have to warn the reader that there will be another notion of free convolution, which will refer to the product of free random variables and which will appear in Lecture 14. In order to distinguish these two notions of free convolutions, $\boxplus$ is also called the "additive free convolution."

Our aim is to find an effective way of calculating the free convolution of two probability measures. According to our general philosophy that free independence is better described in terms of cumulants than in terms of moments, we should check what adding free variables means for its cumulants.

Recall from Notation 11.18 that, for a random variable a we put

$$\kappa_n^a := \kappa_n(a, \dots, a)$$

and call $(\kappa_n^a)_{n \geq 1}$ the "free cumulants of a." Clearly, the free cumulants of a contain the same information as the moments of a. However, free cumulants behave much better with respect to taking sums of free variables. This is a direct consequence of the vanishing of mixed cumulants in free random variables and the multilinearity of our cumulant functionals.

PROPOSITION 12.3. *Let a and b be free random variables in some non-commutative probability space. Then we have*

$$\kappa_n^{a+b} = \kappa_n^a + \kappa_n^b \qquad \forall n \geq 1. \tag{12.1}$$

PROOF. We have

$$\begin{aligned} \kappa_n^{a+b} &= \kappa_n(a+b, \dots, a+b) \\ &= \kappa_n(a, \dots, a) + \kappa_n(b, \dots, b) \\ &= \kappa_n^a + \kappa_n^b, \end{aligned}$$

because cumulants which have both a and b as arguments vanish by Theorem 11.16. □

Thus, the sum of freely independent random variables is easy to describe on the level of cumulants: the cumulants are additive in such a case.

Of course, the main problem has now been shifted to the connection between moments and cumulants. Let us recall what our Definition 11.3 for the free cumulants means in the case of one variable: If $(m_n)_{n\geq 1}$ and $(\kappa_n)_{n\geq 1}$ are the moments and the free cumulants, respectively, of some random variable, then the connection between these two sequences of numbers is given by our moment-cumulant formula

$$m_n = \sum_{\pi \in NC(n)} \kappa_\pi, \tag{12.2}$$

where $\pi \mapsto \kappa_\pi$ is the multiplicative extension of cumulants to non-crossing partitions, i.e.

$$\kappa_\pi := \kappa_{|V_1|} \cdots \kappa_{|V_r|} \qquad \text{for} \qquad \pi = \{V_1, \dots, V_r\} \in NC(n).$$

EXAMPLES 12.4. Let us write down the moment-cumulant formula and its Möbius inversion in this special case of one random variable for small n. Of course, this is just a specialization of the corresponding formulas from Examples 11.6.

(1) For $n = 1$, we have

$$m_1 = \kappa_{|} = \kappa_1$$

and thus

$$\kappa_1 = m_1.$$

(2) For $n = 2$, we have

$$m_2 = \kappa_{\sqcup} + \kappa_{|\,|} = \kappa_2 + \kappa_1^2,$$

and thus

$$\kappa_2 = m_2 - m_1^2.$$

(3) For $n = 3$, we have

$$m_3 = \kappa_{\sqcup\!\sqcup} + \kappa_{|\,\sqcup} + \kappa_{\sqcup\,|} + \kappa_{\sqcup\!\!|\!\!\sqcup} + \kappa_{|\,|\,|} = \kappa_3 + 3\kappa_1\kappa_2 + \kappa_1^3,$$

and thus

$$\kappa_3 = m_3 - 3m_1m_2 + 2m_1^3.$$

For concrete calculations, however, one would prefer to have a more analytical description of the relation between moments and cumulants. This can be achieved by translating the moment-cumulant formula to a formula which involves the corresponding formal power series. Note that we treated this problem in Lecture 10, in the context of multiplicative families of functions on non-crossing partitions. Namely, if we use the moments m_n to build a multiplicative family of functions f_n on NC, by putting $f_n(1_n) := m_n$ for all $n \geq 1$, and the cumulants κ_n to built a multiplicative family of functions g_n on NC, by putting $g_n(1_n) := \kappa_n$ for all $n \geq 1$, then the relation (12.2) between moments and cumulants amounts to $f = g * \zeta$. What this means for the relation between the $(m_n)_{n\geq 1}$ and $(\kappa_n)_{n\geq 1}$ in terms of generating power series was the content of Theorem 10.23. Let us reformulate this in our present language.

THEOREM 12.5. *Let $(m_n)_{n\geq 1}$ and $(\kappa_n)_{n\geq 1}$ be the moments and free cumulants, respectively, of some random variable and consider the corresponding formal power series*

$$M(z) := 1 + \sum_{n=1}^{\infty} m_n z^n \tag{12.3}$$

and

$$C(z) := 1 + \sum_{n=1}^{\infty} \kappa_n z^n. \tag{12.4}$$

Then we have

$$C[zM(z)] = M(z). \tag{12.5}$$

If our random variable is a selfadjoint element in a C^*-probability space, hence its distribution is a probability measure on $\mathbb{R}$, it is advantageous to consider instead of the moment-generating series $M(z)$ the closely related Cauchy transform, because the latter has nice analytic properties and allows in particular recovery of the corresponding probability measure concretely with the help of Stieltjes inversion formula (compare Remark 2.20). Recall that the Cauchy transform for a probability measure μ is defined by

$$G_\mu(z) = \int_{\mathbb{R}} \frac{d\mu(t)}{z - t}$$

and has in the case of a compactly supported μ the following power series expansion about $z = \infty$

$$G_\mu(z) = \sum_{n=0}^{\infty} \frac{m_n}{z^{n+1}}, \tag{12.6}$$

where m_n are the moments of μ. This means of course that we have

$$G_\mu(z) = \frac{1}{z} M(1/z). \tag{12.7}$$

(This equation can be either considered as equality between formal power series, or as an analytic equation for sufficiently large z.) Thus we can reformulate the previous theorem in terms of G_μ. When Voiculescu first discovered these relations, he formulated the results not in terms of the cumulant-generating series $C(z)$, but in terms of a closely related series, which he called "$\mathcal{R}$-transform."

NOTATION 12.6. Let μ be a compactly supported probability measure on $\mathbb{R}$. Denote by $(\kappa_n)_{n\geq 1}$ the free cumulants of μ. The $\mathcal{R}$-**transform** of μ is the formal power series

$$\mathcal{R}_\mu(z) := \sum_{n=0}^{\infty} \kappa_{n+1} z^n. \tag{12.8}$$

Now we can combine the reformulation of our above Theorem 12.5 in terms of Cauchy transform and $\mathcal{R}$-transform together with the $\mathcal{R}$-transform formulation of Proposition 12.3 to obtain the main theorem about the description of additive free convolution.

THEOREM 12.7. (1) *The relation between the Cauchy transform $G_\mu(z)$ and the $\mathcal{R}$-transform $\mathcal{R}_\mu(z)$ of a probability measure μ is given by*

$$G_\mu\big[\mathcal{R}_\mu(z) + 1/z\big] = z. \tag{12.9}$$

(2) *The $\mathcal{R}$-transform linearizes the free convolution, i.e. if μ and ν are compactly supported probability measures on $\mathbb{R}$, then we have*

$$\mathcal{R}_{\mu\boxplus\nu}(z) = \mathcal{R}_\mu(z) + \mathcal{R}_\nu(z). \tag{12.10}$$

PROOF. (1) We just have to note that the formal power series $M(z)$ and $C(z)$ from Theorem 12.5 and $G(z)$, $\mathcal{R}(z)$, and $K(z) = \mathcal{R}(z) + \frac{1}{z}$ are related by (12.7) and

$$C(z) = 1 + z\mathcal{R}(z) = zK(z), \qquad \text{thus} \qquad K(z) := \frac{C(z)}{z}.$$

This gives (where we leave verification of the validity of these formal power series manipulations as a straightforward exercise to the reader)

$$K[G(z)] = \frac{1}{G(z)} C[G(z)] = \frac{1}{G(z)} C\Big[\frac{M(1/z)}{z}\Big] = \frac{1}{G(z)} M(1/z) = z,$$

thus $K[G(z)] = z$ and hence also

$$G[\mathcal{R}(z) + 1/z] = G[K(z)] = z.$$

(2) Since, by Proposition 12.3, the coefficients of the $\mathcal{R}$-transform are additive under free convolution, the same is of course also true for the $\mathcal{R}$-transform itself. □

The $\mathcal{R}$-transform was introduced by Voiculescu as the main tool for dealing with free convolution and the above Theorem 12.7 presents two of his main results about this. His treatment was much more analytical, and showed the existence of the free cumulants of a random variable, but without giving a concrete combinatorial description of them.

Analytic calculation of free convolution

The above Theorem 12.7 provides us with a quite effective machinery for calculating the free convolution. Let μ, ν be probability measures on $\mathbb{R}$, then we can calculate $\mu \boxplus \nu$ as follows. Out of μ and ν we calculate G_μ and G_ν, respectively, then we use the relation between Cauchy transform and $\mathcal{R}$-transform, Theorem 12.7, to calculate the corresponding $\mathcal{R}$-transforms $\mathcal{R}_\mu$ and $\mathcal{R}_\nu$. The free convolution on the level of $\mathcal{R}$-transforms is now quite easily described by $\mathcal{R}_{\mu\boxplus\nu}(z) = \mathcal{R}_\mu(z) + \mathcal{R}_\nu(z)$. It remains to go over to $G_{\mu\boxplus\nu}$ by invoking once again Theorem 12.7 and finally to use the Stieltjes inversion formula to recover $\mu \boxplus \nu$ itself. All these calculations should be done on the level of analytic functions, not just as formal power series manipulations.

Of course, explicit formulas for the transition between the Cauchy transform and the $\mathcal{R}$-transform might not always be obtainable, but the following examples show that non-trivial examples can be treated.

EXAMPLES 12.8. (1) Let

$$\mu = \nu = \frac{1}{2}(\delta_{-1} + \delta_{+1}).$$

Then we have

$$G_\mu(z) = \int \frac{1}{z-t}\, d\mu(t) = \frac{1}{2}\Big(\frac{1}{z+1} + \frac{1}{z-1}\Big) = \frac{z}{z^2-1}.$$

Put

$$K_\mu(z) = \frac{1}{z} + \mathcal{R}_\mu(z).$$

Then $z = G_\mu[K_\mu(z)]$ gives

$$K_\mu(z)^2 - \frac{K_\mu(z)}{z} = 1,$$

which has as solutions

$$K_\mu(z) = \frac{1 \pm \sqrt{1+4z^2}}{2z}.$$

Thus the $\mathcal{R}$-transform of μ is given by

$$\mathcal{R}_\mu(z) = K_\mu(z) - \frac{1}{z} = \frac{\sqrt{1+4z^2}-1}{2z}.$$

(Note: $\mathcal{R}_\mu(0) = k_1(\mu) = m_1(\mu) = 0$, so that we have to choose the plus-sign for the square root.) Hence we get

$$\mathcal{R}_{\mu\boxplus\mu}(z) = 2\mathcal{R}_\mu(z) = \frac{\sqrt{1+4z^2}-1}{z},$$

and

$$K(z) := K_{\mu\boxplus\mu}(z) = \mathcal{R}_{\mu\boxplus\mu}(z) + \frac{1}{z} = \frac{\sqrt{1+4z^2}}{z},$$

which allows us to determine $G := G_{\mu\boxplus\mu}$ via

$$z = K[G(z)] = \frac{\sqrt{1+4G(z)^2}}{G(z)}$$

as

$$G(z) = \frac{1}{\sqrt{z^2-4}}.$$

From this we can calculate the density

$$\frac{d(\mu\boxplus\mu)(t)}{dt} = -\frac{1}{\pi}\lim_{\varepsilon\to 0}\Im\frac{1}{\sqrt{(t+i\varepsilon)^2-4}} = -\frac{1}{\pi}\Im\frac{1}{\sqrt{t^2-4}},$$

so that we finally get

$$\frac{d(\mu\boxplus\mu)(t)}{dt} = \begin{cases} \frac{1}{\pi\sqrt{4-t^2}} & |t|\le 2 \\ 0 & \text{otherwise.} \end{cases} \tag{12.11}$$

Thus $\mu\boxplus\mu$ is the arcsine distribution. Note that in the corresponding classical case, one gets a binomial distribution

$$\mu * \mu = \frac{1}{4}\delta_{-2} + \frac{1}{2}\delta_0 + \frac{1}{4}\delta_{+2}.$$

Thus we are justified in calling $\mu\boxplus\mu$ a "free binomial distribution."

(2) In the same way as above, we can also calculate $\mu^{\boxplus 4}$ instead of $\mu^{\boxplus 2}$. Note that this corresponds to the free convolution of the arcsine distribution with itself and thus should recover the result of Kesten for the moment-generating series for the Laplacian in the free group $\mathbb{F}_2$; see Example 4.5. The calculation proceeds as above; the $\mathcal{R}$-transform for $\mu^{\boxplus 4}$ is given by

$$\mathcal{R}(z) = 2\frac{\sqrt{1+4z^2}-1}{z},$$

which results in a Cauchy transform

$$G(z) = \frac{-z+2\sqrt{z^2-12}}{z^2-16}.$$

If we rewrite this in the moment-generating series $M(z) = \frac{1}{z}G(1/z)$ we get exactly the result mentioned in (4.9),

$$M(z) = \frac{2\sqrt{1-12z^2}-1}{1-16z^2}.$$

More general $\mu^{\boxplus n}$ for all $n\in\mathbb{N}$ will be addressed in Exercise 12.21.

EXERCISE 12.9. For a probability measure μ and a real number $r \in \mathbb{R}$ we denote by $S_r(\mu)$ the probability measure which is the shift of μ by the amount r, i.e. for measurable $A \subset \mathbb{R}$

$$S_r(\mu)(A) := \mu(A - r), \qquad \text{where } A - r := \{t - r \mid t \in A\}. \tag{12.12}$$

Show that the free convolution has the property

$$\mu \boxplus \delta_r = S_r(\mu). \tag{12.13}$$

REMARKS 12.10. The above examples reveal some properties of the free convolution, which are quite surprising compared to the corresponding classical situation.

(1) The free convolution has the property that the convolution of discrete distributions can be an absolutely continuous distribution (i.e. a distribution which has a density with respect to Lebesgue measure).

(2) In particular, we see that $\boxplus$ is not distributive with respect to convex combinations of probability measures. If we put, as before, $\mu := \frac{1}{2}(\delta_{-1} + \delta_{+1})$, then we have seen that $\mu \boxplus \mu$ is the arcsine distribution. However, by invoking also the above Exercise 12.9, we have

$$\frac{1}{2}(\delta_{-1} \boxplus \mu) + \frac{1}{2}(\delta_{+1} \boxplus \mu) = \frac{1}{2}S_{-1}(\mu) + \frac{1}{2}S_{+1}(\mu) = \frac{1}{4}\delta_{-2} + \frac{1}{2}\delta_0 + \frac{1}{4}\delta_2,$$

so that we see:

$$\frac{1}{2}(\delta_{-1} + \delta_{+1}) \boxplus \mu \neq \frac{1}{2}(\delta_{-1} \boxplus \mu) + \frac{1}{2}(\delta_{+1} \boxplus \mu).$$

Proof of the free central limit theorem via $\mathcal{R}$-transform

With the help of the $\mathcal{R}$-transform machinery we can now give a more analytic and condensed proof of the free central limit theorem. Since free cumulants are polynomials in moments and vice versa, the convergence of moments is equivalent to the convergence of cumulants. Consider random variables $a_1, a_2, \ldots$, which are free, identically distributed, centered, and with variance σ^2. In order to prove that the sum $(a_1 + \cdots + a_N)/\sqrt{N}$ converges in distribution to a semicircular variable s of variance σ^2, it suffices thus to show that

$$\mathcal{R}_{(a_1+\cdots+a_N)/\sqrt{N}}(z) \to \mathcal{R}_s(z) = \sigma^2 z$$

in the sense of convergence of the coefficients of the formal power series. It is easy to see that

$$\mathcal{R}_{\lambda a}(z) = \lambda \mathcal{R}_a(\lambda z).$$

Thus we get

$$\begin{aligned}
\mathcal{R}_{(a_1+\cdots+a_N)/\sqrt{N}}(z) &= \frac{1}{\sqrt{N}} \cdot \mathcal{R}_{a_1+\cdots+a_N}\Big(\frac{z}{\sqrt{N}}\Big) \\
&= N\frac{1}{\sqrt{N}} \cdot \mathcal{R}_{a_i}\Big(\frac{z}{\sqrt{N}}\Big) \\
&= \sqrt{N} \cdot \mathcal{R}_{a_i}\Big(\frac{z}{\sqrt{N}}\Big) \\
&= \sqrt{N} \cdot \Big(\kappa_1 + \kappa_2\frac{z}{\sqrt{N}} + \kappa_3\frac{z^2}{N} + \dots\Big) \\
&= \sqrt{N} \cdot \Big(\sigma^2\frac{z}{\sqrt{N}} + \kappa_3\frac{z^2}{N} + \dots\Big) \\
&\to \sigma^2 z,
\end{aligned}$$

since $\kappa_1 = 0$ and $\kappa_2 = \sigma^2$.

Free Poisson distribution

One of the most prominent distributions in classical probability theory beyond the normal distribution is the Poisson distribution. One can get a classical Poisson distribution as the limit in distribution for $N \to \infty$ of

$$\Big(\Big(1 - \frac{\lambda}{N}\Big)\delta_0 + \frac{\lambda}{N}\delta_\alpha\Big)^{*N}.$$

See Exercise 12.22 for a precise formulation. Usually, the parameters λ and α are called the "rate" and the "jump size" of the limiting Poisson distribution (the latter referring to the fact that in the corresponding Poisson process α is the size of the possible jumps; for the distribution it just means that it is concentrated on natural multiples of α.)

Let us look at the free counterpart of that limit theorem. Clearly, we will call a distribution appearing there in the limit a "free Poisson distribution." We will also use the names "rate" and "jump size" for the parameters, although the latter clearly has no real meaning in the non-commutative context.

PROPOSITION 12.11. **(Free Poisson limit theorem)** *Let $\lambda \geq 0$ and $\alpha \in \mathbb{R}$. Then the limit in distribution for $N \to \infty$ of*

$$\Big(\Big(1 - \frac{\lambda}{N}\Big)\delta_0 + \frac{\lambda}{N}\delta_\alpha\Big)^{\boxplus N}$$

has free cumulants $(\kappa_n)_{n\geq 1}$ which are given by

$$\kappa_n = \lambda\alpha^n \qquad (n \geq 1).$$

This corresponds to a probability measure ν on $\mathbb{R}$ which is given by

$$\nu = \begin{cases} (1-\lambda)\delta_0 + \lambda\tilde{\nu} & \text{if } 0 \le \lambda \le 1 \\ \tilde{\nu} & \text{if } \lambda > 1, \end{cases} \tag{12.14}$$

where $\tilde{\nu}$ is the measure supported on the interval

$$[\alpha(1-\sqrt{\lambda})^2, \alpha(1+\sqrt{\lambda})^2]$$

with density

$$d\tilde{\nu}(t) = \frac{1}{2\pi\alpha t}\sqrt{4\lambda\alpha^2 - \big(t - \alpha(1+\lambda)\big)^2}\,dt. \tag{12.15}$$

Note that our choice of the parameters λ and α corresponds exactly to the requirement that, for sufficiently large N, $(1-\lambda/N)\delta_0 + \lambda/N\delta_\alpha$, and thus also its N-fold free convolution power, is a probability measure on $\mathbb{R}$.

DEFINITION 12.12. For given $\lambda \ge 0$ and $\alpha \in \mathbb{R}$, the probability measure given by (12.14) and (12.15) is called **free Poisson distribution with rate $\boldsymbol{\lambda}$ and jump size $\boldsymbol{\alpha}$**.

PROOF. Let us put

$$\nu_N := \Big(1 - \frac{\lambda}{N}\Big)\delta_0 + \frac{\lambda}{N}\delta_\alpha.$$

This is a probability measure on $\mathbb{R}$ for $N \ge \lambda$. In the following we will always assume that N is large enough. We denote by $m_n(\nu)$ and $\kappa_n(\nu)$ the nth moment and nth cumulant, respectively, of a measure ν. Then we have

$$m_n(\nu_N) = \frac{\lambda}{N}\alpha^n$$

and thus by the moment-cumulant formula (note that $\mu(\cdot,\cdot)$ denotes here the Möbius function on NC)

$$\begin{aligned}
\kappa_n(\nu_N) &= \sum_{\pi \in NC(n)} m_\pi(\nu_N)\mu(\pi, 1_n) \\
&= \frac{\lambda}{N}\alpha^n + \sum_{\substack{\pi \in NC(n) \\ \pi \neq 1_n}} m_\pi(\nu_N)\mu(\pi, 1_n) \\
&= \frac{\lambda}{N}\alpha^n + O(1/N^2),
\end{aligned}$$

where $O(1/N^2)$ denotes a term which is bounded by a constant times $1/N^2$. Thus we have

$$\kappa_n(\nu_N^{\boxplus N}) = N \cdot \kappa_n(\nu_N) = \lambda\alpha^n + O(1/N).$$

Letting N go to infinity shows that the nth cumulant of ν_N converges towards $\kappa_n = \lambda\alpha^n$. Since the convergence of all cumulants is equivalent to the convergence of all moments (for a formal proof of this in a more general situation, see Lemma 13.2) it only remains to calculate the explicit form of the limit from knowledge of its free cumulants. For this we will use the $\mathcal{R}$-transform machinery.

The $\mathcal{R}$-transform of the limit is given by

$$\mathcal{R}(z) = \sum_{n=0}^{\infty} \kappa_{n+1} z^n = \sum_{n=0}^{\infty} \lambda\alpha^{n+1} z^n = \lambda\alpha \frac{1}{1-\alpha z}.$$

Thus the corresponding Cauchy transform $G(z)$ has to fulfill the equation

$$\lambda\alpha \frac{1}{1-\alpha G(z)} + \frac{1}{G(z)} = z,$$

which has the solutions

$$G(z) = \frac{z + \alpha - \lambda\alpha \pm \sqrt{\left(z - \alpha(1+\lambda)\right)^2 - 4\lambda\alpha^2}}{2\alpha z}.$$

Since $G(z)$ must behave like $1/z$ for large z we have to choose the minus-sign; application of Stieltjes inversion formula leads finally to the asserted form of the probability measure. □

If we now look back at Example 11.26 then we see that we have found there a quite surprising realization of a free Poisson element. Let us state the result of that example in our present language.

PROPOSITION 12.13. *The square of a semicircular element of variance σ^2 is a free Poisson element of rate $\lambda = 1$ and jump size $\alpha = \sigma^2$.*

REMARKS 12.14. (1) We want to emphasize that the analogous relation in classical probability theory does not hold: the square of a normal variable is, of course, not a classical Poisson variable. In contrast to the free case, there is no direct relation between normal and Poisson variables in the classical world. This is a manifestation of the general observation that the lattice of non-crossing partitions has "more structure" than the lattice of all partitions, which results in relations in free probability theory without classical precedent.

(2) The combinatorial explanation behind the fact that the square of a semicircular element is a free Poisson element is that the number of non-crossing pairings of a set of $2n$ elements is the same as the number of non-crossing pairings of a set of n elements, namely both are equal to the Catalan number C_n. In the classical world, both corresponding numbers have no clear relation. The number of pairings of a set of $2n$

elements (i.e. the even moments of a Gaussian variable of variance 1) is $(2n-1)\cdot(2n-3)\cdots 5\cdot 3\cdot 1$, whereas the number of partitions of a set of n elements (i.e. the moments of a Poisson variable) is counted by the so-called "Bell numbers" B_n. For more information on those, see Exercise 12.23.

Compound free Poisson distribution

There exists a natural generalization of the class of classical Poisson distributions – the so-called "compound Poisson distributions." One possibility for defining them is via a generalization of the limit theorem for the Poisson distribution. Again, we have canonical counterparts of this notion and results in the free world. We present the free version of the limit theorem in the following proposition. The proof of this is left to the reader.

PROPOSITION 12.15. *Let $\lambda \geq 0$ and ν a probability measure on $\mathbb{R}$ with compact support. Then the limit in distribution for $N \to \infty$ of*

$$\left(\left(1-\frac{\lambda}{N}\right)\delta_0 + \frac{\lambda}{N}\nu\right)^{\boxplus N}$$

has free cumulants $(\kappa_n)_{n\geq 1}$ which are given by

$$\kappa_n = \lambda \cdot m_n(\nu) \qquad (n \geq 1), \tag{12.16}$$

and thus an $\mathcal{R}$-transform of the form

$$\mathcal{R}(z) = \lambda \int_{\mathbb{R}} \frac{x}{1-xz}\, d\nu(x).$$

DEFINITION 12.16. A probability measure μ on $\mathbb{R}$ with free cumulants $\kappa_n = \kappa_n(\mu)$ of the form (12.16) for some $\lambda > 0$ and some compactly supported probability measure ν on $\mathbb{R}$, is called a **compound free Poisson distribution** (with rate λ and jump distribution ν).

REMARK 12.17. We can recover, of course, a free Poisson element as a special compound free Poisson element for the choice $\nu = \delta_\alpha$. A general compound free Poisson distribution can be thought of as a superposition of freely independent Poisson distributions; see Exercise 12.25 for more details on this.

Again, there is a relation between semicircular elements and compound Poisson elements.

PROPOSITION 12.18. *Let* $(\mathcal{A}, \varphi)$ *be a non-commutative probability space. Let* $s, a \in \mathcal{A}$ *such that* s *is a semicircular element of variance* 1 *and such that* s *and* a *are free. Then the cumulants of* sas *are given by*

$$\kappa_n(sas, \ldots, sas) = \varphi(a^n) \qquad \forall n \geq 1. \tag{12.17}$$

In particular, if a *is living in a* C^**-probability space and is selfadjoint with distribution* ν*, then* sas *is a compound free Poisson element of rate* $\lambda = 1$ *and jump distribution* ν*.*

PROOF. By Theorem 11.12, we get

$$\kappa_n(sas, \ldots, sas) = \sum_{\substack{\pi \in NC(3n) \\ \pi \vee \sigma = 1_{3n}}} \kappa_\pi[s, a, s, s, a, s, \ldots, s, a, s],$$

where $\sigma \in NC(3n)$ is the partition

$$\sigma = \{(1, 2, 3), (4, 5, 6), \ldots, (3n-2, 3n-1, 3n)\}. \tag{12.18}$$

Since s and a are freely independent, their mixed cumulants vanish and κ_π can only give a non-vanishing contribution if no block of π connects an s with an a. Furthermore, since s is semicircular, each block which connects among the s has to consist of exactly two elements. But then the requirement $\pi \vee \sigma = 1_{3n}$ implies, by our usual arguments (see, for instance, the proof of Proposition 11.25) that the blocks of π which connect among the s must look like this:

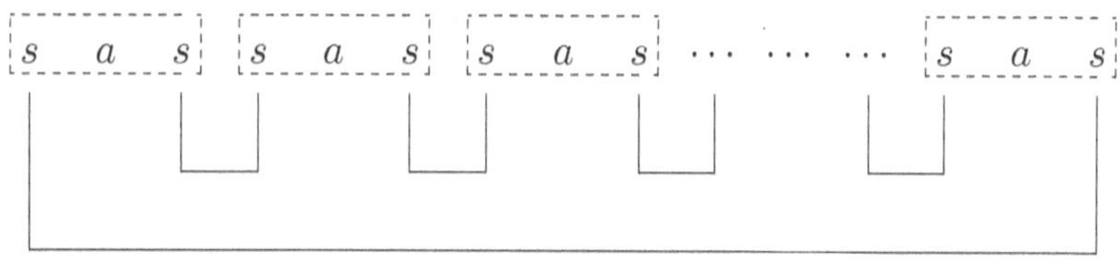

This means that π must be of the form $\pi_s \cup \pi_a$ where π_s is the special partition

$$\pi_s = \{(1, 3n), (3, 4), (6, 7), (9, 10), \ldots, (3n-3, 3n-2)\} \\ \in NC(1, 3, 4, 6, 7, 9, \ldots, 3n-3, 3n-2, 3n),$$

and where π_a is a partition restricted to the position of the a. Since π_s glues the blocks of σ together, π_a does not have to fulfill any constraint and can be an arbitrary element in $NC(2, 5, 8, \ldots, 3n-1)$. Since $k_\pi[s, a, s, s, a, s, \ldots, s, a, s]$ factorizes for $\pi = \pi_s \cup \pi_a$ into

$$\begin{aligned}\kappa_{\pi_s \cup \pi_a}[s, a, s, s, a, s, \ldots, s, a, s] &= \kappa_{\pi_s}[s, s, \ldots, s] \cdot \kappa_{\pi_a}[a, a, \ldots, a] \\ &= \kappa_{\pi_a}[a, a, \ldots, a],\end{aligned}$$

we get finally

$$\kappa_n(sas, \dots, sas) = \sum_{\pi_a \in NC(n)} k_{\pi_a}[a, a, \dots, a] = \varphi(a^n).$$

□

EXAMPLE 12.19. As a generalization of the last proposition, consider now the following situation. Let $(\mathcal{A}, \varphi)$ be a non-commutative probability space. Consider $s, a_1, \dots, a_m \in \mathcal{A}$ such that s is a semicircular element of variance 1 and such that s and $\{a_1, \dots, a_m\}$ are freely independent. Put $P_i := sa_is$. As above we can calculate the joint distribution of these elements as

$$\begin{aligned} \kappa_n(P_{i(1)}, \dots, P_{i(n)}) &= \kappa_n(sa_{i(1)}s, \dots, sa_{i(n)}s) \\ &= \sum_{\substack{\pi \in NC(3n) \\ \pi \vee \sigma = 1_{3n}}} k_\pi[s, a_{i(1)}, s, s, a_{i(2)}, s, \dots, s, a_{i(n)}, s] \\ &= \sum_{\pi_a \in NC(n)} \kappa_{\pi_a}[a_{i(1)}, a_{i(2)}, \dots, a_{i(n)}] \\ &= \varphi(a_{i(1)}a_{i(2)} \cdots a_{i(n)}), \end{aligned}$$

where $\sigma \in NC(3n)$ is as before the partition given in (12.18). Thus we have again the result that the cumulants of $P_1, \dots, P_m$ are given by the moments of $a_1, \dots, a_m$. This contains of course the statement that each of the P_i is a compound Poisson element, but we also get that orthogonality between the a_i is translated into free independence between the P_i. Namely, assume that all a_i are orthogonal in the sense $a_ia_j = 0$ for all $i \neq j$. Consider now a mixed cumulant in the P_i, i.e. $\kappa_n(P_{i(1)}, \dots, P_{i(n)})$, with $i(l) \neq i(k)$ for some l, k. Of course, then there are also two neighboring indices which are different, i.e. we can assume that $k = l + 1$. But then we have

$$\kappa_n(P_{i(1)}, \dots, P_{i(l)}, P_{i(l+1)}, \dots, P_{i(n)}) = \varphi(a_{i(1)} \dots a_{i(l)}a_{i(l+1)} \dots a_{i(n)}) = 0.$$

Thus mixed cumulants in the P_i vanish and, by our Theorem 11.20, $P_1, \dots, P_m$ have to be freely independent.

Exercises

EXERCISE 12.20. Let $(\mathcal{A}, \varphi)$ be a non-commutative probability space and consider a family of random variables ("stochastic process") $(a_t)_{t \geq 0}$ with $a_t \in \mathcal{A}$ for all $t \geq 0$. Consider, for $0 \leq s < t$, the following formal power series

$$G(t,s)=\sum_{n=0}^{\infty}\int\cdots\int_{t\geq t_1\geq\cdots\geq t_n\geq s}\varphi(a_{t_1}\dots a_{t_n})\,dt_1\dots dt_n, \tag{12.19}$$

which can be considered as a kind of replacement for the Cauchy transform. We will now consider a generalization to this case of Voiculescu's formula for the connection between Cauchy transform and $\mathcal{R}$-transform.

(a) Denote by $\kappa_n(t_1,\dots,t_n):=\kappa_n(a_{t_1},\dots,a_{t_n})$ the free cumulants of $(a_t)_{t\geq 0}$. Show that $G(t,s)$ fulfills the following differential equation

$$\begin{aligned}
&\frac{d}{dt}G(t,s) \qquad\qquad (12.20)\\
&=\sum_{n=0}^{\infty}\int\cdots\int_{t\geq t_1\geq\cdots\geq t_n\geq s}\kappa_{n+1}(t,t_1,\dots,t_n)\cdot G(t,t_1)\\
&\qquad\qquad\cdot G(t_1,t_2)\cdots G(t_{n-1},t_n)\cdot G(t_n,s)\,dt_1\dots dt_n\\
&=\kappa_1(t)G(t,s)+\int_s^t\kappa_2(t,t_1)\cdot G(t,t_1)\cdot G(t_1,s)\,dt_1\\
&\quad+\iint_{t\geq t_1\geq t_2\geq s}\kappa_3(t,t_1,t_2)\cdot G(t,t_1)\cdot G(t_1,t_2)\cdot G(t_2,s)\,dt_1dt_2+\cdots.
\end{aligned}$$

(b) Show that in the special case of a constant process, i.e. $a_t=a$ for all $t\geq 0$, the above differential equation goes over, after Laplace transformation, into Voiculescu's formula for the connection between Cauchy transform and $\mathcal{R}$-transform.

EXERCISE 12.21. For $\mu=\frac{1}{2}(\delta_{-1}+\delta_{+1})$ calculate the density of the "free binomial distributions" $\mu^{\boxplus n}$ for all natural n. (The resulting measures were found by Kesten in the context of random walks on free groups.) Does the result also make sense for non-integer n? (We will come back to this question in Example 14.15.)

EXERCISE 12.22. Show (for example, by using Fourier transforms) that the limit of

$$\left(\left(1-\frac{\lambda}{N}\right)\delta_0+\frac{\lambda}{N}\delta_\alpha\right)^{*N},$$

is given by the classical Poisson distribution of rate λ and jump size α, i.e. by the probability distribution

$$e^{-\lambda}\sum_{k=0}^{\infty}\frac{\lambda^k}{k!}\delta_{k\alpha}$$

($\delta_{k\alpha}$ is here the discrete probability measure with mass 1 at $k\alpha$). This is sometimes called the "law of rare events."

EXERCISE 12.23. Let B_n denote the number of partitions of the set $(1,\ldots,n)$. Show that the exponential generating series of these Bell numbers is given by

$$\sum_{n=1}^{\infty} \frac{B_n}{n!} z^n = \exp(e^z - 1) \tag{12.21}$$

Determine the value of B_n for small n.

EXERCISE 12.24. Prove the limit theorem for compound free Poisson distribution, Proposition 12.15.

EXERCISE 12.25. (1) Let a_1 and a_2 be free Poisson elements with rate λ_1 and λ_2 and jump size α_1 and α_2, respectively. Assume that a_1 and a_2 are freely independent. Show that $a_1 + a_2$ is a compound free Poisson element. Show that it is a free Poisson element if and only if $\alpha_1 = \alpha_2$.

(2) Show that any compound free Poisson a can be approximated in distribution by elements a_n where each a_n is the sum of finitely many freely independent free Poisson elements.

LECTURE 13

More about limit theorems and infinitely divisible distributions

The theory of infinitely divisible distributions is an important development in classical probability theory which generalizes the central limit theorem and the Poisson limit theorem. In this lecture we want to develop the basics of the free counterpart of that theory.

Limit theorem for triangular arrays

We will now consider a more general limit theorem, corresponding to triangular arrays of free random variables. In the corresponding classical situation we get in the limit the so-called infinitely divisible distributions. We will see later that we have an analogous characterization in our free situation. In order to keep our considerations simple we restrict to the case where we have identical distributions within the rows of our triangle.

We will formulate our general limit theorem for the case of families of random variables, indexed by some set I. Readers who do not feel comfortable with such a generality should rewrite the theorem for the special case where I consists of one element; for this, see also our Remark 13.4.

THEOREM 13.1. **(Free limit theorem for triangular arrays)** *Let, for each $N \in \mathbb{N}$, $(\mathcal{A}_N, \varphi_N)$ be a non-commutative probability space. Let I be an index set. Consider a triangular field of random variables, i.e. for each $i \in I$, $N \in \mathbb{N}$ and $1 \leq r \leq N$ we have a random variable $a_{N;r}^{(i)} \in \mathcal{A}_N$. Assume that, for each fixed $N \in \mathbb{N}$, the sets $\{a_{N;1}^{(i)}\}_{i\in I}, \dots, \{a_{N;N}^{(i)}\}_{i\in I}$ are free and identically distributed. Then the following statements are equivalent.*

(1) *The sums over the rows of our triangle converge in distribution, i.e. there is a family of random variables $(b_i)_{i\in I}$ in some non-commutative probability space such that*

$$\left(a_{N;1}^{(i)} + \cdots + a_{N;N}^{(i)}\right)_{i\in I} \xrightarrow{\text{distr}} (b_i)_{i\in I}. \tag{13.1}$$

(2) *For all* $n \geq 1$ *and all* $i(1), \dots, i(n) \in I$ *the limits*

$$\lim_{N\to\infty} N \cdot \varphi_N(a_{N;r}^{(i(1))} \cdots a_{N;r}^{(i(n))}) \tag{13.2}$$

(which are independent of r *by the assumption of identical distribution) exist.*

Furthermore, if these conditions are satisfied, then the joint distribution of the limit family $(b_i)_{i\in I}$ *is determined in terms of free cumulants by* ($n \geq 1$, $i(1), \dots, i(n) \in I$)

$$\kappa_n(b_{i(1)}, \dots, b_{i(n)}) = \lim_{N\to\infty} N \cdot \varphi_N(a_{N;r}^{(i(1))} \cdots a_{N;r}^{(i(n))}). \tag{13.3}$$

Since our main tools are cumulants and not moments, it is good to convince oneself that the statement on convergence of moments is equivalent to convergence of cumulants.

LEMMA 13.2. *Let* $(\mathcal{A}_N, \varphi_N)$ *be a sequence of probability spaces and let, for each* $i \in I$, *a random variable* $a_N^{(i)} \in \mathcal{A}_N$ *be given. Denote by* κ^N *the free cumulants corresponding to* φ_N. *Then the following two statements are equivalent.*

(1) *For each* $n \geq 1$ *and each* $i(1), \dots, i(n) \in I$ *the limit*

$$\lim_{N\to\infty} N \cdot \varphi_N(a_N^{(i(1))} \cdots a_N^{(i(n))})$$

exists.

(2) *For each* $n \geq 1$ *and each* $i(1), \dots, i(n) \in I$ *the limit*

$$\lim_{N\to\infty} N \cdot \kappa_n^N(a_N^{(i(1))}, \dots, a_N^{(i(n))})$$

exists.

Furthermore the corresponding limits are the same.

PROOF. (2) $\Longrightarrow$ (1): We have

$$\lim_{N\to\infty} N \cdot \varphi_N(a_N^{(i(1))} \cdots a_N^{(i(n))}) = \lim_{N\to\infty} \sum_{\pi\in NC(n)} N \cdot \kappa_\pi^N[a_N^{(i(1))}, \dots, a_N^{(i(n))}].$$

By assumption (2), all terms for π with more than one block tend to zero and the term for $\pi = 1_n$ tends to the finite limit given by (2). The other direction (1) $\Longrightarrow$ (2) is analogous. □

PROOF OF THEOREM 13.1. We write

$$\varphi_N\big((a_{N;1}^{(i(1))} + \cdots + a_{N;N}^{(i(N))})^n\big) = \sum_{r(1),\dots,r(n)=1}^{N} \varphi_N(a_{N;r(1)}^{(i(1))} \cdots a_{N;r(n)}^{(i(n))})$$

and observe that for fixed N a lot of terms in the sum give the same contribution. Namely, the tuples $(r(1), \dots, r(n))$ and $(r'(1), \dots, r'(n))$

give the same contribution if the indices agree at the same places. As in the case of the central limit theorem (see, in particular, (8.1)), we encode this relevant information by a partition π (which might a priori be a crossing partition). Let $(r(1), \dots, r(n))$ be an index-tuple corresponding to a fixed π, i.e. $p \sim_\pi q$ if and only if $r(p) = r(q)$ $(p, q = 1, \dots, n)$. Then we can write

$$\begin{aligned}\varphi_N(a_{N;r(1)}^{(i(1))} \cdots a_{N;r(n)}^{(i(n))}) &= \sum_{\sigma \in NC(n)} \kappa_\sigma^N [a_{N;r(1)}^{(i(1))}, \dots, a_{N;r(n)}^{(i(n))}] \\ &= \sum_{\substack{\sigma \in NC(n) \\ \sigma \leq \pi}} \kappa_\sigma^N [a_{N;r}^{(i(1))}, \dots, a_{N;r}^{(i(n))}]\end{aligned}$$

(where the latter expression is independent of r). The last equality comes from the fact that elements belonging to different blocks of π are free. The number of tuples $(r(1), \dots, r(n))$ corresponding to π is of order $N^{|\pi|}$, thus we get

$$\begin{aligned}&\lim_{N\to\infty} \varphi_N\big((a_{N;1}^{(i(1))} + \cdots + a_{N;N}^{(i(n))})^n\big) \\ &\qquad = \sum_{\pi \in \mathcal{P}(n)} \sum_{\substack{\sigma \in NC(n) \\ \sigma \leq \pi}} \lim_{N\to\infty} N^{|\pi|} \kappa_\sigma^N [a_{N;r}^{(i(1))}, \dots, a_{N;r}^{(i(n))}]. \qquad (13.4)\end{aligned}$$

By taking suitable polynomials in the left-hand expressions (namely, classical cumulants of them), we see that the convergence in distribution of the sum of our rows is equivalent to the existence of all limits

$$\sum_{\substack{\sigma \in NC(n) \\ \sigma \leq \pi}} \lim_{N\to\infty} N^{|\pi|} \kappa_\sigma^N [a_{N;r}^{(i(1))}, \dots, a_{N;r}^{(i(n))}]$$

for all $\pi \in \mathcal{P}(n)$. By induction, the latter is equivalent to the existence of all limits

$$\lim_{N\to\infty} N^{|\pi|} \kappa_\sigma^N [a_{N;r}^{(i(1))}, \dots, a_{N;r}^{(i(n))}]$$

for any $\pi \in \mathcal{P}(n)$ and $\sigma \in NC(n)$ with $\sigma \leq \pi$. By invoking also Lemma 13.2, this is equivalent to the existence of all limits (13.2).

If the existence of these limits is assumed then we get non-vanishing contributions in (13.4) exactly in those cases where the power of N agrees with the number of factors from the cumulants κ_σ. This means that $|\pi| = |\sigma|$, which can only be the case if π itself is a non-crossing partition and $\sigma = \pi$. But this gives, again by using Lemma 13.2, exactly the assertion on the cumulants of the limiting distribution. □

EXERCISE 13.3. Apply the limit theorem 13.1 to the special situations treated in our limit theorems for free Poisson and compound free

Poisson distributions from the last lecture and check that one recovers the statements from Propositions 12.11 and 12.15.

REMARK 13.4. In the case of C^*-probability spaces and I consisting of just one element, this theorem reduces to the statement that, for given probability measures μ_N on $\mathbb{R}$ with compact support, the convergence in distribution of $\mu_N^{\boxplus N}$ to some compactly supported probability measure μ is equivalent to the fact that

$$\kappa_n^\mu = \lim_{N\to\infty} N \cdot m_n(\mu_N).$$

Our main goal in the following will be to characterize the possible limits in this case.

Cumulants of operators on Fock space

Before we study further the possible limit distributions of triangular arrays, we would like to give a nice application of that limit theorem; it can be used to determine very easily the cumulants of creation, annihilation and gauge operators on a full Fock space. We use here the notations as introduced in Lecture 7 in Definitions 7.13 and 7.24.

PROPOSITION 13.5. *Let $\mathcal{H}$ be a Hilbert space and consider the C^*-probability space $(B(\mathcal{F}(\mathcal{H})), \tau_\mathcal{H})$. Then the cumulants of the random variables $l(f)$, $l^*(g)$, $\Lambda(T)$ $(f, g \in \mathcal{H}, T \in B(\mathcal{H}))$ are of the following form. We have $(n \geq 2, f, g \in \mathcal{H}, T_1, \dots, T_{n-2} \in B(\mathcal{H}))$*

$$\kappa_n(l^*(f), \Lambda(T_1), \dots, \Lambda(T_{n-2}), l(g)) = \langle T_1 \dots T_{n-2} g, f\rangle \tag{13.5}$$

and all other cumulants with arguments from the set $\{l(f) \mid f \in \mathcal{H}\} \cup \{l^*(g) \mid g \in \mathcal{H}\} \cup \{\Lambda(T) \mid T \in B(\mathcal{H})\}$ vanish.

PROOF. For $N \in \mathbb{N}$, put

$$\mathcal{H}_N := \underbrace{\mathcal{H} \oplus \cdots \oplus \mathcal{H}}_{N \text{ times}}$$

and $(f, g \in \mathcal{H}, T \in B(\mathcal{H}))$

Then it is easy to see that the random variables $\{l(f), l^*(g), \Lambda(T) \mid f, g \in \mathcal{H}, T \in B(\mathcal{H})\}$ in $(B(\mathcal{F}(\mathcal{H})), \tau_\mathcal{H})$ have the same joint distribution as the random variables

$$\left\{l\Big(\frac{f \oplus \cdots \oplus f}{\sqrt{N}}\Big), l^*\Big(\frac{g \oplus \cdots \oplus g}{\sqrt{N}}\Big), \Lambda(T \oplus \cdots \oplus T) \mid f, g \in \mathcal{H}, T \in B(\mathcal{H})\right\}$$

in $(B(\mathcal{F}(\mathcal{H}_N)), \tau_{\mathcal{H}_N})$. The latter variables, however, are the sum of N free random variables, the summands having the same joint distribution as $\{l_N(f), l_N^*(g), \Lambda_N(T) \mid f, g \in \mathcal{H}, T \in B(\mathcal{H})\}$ in $(B(\mathcal{F}(\mathcal{H})), \tau_\mathcal{H})$, where

$$l_N(f) := \frac{1}{\sqrt{N}} l(f), \qquad l_N^*(g) := \frac{1}{\sqrt{N}} l^*(g), \qquad \Lambda_N(T) := \Lambda(T).$$

Hence we know from our limit theorem 13.1 that the cumulants $\kappa_n(a^{(1)}, \dots, a^{(n)})$ for $a^{(i)} \in \{l(f), l^*(g), \Lambda(T) \mid f, g \in \mathcal{H}, T \in B(\mathcal{H})\}$ can also be calculated as

$$\kappa_n(a^{(1)}, \dots, a^{(n)}) = \lim_{N\to\infty} N \cdot \varphi_{\mathcal{H}_N}(a_N^{(1)} \cdots a_N^{(n)}).$$

This yields directly the assertion. □

Infinitely divisible distributions

Finally, we wish to characterize the compactly supported probability measures on $\mathbb{R}$ which can arise as limits of triangular arrays. It will turn out that these are exactly the infinitely divisible ones.

DEFINITION 13.6. Let μ be a probability measure on $\mathbb{R}$ with compact support. We say that μ is **infinitely divisible** (in the free sense) if, for each positive integer n, there exists a probability measure μ_n, such that

$$\mu = (\mu_n)^{\boxplus n}. \tag{13.6}$$

REMARK 13.7. (1) Of course, we will denote the probability measure μ_n appearing in the above definition by $\mu^{\boxplus 1/n}$. One should note that the existence of $\mu^{\boxplus 1/n}$ as a linear (not necessarily positive) functional on polynomials is not a problem, the non-trivial requirement for μ being infinitely divisible is the existence of this $\mu^{\boxplus 1/n}$ in the class of probability measures.

(2) Since $\mu^{\boxplus p/q} = (\mu^{\boxplus 1/q})^{\boxplus p}$ for positive integers p, q, it follows that the rational convolution powers are probability measures. By continuity, we then also have that all convolution powers $\mu^{\boxplus t}$ for real $t > 0$ are probability measures. Thus the property "infinitely divisible" is equivalent to the existence of the convolution semigroup $\mu^{\boxplus t}$ in the class of probability measures for all $t > 0$.

In order to get a better understanding of the class of infinitely divisible distributions it would be good to have a sufficiently large supply of operators whose distributions are infinitely divisible. As we will see in the next proposition, sums of creation, annihilation and gauge operators on full Fock spaces are such a class. Even better, in the proof of Theorem 13.16 we will see that every infinitely divisible distribution can be realized in such a way.

PROPOSITION 13.8. *Let $\mathcal{H}$ be a Hilbert space and consider the C^*-probability space $(B(\mathcal{F}((\mathcal{H})), \tau_{\mathcal{H}})$. For $f \in \mathcal{H}$, $T = T^* \in B(\mathcal{H})$ and $\lambda \in \mathbb{R}$, let a be the selfadjoint operator*

$$a := l(f) + l^*(f) + \Lambda(T) + \lambda \cdot 1. \tag{13.7}$$

Then the distribution of a is infinitely divisible.

PROOF. Clearly, it suffices to consider the case where $\lambda = 0$. Let us use the frame which we have set up in the proof of Proposition 13.5. There we have remarked that $a = l(f) + l^*(f) + \Lambda(T)$ has the same distribution as the sum

$$\Big[l\Big(\frac{f \oplus 0 \oplus \cdots \oplus 0}{\sqrt{N}}\Big) + l^*\Big(\frac{f \oplus 0 \oplus \cdots \oplus 0}{\sqrt{N}}\Big) + \Lambda(T \oplus 0 \oplus \cdots \oplus 0)\Big] + \cdots$$
$$\cdots + \Big[l\Big(\frac{0 \oplus \cdots \oplus 0 \oplus f}{\sqrt{N}}\Big) + l^*\Big(\frac{0 \oplus \cdots \oplus 0 \oplus f}{\sqrt{N}}\Big) + \Lambda(0 \oplus \cdots \oplus 0 \oplus T)\Big].$$

The N summands are selfadjoint, all have the same distribution (which is a probability measure), and they are freely independent. Since we have this for each $N \in \mathbb{N}$ this yields the infinite divisibility of the distribution of a. □

EXERCISE 13.9. Realize a free Poisson distribution of rate λ and jump size α in the form $l(f) + l^*(f) + \Lambda(T) + \beta \cdot 1$ for suitably chosen f, T, and β.

Conditionally positive definite sequences

Before looking at the relation between infinitely divisible distributions and limit theorems, we want to determine what infinite divisibility means for the cumulants of the distribution. It will turn out that the following is the right concept for this problem.

NOTATION 13.10. Let $(t_n)_{n\geq 1}$ be a sequence of complex numbers. We say that $(t_n)_{n\geq 1}$ is **conditionally positive definite** if we have for all $r = 1, 2, \ldots$ and all $\alpha_1, \ldots, \alpha_r \in \mathbb{C}$ that

$$\sum_{n,m=1}^{r} \alpha_n \bar{\alpha}_m t_{n+m} \geq 0.$$

REMARKS 13.11. (1) Recall that a sequence $(s_n)_{n\geq 0}$ of complex numbers is called "positive definite" if we have for all $r = 0, 1, \ldots$ and all $\alpha_0, \ldots, \alpha_r \in \mathbb{C}$ that

$$\sum_{n,m=0}^{r} \alpha_n \bar{\alpha}_m s_{n+m} \geq 0.$$

This is equivalent to the fact that $\langle X^n, X^m\rangle := s_{n+m}$ defines a non-negative sesquilinear form on $\mathbb{C}\langle X\rangle$, the vector space of polynomials in one variable. The "conditional" in the notation above refers to the fact that conditionally positive definite sequences provide in the same way a non-negative sesquilinear form restricted to the subspace of polynomials without constant term.

(2) If $(t_n)_{n\geq 1}$ is a conditionally positive definite sequence, then we can define the shifted sequence $(s_n)_{n\geq 0}$ by $s_n := t_{n+2}$ for $n \geq 0$. (Note that we loose the information about t_1 by doing so; on the other hand, the value of t_1 is irrelevant for the question whether $(t_n)_{n\geq 1}$ is conditionally positive definite.) Clearly, the statement that $(t_n)_{n\geq 1}$ is conditionally positive definite is equivalent to the statement that $(s_n)_{n\geq 0}$ is positive definite, because

$$\sum_{n,m=0}^{r} \alpha_n \bar{\alpha}_m s_{n+m} = \sum_{n,m=0}^{r} \alpha_n \bar{\alpha}_m t_{n+m+2} = \sum_{n,m=0}^{r} \beta_{n+1} \bar{\beta}_{m+1} t_{(n+1)+(m+1)},$$

with $\beta_{n+1} = \alpha_n$ for $n = 0, \dots, r$.

(3) Let $(t_n)_{n\geq 1}$ be a conditionally positive definite sequence. What does this mean for the generating power series

$$\mathcal{R}(z) := \sum_{n=0}^{\infty} t_{n+1} z^n?$$

(Let us use $\mathcal{R}$ as notation for this power series, because we will only be interested in the case where the t_n are cumulants of a probability measure μ, and then indeed the considered power series is the $\mathcal{R}$-transform of μ.) It is easier to handle this question in terms of the shifted sequence $(s_n)_{n\geq 0}$ with $s_n := t_{n+2}$ $(n \geq 0)$. This sequence $(s_n)_{n\geq 0}$ is positive definite, which means that we are dealing with moments of a finite measure ρ,

$$s_n = \int x^n \, d\rho(x) \qquad (n \geq 0).$$

Thus our $\mathcal{R}(z)$ is

$$\begin{aligned} \mathcal{R}(z) &= t_1 + \sum_{n=0}^{\infty} s_n z^{n+1} \\ &= t_1 + \sum_{n=0}^{\infty} z \int_{\mathbb{R}} (xz)^n \, d\rho(x) \\ &= t_1 + \int_{\mathbb{R}} \frac{z}{1 - xz} \, d\rho(x). \end{aligned}$$

In order to stay within our frame of measures with compact support and thus also justify the last equation on an analytic level, we only consider the situation that ρ has compact support (in which case it is also uniquely determined by the s_n). In terms of the $(s_n)_{n\geq 0}$ this amounts to the requirement that they do not grow faster than exponentially in n. Let us work this out in the following.

DEFINITION 13.12. We say that a sequence $(s_n)_{n\geq 0}$ of complex numbers **does not grow faster than exponentially** if there exists a constant $c > 0$ such that

$$|s_n| \leq c^n \qquad \forall n \in \mathbb{N}.$$

LEMMA 13.13. *Let ρ be a finite measure on $\mathbb{R}$ and let*

$$s_n := \int_{\mathbb{R}} x^n \, d\rho(x) \qquad (n \in \mathbb{N})$$

be its moments. Then the following statements are equivalent.

(1) *ρ has compact support.*

(2) *The moments s_n exist for all $n \in \mathbb{N}$ and the sequence $(s_n)_{n\in\mathbb{N}}$ does not grow faster than exponentially.*

PROOF. (1) $\Longrightarrow$ (2): Assume ρ has compact support included in the interval $[-R, R]$. Put

$$c := \max(1, \rho(\mathbb{R})) < \infty.$$

Then we have

$$|s_n| \leq \int_{-R}^{R} |x^n| \, d\rho(x) \leq R^n \rho(\mathbb{R}) \leq (Rc)^n.$$

(2) $\Longrightarrow$ (1): This follows from the fact that the support of ρ is bounded by $\lim_{n\to\infty} \sqrt[2n]{s_{2n}}$ (for this, compare the proof of Proposition 3.17). Thus, if we assume an exponential bound $|s_n| \leq c^n$, this yields as bound for the support of ρ

$$\lim_{n\to\infty} \sqrt[2n]{s_{2n}} \leq \lim_{n\to\infty} \sqrt[2n]{c^{2n}} = c.$$

□

Clearly, the condition of exponential growth for a sequence $(s_n)_n$ is the same as for the shifted sequence $(t_n)_n = (s_{n-2})_n$. Thus we can summarize our considerations from the above Remark 13.11 as follows.

PROPOSITION 13.14. *Let $(t_n)_{n\geq 1}$ be a sequence of complex numbers which does not grow faster than exponentially. Then the following statements are equivalent.*

(1) *The sequence $(t_n)_{n\geq 0}$ is conditionally positive definite.*

(2) *There exists a finite measure ρ on $\mathbb{R}$ with compact support such that*

$$\sum_{n=0}^{\infty} t_{n+1} z^n = t_1 + \int_{\mathbb{R}} \frac{z}{1-xz} \, d\rho(x).$$

If we want to apply this to the characterization of infinitely divisible distributions we need one more observation. Our sequence (t_n) will be given by the cumulants of a compactly supported probability measure μ. By the above Lemma 13.13, the moments of μ do not grow faster than exponentially. However, what we really need is that the cumulants have this property. That this is indeed equivalent follows from the fact that the size of $NC(n)$ and also its Möbius function do not grow faster than exponentially.

PROPOSITION 13.15. *Let $(\mathcal{A}, \varphi)$ be a non-commutative probability space and consider a random variable $a \in \mathcal{A}$. Let $m_n := \varphi(a^n)$ $(n \in \mathbb{N})$ and $\kappa_n := \kappa_n^a$ be the moments and free cumulants, respectively, of a. Then the fact that the sequence $(m_n)_n$ of moments does not grow faster than exponentially is equivalent to the fact that the sequence $(\kappa_n)_n$ of cumulants does not grow faster than exponentially.*

PROOF. Let us assume that the cumulants do not grow faster than exponentially. By the moment-cumulant formula, we have $m_n = \sum_{\pi \in NC(n)} \kappa_\pi$. Note that the assumption $|\kappa_n| \leq c^n$ implies that we also have $|\kappa_\pi| \leq c^n$ for all $\pi \in NC(n)$. Thus, we get

$$|m_n| \leq \sum_{\pi \in NC(n)} |\kappa_\pi| \leq \#NC(n) \cdot c^n.$$

If we note now that for all $n \in \mathbb{N}$

$$\#NC(n) = C_n = \frac{1}{n+1}\binom{2n}{n} \leq 4^n$$

then we get the desired growth bound for the moments.

For the other direction, we also need the fact that the Möbius function $\mu(\cdot, \cdot)$ on NC does not grow faster than exponentially. We know that

$$|\mu(0_n, 1_n)| = |(-1)^{n-1} C_{n-1}| \leq 4^n.$$

This implies, for any $\pi \in NC(n)$,

$$|\mu(\pi, 1_n)| = |\mu(0_n, K(\pi))| \leq 4^n,$$

so that from the assumption $m_n \leq c^n$ we get finally

$$|\kappa_n| = |\sum_{\pi \in NC(n)} m_\pi \mu(\pi, 1_n)| \leq \sum_{\pi \in NC(n)} c^n \cdot 4^n \leq 4^n \cdot c^n \cdot 4^n,$$

thus the wanted exponential bound for the growth of the cumulants. □

Characterization of infinitely divisible distributions

Now we are ready to state our final theorem on characterizing infinitely divisible distributions with compact support. The formula (13.8) is the free analog of the classical Levy–Khintchine formula, in the version for compactly supported measures.

THEOREM 13.16. *Let μ be a probability measure on $\mathbb{R}$ with compact support and let $\kappa_n := \kappa_n^\mu$ be the free cumulants of μ. Then the following statements are equivalent.*

(1) μ is infinitely divisible.

(2) The sequence $(\kappa_n)_{n\geq 1}$ of free cumulants of μ is conditionally positive definite.

(3) The $\mathcal{R}$-transform of μ is of the form

$$\mathcal{R}(z) = \kappa_1 + \int_{\mathbb{R}} \frac{z}{1-xz} d\rho(x), \tag{13.8}$$

for some finite measure ρ on $\mathbb{R}$ with compact support.

(4) μ is a possible limit in our triangular array limit theorem: for each $N \in \mathbb{N}$, there exists a compactly supported probability measure μ_N such that

$$\mu_N^{\boxplus N} \xrightarrow{\text{distr}} \mu.$$

PROOF. The equivalence between (2) and (3) follows from Proposition 13.14 together with Proposition 13.15. The remaining equivalences will be proved via the chain of implications (4)$\Longrightarrow$(2)$\Longrightarrow$(1)$\Longrightarrow$(4).

(4) $\Longrightarrow$ (2): Assume that we have $\mu_N^{\boxplus N} \xrightarrow{\text{distr}} \mu$. Let a_N be a self-adjoint random variable in some C^*-probability space $(\mathcal{A}_N, \varphi_N)$ which has distribution μ_N. Then our limit theorem (in the special form considered in Remark 13.4) tells us that we get the cumulants of μ as

$$\kappa_n = \lim_{N\to\infty} N \cdot \varphi_N(a_N^n).$$

Consider now $\alpha_1, \dots, \alpha_r \in \mathbb{C}$. Then we have

$$\begin{aligned}
\sum_{n,m=1}^{k} \alpha_n \bar{\alpha}_m \kappa_{n+m} &= \lim_{N\to\infty} N \cdot \sum_{n,m=1}^{k} \varphi_N(\alpha_n \bar{\alpha}_m a_N^{n+m}) \\
&= \lim_{N\to\infty} N \cdot \varphi_N\Big(\big(\sum_{n=1}^{k} \alpha_n a_N^n\big) \cdot \big(\sum_{m=1}^{k} \alpha_m a_N^m\big)^*\Big) \\
&\geq 0,
\end{aligned}$$

because all φ_N are positive.

(2) $\Longrightarrow$ (1): Denote by $\mathbb{C}_0\langle X\rangle$ the polynomials in one variable X without constant term, i.e.

$$\mathbb{C}_0\langle X\rangle := \mathbb{C}X \oplus \mathbb{C}X^2 \oplus \cdots .$$

We equip this vector space with an inner product by sesquilinear extension of

$$\langle X^n, X^m\rangle := \kappa_{n+m} \qquad (n, m \geq 1). \tag{13.9}$$

The assumption (2) on the sequence of cumulants yields that this is indeed a non-negative sesquilinear form. Thus we get a Hilbert space $\mathcal{H}$ after dividing out the kernel and completion. In the following we will identify elements from $\mathbb{C}_0\langle X\rangle$ with their images in $\mathcal{H}$. We consider now in the C^*-probability space $(B(\mathcal{F}((\mathcal{H})), \tau_{\mathcal{H}})$ the operator

$$a := l(X) + l^*(X) + \Lambda(X) + \kappa_1 \cdot 1, \tag{13.10}$$

where X in $\Lambda(X)$ is considered as the multiplication operator with X. (Note that, by our assumption of compact support of μ, this operator X is indeed bounded.) By Proposition 13.8, we know that the distribution of a is infinitely divisible. We claim that this distribution is the given μ. Indeed, this follows directly from Proposition 13.5. For $n = 1$, we have

$$\kappa_1^a = \kappa_1;$$

for $n = 2$, we get

$$\kappa_2^a = \kappa_2(l^*(X), l(X)) = \langle X, X\rangle = \kappa_2,$$

and for $n > 2$, we have

$$\begin{aligned}\kappa_n^a &= \kappa_n(l^*(X), \Lambda(X), \ldots, \Lambda(X), l(X)\rangle \\ &= \langle X, \Lambda(X)^{n-2}X\rangle \\ &= \langle X, X^{n-1}\rangle \\ &= \kappa_n.\end{aligned}$$

Thus all cumulants of a agree with the corresponding cumulants of μ and hence the two distributions coincide.

(1) $\Longrightarrow$ (4): Just put, for each $N \in \mathbb{N}$, $\mu_N := \mu^{\boxplus 1/N}$ which exists as a probability measure by the definition of "infinitely divisible." $\square$

Exercises

EXERCISE 13.17. Show that $\frac{1}{2}\delta_{-1} + \frac{1}{2}\delta_{+1}$ is not infinitely divisible.

EXERCISE 13.18. In this exercise we will investigate the relation between compound free Poisson distributions (see Definition 12.16) and infinitely divisible distributions.

(1) Show that every compound free Poisson distribution is infinitely divisible.

(2) Show that there exist infinitely divisible distributions with compact support which are not compound free Poisson distributions.

(3) Show that any infinitely divisible distribution with compact support can be approximated in distribution by compound free Poisson distributions.

EXERCISE 13.19. This exercise addresses the relation between infinitely divisible distributions and stationary processes with free increments. A **stationary process with free increments** or a **free Levy process** is a collection of selfadjoint random variables x_t for all real $t \geq 0$ (think of t as time), living in a C^*-probability space such that we have:

(1) $x_0 = 0$;

(2) for any set of times $0 < t_1 < \cdots < t_n$ the increments

$$x_{t_1}, x_{t_2} - x_{t_1}, \ldots, x_{t_n} - x_{t_{n-1}}$$

are freely independent;

(3) for all $0 \leq s < t$ the distribution of $x_t - x_s$ depends only on $t - s$.

(1) Let $(x_t)_{t \geq 0}$ be a stationary process with free increments. Show that the distribution μ of x_1 is infinitely divisible and that the distribution of x_t is given by $\mu^{\boxplus t}$ for any $t > 0$.

(2) Let μ be an infinitely divisible distribution. Show that there exists a stationary process $(x_t)_{t \geq 0}$ with free increments such that the distribution of x_1 is equal to μ.

[Hint: one might adapt the construction in the proof of Theorem 13.16 from the level of one operator to the level of processes by considering a family of operators on $\mathcal{F}(\mathcal{H} \otimes L^2(\mathbb{R}_+))$ instead of one operator on $\mathcal{F}(\mathcal{H})$.]

LECTURE 14

Products of free random variables

In the previous lectures we treated the sum of freely independent variables. In particular, we showed how one can understand and solve from a combinatorial point of view the problem of describing the distribution of $a+b$ in terms of the distributions of a and of b if these variables are freely independent. Now we want to turn to the corresponding problem for the product. Thus we want to understand how we get the distribution of ab out of the distribution of a and of b if a and b are freely independent.

Note that for the corresponding classical problem no new considerations are required, since this can be reduced to the additive problem. Namely, if a and b commute, we have $ab = \exp(\log a + \log b)$ and thus we only need to apply the additive theory to $\log a$ and $\log b$. In the non-commutative situation, however, the functional equation for the exponential function no longer holds, so there is no clear way to reduce the multiplicative problem to the additive one and some new considerations are needed. In our combinatorial treatment it will turn out that the description of the multiplication of freely independent variables is intimately connected with the complementation map K in the lattice of non-crossing partitions. Since there is no counterpart of the complementation map for all partitions, statements concerning the multiplication of freely independent variables might be quite different from what one expects classically. With respect to additive problems, classical and free probability theory are quite parallel (combinatorially this essentially means that one replaces arguments for all partitions by the corresponding arguments for non-crossing partitions); with respect to multiplicative problems the world of free probability is, however, much richer.

Multiplicative free convolution

As in the additive case we want to define a binary operation on probability measures on $\mathbb{R}$ which corresponds to the product of free random variables. However, one has to note the following problem: if x and y are selfadjoint random variables in a C^*-probability space $(\mathcal{A}, \varphi)$, then the product xy (unlike the sum $x + y$) is not selfadjoint in general; as a consequence, the moments of xy will in general not be the moments of a probability measure, even if the moments of x and of y are so. In order to fix this problem, we will assume that at least one of the elements x and y is positive. Indeed, if (say) x is a positive element of $\mathcal{A}$, y is selfadjoint and x and y are free, then xy has the same moments as the selfadjoint random variable $\sqrt{x}y\sqrt{x}$, because we have for all $n \geq 1$ that

$$\begin{aligned}\varphi((\sqrt{x}y\sqrt{x}\,)^n) &= \varphi(\sqrt{x}y(xy)^{n-1} \cdot \sqrt{x}) \\ &= \varphi(\sqrt{x} \cdot \sqrt{x}y(xy)^{n-1}) \\ &= \varphi((xy)^n).\end{aligned}$$

(At the second equality sign we took into account that φ has to be tracial on the unital subalgebra of $\mathcal{A}$ generated by x and y; cf. Proposition 5.19.)

For most of our considerations, we will in fact only use the more symmetric framework where both x and y are positive elements. We thus arrive to the following definition.

DEFINITION 14.1. Let μ and ν be compactly supported probability measures on $\mathbb{R}_+$. Then their **multiplicative free convolution** $\mu \boxtimes \nu$ is defined as the distribution in analytical sense of $\sqrt{x}y\sqrt{x}$ where x and y are positive elements in some C^*-probability space, such that x and y are free, and x and y have μ and ν, respectively, as their distributions in analytical sense.

REMARKS 14.2. (1) As for the additive case, one has to note that one can always find x and y as required and that the result does not depend on the specific choice of x and y, but only on their distributions μ and ν.

(2) Since the moments of $\sqrt{x}y\sqrt{x}$ are the same as those of $\sqrt{y}x\sqrt{y}$ the operation $\boxtimes$ is commutative,

$$\mu \boxtimes \nu = \nu \boxtimes \mu.$$

(3) Note that we need the operator $\sqrt{x}y\sqrt{x}$ just to be sure that we are dealing with moments of a probability distribution. Thus we can also define $\mu \boxtimes \nu$, for μ and ν compactly supported probability measures on $\mathbb{R}_+$, as the probability distribution whose moments are given by $\varphi((ab)^n)$ for any choice of a, b in some non-commutative probability space $(\mathcal{A}, \varphi)$ such that a and b are free and such that the moments of μ and ν are given by $\varphi(a^n)$ and $\varphi(b^n)$, respectively. This probability distribution is then uniquely determined and necessarily compactly supported on $\mathbb{R}_+$.

(4) The previous remark makes the associativity of the multiplicative free convolution obvious. Given r compactly supported probability measures $\mu_1, \dots, \mu_r$ on $\mathbb{R}_+$, $\mu_1 \boxtimes \cdots \boxtimes \mu_r$ is the (necessarily uniquely determined and compactly supported) probability measure on $\mathbb{R}_+$, whose moments are given by the moments of $a_1 \cdots a_r$, whenever $a_1, \dots, a_r$ are free elements in some non-commutative probability space such that the moments of a_i are the same as the moments of μ_i, for all $i = 1, \dots, r$.

(5) Since it is enough to have one of the involved operators x and y positive in order to recognize the moments of xy as giving rise to a probability measure, we can extend the definition of $\boxtimes$ to a mapping

$$\boxtimes : P_c(\mathbb{R}) \times P_c(\mathbb{R}_+) \to P_c(\mathbb{R}), \tag{14.1}$$

where $P_c(A)$ denotes the set of all compactly supported probability measures on $A \subset \mathbb{R}$. We will indicate it explicitly if we consider $\boxtimes$ in this more general context. Usually, we consider it as a binary operation on $P_c(\mathbb{R}_+)$.

(6) There is also a variant of $\boxtimes$ for probability measures on the circle. Note that probability measures on the circle arise naturally as $*$-distributions in analytical sense of unitary operators. Furthermore, the product of any two unitary operators is again unitary, so that, for probability measures μ and ν on the circle, we can define $\mu \boxtimes \nu$ as the distribution in analytical sense of uv, if u and v are unitary operators in a C^*-probability space such that u has distribution μ, v has distribution ν, and u and v are freely independent. Much of the theory of $\boxtimes$ for unitary elements is parallel to the theory of $\boxtimes$ for positive elements. We will restrict ourselves in this book to the latter situation.

As in the additive case, we would now like to understand, in general, how to find the distribution of ab, for a and b free, from the distribution of a and the distribution of b and, in particular, how we can calculate the multiplicative free convolution $\mu \boxtimes \nu$ from μ and ν. In this lecture we will mainly talk about the general problem on the combinatorial level. The translation of this into formal power series (called S-transforms) which will yield an analytical description for $\mu \boxtimes \nu$ will be postponed to Part 3, where we will talk more systematically about transforms.

Combinatorial description of free multiplication

Our combinatorial description of the product of free variables relies crucially on the complementation map $K : NC(n) \to NC(n)$, which was introduced in Definition 9.21. Let us first note in the following exercise another possibility for characterizing $K(\pi)$. We will use this characterization in the proof of the second formula in our main theorem on the description of products of free random variables.

EXERCISE 14.3. Let π be a partition in $NC(n)$. Prove that the Kreweras complement $K(\pi)$ can also be characterized in the following way. It is the only element $\sigma \in NC(\bar{1}, \dots, \bar{n})$ with the properties that $\pi \cup \sigma \in NC(1, \bar{1}, \dots, n, \bar{n}) \cong NC(2n)$ is non-crossing and that

$$(\pi \cup \sigma) \vee \{(1, \bar{1}), (2, \bar{2}), \dots, (n, \bar{n})\} = 1_{2n}. \tag{14.2}$$

THEOREM 14.4. *Let $(\mathcal{A}, \varphi)$ be a non-commutative probability space and consider random variables $a_1, \dots, a_n, b_1, \dots, b_n \in \mathcal{A}$ such that $\{a_1, \dots, a_n\}$ and $\{b_1, \dots, b_n\}$ are freely independent. Then we have*

$$\varphi(a_1 b_1 a_2 b_2 \dots a_n b_n) = \sum_{\pi \in NC(n)} \kappa_\pi[a_1, a_2, \dots, a_n] \cdot \varphi_{K(\pi)}[b_1, b_2, \dots, b_n] \tag{14.3}$$

and

$$\kappa_n(a_1 b_1, a_2 b_2, \dots, a_n b_n) = \sum_{\pi \in NC(n)} \kappa_\pi[a_1, a_2, \dots, a_n] \cdot \kappa_{K(\pi)}[b_1, b_2, \dots, b_n]. \tag{14.4}$$

PROOF. By using the vanishing of mixed cumulants in free variables we obtain

$$\varphi(a_1 b_1 a_2 b_2 \dots a_n b_n) = \sum_{\pi \in NC(2n)} \kappa_\pi[a_1, b_1, a_2, b_2, \dots, a_n, b_n]$$

$$= \sum_{\substack{\pi_a \in NC(1,3,\dots,2n-1), \pi_b \in NC(2,4,\dots,2n) \\ \pi_a \cup \pi_b \in NC(2n)}} \kappa_{\pi_a}[a_1, a_2, \dots, a_n] \cdot \kappa_{\pi_b}[b_1, b_2, \dots, b_n]$$

$$= \sum_{\pi_a \in NC(1,3,\dots,2n-1)} \kappa_{\pi_a}[a_1, a_2, \dots, a_n] \cdot \Big(\sum_{\substack{\pi_b \in NC(2,4,\dots,2n) \\ \pi_a \cup \pi_b \in NC(2n)}} \kappa_{\pi_b}[b_1, b_2, \dots, b_n] \Big).$$

Now note that, for fixed $\pi_a \in NC(1, 3, \dots, 2n-1) \cong NC(n)$, the condition $\pi_a \cup \pi_b \in NC(2n)$ for $\pi_b \in NC(2, 4, \dots, 2n) \cong NC(n)$ means nothing but $\pi_b \leq K(\pi_a)$ (since $K(\pi_a)$ is by definition the biggest element with this property). Thus we can continue

$$\varphi(a_1 b_1 a_2 b_2 \dots a_n b_n)$$

$$= \sum_{\pi_a \in NC(n)} \kappa_{\pi_a}[a_1, a_2, \dots, a_n] \cdot \Big(\sum_{\pi_b \leq K(\pi_a)} \kappa_{\pi_b}[b_1, b_2, \dots, b_n] \Big)$$

$$= \sum_{\pi_a \in NC(n)} \kappa_{\pi_a}[a_1, a_2, \dots, a_n] \cdot \varphi_{K(\pi_a)}[b_1, b_2, \dots, b_n].$$

One can get (14.4) from (14.3) by convolution with the Möbius function; however, by invoking the simple observation made in Exercise 14.3, one can also give a nice direct proof as follows. By using Theorem 11.12 for cumulants with products as entries we get

$$\kappa_n(a_1b_1,\dots,a_nb_n) = \sum_{\substack{\pi\in NC(2n)\\ \pi\vee\sigma=1_{2n}}} \kappa_\pi[a_1,b_1,\dots,a_n,b_n],$$

where $\sigma = \{(1,2),(3,4),\dots,(2n-1,2n)\}$. By the vanishing of mixed cumulants only such π contribute in the sum which do not couple a with b, thus they are of the form $\pi = \pi_a \cup \pi_b$ with $\pi_a \in NC(a_1,a_2,\dots,a_n)$ and $\pi_b \in NC(b_1,b_2,\dots,b_n)$. Fix now an arbitrary π_a. Then, by Exercise 14.3, there is exactly one π_b which fulfills the requirements that $\pi_a\cup\pi_b$ is non-crossing and that $(\pi_a\cup\pi_b)\vee\sigma = 1_{2n}$, namely π_b has to be the complement $K(\pi_a)$ of π_a. But then the above sum reduces to the right-hand side of (14.4). □

REMARK 14.5. One should note that in the case $a_1 = \cdots = a_n =: a$ and $b_1 = \cdots = b_n =: b$ the structure of the formula (14.4) is

$$\kappa_n^{ab} = \sum_{\pi\in NC(n)} \kappa_\pi^a \cdot \kappa_{K(\pi)}^b,$$

which, in the language of Lecture 10, says that the multiplicative function determined by the cumulants of ab is the convolution of the multiplicative functions determined by the cumulants of a with the multiplicative function determined by the cumulants of b. We will examine this more systematically in Part 3.

EXAMPLES 14.6. (1) Let us write down explicitly the formulas (14.3) and (14.4) for small n.
For $n = 1$ we get

$$\varphi(ab) = \kappa_1(a)\varphi(b) \qquad \text{and} \qquad \kappa_1(ab) = \kappa_1(a)\kappa_1(b),$$

which are just versions of the factorization rule $\varphi(ab) = \varphi(a)\varphi(b)$.
For $n = 2$ we get

$$\varphi(a_1b_1a_2b_2) = \kappa_1(a_1)\kappa_1(a_2)\varphi(b_1b_2) + \kappa_2(a_1,a_2)\varphi(b_1)\varphi(b_2)$$

and

$$\kappa_2(a_1b_1,a_2b_2) = \kappa_1(a_1)\kappa_1(a_2)\kappa_2(b_1,b_2) + \kappa_2(a_1,a_2)\kappa_1(b_1)\kappa_1(b_2)$$

which are both rephrasings of the formula

$$\begin{aligned}\varphi(a_1b_1a_2b_2) = {}& \varphi(a_1)\varphi(a_2)\varphi(b_1b_2)\\ &+ \varphi(a_1a_2)\varphi(b_1)\varphi(b_2) - \varphi(a_1)\varphi(a_2)\varphi(b_1)\varphi(b_2).\end{aligned}$$

(2) Let us specialize the first formula (14.3) to the case where $a_1, \ldots, a_n$ are elements chosen from free semicircular elements s_i. The only non-trivial cumulants are then $\kappa_2(s_i, s_j) = \delta_{ij}$, and we have

$$\varphi(s_{p(1)}b_1 \cdots s_{p(n)}b_n) = \sum_{\pi \in NC_2^{(p)}(n)} \varphi_{K(\pi)}[b_1, \ldots, b_n], \tag{14.5}$$

where $NC_2^{(p)}(n)$ denotes those non-crossing pairings of n elements whose blocks connect only the same p-indices, i.e. only the same semicircular elements. An example is

$$\varphi(s_1b_1s_1b_2s_2b_3s_2b_4s_2b_5s_2b_6s_3b_7s_3b_8) =$$
$$\varphi(b_1)\varphi(b_2b_6b_8)\varphi(b_3b_5)\varphi(b_4)\varphi(b_7) + \varphi(b_1)\varphi(b_2b_4b_6b_8)\varphi(b_3)\varphi(b_5)\varphi(b_7),$$

where we have two contributing $\pi \in NC(6)$ according to the pictures

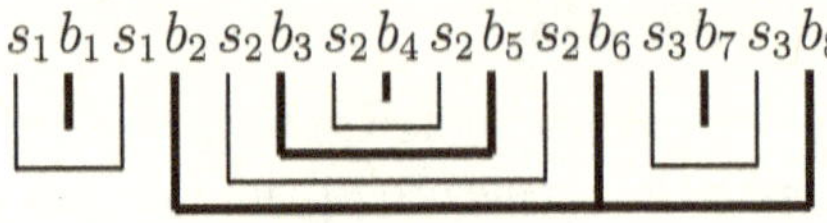

and

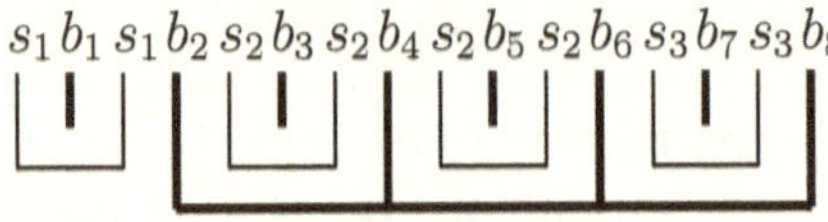

Formula (14.5) will become relevant in Lecture 22, in the context of asymptotic freeness of random matrices.

Theorem 14.4 is the basic combinatorial result about the product of free variables. By translating this into generating power series one can get Voiculescu's description of multiplicative free convolution via the so-called S-transform. However, this translation is not as obvious as in the case of the $\mathcal{R}$-transform and thus we will postpone this to Part 3, where we talk more systematically about transforms.

For the moment we want to show that even without running through analytic calculations the above description can be used quite effectively to handle some special important situations. In the rest of this lecture, we will apply Theorem 14.4 to calculate the distribution of free compressions of random variables.

Compression by a free projection

There is a general way of producing new non-commutative probability spaces out of given ones, namely by compressing with projections. Let us first introduce this general concept.

NOTATION 14.7. If $(\mathcal{A}, \varphi)$ is a non-commutative probability space and $p \in \mathcal{A}$ a projection (i.e. $p^2 = p$) such that $\varphi(p) \neq 0$, then we can consider the **compression** $(p\mathcal{A}p, \varphi^{p\mathcal{A}p})$, where

$$p\mathcal{A}p := \{pap \mid a \in \mathcal{A}\} \tag{14.6}$$

and

$$\varphi^{p\mathcal{A}p}(\cdot) := \frac{1}{\varphi(p)}\varphi(\cdot) \qquad \text{restricted to } p\mathcal{A}p. \tag{14.7}$$

We will denote the cumulants corresponding to $\varphi^{p\mathcal{A}p}$ by $\kappa^{p\mathcal{A}p}$, whereas κ refers as usual to the cumulants corresponding to φ.

REMARKS 14.8. (1) Note that the compression $(p\mathcal{A}p, \varphi^{p\mathcal{A}p})$ is indeed a non-commutative probability space: $p\mathcal{A}p$ is an algebra, whose unit element is $p = p \cdot 1 \cdot p$; and we have rescaled φ just to get $\varphi^{p\mathcal{A}p}(p) = 1$.

(2) Additional properties of $(\mathcal{A}, \varphi)$ will usually pass over to the compression. In particular, if $(\mathcal{A}, \varphi)$ is a C^*-probability space and p is selfadjoint, then it is immediately checked that $(p\mathcal{A}p, \varphi^{p\mathcal{A}p})$ is also a C^*-probability space.

EXAMPLE 14.9. If $\mathcal{A} = M_4(\mathbb{C})$ are the 4×4 matrices equipped with the normalized trace $\varphi = \mathrm{tr}_4$ and p is the projection

$$p = \begin{pmatrix} 1 & 0 & 0 & 0 \\ 0 & 1 & 0 & 0 \\ 0 & 0 & 0 & 0 \\ 0 & 0 & 0 & 0 \end{pmatrix},$$

then

$$p \begin{pmatrix} \alpha_{11} & \alpha_{12} & \alpha_{13} & \alpha_{14} \\ \alpha_{21} & \alpha_{22} & \alpha_{23} & \alpha_{24} \\ \alpha_{31} & \alpha_{32} & \alpha_{33} & \alpha_{34} \\ \alpha_{41} & \alpha_{42} & \alpha_{43} & \alpha_{44} \end{pmatrix} p = \begin{pmatrix} \alpha_{11} & \alpha_{12} & 0 & 0 \\ \alpha_{21} & \alpha_{22} & 0 & 0 \\ 0 & 0 & 0 & 0 \\ 0 & 0 & 0 & 0 \end{pmatrix},$$

and going over to the compressed space just means that we throw away the zeros and identify $pM_4(\mathbb{C})p$ with the 2×2 matrices $M_2(\mathbb{C})$. Of course, the renormalized state $\mathrm{tr}_4^{p\mathcal{A}p}$ coincides with the state tr_2 on $M_2(\mathbb{C})$.

If we have some random variables $a_1, \ldots, a_m$ in the original space, not much can be said about the compressed variables $pa_1p, \ldots, pa_mp$ in the compressed space in general. However, in the case that p is free from the considered variables, we can apply our machinery and relate the distribution of the compressed variables with the distribution of the original ones.

THEOREM 14.10. *Consider a non-commutative probability space* $(\mathcal{A}, \varphi)$ *and random variables* $p, a_1, \dots, a_m \in \mathcal{A}$ *such that* p *is a projection with* $\varphi(p) \neq 0$ *and such that* p *is freely independent from* $\{a_1, \dots, a_m\}$. *Put* $\lambda := \varphi(p)$. *Then we have the following relation between the cumulants of* $a_1, \dots, a_m \in \mathcal{A}$ *and the cumulants of the compressed variables* $pa_1p, \dots, pa_mp \in p\mathcal{A}p$*:*

$$\kappa_n^{p\mathcal{A}p}(pa_{i(1)}p, \dots, pa_{i(n)}p) = \frac{1}{\lambda}\kappa_n(\lambda a_{i(1)}, \dots, \lambda a_{i(n)}) \tag{14.8}$$

for all $n \geq 1$ *and all* $1 \leq i(1), \dots, i(n) \leq m$.

REMARK 14.11. The fact that p is a projection implies

$$\varphi(pa_{i(1)}ppa_{i(2)}p \cdots pa_{i(n)}p) = \varphi(pa_{i(1)}pa_{i(2)}p \cdots a_{i(n)}p),$$

so that apart from the first p we are in the situation where we have a free product of a's with p. If we assumed a tracial situation, then of course the first p could be absorbed by the last one. However, we want to treat the theorem in full generality. But even without traciality we can arrive at the situation treated in Theorem 14.4, just by enlarging $\{a_1, \dots, a_m\}$ to $\{1, a_1, \dots, a_m\}$ (which does not interfere with the assumption on free independence because 1 is freely independent from everything) and reading $\varphi(pa_{i(1)}pa_{i(2)}p \dots a_{i(n)}p)$ as $\varphi(1pa_{i(1)}pa_{i(2)}p \dots a_{i(n)}p)$.

PROOF. By using Theorem 14.4 in the form indicated in the above remark, we get

$$\begin{aligned}\varphi^{p\mathcal{A}p}(pa_{i(1)}p \cdots pa_{i(n)}p) &= \frac{1}{\lambda}\varphi(pa_{i(1)}p \cdots pa_{i(n)}p) \\ &= \frac{1}{\lambda}\varphi_{n+1}(1p, a_{i(1)}p, \dots, a_{i(n)}p) \\ &= \frac{1}{\lambda} \sum_{\sigma \in NC(n+1)} \kappa_\sigma[1, a_{i(1)}, \dots, a_{i(n)}] \cdot \varphi_{K(\sigma)}[p, p, \dots, p].\end{aligned}$$

Now we observe that $k_\sigma[1, a_{i(1)}, \dots, a_{i(n)}]$ can only be different from zero if σ does not couple the random variable 1 with anything else, i.e. $\sigma \in NC(n+1) = NC(0, 1, \dots, n)$ must be of the form $\sigma = (0) \cup \pi$ with $\pi \in NC(1, \dots, n)$. So in fact the sum runs over $\pi \in NC(n)$ and, since $\kappa_1(1) = 1$, $\kappa_\sigma[1, a_{i(1)}, \dots, a_{i(n)}]$ is nothing but $\kappa_\pi[a_{i(1)}, \dots, a_{i(n)}]$. Furthermore, $p^2 = p$ implies that all moments of p are equal to $\varphi(p) = \lambda$, which gives

$$\varphi_{K(\sigma)}[p, p, \dots, p] = \lambda^{|K(\sigma)|}.$$

Using the easily checked fact (see Exercise 9.23) that $|\sigma| + |K(\sigma)| = n+2$ for all $\sigma \in NC(n+1)$ we can rewrite $|K(\sigma)|$ in terms of $|\pi| = |\sigma| - 1$

and continue our above calculation as follows.

$$\begin{aligned}\varphi_n^{pAp}(pa_{i(1)}p,\dots,pa_{i(n)}p) &= \frac{1}{\lambda}\sum_{\pi\in NC(n)} \kappa_\pi[a_{i(1)},\dots,a_{i(n)}]\lambda^{n+1-|\pi|}\\ &= \sum_{\pi\in NC(n)} \frac{1}{\lambda^{|\pi|}}\kappa_\pi[\lambda a_{i(1)},\dots,\lambda a_{i(n)}].\end{aligned}$$

Since the function $\pi \mapsto \frac{1}{\lambda^{|\pi|}}\kappa_\pi(\lambda a_{i(1)},\dots,\lambda a_{i(n)})$ is multiplicative, we see that according to Remarks 11.19 the cumulants of the compressed random variables are given by

$$\begin{aligned}\kappa_n^{pAp}[pa_{i(1)}p,\dots,pa_{i(n)}p] &= \frac{1}{\lambda^{|1_n|}}\kappa_{1_n}[\lambda a_{i(1)},\dots,\lambda a_{i(n)}]\\ &= \frac{1}{\lambda}\kappa_n(\lambda a_{i(1)},\dots,\lambda a_{i(n)}).\end{aligned}$$

□

This theorem has two interesting corollaries. The first states that free independence is preserved under taking free compressions, whereas the second is a very surprising statement about free harmonic analysis.

COROLLARY 14.12. *Let $(\mathcal{A},\varphi)$ be a non-commutative probability space and $p\in\mathcal{A}$ a projection such that $\varphi(p)\neq 0$. Consider unital subalgebras $\mathcal{A}_1,\dots,\mathcal{A}_m\subset\mathcal{A}$ such that p is freely independent from $\mathcal{A}_1\cup\dots\cup\mathcal{A}_m$. Then the following two statements are equivalent.*

(1) *The subalgebras $\mathcal{A}_1,\dots,\mathcal{A}_m\subset\mathcal{A}$ are freely independent in the original non-commutative probability space $(\mathcal{A},\varphi)$.*

(2) *The compressed subalgebras $p\mathcal{A}_1p,\dots,p\mathcal{A}_mp\subset p\mathcal{A}p$ are freely independent in the compressed probability space $(p\mathcal{A}p,\varphi^{p\mathcal{A}p})$.*

PROOF. Since the cumulants of the $\mathcal{A}_i$ coincide with the cumulants of the compressions $p\mathcal{A}_ip$ up to some power of λ, the vanishing of mixed cumulants in the $\mathcal{A}_i$ is equivalent to the vanishing of mixed cumulants in the $p\mathcal{A}_ip$. □

Convolution semigroups $(\mu^{\boxplus t})_{t\geq 1}$

For the second consequence of our Theorem 14.10 let us see what it tells us in the case of one variable. Consider a random variable x in a C^*-probability space whose distribution is a probability measure μ on $\mathbb{R}$. Then the above theorem tells us that, for p a projection with $\varphi(p)=\lambda$ and such that x and p are free, the distribution of pxp in the compressed space is given by a distribution whose cumulants are $1/\lambda$ times the corresponding cumulants of λx. Going over from x to λx is of course just a rescaling of our distribution μ by a factor λ – let us denote this by $D_\lambda(\mu)$. Multiplying cumulants by a factor $1/\lambda$, on the other hand, corresponds to taking the $1/\lambda$-fold free convolution of the given distribution – which, a priori, only makes sense for integer $1/\lambda$. Thus, for $\lambda = 1/n$, we get that the distribution of pap is $(D_\lambda(\mu))^{\boxplus n}$, and hence compressing by free projections has, up to trivial rescalings, the same effect as taking convolution powers. Since in the compression picture we are not restricted to λ of the form $1/n$, this gives us the possibility of also obtaining non-integer free convolution powers.

COROLLARY 14.13. *Let μ be a compactly supported probability measure on $\mathbb{R}$. Then there exists a semigroup (μ_t) $(t \in \mathbb{R},\ t \geq 1)$ of compactly supported probability measures on $\mathbb{R}$ such that*

$$\mu_1 = \mu,$$

$$\mu_{s+t} = \mu_s \boxplus \mu_t \qquad (s, t \geq 1)$$

and the mapping $t \mapsto \mu_t$ is continuous with respect to the weak topology on probability measures (i.e. all moments of μ_t are continuous in t).*

PROOF. Let x be a selfadjoint random variable and p a selfadjoint projection in some C^*-probability space $(\mathcal{A}, \varphi)$ such that $\varphi(p) = \frac{1}{t}$, the distribution of x is equal to μ, and x and p are freely independent. (It is no problem to realize such a situation with the usual free product constructions as described in Lecture 7.) Put now $x_t := p(tx)p$ and consider this as an element in the compressed space $(p\mathcal{A}p, \varphi^{p\mathcal{A}p})$. As we noted in Remarks 14.8, this compressed space is again a C^*-probability space, thus the distribution μ_t of $x_t \in p\mathcal{A}p$ is a compactly supported probability measure. Furthermore, by Theorem 14.10, we know that the cumulants of x_t are given by

$$\kappa_n^{\mu_t} = \kappa_n^{p\mathcal{A}p}(x_t, \ldots, x_t) = t\kappa_n\big(\frac{1}{t}tx, \ldots, \frac{1}{t}tx\big) = t\kappa_n(x, \ldots, x) = t\kappa_n^{\mu}.$$

This implies that for all $n \geq 1$

$$\kappa_n^{\mu_{s+t}} = (s+t)\kappa_n^{\mu} = s\kappa_n^{\mu} + t\kappa_n^{\mu} = \kappa_n^{\mu_s} + \kappa_n^{\mu_t},$$

which just means that $\mu_{s+t} = \mu_s \boxplus \mu_t$. Since $\kappa_n^{\mu_1} = \kappa_n^{\mu}$, we also have $\mu_1 = \mu$. Furthermore, the mapping $t \mapsto t\kappa_n^{\mu}$ is clearly continuous, thus all cumulants, and hence also all moments, of μ_t are continuous in t. □

REMARKS 14.14. (1) For $t = n \in \mathbb{N}$, we have of course the convolution powers $\mu_n = \mu^{\boxplus n}$. The corollary states that we can interpolate between these for non-natural powers. Of course, the crucial fact is that we claim the μ_t to be always probability measures. As linear functionals these objects exist trivially, the non-trivial fact is positivity.

(2) Note that the corollary claims the existence of μ_t only for $t \geq 1$. For $0 < t < 1$, μ_t does not exist as a probability measure in general. In particular, the existence of the semigroup μ_t for all $t > 0$ is, as we have seen in the last lecture, equivalent to μ being infinitely divisible in the free sense.

(3) As we have seen in the proof of Theorem 14.10, the corollary relies very much on the complementation map and its properties. Thus it is no surprise that there is no classical analog of this result. In the classical case one usually cannot interpolate between the natural convolution powers. For example, if $\mu = \frac{1}{2}(\delta_{-1} + \delta_1)$ is a symmetric Bernoulli distribution, we have $\mu * \mu * \mu = \frac{1}{8}\delta_{-3} + \frac{3}{8}\delta_{-1} + \frac{3}{8}\delta_1 + \frac{1}{8}\delta_3$ and it is trivial to check that it is not possible to write $\mu * \mu * \mu$ as $\nu * \nu$ for some other probability measure $\nu = \mu^{*3/2}$.

EXAMPLE 14.15. Consider the symmetric Bernoulli distribution $\mu = \frac{1}{2}(\delta_{-1} + \delta_{+1})$. Then, the compression with a projection having $\varphi(p) = 1/n$ gives a distribution which is, up to rescaling, the n-fold free convolution of μ with itself. For example, for $\varphi(p) = 1/2$, we get that the compression of Bernoulli has the arcsine distribution. More generally, we get rescalings of the Kesten measures $\mu^{\boxplus n}$, which were calculated in Exercise 12.21. Furthermore, our above corollary explains the fact that the calculations of $\mu^{\boxplus n}$ in that exercise also make sense for non-integer $n \geq 1$.

Compression by a free family of matrix units

DEFINITION 14.16. Let $(\mathcal{A}, \varphi)$ be a non-commutative probability space. A **family of matrix units** is a set $\{e_{ij}\}_{i,j=1,\ldots,d} \subset \mathcal{A}$ (for some $d \in \mathbb{N}$) with the properties

$$e_{ij}e_{kl} = \delta_{jk}e_{il} \qquad \forall i, j, k, l = 1, \ldots, d, \tag{14.9}$$

$$\sum_{i=1}^{d} e_{ii} = 1. \tag{14.10}$$

REMARKS 14.17. (1) Consider the non-commutative probability space $(M_d(\mathbb{C}), \mathrm{tr}_d)$ of $d \times d$ matrices equipped with the normalized trace. Then the canonical family of matrix units is $\{E_{ij}\}_{i,j=1,\ldots,d}$, where E_{ij} is the matrix

$$E_{ij} = (\delta_{ik}\delta_{jl})_{k,l=1}^{d}.$$

(2) For a non-commutative probability space $(\mathcal{A}, \varphi)$ we considered in Exercise 1.23 the non-commutative probability space $(M_d(\mathcal{A}), \mathrm{tr} \otimes \varphi)$ of $d \times d$ matrices over $\mathcal{A}$. The $\{E_{ij}\}_{i,j=1,\ldots,d}$ from above are sitting inside

such an $M_d(\mathcal{A})$ via

$$M_d(\mathbb{C}) \hat{=} M_d(\mathbb{C}) \otimes 1_{\mathcal{A}} \subset M_d(\mathbb{C}) \otimes \mathcal{A} \hat{=} M_d(\mathcal{A}),$$

and form there a family of matrix units. In such a situation we will be interested in the question when the family of matrix units is free from a matrix $A \in M_d(\mathcal{A})$. (Note that the algebra generated by the matrix units is just $M_d(\mathbb{C}) \otimes 1_{\mathcal{A}}$.) We will present the solution to this problem in Theorem 14.20.

THEOREM 14.18. *Consider a non-commutative probability space $(\mathcal{A}, \varphi)$ and random variables $a^{(1)}, \dots, a^{(m)} \in \mathcal{A}$. Furthermore, let $\{e_{ij}\}_{i,j=1,\dots,d} \subset \mathcal{A}$ be a family of matrix units which satisfies*

$$\varphi(e_{ij}) = \delta_{ij}\frac{1}{d} \qquad \forall i, j = 1, \dots, d, \tag{14.11}$$

and such that $\{a^{(1)}, \dots, a^{(m)}\}$ is freely independent from $\{e_{ij}\}_{i,j=1,\dots,d}$. Put now $a_{ij}^{(r)} := e_{1i}a^{(r)}e_{j1}$ and $p := e_{11}$, $\lambda := \varphi(p) = 1/d$. Then we have the following relation between the cumulants of $a^{(1)}, \dots, a^{(m)} \in \mathcal{A}$ and the cumulants of the compressed variables $a_{ij}^{(r)} \in p\mathcal{A}p$ ($i, j = 1, \dots, d$; $r = 1, \dots, m$). For all $n \geq 1$ and all $1 \leq r(1), \dots, r(n) \leq m$, $1 \leq i(1), j(1), \dots, i(n), j(n) \leq d$ we have

$$\kappa_n^{p\mathcal{A}p}(a_{i(1)j(1)}^{(r(1))}, \dots, a_{i(n)j(n)}^{(r(n))})$$
$$= \begin{cases} \frac{1}{\lambda}\kappa_n(\lambda a^{(r(1))}, \dots, \lambda a^{(r(n))}) & \text{if } j(k) = i(k+1) \text{ for all } k = 1, \dots, n \\ 0 & \text{otherwise} \end{cases} \tag{14.12}$$

(where we put $i(n+1) := i(1)$).

NOTATION 14.19. Let a partition $\pi \in NC(n)$ and an n-tuple of double-indices $(i(1)j(1), i(2)j(2), \dots, i(n)j(n))$ be given. Then we say that π **couples in a cyclic way** (c.c.w., for short) the indices $(i(1)j(1), \dots, i(n)j(n))$ if we have for each block $(r_1 < r_2 < \dots < r_s) \in \pi$ that $j(r_k) = i(r_{k+1})$ for all $k = 1, \dots, s$ (where we put $r_{s+1} := r_1$).

PROOF. Let us denote in the following the fixed tuple of indices by

$$(\vec{i}, \vec{j}) := (i(1)j(1), i(2)j(2), \dots, i(n)j(n)).$$

As in the case of one free projection we calculate

$$\varphi_n^{p\mathcal{A}p}(a_{i(1)j(1)}^{(r(1))}, \dots, a_{i(n)j(n)}^{(r(n))})$$
$$= \frac{1}{\lambda} \sum_{\sigma \in NC(n+1)} \kappa_\sigma[1, a^{(r(1))}, \dots, a^{(r(n))}] \cdot \varphi_{K(\sigma)}[e_{1,i(1)}, e_{j(1)i(2)}, \dots, e_{j(n)1}].$$

Again σ has to be of the form $\sigma = (0) \cup \pi$ with $\pi \in NC(n)$. The factor $\varphi_{K(\sigma)}$ gives

$$\varphi_{K(\sigma)}[e_{1,i(1)}, e_{j(1)i(2)}, \dots, e_{j(n)1}] = \varphi_{K(\pi)}[e_{j(1)i(2)}, e_{j(2),j(3)}, \dots, e_{j(n)i(1)}]$$
$$= \begin{cases} \lambda^{|K(\pi)|} & \text{if } K(\pi) \text{ c.c.w. } (\vec{j}, \vec{i}) \\ 0 & \text{otherwise,} \end{cases}$$

where

$$(\vec{j}, \vec{i}) := (j(1)i(2), j(2)i(3), \dots, j(n)i(1)).$$

Now one has to observe that cyclicity of $K(\pi)$ in $(\vec{j}, \vec{i})$ is equivalent to cyclicity of π in $(\vec{i}, \vec{j})$. (The proof of this is left to the reader.) Then one can continue the above calculation as follows.

$$\varphi_n^{pAp}(a_{i(1)j(1)}^{(r(1))}, \dots, a_{i(n)j(n)}^{(r(n))}) = \frac{1}{\lambda} \sum_{\substack{\pi \in NC(n) \\ \pi \text{ c.c.w. } (\vec{i}, \vec{j})}} \kappa_\pi[a^{(r(1))}, \dots, a^{(r(n))}] \cdot \lambda^{|K(\pi)|}$$
$$= \sum_{\substack{\pi \in NC(n) \\ \pi \text{ c.c.w. } (\vec{i}, \vec{j})}} \frac{1}{\lambda^{|\pi|}} \kappa_\pi[\lambda a^{(r(1))}, \dots, \lambda a^{(r(n))}],$$

where we sum only over those π which couple in a cyclic way $(\vec{i}, \vec{j})$. Noticing that the function

$$\pi \mapsto \begin{cases} \frac{1}{\lambda^{|\pi|}} \kappa_\pi[\lambda a^{(r(1))}, \dots, \lambda a^{(r(n))}] & \text{if } \pi \text{ c.c.w. } (\vec{i}, \vec{j}) \\ 0 & \text{otherwise} \end{cases}$$

is multiplicative, gives the statement. □

We can now come back to the question, raised in Remark 14.17, when $M_d(\mathbb{C}) \hat{=} M_d(\mathbb{C}) \otimes 1_{\mathcal{A}} \subset M_d(\mathcal{A})$ is free from a matrix $A \in M_d(\mathcal{A})$.

THEOREM 14.20. *Consider random variables a_{ij} $(i, j = 1, \dots, d)$ in some non-commutative probability space $(\mathcal{A}, \varphi)$. Then the following two statements are equivalent.*

(1) *The matrix $A := (a_{ij})_{i,j=1}^d$ is freely independent from $M_d(\mathbb{C})$ in the non-commutative probability space $(M_d(\mathcal{A}), \mathrm{tr}_d \otimes \varphi)$.*

(2) *Cumulants of $\{a_{ij} \mid i, j = 1, \dots, d\}$ in $(\mathcal{A}, \varphi)$ have the property that only cyclic cumulants $\kappa_n(a_{i(1)i(2)}, a_{i(2)i(3)}, \dots, a_{i(n)i(1)})$ are different from zero and the value of such a cumulant depends only on n, but not on the tuple $(i(1), \dots, i(n))$.*

PROOF. That (1) implies (2) follows directly from Theorem 14.18 (for the case $m = 1$), because we can identify the entries of the matrix A with the compressions by the matrix units.

For the other direction, note that joint moments (with respect to $\mathrm{tr}_d \otimes \varphi$) of the matrix $A := (a_{ij})_{i,j=1}^d$ and elements from $M_d(\mathbb{C})$ can be expressed in terms of moments of the entries of A. Thus the free independence between A and $M_d(\mathbb{C})$ depends only on the joint distribution (i.e. on the cumulants) of the a_{ij}. This implies that if we can present a realization of the joint distribution of the a_{ij} $(i,j = 1,\dots,d)$ in which the corresponding matrix $A = (a_{ij})_{i,j=1}^d$ is free from $M_d(\mathbb{C})$, then we are done. But this representation is given by Theorem 14.18. Namely, let a be a random variable whose cumulants are given, up to a factor, by the cyclic cumulants of the a_{ij}, i.e. $k_n^a = d^{n-1}k_n(a_{i(1)i(2)}, a_{i(2)i(3)}, \dots, a_{i(n)i(1)})$. Let furthermore $\{e_{ij}\}_{i,j=1,\dots,d}$ be a family of matrix units which are freely independent from a in some probability space $(\tilde{\mathcal{A}}, \tilde{\varphi})$. Then we compress a with the free matrix units as in Theorem 14.18 and denote the compressions by $\tilde{a}_{ij} := e_{1i}ae_{j1}$. By Theorem 14.18 and the choice of the cumulants for a, we have that the joint distribution of the $\tilde{a}_{ij}$ in $(e_{11}\tilde{\mathcal{A}}e_{11}, \tilde{\varphi}^{e_{11}\tilde{\mathcal{A}}e_{11}})$ coincides with the joint distribution of the a_{ij}. Furthermore, the matrix $\tilde{A} := (\tilde{a}_{ij})_{i,j=1}^d$ is freely independent from $M_d(\mathbb{C})$ in $(M_d(e_{11}\tilde{\mathcal{A}}e_{11}), \tilde{\varphi}^{e_{11}\tilde{\mathcal{A}}e_{11}} \otimes \mathrm{tr}_d)$, because the mapping

$$\tilde{\mathcal{A}} \to M_d(e_{11}\tilde{\mathcal{A}}e_{11}), \qquad y \mapsto (e_{1i}ye_{j1})_{i,j=1}^d$$

is an isomorphism which sends a into $\tilde{A}$ and e_{kl} into the canonical matrix units E_{kl} in $M_d(\mathbb{C}) \otimes 1$. □

Exercises

EXERCISE 14.21. Show that we have for all compactly supported probability measures μ on $\mathbb{R}$ and for all t with $t \geq 1$ that

$$\mu \boxtimes \Big((1 - \frac{1}{t})\delta_0 + \frac{1}{t}\delta_t \Big) = (1 - \frac{1}{t})\delta_0 + \frac{1}{t}\mu^{\boxplus t}. \tag{14.13}$$

(In this formulation we use the extended definition (14.1) of $\boxtimes$, where only one of the involved probability measures has to be supported on $\mathbb{R}_+$.)

EXERCISE 14.22. Let $(\mathcal{A}, \varphi)$ be a non-commutative probability space. Consider a family of matrix units $\{e_{ij}\}_{i,j=1,\dots,d} \subset \mathcal{A}$, which satisfies (14.11), and a subset $\mathcal{X} \subset \mathcal{A}$ such that $\{e_{ij}\}_{i,j=1,\dots,d}$ and $\mathcal{X}$ are freely independent. Consider now, for $i = 1,\dots,d$, the compressed subsets $\mathcal{X}_i := e_{1i}\mathcal{X}e_{i1} \subset e_{11}\mathcal{A}e_{11}$. Show that $\mathcal{X}_1,\dots,\mathcal{X}_d$ are freely independent in the compressed space $(e_{11}\mathcal{A}e_{11}, \varphi^{e_{11}\mathcal{A}e_{11}})$.

LECTURE 15

R-diagonal elements

There is a substantial difference in our understanding of normal operators on the one hand and non-normal operators on the other hand. Whereas the first case takes place in the classical commutative world, where we have available the sophisticated tools of analytic function theory, the second case is really non-commutative in nature and is much harder to analyze. It is therefore quite important to have sufficiently large classes of non-normal operators which can be treated.

In this lecture we present one of the most prominent class of non-normal operators arising from free probability – the class of R-diagonal operators. It will turn out that these operators are simple enough to allow concrete calculations, but also this class is big enough to appear quite canonically in a lot of situations.

Motivation: cumulants of Haar unitary elements

The motivation for the introduction of R-diagonal elements was the observation that the two most prominent non-selfadjoint elements in free probability theory – the circular element and the Haar unitary element – show similar structures of their $*$-cumulants. The $*$-cumulants for a circular element c are very easy to determine: the only non-vanishing $*$-cumulants are

$$\kappa_2(c, c^*) = 1 = \kappa_2(c^*, c).$$

For a Haar unitary element u, however, calculation of the $*$-cumulants is quite non-trivial. Determination of these $*$-cumulants will be our goal in this section.

So let u be a Haar unitary. Recall that this means that u is unitary and that all $*$-moments of the form $\varphi(u^k)$ ($k \in \mathbb{Z}$) vanish unless $k = 0$; for $k = 0$ we have of course $\varphi(u^0) = \varphi(1) = 1$. This clearly gives complete information about the $*$-distribution of u because any $*$-moment of u can, by the unitarity condition, be reduced to a moment of the form $\varphi(u^k)$ for $k \in \mathbb{Z}$.

We want to calculate $\kappa_n(u_1, \dots, u_n)$, where $u_1, \dots, u_n \in \{u, u^*\}$. First we note that such a cumulant can only be different from zero if the number of u among $u_1, \dots, u_n$ is the same as the number of u^*

among $u_1, \ldots, u_n$. This follows directly by the formula $\kappa_n(u_1, \ldots, u_n) = \sum_{\pi \in NC(n)} \varphi_\pi[u_1, \ldots, u_n]\mu(\pi, 1_n)$. Indeed, if the number of u is not equal to the number of u^*, then for all $\pi \in NC(n)$ there exists a block $V \in \pi$ which contains an unequal number of u and u^*; but then $\varphi(V)[u_1, \ldots, u_n]$, and thus also $\varphi_\pi[u_1, \ldots, u_n]$, vanishes. This means, in particular, that only cumulants of even length are different from zero.

Consider now a cumulant where the number of u and the number of u^* are the same. We claim that only such cumulants are different from zero where the entries are alternating in u and u^*. We will prove this by induction on the length of the cumulant. For length 2 this is clear, because $\kappa_2(u, u) = \varphi(uu) - \varphi(u)\varphi(u) = 0$ and in the same way $\kappa_2(u^*, u^*) = 0$. Assume now that we have proved the vanishing of all non-alternating cumulants of length smaller than n and consider a non-alternating cumulant of length n. Non-alternating means that we find in the sequence of arguments at least one of the patterns $\kappa_n(\ldots, u^*, u, u, \ldots)$, $\kappa_n(\ldots, u, u, u^*, \ldots)$, $\kappa_n(\ldots, u, u^*, u^*, \ldots)$, or $\kappa_n(\ldots, u^*, u^*, u, \ldots)$. Note that actually it suffices to consider the first two cases because we can get the other two from those by replacing u by u^*, and if u is a Haar unitary then so is u^*. Let us only treat the first case, the second is similar. So let us consider $\kappa_n(\ldots, u^*, u, u, \ldots)$ and say that the positions of $\ldots, u^*, u, u, \ldots$ are $\ldots, m, m+1, m+2, \ldots$. By Proposition 11.15, we have that $\kappa_n(\ldots, 1, u, \ldots) = 0$. On the other hand, the latter cumulant is the same as $\kappa_n(\ldots, u^* \cdot u, u, \ldots)$, and then we can use Theorem 11.12 to write this as

$$0 = \kappa(\ldots, u^* \cdot u, u, \ldots) = \sum_{\substack{\pi \in NC(n) \\ \pi \vee \sigma = 1_n}} \kappa_\pi[\ldots, u^*, u, u, \ldots].$$

Here

$$\sigma = \{(1), (2), \ldots, (m, m+1), (m+2), \ldots, (n)\}$$

is the partition which glues together the elements m and $m+1$. Which partitions $\pi \in NC(n)$ have the property $\pi \vee \sigma = 1_n$? Of course, we have the possibility $\pi = 1_n$. The only other possibilities are π which consist of exactly two blocks, one of them containing m and the other containing $m+1$. For these π, the summand $\kappa_\pi[\ldots, u^*, u, u, \ldots]$ factorizes into a product of two cumulants of smaller length, so, by the induction hypothesis, each of the two blocks of π must connect alternatingly u and u^*. This implies that such a π cannot connect $m+1$ with $m+2$, and hence it must connect m with $m+2$. But this forces $m+1$ to give rise to a singleton of π, i.e. such a π looks like this:

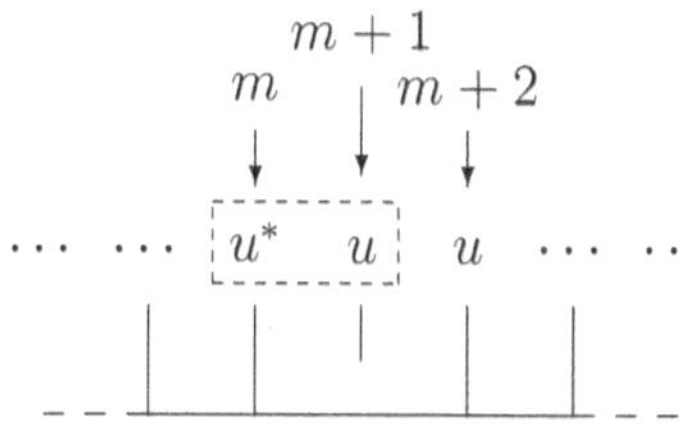

Hence one factor of κ_π for such a π is just $\kappa_1(u) = 0$. This implies that only $\pi = 1_n$ makes a contribution to the above sum, i.e. we get

$$0 = \sum_{\substack{\pi \in NC(n)\\ \pi \vee \sigma = 1_n}} \kappa_\pi[\dots, u^*, u, u, \dots] = \kappa_n(\dots, u^*, u, u, \dots),$$

which proves our claim on the vanishing of non-alternating cumulants.

Finally, it remains to determine the value of the alternating cumulants. Let us denote by α_n the value of such a cumulant of length $2n$, i.e.

$$\kappa_{2n}(u, u^*, \dots, u, u^*) =: \alpha_n = \kappa_{2n}(u^*, u, \dots, u^*, u).$$

The last equality comes from the fact that with u also u^* is a Haar unitary. We now use again Proposition 11.15 and Theorem 11.12:

$$\begin{aligned} 0 = \kappa_{2n-1}(1, u, u^*, \dots, u, u^*) &= \kappa_{2n-1}(u \cdot u^*, u, u^*, \dots, u, u^*) \\ &= \sum_{\substack{\pi \in NC(2n)\\ \pi \vee \sigma = 1_{2n}}} \kappa_\pi[u, u^*, u, u^*, \dots, u, u^*], \end{aligned}$$

where $\sigma = \{(1,2), (3), (4), \dots, (2n)\} \in NC(2n)$ is the partition which couples the first two elements. Again, π fulfilling the condition $\pi \vee \sigma = 1_{2n}$ are either $\pi = 1_{2n}$ or consist of exactly two blocks, one containing the element 1 and the other containing the element 2. Note that in the latter case the next element in the block containing 1 must correspond to an u^*, hence be of the form $2(p+1)$ for some $1 \le p \le n-1$, and such a π must look like this:

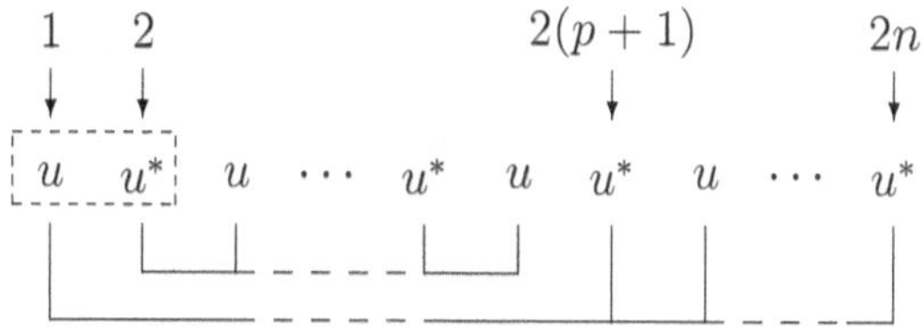

Then we can continue the above calculation as follows:

$$\begin{aligned} 0 &= \sum_{\substack{\pi\in NC(2n)\\ \pi\vee\sigma=1_{2n}}} \kappa_\pi[u,u^*,u,u^*,\dots,u,u^*] \\ &= k_{2n}(u,u^*,u,u^*,\dots,u,u^*) \\ &\quad + \sum_{p=1}^{n-1} k_{2n-2p}(u,u^*,\dots,u,u^*)\cdot\kappa_{2p}(u^*,u,\dots,u^*,u) \\ &= \alpha_n + \sum_{p=1}^{n-1}\alpha_{n-p}\alpha_p. \end{aligned}$$

Thus we have the recursion

$$\alpha_n = -\sum_{p=1}^{n-1}\alpha_{n-p}\alpha_p, \tag{15.1}$$

which is up to the minus-sign and a shift in the indices by 1 the recursion relation (2.8) for the Catalan numbers. Since also $\alpha_1 = \kappa_2(u,u^*) = 1 = C_0$, we have finally proved the following statement.

PROPOSITION 15.1. *The alternating $*$-cumulants of a Haar unitary u are given by*

$$\kappa_{2n}(u,u^*,\dots,u,u^*) = \kappa_{2n}(u^*,u,\dots,u^*,u) = (-1)^{n-1}C_{n-1}. \tag{15.2}$$

All other $*$-cumulants of u vanish.

Definition of R-diagonal elements

We see that for both the circular and the Haar unitary elements a lot of $*$-cumulants vanish, namely those for which the arguments are not alternating between the element and its adjoint. We will take this as the defining property of R-diagonal elements, thus providing a class of (in general, non-normal) elements which contain circular and Haar unitary elements as special cases.

NOTATION 15.2. Let a be a random variable in a $*$-probability space. A cumulant $\kappa_{2n}(a_1,\dots,a_{2n})$ with arguments from $\{a,a^*\}$ is said to have **alternating arguments** or is **alternating**, if there does not exist any a_i $(1\le i\le 2n-1)$ with $a_{i+1}=a_i$. Cumulants with an odd number of arguments will always be considered as not alternating.

For example, $\kappa_8(a,a^*,a^*,a,a,a^*,a,a^*)$ or $\kappa_5(a,a^*,a,a^*,a)$ are not alternating, whereas $\kappa_6(a,a^*,a,a^*,a,a^*)$ is alternating.

DEFINITION 15.3. Let $(\mathcal{A}, \varphi)$ be a $*$-probability space.

(1) A random variable $a \in \mathcal{A}$ is called ***R*-diagonal** if for all $n \in \mathbb{N}$ we have that $\kappa_n(a_1, \ldots, a_n) = 0$ whenever the arguments $a_1, \ldots, a_n \in \{a, a^*\}$ are not alternating in a and a^*.

(2) If $a \in \mathcal{A}$ is R-diagonal we denote the non-vanishing cumulants by $\alpha_n := \kappa_{2n}(a, a^*, a, a^*, \ldots, a, a^*)$ and $\beta_n := \kappa_{2n}(a^*, a, a^*, a, \ldots, a^*, a)$ $(n \geq 1)$. The sequences $(\alpha_n)_{n \geq 1}$ and $(\beta_n)_{n \geq 1}$ are called the **determining sequences** of a.

(3) If the state φ restricted to the $*$-algebra generated by an R-diagonal $a \in \mathcal{A}$ is tracial – which means that the two determining sequences coincide, $\alpha_n = \beta_n$ for all n – then we call a a **tracial *R*-diagonal** element.

EXAMPLES 15.4. (1) The only non-vanishing cumulants for a circular element are $\kappa_2(c, c^*) = \kappa_2(c^*, c) = 1$. Thus a circular element is a tracial R-diagonal element with determining sequence

$$\alpha_n = \beta_n = \begin{cases} 1 & n = 1 \\ 0 & n > 1. \end{cases} \tag{15.3}$$

(2) Let u be a Haar unitary. We calculated its cumulants in Proposition 15.1. In our present language, we showed there that u is a tracial R-diagonal element with determining sequence

$$\alpha_n = \beta_n = (-1)^{n-1} C_{n-1}. \tag{15.4}$$

NOTATION 15.5. Note that many R-diagonal elements a (e.g. circular or Haar unitary elements) are tracial. In this case all information about the $*$-distribution of a is contained in one sequence $(\alpha_n)_{n \geq 1}$ and many formulas can then be formulated quite compactly in terms of the corresponding multiplicative function on non-crossing partitions. We will denote this function also by α, thus in analogy to our moment and cumulant functions we will then use the notation

$$\alpha_\pi := \prod_{V \in \pi} \alpha_{|V|} \qquad (\pi \in NC(n)). \tag{15.5}$$

It is clear that all information on the $*$-distribution of an R-diagonal element a is contained in its determining sequences. In the next proposition we will connect the determining sequences with the distribution of aa^* and the distribution of a^*a.

PROPOSITION 15.6. *Let a be an R-diagonal random variable and*

$$\alpha_n := \kappa_{2n}(a, a^*, a, a^*, \ldots, a, a^*),$$
$$\beta_n := \kappa_{2n}(a^*, a, a^*, a, \ldots, a^*, a),$$

the determining sequences of a.

(1) *Then we have*

$$\kappa_n(aa^*, \dots, aa^*) = \sum_{\substack{\pi \in NC(n) \\ \pi = \{V_1, \dots, V_r\}}} \alpha_{|V_1|}\beta_{|V_2|}\cdots\beta_{|V_r|} \tag{15.6}$$

and

$$\kappa_n(a^*a, \dots, a^*a) = \sum_{\substack{\pi \in NC(n) \\ \pi = \{V_1, \dots, V_r\}}} \beta_{|V_1|}\alpha_{|V_2|}\cdots\beta_{|V_r|} \tag{15.7}$$

where V_1 *denotes that block of* $\pi \in NC(n)$ *which contains the first element* 1.

(2) *In the tracial case (i.e. if* $\alpha_n = \beta_n$ *for all* n*) we have*

$$\kappa_n(aa^*, \dots, aa^*) = \kappa_n(a^*a, \dots, a^*a) = \sum_{\pi \in NC(n)} \alpha_\pi. \tag{15.8}$$

PROOF. (1) Applying Theorem 11.12 yields

$$\kappa_n(aa^*, \dots, aa^*) = \sum_{\substack{\pi \in NC(2n) \\ \pi \vee \sigma = 1_{2n}}} \kappa_\pi[a, a^*, \dots, a, a^*]$$

with $\sigma = \{(1,2), (3,4), \dots, (2n-1, 2n)\} \in NC(2n)$.

Note that this is exactly the same σ which appeared in the proof of Proposition 11.25. In that proof it was shown that a partition $\pi \in NC(2n)$ fulfills the condition $\pi \vee \sigma = 1_{2n}$ if and only if it has the following properties: the block of π which contains the element 1 also contains the element $2n$ and, for each $1 \leq k \leq n-1$, the block of π which contains the element 2k also contains the element $2k+1$. Moreover, in the proof of Proposition 11.25 it was observed that the set of partitions $\pi \in NC(2n)$ which have these properties is in canonical bijection with $NC(n)$. It is immediate that, under this bijection, $\kappa_\pi[a, a^*, \dots, a, a^*]$ is transformed into the product appearing on the right-hand side of Equation (15.6). This proves (15.6), and the proof of (15.7) is similar.

(2) This is a direct consequence from the first part, if the α and β are the same. □

COROLLARY 15.7. (1) *The* $*$*-distribution of an* R*-diagonal element* a *is uniquely determined by the distribution of* aa^* *and the distribution of* a^*a.

(2) *The* $*$*-distribution of a tracial* R*-diagonal element* a *is uniquely determined by the distribution of* a^*a.

PROOF. This follows immediately from the observation that the formulas (15.6) and (15.7) can inductively be resolved for $(\alpha_n)_{n\geq 1}$ and $(\beta_n)_{n\geq 1}$ in terms of the cumulants of aa^* and the cumulants of a^*a. □

PROPOSITION 15.8. *Let a and b be elements in a $*$-probability space $(\mathcal{A}, \varphi)$ such that a is R-diagonal and such that $\{a, a^*\}$ and $\{b, b^*\}$ are freely independent. Then ab is R-diagonal.*

PROOF. We have to show that non-alternating cumulants in ab and b^*a^* vanish. Thus we have to look at situations like $\kappa_n(\ldots, ab, ab, \ldots)$ or $\kappa_n(\ldots, b^*a^*, b^*a^*, \ldots)$. We will only consider the first case, the latter is similar. In order to be able to distinguish the various a appearing in the cumulant, we will put some indices on them. So we consider $\kappa_n(\ldots, a_1b, a_2b, \ldots)$, where $a_1 = a_2 = a$.

By Theorem 11.12, we have

$$\kappa_n(\ldots, a_1b, a_2b, \ldots) = \sum_{\substack{\pi\in NC(2n)\\ \pi\vee\sigma=1_{2n}}} \kappa_\pi[\ldots, a_1, b, a_2, b, \ldots], \tag{15.9}$$

where $\sigma = \{(1, 2), (3, 4), \ldots, (2n-1, 2n)\}$.

The fact that a and b are $*$-freely independent implies, by Theorem 11.20 on the vanishing of mixed cumulants, that only such partitions $\pi \in NC(2n)$ contribute to the sum for which each of their blocks contains elements only from $\{a, a^*\}$ or only from $\{b, b^*\}$. Let V be the block containing a_2. We have to examine two situations.

On the one hand, it might happen that a_2 is the first element in the block V. Then the last element in V must be an a^*. This situation can be sketched in the following way:

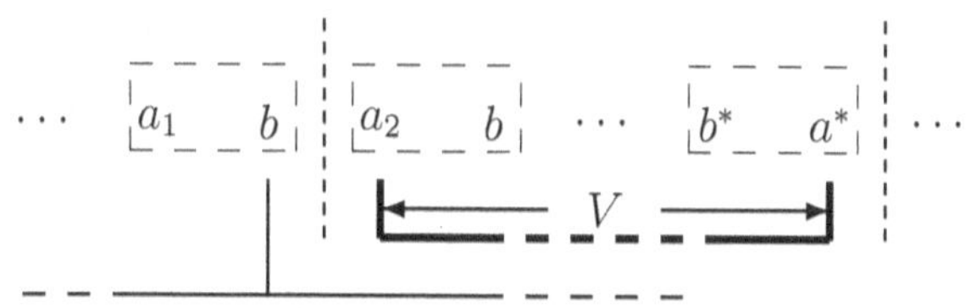

In this case π does not couple the block of σ which contains a_1 with the block of σ which contains a_2; thus $\pi \vee \sigma$ cannot be equal to 1_{2n}.

On the other hand, it can happen that a_2 is not the first element of V. Since a is R-diagonal, the element preceding a_2 in V is an a^*.

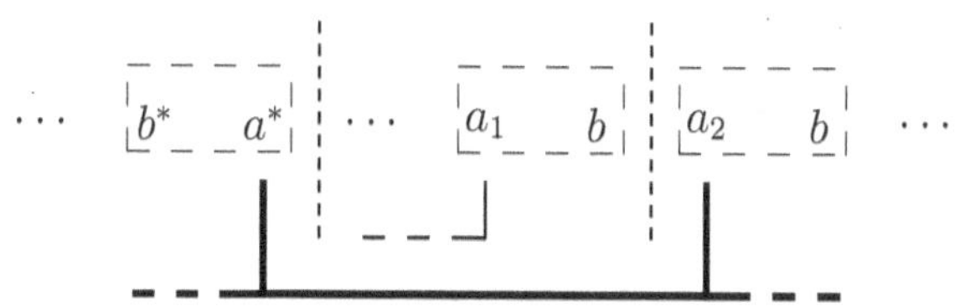

But then again the block of σ containing a_1 and the block of σ containing a_2 will not be coupled by π, and thus $\pi \vee \sigma$ cannot be equal to 1_{2n}.

As in both cases we do not find any partition contributing to the sum (15.9), this has to vanish and thus we get the assertion. □

If we take a to be a Haar unitary then we get the following corollary.

COROLLARY 15.9. *Let u and b be elements in some $*$-probability space such that u is a Haar unitary and such that u and b are $*$-free. Then ub is a R-diagonal.*

The previous results yield now the following characterization of R-diagonal elements by an "invariance of the $*$-distribution under multiplication with a free Haar unitary."

THEOREM 15.10. *Let a be an element in a $*$-probability space $(\mathcal{A}, \varphi)$. Furthermore, let u be a Haar unitary in $(\mathcal{A}, \varphi)$ such that u and a are $*$-free. Then a is R-diagonal if and only if a has the same $*$-distribution as ua.*

PROOF. $\Longrightarrow$: We assume that a is R-diagonal and, by Corollary 15.9, we know that ua is R-diagonal, too. In order to see that a and ua have the same $*$-distribution we invoke Corollary 15.7, which tells us that the distribution of an R-diagonal element a is determined by the distribution of aa^* and the distribution of a^*a. So we have to show that the distribution of aa^* agrees with the distribution of $ua(ua)^*$ and that the distribution of a^*a agrees with the distribution of $(ua)^*ua$. For the latter this is directly clear, whereas for the former one only has to observe that, for a selfadjoint random variable y and an unitary u, uyu^* has the same distribution as y if u is $*$-free from y. Note that in the non-tracial case one really needs the freeness assumption in order to get the first u to cancel the last u^* via

$$\varphi((uyu^*)^n) = \varphi(uy^n u^*) = \varphi(uu^*)\varphi(y^n) = \varphi(y^n).$$

$\Longleftarrow$: We assume that the $*$-distribution of a is the same as the $*$-distribution of ua. As, by Corollary 15.9, ua is R-diagonal it follows that a is R-diagonal, too. □

COROLLARY 15.11. *Let a be R-diagonal. Then aa^* and a^*a are freely independent.*

PROOF. By enlarging $(\mathcal{A}, \varphi)$, if necessary, we may assume that there exists a Haar unitary u in our $*$-probability space, such that u and a are $*$-free (see Exercise 6.17). Since a has the same $*$-distribution as ua it suffices to prove the statement for ua. But there it just says that uaa^*u^* and $a^*u^*ua = a^*a$ are freely independent, which follows easily from the definition of free independence (see Exercise 5.24). $\square$

Special realizations of tracial R-diagonal elements

Let us recall from Notations 11.24 that we say an element x in a non-commutative probability space is even if all its odd moments vanish (which is equivalent to the fact that all odd cumulants vanish) and that we find it convenient to record all non-trivial information about such an element x in its determining sequence $(\alpha_n^x)_{n\geq 1}$ with $\alpha_n^x := \kappa_{2n}^x$. In this section all considered even elements will live in a $*$-probability space and be selfadjoint, so that their $*$-distribution is determined by their determining sequence. It is no accident that we denote the non-trivial information (cumulants of even length) for even elements in the same way as the non-trivial information (alternating cumulants) for tracial R-diagonal elements . Selfadjoint even elements and tracial R-diagonal elements show quite a lot of similarities, in a sense the latter can be seen as the non-normal relatives of the former. Let us in particular point out that in both cases the information about the distribution of the square of the variable is calculated in the same way from the determining sequence; namely, by Proposition 15.6 we have for a tracial R-diagonal element a with determining sequence $(\alpha_n^a)_{n\geq 1}$ that

$$\kappa_n(a^*a,\dots,a^*a) = \sum_{\pi\in NC(n)} \alpha_\pi^a, \tag{15.10}$$

whereas, by Proposition 11.25, we have for an even element x with determining sequence $(\alpha_n^x)_{n\geq 1}$ that

$$\kappa_n(x^2,\dots,x^2) = \sum_{\pi\in NC(n)} \alpha_\pi^x. \tag{15.11}$$

This suggests that for each tracial R-diagonal distribution there should be a corresponding even distribution with the same determining sequence. This is indeed the case, and will be made precise in the next proposition.

PROPOSITION 15.12. (1) *Let a be a tracial R-diagonal element in a $*$-probability space $(\mathcal{A},\varphi)$. Consider in the $*$-probability space $(M_2(\mathcal{A}),\varphi\otimes \mathrm{tr}_2)$ of 2×2 matrices over $\mathcal{A}$ the random variable*

$$X = \begin{pmatrix} 0 & a \\ a^* & 0 \end{pmatrix}.$$

Then X is selfadjoint and even, and its determining sequence is the same as the determining sequence of a.

(2) *If x is a selfadjoint even element in some $*$-probability space and u is a Haar unitary which is $*$-free from x, then ux is a tracial R-diagonal element with the same determining sequence as x.*

(3) *Any tracial R-diagonal distribution arising in a $*$-probability space can be realized in the form ux where u is a Haar unitary, x is selfadjoint and even, and u and x are $*$-free. The determining sequence of the R-diagonal distribution and the determining sequence of the selfadjoint even element are necessarily the same.*

PROOF. (1) X is clearly selfadjoint and it is even because odd powers of X have zero entries on the diagonal. Since the determining sequences $(\alpha_n^a)_{n\geq 1}$ of a and $(\alpha_n^X)_{n\geq 1}$ of X can be calculated out of the moments of a^*a and X^2, respectively, in the same way (namely by Möbius inversion of formulas (15.10) and (15.11)), it is enough to see that the moments of a^*a are the same as the moments of X^2 in order to infer that both a and X have the same determining sequences. However,

$$(X^2)^n = \begin{pmatrix} (aa^*)^n & 0 \\ 0 & (a^*a)^n \end{pmatrix}$$

and thus the moments of X^2 are clearly the same as the corresponding moments of a^*a.

(2) Put $a := ux$. Then, by Corollary 15.9, a is R-diagonal. Since both u and x are normal, our state restricted to the $*$-algebra generated by u and restricted to the $*$-algebra generated by x is a trace. Thus, by Proposition 5.19, it is also a trace restricted to the $*$-algebra generated by u and x, and thus a is a tracial R-diagonal element. Now note that $a^*a = x^2$ and thus the moments of a^*a coincide with the moments of x^2. As before, this implies that their determining sequences are also the same.

(3) This is just a combination of the first and second parts; take for x the matrix X from part (1). □

The above correspondence between tracial R-diagonal elements and selfadjoint even elements goes of course over to a C^*-probability framework, i.e. if the R-diagonal element lives in a C^*-probability space, then the corresponding even element does so, and the other way around. However, in the C^*-context, we can also get other canonical realizations of R-diagonal elements which involve positive elements instead of even ones.

PROPOSITION 15.13. *Let a be a tracial R-diagonal element in a C^*-probability space. Then the $*$-distribution of a can be realized in the form uq, where u and q are elements in some C^*-probability space such* *that u is a Haar unitary, q is a positive element, and u and q are $*$-free. The distribution of q is then necessarily the same as the distribution of $\sqrt{a^*a}$.*

PROOF. Note first that if we have a realization of the $*$-distribution of a in the form $b := uq$ as in the proposition, then $b^*b = qu^*uq = q^2$. Since both b^*b and q^2 are positive we can take the square root (which is possible in a C^*-framework), yielding $\sqrt{b^*b} = \sqrt{q^2} = q$ (because q is positive). Since $\sqrt{b^*b}$ has the same distribution as $\sqrt{a^*a}$, we get the assertion on the distribution of q.

Let us now construct the asserted realization. By invoking the free product construction for C^*-probability space we can construct elements u and q in some C^*-probability space such that

- u is a Haar unitary,
- q is positive and has the same distribution as $|a| = \sqrt{a^*a}$ (which is, by traciality, the same as the distribution of $\sqrt{aa^*}$),
- u and q are $*$-free.

Then, by Corollary 15.9, uq is R-diagonal; since u and q are normal we have again that uq is a tracial R-diagonal element. However, $(uq)^*(uq) = q^2$ has the same distribution as a^*a, and thus the tracial R-diagonal element uq and the tracial R-diagonal element a have the same $*$-distribution, by Corollary 15.7. □

The previous result can actually be refined to a statement about the polar decomposition of a tracial R-diagonal element – which has the nice feature that we find the elements u and q not just in some other non-commutative probability space, but in the von Neumann algebra generated by a. Since we are not using von Neumann algebras in this book, we give that result just as an additional statement for the interested reader without elaborating on the relevant facts about von Neumann algebras. Let us just recall that any bounded operator a on a Hilbert space admits a unique polar decomposition in the form $a = u|a|$, where $|a| := \sqrt{a^*a}$ and where u is a partial isometry (as defined in Definition 7.21) such that $\ker(u) = \ker(a)$. Whereas the absolute value $|a|$ lies, by continuous functional calculus, always in the C^*-algebra generated by a, this is not true in general for u. However, what is true in general is that u lies in the von Neumann algebra generated by a. Thus, for a meaningful formulation of a polar decomposition result for R-diagonal operators one needs a non-commutative probability space which is a von Neumann algebra (usually this is called a W^*-probability space).

COROLLARY 15.14. *Let* $(\mathcal{A}, \varphi)$ *be a* W^**-probability space (i.e.* $\mathcal{A}$ *is a von Neumann algebra) with* φ *a faithful trace and let* $a \in \mathcal{A}$ *be such that* $\ker(a) = \{0\}$. *Then the following two statements are equivalent.*

(1) a *is* R*-diagonal.*

(2) a *has a polar decomposition of the form* $a = uq$, *where* u *is a Haar unitary and* u, q *are* $*$*-free.*

PROOF. The implication (2) $\Longrightarrow$ (1) is just an application of Corollary 15.9.

(1) $\Longrightarrow$ (2): Let $\tilde{a} = \tilde{u}\tilde{q}$ be the R-diagonal element, which we constructed in Proposition 15.13 (where we write now $\tilde{u}$ and $\tilde{q}$ for the u and q appearing in that proposition) and which has the same $*$-distribution as a. But this means that the von Neumann algebra generated by a is isomorphic to the von Neumann algebra generated by $\tilde{a}$ via the mapping $a \mapsto \tilde{a}$ (for this we need the faithfulness of the trace, see the corresponding statement on the level of C^*-algebras in Theorem 4.11). Since the polar decomposition takes places inside the von Neumann algebras, the polar decomposition of a is mapped to the polar decomposition of $\tilde{a}$ under this isomorphism. But the polar decomposition of $\tilde{a}$ is by construction just $\tilde{a} = \tilde{u}\tilde{q}$ (note that we need the condition on the kernel of a for this) and thus has the stated properties. Hence these properties (which rely only on the $*$-distributions of the elements involved in the polar decomposition) are also true for the elements in the polar decomposition of a. □

EXAMPLE 15.15. A circular element can be realized in (or has polar decomposition of) the form $c = uq$ where u is Haar unitary and $*$-free from q and where q has the distribution of $\sqrt{c^*c}$. The latter is the same as the distribution of $\sqrt{s^2}$ where s is a semicircular element of radius 1. This distribution of q is the so-called **quarter-circular distribution** and is given by a density

$$\frac{1}{\pi}\sqrt{4 - t^2}\, dt \qquad \text{on } [0, 2].$$

REMARKS 15.16. (1) The above polar decomposition was only for the case when a has trivial kernel. What happens in the case of non-trivial kernel? We restrict the problem for the moment to a tracial situation. Then it is still true that we can realize the given distribution in the form $b = uq$, where u is a Haar unitary, q is positive, and u and q are $*$-free. This was the content of Proposition 15.13. However, it is not true any more that uq is the polar decomposition of b, because a Haar unitary u always has trivial kernel. Thus there is (in a W^*-setting) a polar decomposition of b of the form $b = vq$, where v is a partial isometry. In this representation it will not be true that v and q are $*$-free because they share a common kernel.

(2) One might wonder whether one also has analogs of the representations from this section for non-tracial R-diagonal elements. Note that in such a case it is not possible to realize such elements in the form uq with u a Haar unitary, q positive, and u and q $*$-free. Because u and q are normal elements, our state restricted to the $*$-algebra generated by uq is necessarily tracial; the same is true with ux for x selfadjoint and even. What one can expect is to realize non-tracial R-diagonal elements in the form vq or vx, where v is an R-diagonal partial isometry. There are some results in this direction, the general situation however is not so clear. We will address this kind of question in Exercises 15.27 and 15.28.

Product of two free even elements

THEOREM 15.17. *Let x, y be two selfadjoint even random variables in some $*$-probability space. If x and y are freely independent then xy is a tracial R-diagonal element. Furthermore, the determining sequence of xy is given in terms of the determining sequence of x and the determining sequence of y as follows.*

$$\alpha_n^{xy} = \sum_{\substack{\pi,\sigma\in NC(n)\\ \sigma\leq K(\pi)}} \alpha_\pi^x \cdot \alpha_\sigma^y. \tag{15.12}$$

PROOF. Put $a := xy$. We have to see that non-alternating cumulants in $a = xy$ and $a^* = yx$ vanish. Since it is clear that cumulants of odd length in xy and yx vanish always, it remains to check the vanishing of cumulants of the form $\kappa_n(\dots, xy, xy, \dots)$. (Because of the symmetry of our assumptions in x and y this will also yield the case $\kappa_n(\dots, yx, yx, \dots)$.) By Theorem 11.12, we can write this cumulant as

$$\kappa_n(\dots, xy, xy, \dots) = \sum_{\substack{\pi\in NC(2n)\\ \pi\vee\sigma=1_{2n}}} \kappa_\pi[\dots, x, y, x, y, \dots],$$

where $\sigma = \{(1,2), (3,4), \dots, (2n-1, 2n)\}$. In order to be able to distinguish y appearing at different positions we will label them by indices (i.e. $y_i = y$ for all appearing i). Thus we have to look at $\kappa_\pi[\dots, x, y_1, x, y_2, \dots]$ for $\pi \in NC(2n)$. Because of the free independence of x and y, π only gives a contribution if it does not couple x with y. Furthermore all blocks of π have to be of even length, by our assumption that x and y are even. Let V be that block of π which contains y_1. Then there are two possibilities.

(i) y_1 is not the last element in V. Let y_3 be the next element in V, then we must have a situation like this

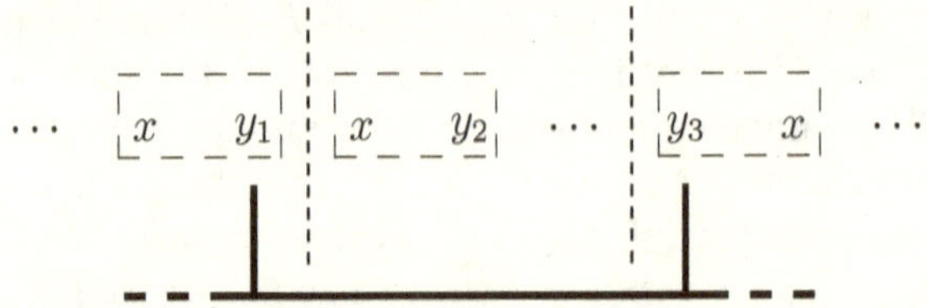

Note that y_3 has to belong to a product yx as indicated, because both the number of x and the number of y lying between y_1 and y_3 have to be even. But then everything lying between y_1 and y_3 is not connected to the rest (neither by π nor by σ), and thus the condition $\pi \vee \sigma = 1_{2n}$ cannot be fulfilled.

(ii) y_1 is the last element in the block V. Let y_0 be the first element in V. Then we have a situation as follows

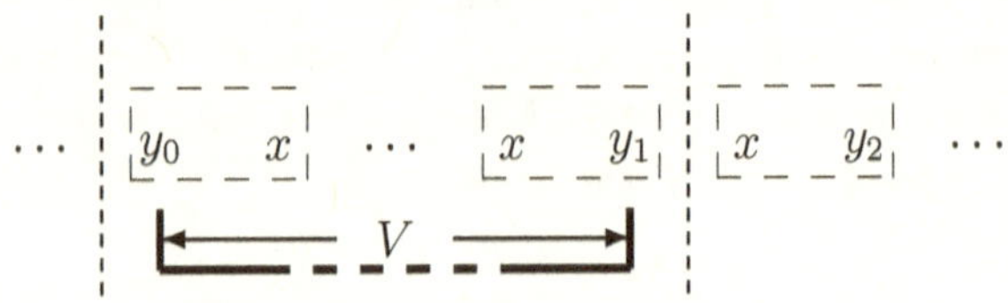

Again we have that y_0 must come from a product yx, because the number of x and the number of y lying between y_0 and y_1 both have to be even (although now some of the y from that interval might be connected to V, too, but that also has to be an even number). But then everything lying between y_0 and y_1 is separated from the rest and we cannot fulfill the condition $\pi \vee \sigma = 1_{2n}$.

Thus in any case there is no π which fulfills $\pi \vee \sigma = 1_{2n}$ and has also $k_\pi[\dots, x, y_1, x, y_2, \dots]$ different from zero. Hence $\kappa(\dots, xy, xy, \dots)$ vanishes.

So it remains to calculate $\kappa_{2n}(xy, yx, \dots, xy, yx)$. This is quite similar to the proof of Proposition 11.25. We will be quite brief and leave the details to the reader.

By Theorem 11.12, we get

$$\kappa_{2n}(xy, yx, \dots, xy, yx) = \sum_{\substack{\pi \in NC(4n) \\ \pi \vee \sigma = 1_{4n}}} \kappa_\pi[x, y, y, x, \dots, x, y, y, x],$$

where $\sigma = \{(1,2), (3,4), \dots, (4n-1, 4n)\} \in NC(4n)$. As in the proof of Proposition 11.25 one can show that the requirement $\pi \vee \sigma = 1_{4n}$ is equivalent to the following properties of π. The block containing 1 must also contain $4n$ and, for each $k = 1, \ldots, 2n-1$, the block containing $2k$ must also contain $2k+1$. The set of partitions in $NC(4n)$ fulfilling these properties are in canonical bijection with $NC(2n)$. Furthermore we have to take into account that each block of $\pi \in NC(4n)$ couples either only x or only y. For the image of π in $NC(2n)$ this means that it splits into blocks living on the odd numbers – corresponding to a $\pi_1 \in NC(1, 3, \ldots, 2n-1)$ – and blocks living on the even numbers – corresponding to a $\pi_2 \in NC(2, 4, \ldots, 2n)$. Under this identification the quantity $\kappa_\pi[x, y, y, x, \ldots, x, y, y, x]$ goes over to $\alpha^x_{\pi_1} \cdot \alpha^y_{\pi_2}$. The fact that the union of π_1 and π_2 must be non-crossing amounts to the requirement that $\pi_2 \leq K(\pi_1)$. Renaming π_1 as π and π_2 as σ gives (15.12). □

As we have seen in the last section we can make a transition from a selfadjoint even element to an R-diagonal element with the same determining sequence by multiplying the given even element with a free Haar unitary. Our present considerations show that instead of a Haar unitary we could also take a symmetric Bernoulli element. Recall that a symmetric Bernoulli variable is a selfadjoint even element b with $b^2 = 1$.

COROLLARY 15.18. *Let x be a selfadjoint even element and b be a symmetric Bernoulli variable, such that x and b are free. Then xb is R-diagonal and has the same determining sequence as the even element x.*

PROOF. By Theorem 15.17, $a := xb$ is a tracial R-diagonal element. Since $aa^* = xb^2x = x^2$, all moments of aa^* are the same as the corresponding moments of x^2 which implies that the R-diagonal element a and the even element x have the same determining sequence. □

EXERCISE 15.19. Let s be a semicircular element of variance 1 and b be a symmetric Bernoulli variable which is free from s. Prove that sb is a circular element.

The free anti-commutator of even elements

If we understand the $*$-distribution of xy, then we can of course also make some statements about the distribution of the corresponding commutator $xy - yx$ or anti-commutator $xy + yx$.

THEOREM 15.20. *Let x and y be two selfadjoint even elements which are freely independent. Then their anti-commutator $xy + yx$*

is also selfadjoint and even, and its determining sequence is given by

$$\alpha_n^{xy+yx} = 2 \sum_{\substack{\pi,\sigma \in NC(n) \\ \sigma \leq K(\pi)}} \alpha_\pi^x \cdot \alpha_\sigma^y. \tag{15.13}$$

PROOF. Since, by Theorem 15.17, xy is R-diagonal it is clear that cumulants in $xy + yx$ of odd length vanish and that for even length we get

$$\begin{aligned}\alpha_n^{xy+yx} &= \kappa_{2n}^{xy+yx} \\ &= \kappa_{2n}(xy + yx, \dots, xy + yx) \\ &= \kappa_{2n}(xy, yx, \dots, xy, yx) + \kappa_{2n}(yx, xy, \dots, yx, xy).\end{aligned}$$

Since xy is tracial we have actually that the last two summands coincide and thus

$$\alpha_n^{xy+yx} = 2\alpha_n^{xy}.$$

The assertion follows then from formula (15.12). □

REMARKS 15.21. (1) Instead of the anti-commutator one can also consider the selfadjoint version $i(xy - yx)$ of the commutator. One sees easily that in our case where x and y are even the distribution of this commutator is the same as the distribution of the anti-commutator.

(2) If one wants to consider the case of the free commutator or anti-commutator for general selfadjoint x and y then the situation becomes much more involved. In such a situation xy is of course not R-diagonal, so we have no good tools for calculating the joint moments in xy and yx. (Note that if we only consider moments in xy, then we are back to the problem of the product of free variables, which we treated in the last lecture. The point of the commutator or anti-commutator is that one needs to understand the $*$-moments of xy, not just the moments.)

(3) Even though we have no useful general formulas for the $*$-moments of xy in the general case, the commutator can nevertheless be treated in full generality, due to some remarkable cancelations which allow the general situation to be reduced to the case of even x and y. We will come back to this in Lecture 19.

(4) As is clear from our above result about the free anti-commutator, the combinatorial formulas are becoming more and more involved and one might start to wonder how much insight such formulas provide. What is really needed for presenting these solutions in a useful way is a machinery which allows us to formalize the proofs and manipulate the results in an algebraic way without having to give too much consideration on the actual kind of summation. Such a machinery will be presented in Part 3, and it will be only with the help of that

apparatus that one can really formulate the results in a form which is also suitable for concrete analytic calculations.

Powers of R-diagonal elements

We have seen that R-diagonality is preserved under several operations (like taking the sum or product of free elements). We will see now that powers of R-diagonal elements are R-diagonal, too.

PROPOSITION 15.22. *Let a be an R-diagonal element and let r be a positive integer. Then a^r is R-diagonal, too.*

PROOF. For notational convenience we deal with the case $r = 3$. General r can be treated analogously.

We have to show that cumulants $\kappa_n(\dots, a^*a^*a^*, a^*a^*a^*, \dots)$ vanish. (Cumulants $\kappa_n(\dots, aaa, aaa, \dots)$ are then covered by the observation that a being R-diagonal is the same as a^* being R-diagonal.) In order to be able to distinguish the relevant a^* we will index them from a_1^* to a_6^*, i.e. we are looking at $\kappa_n(\dots, a_1^*a_2^*a_3^*, a_4^*a_5^*a_6^*, \dots)$.

Theorem 11.12 yields in this case

$$\kappa_n(\dots, a_1^*a_2^*a_3^*, a_4^*a_5^*a_6^*, \dots) = \sum_{\substack{\pi \in NC(3n) \\ \pi \vee \sigma = 1_{3n}}} \kappa_\pi[\dots, a_1^*, a_2^*, a_3^*, a_4^*, a_5^*, a_6^*, \dots], \tag{15.14}$$

where $\sigma := \{(1,2,3), (4,5,6), \dots, (3n-2, 3n-1, 3n)\} \in NC(3n)$. In order to find out which partitions $\pi \in NC(3n)$ contribute to the sum we look at the structure of the block of π containing the element a_4^*; in the following we will call this block V.

There are two situations which can occur. The first possibility is that a_4^* is the first component of V; in this case the last component of V must be an a and, since each block has to contain the same number of a and a^*, this a has to be the third a of an argument a^3.

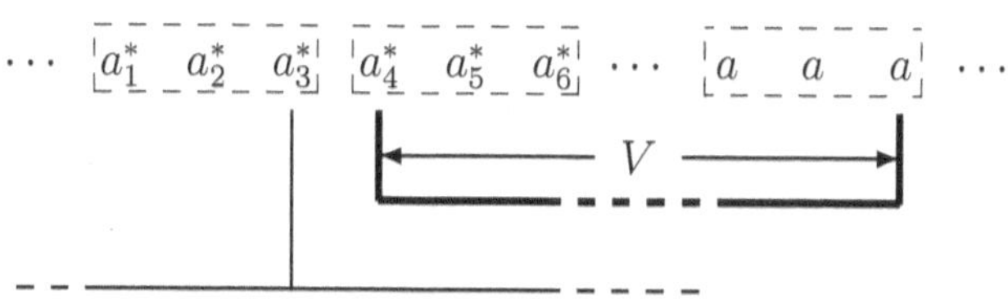

But then the block V is in $\pi \vee \sigma$ not connected with the block containing a_3^* and hence the requirement $\pi \vee \sigma = 1_{3n}$ cannot be fulfilled in such a situation.

The second situation that might happen is that a_4^* is not the first component of V. Then the preceding element in this block must be an a and again it must be the third a of an argument a^3.

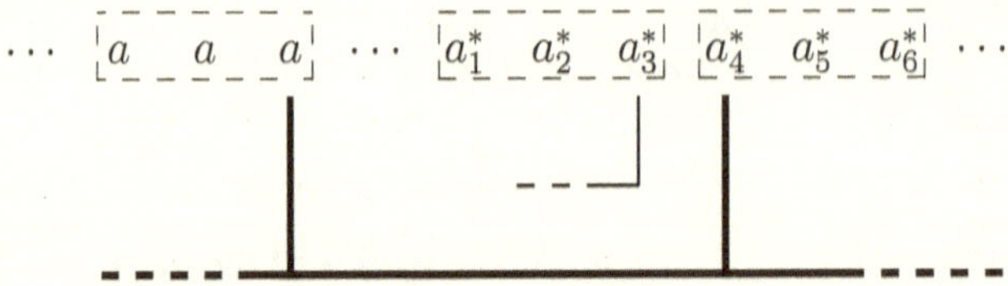

But then the block containing a_3^* is again not connected with V in $\pi \vee \sigma$.

Thus, there exists no π which gives a non-vanishing contribution in (15.14) and we get that $\kappa_n(\dots, a^*a^*a^*, a^*a^*a^*, \dots)$ is zero. $\square$

Exercises

EXERCISE 15.23. Let a and b be two R-diagonal elements which are $*$-free. Show that the sum $a+b$ is also R-diagonal and express its determining sequence in terms of the determining sequence of a and the determining sequence of b.

EXERCISE 15.24. Let a and b be R-diagonal random variables such that $\{a, a^*\}$ is free from $\{b, b^*\}$. By Proposition 15.8 we know that ab is R-diagonal. In this exercise we want to see how we can express the determining sequence of ab in terms of the determining sequence of a and of b. We put

$$\alpha_n^a := \kappa_{2n}(a, a^*, a, a^*, \dots, a, a^*), \qquad \beta_n^a := \kappa_{2n}(a^*, a, a^*, a, \dots, a^*, a),$$
$$\alpha_n^b := \kappa_{2n}(b, b^*, b, b^*, \dots, b, b^*), \qquad \alpha_n^{ab} := \kappa_{2n}(ab, b^*a^*, \dots, ab, b^*a^*).$$

(1) Show that we have

$$\alpha_n^{ab} = \sum_{\substack{\pi=\pi_a\cup\pi_b\in NC(2n)\\ \pi_a=\{V_1,\dots,V_k\}\in NC(1,3,\dots,2n-1)\\ \pi_b=\{V_1',\dots,V_l'\}\in NC(2,4,\dots,2n)}} \alpha^a_{|V_1|}\beta^a_{|V_2|}\cdots\beta^a_{|V_k|}\alpha^b_{|V_1'|}\cdots\alpha^b_{|V_l'|}, \tag{15.15}$$

where V_1 is that block of π which contains the first element 1.

(2) Show that in the tracial case the statement reduces to

$$\alpha_n^{ab} = \sum_{\substack{\pi,\sigma\in NC(n)\\ \sigma\le K(\pi)}} \alpha_\pi^a \cdot \alpha_\sigma^b. \tag{15.16}$$

EXERCISE 15.25. (1) Prove the following statement. Let a be an R-diagonal element and r a positive integer. Then the $*$-distribution of a^r is the same as the $*$-distribution of $a_1 \cdots a_r$ where each a_i ($i =$

$1, \ldots, r)$ has the same $*$-distribution as a and where $a_1, \ldots, a_r$ are $*$-freely independent.

(2) Let $a_1, \ldots, a_n$ be n R-diagonal elements which are $*$-free. Consider a matrix $(\gamma_{ij})_{i,j=1}^n$ of complex numbers. Define $a := \sum_{i,j=1}^n \gamma_{ij} a_i a_j$. Show that a is R-diagonal.

EXERCISE 15.26. Let c be a circular element and r a positive integer.

(1) Calculate the determining sequence of c^r and the moments of $c^{*r}c^r$.

(2) Calculate the norms of powers of circular elements (assuming they live in a C^*-probability space with faithful state φ) via the formula

$$\|c^r\| = \lim_{n\to\infty} \sqrt[2n]{\varphi\big((c^{*r}c^r)^n\big)}.$$

EXERCISE 15.27. (1) Show that a Haar unitary element is the only unitary element that is R-diagonal.

(2) Let v be a partial isometry (as defined in Definition 7.21) which is also R-diagonal. Show that all $*$-moments of v are determined by the knowledge of $\alpha := \varphi(v^*v)$ and $\beta := \varphi(vv^*)$. Let us call such a v an (α, β)-**Haar partial isometry** in the following. Show that such an (α, β)-Haar partial isometry is tracial if and only if $\alpha = \beta$.

(3) Show that for a (α, β)-Haar partial isometry in a C^*-probability space we have necessarily $0 \le \alpha, \beta \le 1$.

EXERCISE 15.28. Let $l_1 := l(f)$ and $l_2 := l(g)$ be two creation operators on a full Fock space $\mathcal{F}(\mathcal{H})$, such that f and g are two orthogonal unit vectors, i.e. l_1 and l_2 are $*$-free with respect to the vacuum expectation state. For a fixed $0 \le \lambda \le 1$ we put $c := l_1 + \sqrt{\lambda} l_2^*$.

(1) Show that c is R-diagonal and that it is tracial if and only if $\lambda = 1$, in which case it is circular. We will call the c for general λ **generalized circular elements**.

(2) Show that the vacuum expectation state restricted to the $*$-algebra generated by c is faithful for $0 < \lambda \le 1$.

(3) Consider the polar decomposition $c = vq$ of a generalized circular element (in a W^*-probability space). Show that v and q are $*$-free and that the polar part is a $(1, \beta)$-Haar partial isometry for suitable β.

(4) Let v_1 be a $(1, \beta)$-Haar isometry and v_2 a $(1, \alpha)$-Haar partial isometry, and assume that v_1 and v_2 are $*$-free. Show that then $v := v_1 v_2^*$ is an (α, β)-Haar partial isometry.

Part 3

Transforms and models

LECTURE 16

The R-transform

In this lecture (and in general in the lectures on transforms in Part 3) we will take a point of view on free cumulants which emphasizes formal power series. This leads us to the concept of *R-transform* for a tuple of non-commutative random variables. The R-transform contains essentially the same information as the free cumulants of the random variables in question, the difference is in the point of view:

free cumulants	$\leftrightarrow$	the R-transform
("coefficients")		("power series").

The R-transform of one variable has already appeared in Lecture 12, where it was used to study the operation of addition of freely independent random variables. In this lecture we will introduce the multi-variable version of the R-transform, and point out the analogy with its counterpart in classical probability, the logarithm of the Fourier transform.

The multi-variable R-transform

We start by introducing the space of series that we want to use, and the operation of extracting a coefficient from such a series.

NOTATIONS 16.1. Let s be a positive integer.

(1) We denote by Θ_s the set of all formal power series of the form

$$f(z_1, \ldots, z_s) = \sum_{n=1}^{\infty} \sum_{i_1,\ldots,i_n=1}^{s} \alpha_{i_1,\ldots i_n} z_{i_1} \cdots z_{i_n}, \tag{16.1}$$

with $\alpha_{i_1,\ldots,i_n} \in \mathbb{C}$ ($\forall\, n \geq 1$, $\forall\, 1 \leq i_1, \ldots, i_n \leq s$), and where $z_1, \ldots, z_s$ are non-commuting indeterminates.

(2) Let $f \in \Theta_s$ be as in Equation (16.1). For every $n \geq 1$ and $1 \leq i_1, \ldots, i_n \leq s$ we denote

$$\alpha_{i_1,\ldots,i_n} =: \mathrm{Cf}_{(i_1,\ldots,i_n)}(f) \tag{16.2}$$

("the coefficient of order $(i_1, \ldots, i_n)$" of f).

We will need, moreover, the following notation for "generalized coefficients" of a power series in Θ_s.

NOTATION 16.2. Let s be a positive integer. For $n \geq 1$, $1 \leq i_1, \ldots, i_n \leq s$ and $\emptyset \neq V \subset \{1, \ldots, n\}$ we will use "$(i_1, \ldots, i_n)|V$" to denote the tuple in $\{1, \ldots, s\}^{|V|}$ obtained from $(i_1, \ldots, i_n)$ by retaining only those i_j with $j \in V$. (For instance if $n = 6$ and $V = \{2, 3, 5\}$ then $(i_1, \ldots, i_6)|V = (i_2, i_3, i_5)$.)

Now, let f be a power series in Θ_s. For every $n \geq 1$, every $1 \leq i_1, \ldots, i_n \leq s$, and every non-crossing partition $\pi = \{V_1, \ldots, V_r\} \in NC(n)$ we denote

$$\mathrm{Cf}_{(i_1,\ldots,i_n);\pi}(f) := \alpha_{(i_1,\ldots,i_n)|V_1} \cdots \alpha_{(i_1,\ldots,i_n)|V_r}. \tag{16.3}$$

Note that in general $\mathrm{Cf}_{(i_1,\ldots,i_n);\pi}(f)$ is not a true coefficient of f, but rather a product of such coefficients. (Of course, if π happens to be the partition with only one block, 1_n, then $\mathrm{Cf}_{(i_1,\ldots,i_n);\pi}(f)$ reduces to the regular coefficient $\mathrm{Cf}_{(i_1,\ldots,i_n)}(f)$.)

DEFINITION 16.3. Let $(\mathcal{A}, \varphi)$ be a non-commutative probability space, and let $a_1, \ldots, a_s$ be an s-tuple of elements of $\mathcal{A}$.

(1) Consider the family of joint moments of $a_1, \ldots, a_s$,

$$\Big\{\varphi(a_{i_1} \cdots a_{i_n}) \mid n \geq 1,\ 1 \leq i_1, \ldots, i_n \leq s\Big\}.$$

We can use all these numbers to make up a power series in Θ_s which will be denoted by $M_{a_1,\ldots,a_s}$, and is called the **moment series** of $a_1, \ldots, a_s$:

$$M_{a_1,\ldots,a_s}(z_1, \ldots, z_s) = \sum_{n=1}^{\infty} \sum_{i_1,\ldots,i_n=1}^{s} \varphi(a_{i_1} \cdots a_{i_n}) z_{i_1} \cdots z_{i_n}. \tag{16.4}$$

(2) Consider on the other hand the family of all free cumulants of $a_1, \ldots, a_s$,

$$\Big\{\kappa_n(a_{i_1}, \ldots, a_{i_n}) \mid n \geq 1,\ 1 \leq i_1, \ldots, i_n \leq s\Big\}.$$

With these numbers we make up a series in Θ_s, denoted by $R_{a_1,\ldots,a_s}$ and called the ***R*-transform** of $a_1, \ldots, a_s$:

$$R_{a_1,\ldots,a_s}(z_1, \ldots, z_s) = \sum_{n=1}^{\infty} \sum_{i_1,\ldots,i_n=1}^{s} \kappa_n(a_{i_1}, \ldots, a_{i_n}) z_{i_1} \cdots z_{i_n}. \tag{16.5}$$

REMARKS 16.4. Let us make some comments on how the notations introduced above relate to some other notations used in the preceding lectures.

(1) When applied to the R-transform $f = R_{a_1,\ldots,a_s} \in \Theta_s$, the notation $\mathrm{Cf}_{(i_1,\ldots,i_n);\pi}(f)$ from Equation (16.3) matches the notations of the type "$\kappa_\pi[a_1, \ldots, a_n]$" used in the lectures from Part 2. Indeed, it is

clear that given a non-commutative probability space $(\mathcal{A}, \varphi)$ and the elements $a_1, \ldots, a_s \in \mathcal{A}$, we have

$$\mathrm{Cf}_{(i_1,\ldots,i_n)}(R_{a_1,\ldots,a_s}) = \kappa_n(a_{i_1}, \ldots, a_{i_n}), \tag{16.6}$$

and more generally

$$\mathrm{Cf}_{(i_1,\ldots,i_n);\pi}(R_{a_1,\ldots,a_s}) = \kappa_\pi[a_{i_1}, \ldots, a_{i_n}] \tag{16.7}$$

for every $n \geq 1$, every $1 \leq i_1, \ldots, i_n \leq s$, and every $\pi \in NC(n)$.

(2) The particular case $s = 1$ of Equation (16.5) gives us a series

$$R_a(z) = \sum_{n=1}^{\infty} \kappa_n(a, \ldots, a) z^n,$$

where a is an element in a non-commutative probability space $(\mathcal{A}, \varphi)$. This is, of course, very closely related to the series $\mathcal{R}_a$ introduced in Notation 12.6. More precisely, the definitions of R_a and of $\mathcal{R}_a$ are made in such a way that the two series only differ by a shift in the powers of z:

$$R_a(z) = z\mathcal{R}_a(z). \tag{16.8}$$

In the free probability literature (and in particular in this book) both the series $\mathcal{R}_a$ and R_a are referred to as "the R-transform of a." We hope the reader will have no difficulty in taking this minor detail into account in the various R-transform calculations which are being shown throughout the book.

REMARK 16.5. As explained in Lecture 11 (see in particular the Appendix to that lecture), the free cumulants represent the free probabilistic analog for the concept of cumulants from classical probability – in the respect that they are obtained from joint moments by the same kind of formulas, but where one only looks at non-crossing partitions (instead of arbitrary partitions) of finite sets. Let us now revisit this fact, from the point of view of power series.

Let $X_1, \ldots, X_s$ be an s-tuple of real random variables in the classical (commutative) sense. It is well known (cf. Exercise 11.37) that the classical joint cumulants of $X_1, \ldots, X_s$ can be retrieved as coefficients of the power series (in s *commuting* indeterminates) $\log \mathcal{F}(\nu)$, where ν is the joint distribution of $X_1, \ldots, X_s$ (ν is a probability measure on $\mathbb{R}^s$ – cf. Example 4.4.1 of Lecture 4). Here $\mathcal{F}(\nu)$ denotes the Fourier transform (also known as the characteristic function) of ν, and the log can be viewed in the formal power series sense.

On the other hand, free cumulants are coefficients of R-transforms for s-tuples of non-commutative random variables; hence the analogy from the level of cumulants leads to the following important statement:

The R-transform is the analog in free probability for the logarithm of the Fourier transform.

This statement is illustrated by the following theorem ("a free independence criterion in terms of R-transforms"), which is a fundamental result in the theory of the R-transform.

THEOREM 16.6. *Let $(\mathcal{A}, \varphi)$ be a non-commutative probability space, and let $a_1, \ldots, a_s$ be elements of $\mathcal{A}$. Then the following two statements are equivalent:*

(1) *$a_1, \ldots, a_s$ are freely independent;*

(2) *we have that*

$$R_{a_1,\ldots,a_s}(z_1, \ldots, z_s) = R_{a_1}(z_1) + \cdots + R_{a_s}(z_s). \tag{16.9}$$

PROOF. This is a restatement of Theorem 11.20 (which said that "free independence is equivalent to the vanishing of all the mixed cumulants"), in the context where the free cumulants of $a_1, \ldots, a_s$ are viewed as coefficients of $R_{a_1,\ldots,a_s}$. □

REMARKS 16.7. (1) Equation (16.9) in the preceding theorem is the free analog for the following basic property of the Fourier transform. Suppose that $X_1, \ldots, X_s$ are random variables on a probability space (in classical, commutative sense); let $\mathcal{F}$ be the Fourier transform of the joint distribution of $X_1, \ldots, X_s$, and let $\mathcal{F}_1, \ldots, \mathcal{F}_s$ denote the Fourier transforms of the individual distributions of $X_1, \ldots, X_s$, respectively (so that $\mathcal{F}$ is a series in s commuting complex variables, while each of $\mathcal{F}_1, \ldots, \mathcal{F}_s$ is a series of one variable). Then $X_1, \ldots, X_s$ are classically independent precisely when

$$\mathcal{F}(z_1, \ldots, z_s) = \mathcal{F}_1(z_1) \cdots \mathcal{F}_s(z_s). \tag{16.10}$$

The analogy between the Equations (16.9) and (16.10) becomes obvious when one applies the log to both sides of (16.10).

(2) The result of the preceding theorem also holds in a version where instead of s elements we deal with s *families* of elements. For example for $s = 2$ this would be stated as follows. Let $(\mathcal{A}, \varphi)$ be a non-commutative probability space, and let $a_1, \ldots, a_p, b_1, \ldots, b_q$ be elements of $\mathcal{A}$. Then the family $\{a_1, \ldots, a_p\}$ is freely independent from $\{b_1, \ldots, b_q\}$ if and only if we have

$$\begin{aligned} &R_{a_1,\ldots,a_p,b_1,\ldots,b_q}(z_1, \ldots, z_p, w_1, \ldots, w_q) \\ &= R_{a_1,\ldots,a_p}(z_1, \ldots, z_p) + R_{b_1,\ldots,b_q}(w_1, \ldots, w_q). \end{aligned} \tag{16.11}$$

This too follows from the fact that free independence is equivalent to the vanishing of the mixed free cumulants (one has to prove and then apply the suitable version of Theorem 11.16).

(3) The phenomenon of vanishing of mixed free cumulants has an important consequence about the addition of free s-tuples, which has already been observed in the case $s = 1$ in Lecture 12 (Proposition 12.3).

PROPOSITION 16.8. *Let $(\mathcal{A}, \varphi)$ be a non-commutative probability space, and let $a_1, \ldots, a_s, b_1, \ldots, b_s$ be elements of $\mathcal{A}$ such that $\{a_1, \ldots, a_s\}$ is freely independent from $\{b_1, \ldots, b_s\}$. Then*

$$R_{a_1+b_1,\ldots,a_s+b_s} = R_{a_1,\ldots,a_s} + R_{b_1,\ldots,b_s}. \tag{16.12}$$

PROOF. For $n \geq 1$ and $1 \leq i_1, \ldots, i_n \leq s$ we have that

$$\mathrm{Cf}_{(i_1,\ldots,i_n)}(R_{a_1+b_1,\ldots,a_s+b_s}) = \kappa_n(a_{i_1} + b_{i_1}, \ldots, a_{i_n} + b_{i_n}).$$

By using the multilinearity of κ_n we expand the latter cumulant as a sum of 2^n terms; and after that we notice that $2^n - 2$ of the 2^n terms are mixed free cumulants of $\{a_1, \ldots, a_s\}$ and $\{b_1, \ldots, b_s\}$, therefore must vanish. We are thus left with a sum of two terms:

$$\begin{aligned} \mathrm{Cf}_{(i_1,\ldots,i_n)}(R_{a_1+b_1,\ldots,a_s+b_s}) &= \kappa_n(a_{i_1}, \ldots, a_{i_n}) + \kappa_n(b_{i_1}, \ldots, b_{i_n}) \\ &= \mathrm{Cf}_{(i_1,\ldots,i_n)}(R_{a_1,\ldots,a_s}) + \mathrm{Cf}_{(i_1,\ldots,i_n)}(R_{b_1,\ldots,b_s}), \end{aligned}$$

and the assertion follows. □

EXAMPLE 16.9. The use of R-transforms can sometimes help our terminology become more concise. As an example let us look at the R-diagonal elements studied in Lecture 15. In the language of the R-transform, an element a in a $*$-probability space $(\mathcal{A}, \varphi)$ is R-diagonal precisely when the R-transform R_{a,a^*} is of the form:

$$R_{a,a^*}(z_1, z_2) = f(z_1 z_2) + g(z_2 z_1) \tag{16.13}$$

where f and g are series of one variable,

$$f(z) = \sum_{n=1}^{\infty} \alpha_n z^n \qquad \text{and} \qquad g(z) = \sum_{n=1}^{\infty} \beta_n z^n.$$

The sequences of coefficients $(\alpha_n)_{n\geq 1}$ and $(\beta_n)_{n\geq 1}$ were called in Lecture 15 the *determining sequences* of the R-diagonal element a; in the same vein, the two power series f and g appearing above will be termed in what follows as the **determining series** of a.

The interpretation in terms of R-transforms is actually the one which explains the name "R-diagonal element": the requirement for a to be R-diagonal is that the joint R-transform R_{a,a^*} is in a certain sense supported on the diagonal of the set which indexes its coefficients.

Another property of the multivariable R-transform which is worth recording is that it behaves very nicely under linear transformations. In order to state this, let us first introduce a notation, concerning linear changes of variables in formal power series.

NOTATION 16.10. Let r and s be positive integers, let f be a series in Θ_s, and let $L = (\lambda_{ij})_{i,j}$ be a complex $s \times r$ matrix. We denote as "$f \circ L$" the series in Θ_r with coefficients defined as follows:

$$\text{Cf}_{j_1,\dots,j_n}(f \circ L) = \sum_{i_1,\dots i_n=1}^{s} \text{Cf}_{i_1,\dots,i_n}(f)\lambda_{i_1j_1} \cdots \lambda_{i_nj_n}, \tag{16.14}$$

for $n \geq 1$ and $1 \leq j_1, \dots, j_n \leq r$.

REMARK 16.11. The explanation for the notation "$f \circ L$" is that, with a small notational abuse, the relation between f and $f \circ L$ can be written as

$$(f \circ L)(w_1, \dots, w_r) = f\Big(\sum_{j=1}^{r} \lambda_{1j}w_j, \dots, \sum_{j=1}^{r} \lambda_{sj}w_j\Big). \tag{16.15}$$

The meaning of Equation (16.15) is the following. Take the expanded form of f,

$$f(z_1, \dots, z_s) = \sum_{n=1}^{\infty} \sum_{i_1,\dots,i_n=1}^{s} \text{Cf}_{(i_1,\dots i_n)}(f)z_{i_1} \cdots z_{i_n}.$$

In this expanded form perform the substitutions

$$z_i = \sum_{j=1}^{r} \lambda_{ij}w_j, \quad 1 \leq i \leq s,$$

then multiply out the monomials $z_{i_1} \cdots z_{i_n}$ and re-group terms to get a series in the indeterminates $w_1, \dots, w_r$. The resulting series will be precisely $f \circ L \in \Theta_r$.

The justification for the notation "$f \circ L$" is even clearer if Equation (16.15) is written in the more concise form

$$(f \circ L)(w_1, \dots, w_r) = f(\, L(w_1, \dots, w_r)\,).$$

PROPOSITION 16.12. *Let $(\mathcal{A}, \varphi)$ be a non-commutative probability space, let $a_1, \dots, a_s$ be elements of $\mathcal{A}$, and let $L = (\lambda_{ij})_{i,j}$ be a complex $s \times r$ matrix. Consider the elements $b_1, \dots, b_r \in \mathcal{A}$ defined by*

$$b_j = \sum_{i=1}^{s} \lambda_{ij}a_i, \quad 1 \leq j \leq r.$$

Then we have

$$R_{b_1,\dots,b_r} = R_{a_1,\dots,a_s} \circ L. \tag{16.16}$$

PROOF. For $n \geq 1$ and $1 \leq j_1, \dots, j_n \leq r$ we have that

$$\begin{aligned}
\mathrm{Cf}_{(j_1,\dots,j_n)}(R_{b_1,\dots,b_r}) &= \kappa_n(b_{j_1}, \dots, b_{j_n}) \\
&= \kappa_n\Big(\sum_{i=1}^{s} \lambda_{i,j_1} a_i, \dots, \sum_{i=1}^{s} \lambda_{i,j_n} a_i\Big) \\
&= \sum_{i_1,\dots,i_n=1}^{s} \lambda_{i_1 j_1} \cdots \lambda_{i_n j_n} \kappa_n(a_{i_1}, \dots, a_{i_n}) \\
&= \sum_{i_1,\dots,i_n=1}^{s} \lambda_{i_1 j_1} \cdots \lambda_{i_n j_n} \mathrm{Cf}_{(i_1,\dots,i_n)}(R_{a_1,\dots,a_s}) \\
&= \mathrm{Cf}_{(j_1,\dots,j_n)}(R_{a_1,\dots,a_s} \circ L).
\end{aligned}$$

□

REMARKS 16.13. (1) If one uses the more suggestive (though somewhat unrigorous) notations from Remark 16.11, then the statement of the preceding proposition can be summarized by the formula

$$R_{L^t(a_1,\dots,a_s)}(w_1, \dots, w_r) = R_{a_1,\dots,a_s}(\, L(w_1, \dots, w_r)\,),$$

where L^t is the transpose of L.

(2) A well-known fact from classical probability theory is that independent Gaussian random variables are characterized by the property of remaining independent under rotations. The formula for the behavior of the R-transform under linear transformations can be used to obtain the free analog of this fact, where instead of independent Gaussian variables we are now dealing with freely independent semicircular variables (in a non-commutative context). See Exercise 16.23 at the end of the lecture for the precise formulation of how this goes.

The functional equation for the R-transform

In this section we derive an important functional equation which is always satisfied by the series $M_{a_1,\dots,a_s}$ and $R_{a_1,\dots,a_s}$, when $a_1, \dots, a_s$ are random variables in a non-commutative probability space. This will extend the functional equation observed for $s = 1$ in Theorem 12.5.

We fix for the section a positive integer s, and we consider the space of power series Θ_s, as in Notations 16.1. It is obvious that Θ_s is an algebra (non-unital, though) under the usual operations of addition, multiplication and scalar multiplication for power series. Besides this it also makes sense to consider compositions of series from Θ_s, as follows.

NOTATION 16.14. Let $f, h_1, \ldots, h_s$ be in Θ_s, and suppose that f is written explicitly,

$$f(z_1, \ldots, z_s) = \sum_{n=1}^{\infty} \sum_{i_1,\ldots,i_n=1}^{s} \alpha_{i_1,\ldots,i_n} z_{i_1} \cdots z_{i_n}.$$

We denote by $f(h_1, \ldots, h_s)$ the series defined as:

$$f(h_1, \ldots, h_s) := \sum_{n=1}^{\infty} \sum_{i_1,\ldots,i_n=1}^{s} \alpha_{i_1,\ldots,i_n} h_{i_1} \cdots h_{i_n}. \tag{16.17}$$

The infinite sum on the right-hand side of Equation (16.17) does not raise convergence problems – indeed, (16.17) can also be written by saying that:

$$\mathrm{Cf}_{(j_1,\ldots,j_m)}\big(f(h_1, \ldots, h_s)\big) := \tag{16.18}$$

$$\sum_{n=1}^{m} \sum_{i_1,\ldots,i_n=1}^{s} \alpha_{i_1,\ldots,i_n} \cdot \mathrm{Cf}_{(j_1,\ldots,j_m)}(h_{i_1} \cdots h_{i_n}),$$

for every $m \geq 1$ and every $1 \leq j_1, \ldots, j_m \leq s$.

The functional equation announced in the title of the section goes as follows.

THEOREM 16.15. *For $f, g \in \Theta_s$ the following two conditions are equivalent:*

(1) $\mathrm{Cf}_{(i_1,\ldots,i_n)}(g) = \sum_{\pi \in NC(n)} \mathrm{Cf}_{(i_1,\ldots,i_n);\pi}(f), \quad \forall\, n \geq 1,\ 1 \leq i_1, \ldots, i_n \leq s.$

(2) $g = f\big(z_1(1+g), \ldots, z_s(1+g)\big).$

PROOF. We fix a series $f \in \Theta_s$,

$$f(z_1, \ldots, z_s) = \sum_{n=1}^{\infty} \sum_{i_1,\ldots,i_n=1}^{s} \alpha_{i_1,\ldots,i_n} z_{i_1} \cdots z_{i_n}.$$

For this f we consider the functional equation

$$\xi = f\big(z_1(1+\xi), \ldots, z_s(1+\xi)\big), \tag{16.19}$$

in the unknown $\xi \in \Theta_s$. The equivalence of the conditions (1) and (2) in the theorem will clearly follow if we show that

(a) Equation (16.19) has a unique solution in Θ_s, and

(b) the series $g \in \Theta_s$ with coefficients defined as in condition (1) is a solution of Equation (16.19).

We show (a) by pointing out that (16.19) is equivalent to a recurrence relation for the coefficients of the series ξ. Indeed, by taking (16.18) into account, we get that (16.19) is equivalent to:

$$\mathrm{Cf}_{(j_1,\ldots,j_m)}(\xi) = \tag{16.20}$$

$$\sum_{n=1}^{m} \sum_{i_1,\ldots,i_n=1}^{s} \alpha_{i_1,\ldots,i_n} \cdot \mathrm{Cf}_{(j_1,\ldots,j_m)}\big(z_{i_1}(1+\xi)\cdots z_{i_n}(1+\xi)\big),$$

for $m \geq 1$ and $1 \leq j_1, \ldots j_m \leq s$. For $m = 1$, Equation (16.20) simply says that the coefficient of z_j in ξ is equal to α_j, for $1 \leq j \leq s$. Now let us consider an $m \geq 2$ and some $1 \leq j_1, \ldots, j_m \leq s$. Then the coefficients appearing on the right-hand side of (16.20) keep track of all the possibilities of producing the word

$$z_{j_1} \cdots z_{j_m} \tag{16.21}$$

as part of the expansion of an expression

$$z_{i_1}(1+\xi)\cdots z_{i_n}(1+\xi), \tag{16.22}$$

with $1 \leq n \leq m$ and $1 \leq i_1, \ldots, i_n \leq s$. However, this can only be done when $i_1, \ldots, i_n$ of (16.22) are sitting among $j_1, \ldots, j_n$ of (16.21), i.e. we have

$$i_1 = j_{b(1)}, \ldots, i_n = j_{b(n)}$$

for some $1 = b(1) < b(2) < \cdots < b(n) \leq m$. If we enumerate $i_1, \ldots, i_n$ in terms of $b(1), \ldots, b(n)$, we get that the right-hand side of (16.20) is equal to:

$$\sum_{n=1}^{m} \sum_{1=b(1)<\cdots<b(n)\leq m} \alpha_{j_{b(1)},j_{b(2)},\ldots,j_{b(n)}} \cdot \mathrm{Cf}_{(j_{b(1)+1},\ldots,j_{b(2)-1})}(\xi)\cdots$$

$$\cdots \mathrm{Cf}_{(j_{b(n)+1},\ldots,j_m)}(\xi) \tag{16.23}$$

(with the appropriate convention that $\mathrm{Cf}_{(j_p,\ldots,j_q)}(\xi) = 1$ when $p > q$). Hence in (16.23) we obtained an expression for the coefficient of order $(j_1, \ldots, j_m)$ of ξ, in terms of some shorter coefficients of the same series (the lengths of the coefficients listed in (16.23) add up to $m - n < m$, so each of them is indeed "shorter"). This is a recurrence relation for the coefficients of ξ, which determines ξ uniquely.

We now go to statement (b), that the series g defined by condition (1) in the theorem is a solution of Equation (16.19). Clearly, this will follow if we show that the coefficients of g satisfy the recurrence from

(16.23). To obtain this recurrence, we write $\mathrm{Cf}_{(j_1,\ldots,j_m)}(g)$ in terms of the coefficients of f:

$$\begin{aligned}\mathrm{Cf}_{(j_1,\ldots,j_m)}(g) &= \sum_{\pi\in NC(m)} \mathrm{Cf}_{(j_1,\ldots,j_m);\pi}(f)\\ &= \sum_{n=1}^{m} \sum_{1=b(1)<\cdots<b(n)\le m} \sum_{\substack{\pi\in NC(m) \text{ s.t.}\\ \{b(1),\ldots,b(n)\}\\ \text{is block of } \pi}} \mathrm{Cf}_{(j_1,\ldots,j_m);\pi}(f)\end{aligned} \tag{16.24}$$

(by enumerating the partitions from $NC(m)$ in terms of their block which contains the number 1). But if a partition $\pi \in NC(m)$ is subjected to the condition of having a prescribed block $B = \{b(1),\ldots,b(n)\}$, with $1 = b(1) < \cdots < b(n) \le m$, then knowing π is equivalent to knowing its restrictions $\pi_1 \in NC(b(2)-b(1)-1)$, $\pi_2 \in NC(b(3)-b(2)-1),\ldots,\pi_n \in NC(m-b(n))$ to the spaces left between the consecutive elements of B; moreover, if $\pi_1,\ldots,\pi_n$ correspond to π in this way, then it is immediate that:

$$\mathrm{Cf}_{(j_1,\ldots,j_m);\pi}(f) = \tag{16.25}$$
$$\alpha_{j_{b(1)},\ldots,j_{b(n)}} \cdot \mathrm{Cf}_{(j_{b(1)+1},\ldots,j_{b(2)-1});\pi_1}(f)\cdots \mathrm{Cf}_{(j_{b(n)+1},\ldots,j_m);\pi_n}(f).$$

By substituting (16.25) in (16.24) we obtain that

$$\begin{aligned}\mathrm{Cf}_{(j_1,\ldots,j_m);\pi}(g) &= \sum_{n=1}^{m} \sum_{1=b(1)<\cdots<b(n)\le m} \alpha_{j_{b(1)},\ldots,j_{b(n)}}\\ &\quad \cdot\Big(\sum_{\pi_1\in NC(b(2)-b(1)-1)} \mathrm{Cf}_{(j_{b(1)+1},\ldots,j_{b(2)-1});\pi_1}(f)\Big)\cdots\\ &\quad \cdots\Big(\sum_{\pi_n\in NC(m-b(n))} \mathrm{Cf}_{(j_{b(n)+1},\ldots,j_m);\pi_n}(f)\Big)\\ &= \sum_{n=1}^{m} \sum_{1=b(1)<\cdots<b(n)\le m} \alpha_{j_{b(1)},j_{b(2)},\ldots,j_{b(n)}} \qquad (16.26)\\ &\quad \cdot \mathrm{Cf}_{(j_{b(1)+1},\ldots,j_{b(2)-1})}(g)\cdots \mathrm{Cf}_{(j_{b(n)+1},\ldots,j_m)}(g),\end{aligned}$$

where at the last equality sign we used again the connection between f and g, given by condition (1) in the theorem. But Equation (16.26) is just a repetition of (16.23) with "g" appearing in the place of "ξ". □

COROLLARY 16.16. *Let $(\mathcal{A},\varphi)$ be a non-commutative probability space, and let $a_1,\ldots,a_s$ be in $\mathcal{A}$. Consider the moment series $M :=$*

$M_{a_1,\ldots,a_s}$ and the R-transform $R := R_{a_1,\ldots,a_s}$. Then M and R satisfy the equation

$$M = R\big(z_1(1+M),\ldots,z_s(1+M)\big). \tag{16.27}$$

PROOF. From the lectures about free cumulants we know that M and R satisfy condition (1) of Theorem 16.15. Therefore they must also satisfy condition (2) of the same theorem, which is (16.27). □

EXAMPLE 16.17. Let $(\mathcal{A},\varphi)$ be a $*$-probability space, and let $c \in \mathcal{A}$ be a circular element. It was observed in Example 11.23 that the only non-vanishing joint cumulants of c and c^* are $k_2(c,c^*) = k_2(c^*,c) = 1$. In the language of the R-transform, this says:

$$R_{c,c^*}(z_1,z_2) = z_1z_2 + z_2z_1.$$

What about the moment series of c and c^*? There does not appear to be any nice formula for the joint moments of c, c^*, but at least the above corollary gives a "non-commutative quadratic equation" for the series $M := M_{c,c^*}$, namely:

$$M = z_1(1+M)z_2(1+M) + z_2(1+M)z_1(1+M). \tag{16.28}$$

More about the one-dimensional case

REMARK 16.18. Let us consider a non-commutative probability space $(\mathcal{A},\varphi)$, and let a be an element of $\mathcal{A}$. The particular case $s=1$ of Corollary 16.16 gives us the functional equation

$$M_a(z) = R_a\big(z(1+M_a(z))\big), \tag{16.29}$$

where

$$M_a(z) = \sum_{n=1}^{\infty} \varphi(a^n)z^n$$

and

$$R_a(z) = \sum_{n=1}^{\infty} \kappa_n(a,a,\ldots,a)z^n.$$

We will look at some other ways in which Equation (16.29) can be reformulated. Clearly, we can start by writing (16.29) in the form

$$M_a = R_a \circ \big(z(1+M_a)\big), \tag{16.30}$$

where $\circ$ denotes composition of power series. Suppose now that the element $a \in \mathcal{A}$ we are working with is such that $\varphi(a) \neq 0$. Since the linear term of both the series M_a and R_a is equal to $\varphi(a)$, it follows that these series are invertible under composition; we will denote their inverses under composition by $M_a^{<-1>}$ and $R_a^{<-1>}$, respectively. In

(16.30) let us compose with $R_a^{<-1>}$ on the left, and with $M_a^{<-1>}$ on the right. We obtain that:

$$\begin{aligned} R_a^{<-1>} &= \big(z(1+M_a)\big) \circ M_a^{<-1>} \\ &= (z \circ M_a^{<-1>}) \cdot (1 + M_a \circ M_a^{<-1>}) \\ &= M_a^{<-1>} \cdot (1+z). \end{aligned}$$

So it is interesting that while the series M_a and R_a are in general quite different from each other, their inverses under composition differ only by a multiplication with $1+z$:

$$R_a^{<-1>}(z) = (1+z)M_a^{<-1>}(z). \tag{16.31}$$

This in particular gives a "practical" method for passing between M_a and R_a. For instance from M_a to R_a, what one has to do is take inverse under composition, then multiply by $1+z$, then take inverse under composition again.

Modulo a shift in the coefficients, the series which appears in Equation (16.31) is another important "transform" of free probability, the *S-transform*, and will be studied in more detail in Lecture 18.

REMARK 16.19. We will conclude the lecture with a discussion of the connection between the functional equation of the R-transform (in the case $s = 1$) and the *Lagrange inversion formula.* The latter formula concerns the implicit equation:

$$\xi(z) = z \cdot u\big(\xi(z)\big), \tag{16.32}$$

where $\xi \in \Theta_1$ is the unknown, and where u is a given power series, $u(t) = u_0 + u_1 t + \cdots + u_n t^n + \cdots$. Equation (16.32) has a unique solution, and Lagrange inversion says that the coefficients of the solution can be determined as follows, for all $n \geq 1$:

$$[\text{ coef. of order } n \text{ of } \xi\ ] = \frac{1}{n} \cdot [\text{ coef. of order } n-1 \text{ of } u^n\ ]. \tag{16.33}$$

There is a striking resemblance between (16.32) and the particular case $s = 1$ of the functional equation of Theorem 16.15, which was:

$$g(z) = f\big(z(1+g(z))\big). \tag{16.34}$$

Indeed, if in (16.34) we add 1 and then multiply by z on both sides, we will get exactly (16.32) for the situation when $u = 1+f$ and $\xi(z) = z(1+g(z))$. But since (16.34) is equivalent to the relation between the moment series and the R-transform of a random variable, this means that the Lagrange inversion formula can also be used to describe this relation. More precisely, the recipe from (16.33) is converted into the following.

PROPOSITION 16.20. *Let $(\mathcal{A}, \varphi)$ be a non-commutative probability space, let a be an element of $\mathcal{A}$, and let R_a be the R-transform of a. Then for every $n \geq 1$ we have:*

$$\varphi(a^n) = \frac{1}{n+1} \cdot [\text{ coef. of order } n \text{ of } (1 + R_a)^{n+1} \,]. \tag{16.35}$$

It is instructive to see directly how Equation (16.35) can be obtained from the relation between moments and free cumulants via summations over non-crossing partitions, as introduced in Lecture 11. This must of course come pretty close to re-proving the Lagrange inversion formula. In fact one of the standard combinatorial proofs of the Lagrange inversion formula is in terms of Lukasiewicz paths, so we only have to look at that one, and use the bijection between non-crossing partitions and Lukasiewicz paths observed in Lecture 9.

PROOF. Let us write explicitly

$$(1 + R_a)(z) := \sum_{n=0}^{\infty} \alpha_n z^n,$$

where $\alpha_0 = 1$ and $\alpha_n = \kappa_n(a, \ldots, a)$ for $n \geq 1$. The coefficient of order n of $(1 + R_a)^{n+1}$ is then spelled out as

$$\sum_{\substack{i_1, \ldots, i_{n+1} \geq 0 \\ i_1 + \cdots + i_{n+1} = n}} \alpha_{i_1} \cdots \alpha_{i_{n+1}} = \sum_{\substack{j_1, \ldots, j_{n+1} \geq -1 \\ j_1 + \cdots + j_{n+1} = -1}} \alpha_{1+j_1} \cdots \alpha_{1+j_{n+1}}$$

(via the obvious change of variable $j_1 = i_1 - 1, \ldots, j_{n+1} = i_{n+1} - 1$).

Consider now the concept of almost-rising path on $\mathbb{Z}^2$ which appeared in Lecture 9 (cf. Definition 9.6.1). We use the coefficients $(\alpha_n)_{n \geq 0}$ of the series $1 + R_a$ in order to define a weight for an almost-rising path γ, as follows: if the steps of γ are $(1, j_1), \ldots, (1, j_m)$ (with $m \geq 1$ and $j_1, \ldots, j_m \in \mathbb{N} \cup \{-1, 0\}$), then the weight of γ is

$$\mathrm{wt}(\gamma) := \alpha_{1+j_1} \alpha_{1+j_2} \cdots \alpha_{1+j_m}.$$

Let Γ_n denote the set of all almost-rising paths going from $(0, 0)$ to $(n + 1, -1)$. Then, clearly, the last expression found for the coefficient of order n of $(1 + R_a)^{n+1}$ is just $\sum_{\gamma \in \Gamma_n} \mathrm{wt}(\gamma)$. On the other hand let us recall that Proposition 9.11 of Lecture 9 gives us an explicit bijection between Γ_n and $Luk(n) \times \{1, \ldots, n + 1\}$, where $Luk(n)$ is the set of Lukasiewicz paths with n steps. If $\gamma \in \Gamma_n$ corresponds via this bijection to $(\gamma_0, m) \in Luk(n) \times \{1, \ldots, n+1\}$, then we have that $\mathrm{wt}(\gamma) = \mathrm{wt}(\gamma_0)$. (Indeed, γ_0 is obtained from γ by suppressing one falling step and then by cyclically permuting the remaining steps, and this does not affect the weight.) By putting all these observations together we get that:

$$\text{coef. of order } n \text{ of } (1+R_a)^{n+1} = (n+1) \cdot \sum_{\gamma_0 \in Luk(n)} \text{wt}(\gamma_0). \quad (16.36)$$

Finally, let us look at the explicit bijection between $Luk(n)$ and $NC(n)$ which is put into evidence in Proposition 9.8. If $\gamma_0 \in Luk(n)$ corresponds via this bijection to $\pi = \{V_1, \ldots, V_r\} \in NC(n)$, then we have that $\text{wt}(\gamma_0) = \alpha_{|V_1|} \cdots \alpha_{|V_r|}$. (Indeed, from the description of the bijection we see that γ_0 has precisely r non-falling steps, and the rises of these steps are $|V_1| - 1, \ldots, |V_r| - 1$.) As a consequence, we can rewrite (16.36) in the form

$$\text{coef. of order } n \text{ of } (1+R_a)^{n+1} = (n+1) \cdot \sum_{\substack{\pi \in NC(n) \\ \pi = \{V_1, \ldots, V_r\}}} \alpha_{|V_1|} \cdots \alpha_{|V_r|}.$$

But in view of the relation between the moments and the free cumulants of a, the sum on the right-hand side is precisely equal to $\varphi(a^n)$. The desired formula (16.35) follows. □

Exercises

EXERCISE 16.21. Let s be a positive integer.

(1) Prove that for every series $f \in \Theta_s$ one can find a non-commutative probability space $(\mathcal{A}, \varphi)$ and some elements $a_1, \ldots, a_s \in \mathcal{A}$ such that $M_{a_1,\ldots,a_s} = f$.

(2) Prove that for every series $g \in \Theta_s$ one can find a non-commutative probability space $(\mathcal{A}, \varphi)$ and some elements $a_1, \ldots, a_s \in \mathcal{A}$ such that $R_{a_1,\ldots,a_s} = g$.

EXERCISE 16.22. Let q, r, s be positive integers. Let L and M be complex matrices of sizes $s \times r$ and respectively $r \times q$, and let f be a series in Θ_s. Verify that

$$(f \circ L) \circ M = f \circ (LM).$$

EXERCISE 16.23. Let $(\mathcal{A}, \varphi)$ be a $*$-probability space and let a_1, a_2 be two selfadjoint elements of $\mathcal{A}$, such that a_1 is free from a_2.

(1) Suppose that a_1 and a_2 are semicircular elements, and that they have the same radius. Prove that $(\cos\theta)a_1 + (\sin\theta)a_2$ is free from $(-\sin\theta)a_1 + (\cos\theta)a_2$ for every $\theta \in \mathbb{R}$.

(2) Conversely, suppose there exists an angle θ which is not an integer multiple of $\pi/2$, such that $(\cos\theta)a_1 + (\sin\theta)a_2$ is free from $(-\sin\theta)a_1 + (\cos\theta)a_2$. Prove that a_1 and a_2 are semicircular elements and that they have the same radius.

LECTURE 17

The operation of boxed convolution

The operation of boxed convolution, $\boxed{\star}$, is a binary operation on the space Θ_s of formal power series considered in the preceding lecture. Its free probabilistic interpretation is that it encodes the multiplication of free tuples of non-commutative random variables, when one keeps track of these tuples by using R-transforms.

On the other hand, the same operation $\boxed{\star}$ can be viewed as a distillation (taking place in the power series framework) for the operation of convolution of multiplicative functions on NC, as encountered in Lecture 10.

Thus $\boxed{\star}$ is a natural object to consider, both from the free probability angle and from a strictly combinatorial point of view. Because of this, $\boxed{\star}$ is a very useful tool in the combinatorics of free probability. In this lecture we develop its basic theory, and show how it can be used in computations with R-transforms.

The definition of boxed convolution, and its motivation

DEFINITION 17.1. Let s be a positive integer, and let Θ_s be the space of power series in s non-commuting indeterminates which was considered in Lecture 16. On Θ_s we define a binary operation $\boxed{\star}$, by the following rule: for every $f, g \in \Theta_s$ and for every $n \geq 1$, $1 \leq i_1, \ldots, i_n \leq s$, the coefficient of order $(i_1, \ldots, i_n)$ of $f \boxed{\star} g$ is:

$$\mathrm{Cf}_{(i_1,\ldots,i_n)}(f \boxed{\star} g) := \sum_{\pi \in NC(n)} \mathrm{Cf}_{(i_1,\ldots,i_n);\pi}(f) \cdot \mathrm{Cf}_{(i_1,\ldots,i_n);K(\pi)}(g) \quad (17.1)$$

(where "$K(\pi)$" stands for the Kreweras complement of a partition $\pi \in NC(n)$, as in Definition 9.21). In the cases when it can be ambiguous what is the number s of indeterminates we work with, we will write "$\boxed{\star}_s$" instead of just "$\boxed{\star}$".

In this section we explain the motivation for introducing $\boxed{\star}$, from the point of view of free probability. The operation $\boxed{\star}$ has in some sense already appeared in this monograph in Lecture 14, in connection to the multiplication of two free tuples of elements in a non-commutative probability space. More precisely, if $(\mathcal{A}, \varphi)$ is a non-commutative probability space and if $a_1, \ldots, a_s, b_1, \ldots, b_s \in \mathcal{A}$ are such that $\{a_1, \ldots, a_s\}$ is free from $\{b_1, \ldots, b_s\}$, then Equation (14.4) in Theorem 14.4 gives us that

$$\kappa_n(a_{i_1}b_{i_1}, \ldots, a_{i_n}b_{i_n}) = \sum_{\pi \in NC(n)} \kappa_\pi[a_{i_1}, \ldots, a_{i_n}] \cdot \kappa_{K(\pi)}[b_{i_1}, \ldots, b_{i_n}],$$

for every $n \geq 1$ and every $1 \leq i_1, \ldots, i_n \leq s$. When the above free cumulants are interpreted as coefficients of the R-transforms of the various s-tuples involved, we obtain the following neat statement.

PROPOSITION 17.2. *Let $(\mathcal{A}, \varphi)$ be a non-commutative probability space and let $a_1, \ldots, a_s, b_1, \ldots, b_s \in \mathcal{A}$ be such that $\{a_1, \ldots, a_s\}$ is free from $\{b_1, \ldots, b_s\}$. Then the R-transform of the s-tuple $(a_1b_1, \ldots, a_sb_s)$ is*

$$R_{a_1b_1,\ldots,a_sb_s} = R_{a_1,\ldots,a_s} \boxed{\star} R_{b_1,\ldots,b_s}. \tag{17.2}$$

The implicit presence of $\boxed{\star}$ in the preceding lectures can in fact be already spotted in the fundamental relation which connects the moments and the free cumulants of a tuple of non-commutative random variables. Let us introduce the following notation.

NOTATION 17.3. Let s be a positive integer. We denote by Zeta (or by Zeta_s, if the specification of s is necessary) the series in Θ_s which has all the coefficients equal to 1:

$$\text{Zeta}(z_1, \ldots, z_s) := \sum_{n=1}^{\infty} \sum_{i_1,\ldots,i_n=1}^{s} z_{i_1} \cdots z_{i_n}. \tag{17.3}$$

Then we have the following.

PROPOSITION 17.4. *Let $(\mathcal{A}, \varphi)$ be a non-commutative probability space, and let $a_1, \ldots, a_s$ be elements of $\mathcal{A}$. Then the moment series $M := M_{a_1,\ldots,a_s}$ and the R-transform $R := R_{a_1,\ldots,a_s}$ are related by the equation*

$$M = R \boxed{\star} \text{Zeta}. \tag{17.4}$$

PROOF. The relation expressing the joint moments of $a_1, \ldots, a_s$ in terms of free cumulants is

$$\varphi(a_{i_1} \cdots a_{i_n}) = \sum_{\pi \in NC(n)} \kappa_\pi[a_{i_1}, \ldots, a_{i_n}],$$

for $n \geq 1$ and $1 \leq i_1, \ldots, i_n \leq s$ (cf. Proposition 11.4.3). This can be rewritten as

$$\text{Cf}_{(i_1,\ldots,i_n)}(M) = \sum_{\pi \in NC(n)} \text{Cf}_{(i_1,\ldots,i_n);\pi}(R) \cdot \text{Cf}_{(i_1,\ldots,i_n);K(\pi)}(\text{Zeta}),$$

which leads to (17.4). □

Basic properties of boxed convolution

PROPOSITION 17.5. *Let s be a positive integer.*

(1) *The operation* $\boxed{\star}$ *on* Θ_s *is associative.*

(2) *Let us denote by* Δ *(or* Δ_s*, if the specification of* s *is necessary) the series in* Θ_s *defined by*

$$\Delta(z_1, \dots, z_s) = z_1 + \cdots + z_s. \tag{17.5}$$

Then Δ *is the unit for* $\boxed{\star}$.

These two properties of $\boxed{\star}$ can be proved either by basic combinatorics (directly from Definition 17.1) or by using the connection to the multiplication of free s-tuples which was recorded in Proposition 17.2. In the proof written below we illustrate both methods (see also Exercise 17.23 at the end of the lecture for the alternative choices of method in the parts (1) and (2) of the proof).

PROOF. (1) Consider three series $f, g, h \in \Theta_s$, about which we will prove that $(f \boxed{\star} g) \boxed{\star} h = f \boxed{\star} (g \boxed{\star} h)$. By Exercise 16.21 in the preceding lecture, one can find non-commutative probability spaces $(\mathcal{A}_i, \varphi_i)$, $1 \le i \le 3$, and elements $a_1, \dots, a_s \in \mathcal{A}_1$, $b_1, \dots, b_s \in \mathcal{A}_2$, $c_1, \dots, c_s \in \mathcal{A}_3$, such that $R_{a_1,\dots,a_s} = f$, $R_{b_1,\dots,b_s} = g$, and $R_{c_1,\dots,c_s} = h$. By considering the free product $(\mathcal{A}, \varphi)$ of the $(\mathcal{A}_i, \varphi_i)$, $1 \le i \le 3$, one can in fact assume that all the $3s$ elements $a_1, \dots, c_s$ belong to the same non-commutative probability space $(\mathcal{A}, \varphi)$, and moreover that the three sets $\{a_1, \dots, a_s\}$, $\{b_1, \dots, b_s\}$ and $\{c_1, \dots, c_s\}$ are freely independent in $(\mathcal{A}, \varphi)$. We then repeatedly apply Proposition 17.2, by also taking into account the associativity of free independence (cf. Lecture 5, Remark 5.20). What we obtain is that both $(f \boxed{\star} g) \boxed{\star} h$ and $f \boxed{\star} (g \boxed{\star} h)$ are equal to the R-transform $R_{a_1b_1c_1,\dots,a_sb_sc_s}$.

(2) Let f be a series in Θ_s. For $n \ge 1$ and $1 \le i_1, \dots, i_n \le s$ we have

$$\mathrm{Cf}_{(i_1,\dots,i_n)}(f \boxed{\star} \Delta) = \sum_{\pi \in NC(n)} \mathrm{Cf}_{(i_1,\dots,i_n);\pi}(f) \cdot \mathrm{Cf}_{(i_1,\dots,i_n);K(\pi)}(\Delta). \tag{17.6}$$

But obviously, $\mathrm{Cf}_{(i_1,\dots,i_n);\rho}(\Delta) = 0$ for every $\rho \ne 0_n$ in $NC(n)$; or equivalently, we have that $\mathrm{Cf}_{(i_1,\dots,i_n);K(\pi)}(\Delta) = 0$ for every $\pi \ne 1_n$ in $NC(n)$. This shows that the right-hand side of (17.6) is in fact equal to

$$\mathrm{Cf}_{(i_1,\dots,i_n);1_n}(f) \cdot \mathrm{Cf}_{(i_1,\dots,i_n);0_n}(\Delta),$$

which is nothing but $\mathrm{Cf}_{(i_1,\dots,i_n)}(f)$.

In this way we obtained that $f \boxstar \Delta = f$, for every $f \in \Theta_s$. The equality $\Delta \boxstar f = f$ is verified in exactly the same way. $\square$

REMARK 17.6. A property which $\boxstar$ does *not* have is distributivity with respect to the addition and/or scalar multiplication of power series. This is because the functionals $\mathrm{Cf}_{(i_1,\ldots,i_n);\pi} : \Theta_s \to \mathbb{C}$ (defined for $n \geq 1$, $1 \leq i_1, \ldots, i_n \leq s$, and $\pi \in NC(n)$) are not linear, except for the case when $\pi = 1_n$.

Also, it is easy to see by example that the boxed convolution $\boxstar_s$ on Θ_s is non-commutative for $s \geq 2$ (cf. Exercise 17.24 at the end of the lecture). For $s = 1$ we do have that $\boxstar_1$ is commutative, as shown in Corollary 17.10 below.

We next describe the series in Θ_s which are invertible with respect to $\boxstar$.

PROPOSITION 17.7. *Let s be a positive integer, and consider the operation $\boxstar$ on Θ_s. Let f be a series in Θ_s. We have that f is invertible with respect to $\boxstar$ if and only if $\mathrm{Cf}_{(i)}(f) \neq 0$ for all $1 \leq i \leq s$.*

PROOF. Let us write f explicitly,

$$f(z_1, \ldots, z_s) = \sum_{n=1}^{\infty} \sum_{i_1,\ldots,i_n=1}^{s} \alpha_{i_1,\ldots,i_n} z_{i_1} \cdots z_{i_n}.$$

"$\Longrightarrow$" The hypothesis is that there exists $g \in \Theta_s$,

$$g(z_1, \ldots, z_s) = \sum_{n=1}^{\infty} \sum_{i_1,\ldots,i_n=1}^{s} \beta_{i_1,\ldots,i_n} z_{i_1} \cdots z_{i_n},$$

such that $f \boxstar g = \Delta$. For every $1 \leq i \leq s$ we then have:

$$\begin{aligned} 1 = \mathrm{Cf}_{(i)}(\Delta) &= \mathrm{Cf}_{(i)}(\, f \boxstar g\,) \\ &= \sum_{\pi \in NC(1)} \mathrm{Cf}_{(i);\pi}(f) \cdot \mathrm{Cf}_{(i);K(\pi)}(g) = \alpha_i \beta_i. \end{aligned}$$

Hence $\alpha_i \beta_i = 1$, which implies $\alpha_i \neq 0$.

"$\Longleftarrow$" Now the hypothesis is that $\alpha_i \neq 0$, $\forall\, 1 \leq i \leq s$. We show how one can construct a series $g \in \Theta_s$, such that $f \boxstar g = \Delta$. The coefficients of g are constructed by induction on their length, n. For $n = 1$ we set $\mathrm{Cf}_{(i)}(g) := \alpha_i^{-1}$, $1 \leq i \leq s$. Suppose next that the coefficients of g have been constructed up to length $n-1$, for some $n \geq 2$. For every $i_1, \ldots, i_n \in \{1, \ldots, s\}$ we define $\mathrm{Cf}_{(i_1,\ldots,i_n)}(g)$ as being equal to

$$-(\alpha_{i_1} \cdots \alpha_{i_n})^{-1} \cdot \sum_{\substack{\pi \in NC(n) \\ \pi \neq 0_n}} \mathrm{Cf}_{(i_1,\ldots,i_n);\pi}(f) \cdot \mathrm{Cf}_{(i_1,\ldots,i_n);K(\pi)}(g). \qquad (17.7)$$

It is immediate that the expression in (17.7) only uses coefficients of g which have length $\leq n-1$, hence it makes sense to use (17.7) as definition for $\mathrm{Cf}_{(i_1,\ldots,i_n)}(g)$. By comparing Equations (17.1) and (17.7) one readily sees that the series g obtained as the result of the inductive construction will satisfy

$$\mathrm{Cf}_{(i_1,\ldots,i_n)}(\, f \boxed{\star}\, g\,) \;=\; 0,$$

for every $n \geq 2$ and $1 \leq i_1, \ldots, i_n \leq s$. Together with the assignment for the coefficients of length 1 of g, this leads to the fact that $f \boxed{\star}\, g = \Delta$.

In a similar way one can construct a series $h \in \Theta_s$ such that $h \boxed{\star}\, f = \Delta$. Then $g = h$ because of the associativity of $\boxed{\star}$, and we conclude that f is invertible. □

Radial series

DEFINITION 17.8. Let s be a positive integer, and let f be a series in Θ_s. If $\mathrm{Cf}_{(i_1,\ldots,i_n)}(f)$ only depends on n (but not on the specific choices of $1 \leq i_1, \ldots, i_n \leq s$), then we will say that f is **radial**. In other words, f is radial when it is of the form:

$$\begin{aligned} f(z_1, \ldots, z_s) &= \sum_{n=1}^{\infty} \sum_{i_1,\ldots,i_n=1}^{s} \alpha_n z_{i_1} \cdots z_{i_n} \qquad (17.8) \\ &= \sum_{n=1}^{\infty} \alpha_n (z_1 + \cdots + z_s)^n \end{aligned}$$

for some complex coefficients $(\alpha_n)_{n=1}^{\infty}$.

For example, both the "special" series Zeta and Δ encountered above (in Equations (17.3) and (17.5), respectively) are radial.

PROPOSITION 17.9. *Let s be a positive integer, and consider the operation $\boxed{\star}$ on Θ_s. Let f be a series in Θ_s. If f is radial, then f is in the center of $\boxed{\star}$, i.e. it satisfies $f \boxed{\star}\, g = g \boxed{\star}\, f$ for all $g \in \Theta_s$.*

PROOF. We denote the coefficients of f by $\alpha_1, \alpha_2, \ldots, \alpha_n, \ldots$ (as in Equation (17.8)). Let $g \in \Theta_s$ be arbitrary. For any $n \geq 1$ and

$1 \leq i_1, \ldots, i_n \leq s$ we write:

$$\begin{aligned} \mathrm{Cf}_{(i_1,\ldots,i_n)}(\, f \boxed{\star}\, g\,) &= \sum_{\pi \in NC(n)} \mathrm{Cf}_{(i_1,\ldots,i_n);\pi}(f) \cdot \mathrm{Cf}_{(i_1,\ldots,i_n);K(\pi)}(g) \\ &= \sum_{\rho \in NC(n)} \mathrm{Cf}_{(i_1,\ldots,i_n);\rho}(g) \cdot \mathrm{Cf}_{(i_1,\ldots,i_n);K^{-1}(\rho)}(f), \end{aligned} \tag{17.9}$$

where the latter equality is obtained by doing the substitution "$\rho = K(\pi)$", and by reversing the order in the product of coefficients of f and g.

The point is now to remark that for every $\rho \in NC(n)$ we have:

$$\mathrm{Cf}_{(i_1,\ldots,i_n);K^{-1}(\rho)}(f) = \mathrm{Cf}_{(i_1,\ldots,i_n);K(\rho)}(f). \tag{17.10}$$

Indeed, the partitions $K(\rho)$ and $K^{-1}(\rho)$ are obtained from each other by a cyclic permutation of $\{1, \ldots, n\}$ (since $K(\rho) = K^2(\,K^{-1}(\rho)\,)$ and by Exercise 9.23.1). As a consequence we can write these two partitions as

$$K(\rho) = \{V_1, \ldots, V_r\}, \qquad K^{-1}(\rho) = \{V_1', \ldots, V_r'\},$$

where $|V_1| = |V_1'| =: m_1, \ldots, |V_p| = |V_r'| =: m_r$. But then the radiality of f implies that both sides of (17.10) are equal to $\alpha_{m_1} \cdots \alpha_{m_r}$.

By substituting (17.10) in (17.9) we obtain exactly the expression defining the coefficient of order $(i_1, \ldots, i_n)$ of $g \boxed{\star}\, f$, and the assertion follows. □

An immediate consequence of the above proposition is the following.

COROLLARY 17.10. *The semigroup* $(\Theta_1, \boxed{\star}_1)$ *is commutative.*

PROOF. Every $f \in \Theta_1$ is radial, hence in the center of $\boxed{\star}_1$. □

Let us note, moreover, that the series from Θ_1 are in some sense showing up in the center of every Θ_s.

PROPOSITION 17.11. *Let s be a positive integer. Consider the map from Θ_1 to Θ_s defined by*

$$\sum_{n=1}^{\infty} \alpha_n z_1^n \;\mapsto\; \sum_{n=1}^{\infty} \alpha_n (z_1 + \cdots + z_s)^n. \tag{17.11}$$

Then this map is a homomorphism between the operations $\boxed{\star}_1$ on Θ_1 and $\boxed{\star}_s$ on Θ_s, and is thus an embedding of $(\,\Theta_1, \boxed{\star}_1\,)$ into the center of $(\,\Theta_s, \boxed{\star}_s\,)$.

PROOF. Let f, g be in Θ_1, and set $f \boxed{\star}_1 g =: h$. Let us denote the images of f, g, h via the map (17.11) by $\widetilde{f}, \widetilde{g}, \widetilde{h}$, respectively. This means that we have the relation

$$\mathrm{Cf}_{(i_1,\ldots,i_n)}(\widetilde{f}) = \mathrm{Cf}_{(1,\ldots,1)}(f), \quad n \geq 1,\ 1 \leq i_1, \ldots, i_n \leq s, \tag{17.12}$$

and similar relations involving g and h. Note that (17.12) immediately extends to:

$$\mathrm{Cf}_{(i_1,\ldots,i_n);\pi}(\widetilde{f}) = \mathrm{Cf}_{(1,\ldots,1);\pi}(f), \tag{17.13}$$

where π is an arbitrary partition in $NC(n)$.

We have to show that $\widetilde{f} \boxed{\star}_s \widetilde{g} = \widetilde{h}$. And indeed, for arbitrary $n \geq 1$ and $1 \leq i_1, \ldots, i_n \leq s$ we have

$$\begin{aligned} \mathrm{Cf}_{(i_1,\ldots,i_n)}(\widetilde{f} \boxed{\star}_s \widetilde{g}) &= \sum_{\pi \in NC(n)} \mathrm{Cf}_{(i_1,\ldots,i_n);\pi}(\widetilde{f}) \cdot \mathrm{Cf}_{(i_1,\ldots,i_n);K(\pi)}(\widetilde{g}) \\ &= \sum_{\pi \in NC(n)} \mathrm{Cf}_{(1,\ldots,1);\pi}(f) \cdot \mathrm{Cf}_{(1,\ldots,1);K(\pi)}(g) \\ &= \mathrm{Cf}_{(1,\ldots,1)}(h) \\ &= \mathrm{Cf}_{(i_1,\ldots,i_n)}(\widetilde{h}). \end{aligned}$$

In this chain of equalities we used successively the definition of $\boxed{\star}_s$, Equation (17.13) and its analog written for g, the definition of $\boxed{\star}_1$, and the analog of (17.12) written for h. □

The next corollary is an obvious consequence of Proposition 17.11.

COROLLARY 17.12. *Let s be a positive integer, and consider the operation $\boxed{\star}_s$ on Θ_s.*

(1) *If $f, g \in \Theta_s$ are radial, then so is $f \boxed{\star}_s g$.*

(2) *Consider a radial series in Θ_s,*

$$f(z_1, \ldots, z_s) = \sum_{n=1}^{\infty} \alpha_n (z_1 + \cdots + z_s)^n.$$

Then f is invertible with respect to $\boxed{\star}_s$ if and only if $\alpha_1 \neq 0$. In the case when $\alpha_1 \neq 0$, the inverse g of f under $\boxed{\star}_s$ is also radial, and is described as follows:

$$g(z_1, \ldots, z_s) = \sum_{n=1}^{\infty} \beta_n (z_1 + \cdots + z_s)^n,$$

where $(\beta_n)_{n=1}^{\infty}$ are such that $\sum_{n=1}^{\infty} \beta_n z_1^n$ is the inverse of $\sum_{n=1}^{\infty} \alpha_n z_1^n$ with respect to the operation $\boxed{\star}_1$ on Θ_1.

REMARK 17.13. In the setting of moment series and R-transforms for s-tuples in a non-commutative probability space, the embedding of Θ_1 into Θ_s from the preceding proposition amounts to repeating the random variables of our s-tuples. More precisely, it is immediate from the definitions that if $(\mathcal{A}, \varphi)$ is a non-commutative probability space and if $a \in \mathcal{A}$, then the s-tuple $(a, \ldots, a) \in \mathcal{A}^s$ will have

$$M_{a,\ldots,a}(z_1, \ldots, z_s) = M_a(z_1 + \cdots + z_s)$$

and

$$R_{a,\ldots,a}(z_1, \ldots, z_s) = R_a(z_1 + \cdots + z_s).$$

The Möbius series and its use

DEFINITION 17.14. Let s be a positive integer. Consider the operation $\boxed{\star}$ on Θ_s, and the series Zeta $\in \Theta_s$ introduced in Notation 17.3. The inverse of Zeta under $\boxed{\star}$ is called the **Möbius series**, and is denoted by Möb (or by Möb_s if the specification of s is necessary).

Note that the above definition of the Möb series indeed makes sense (the series Zeta really is invertible with respect to $\boxed{\star}$, by Proposition 17.7). Alternatively, one can also describe Möb by explicitly indicating its coefficients.

PROPOSITION 17.15. *The explicit formula for the Möbius series in* Θ_s *is:*

$$\text{Möb}_s(z_1, \ldots, z_s) = \sum_{n=1}^{\infty} (-1)^{n-1} C_{n-1} \cdot (z_1 + \cdots + z_s)^n, \tag{17.14}$$

where $(C_n)_{n=1}^{\infty}$ *is the sequence of Catalan numbers.*

PROOF. Due to Corollary 17.12, it will suffice to verify Equation (17.14) in the particular case when $s = 1$.

The particular case $s = 1$ can in turn be easily inferred by using the functional equation of the R-transform. Indeed, we have $\text{Möb}_1 \boxed{\star}_1 \text{Zeta}_1 = \Delta_1$, hence the series $f := \text{Möb}_1$ and $g := \Delta_1$ satisfy condition (1) of Theorem 16.15. Thus Möb_1 and Δ_1 must also satisfy condition (2) of the same theorem, which amounts in this case to:

$$\text{Möb}_1(\, z(1+z)\,) = z. \tag{17.15}$$

This equation is equivalent to

$$\text{Möb}_1(z) + \text{Möb}_1^2(z) = z \tag{17.16}$$

(indeed, both (17.15) and (17.16) state that Möb_1 is the inverse under composition for the series $z + z^2$).

When Möb_1 is written explicitly,

$$\text{Möb}_1(z) = \sum_{n=1}^{\infty} \gamma_n z^n,$$

Equation (17.16) becomes a recurrence for the coefficients γ_n, namely

$$\gamma_1 = 1 \qquad \text{and} \qquad \gamma_n + \sum_{k=1}^{n-1} \gamma_k \gamma_{n-k} = 0, \quad n \geq 2. \tag{17.17}$$

But this is the well-known recurrence of the signed Catalan numbers (as also encountered in the Lecture 10, while proving Proposition 10.15); so we get that the explicit writing of Möb_1 is indeed

$$\text{Möb}_1(z) = \sum_{n=1}^{\infty} (-1)^{n-1} C_{n-1} z^n,$$

and the result follows. □

In the remaining part of this section we will give a couple of illustrations for how the series Möb and Zeta can be used in computations related to R-transforms.

For the first illustration let us look again at the discussion about the square of an even element in a non-commutative probability space (cf. Lecture 11, Proposition 11.25). In order to rephrase that discussion in terms of power series it will be convenient to use the following notation (analogous to the notation "$f \circ L$" which was used in Lecture 16).

NOTATION 17.16. Let f be a series in Θ_1. We will denote by $f \circ \text{Sq}$ the new series in Θ_1 which has all the coefficients of odd order equal to 0 and has

$$\text{coef. of order } 2n \text{ of } f \circ \text{Sq} = \text{coef. of order } n \text{ of } f, \quad \forall\, n \geq 1.$$

In other words, if $f(z) = \sum_{n=1}^{\infty} \alpha_n z^n$ then we have

$$(f \circ \text{Sq})(z) := \sum_{n=1}^{\infty} \alpha_n z^{2n}.$$

Written more succinctly (though somewhat less rigorously), the equation defining $f \circ \text{Sq}$ is thus saying that

$$(f \circ \text{Sq})(z) := f(z^2). \tag{17.18}$$

REMARK 17.17. Let $(\mathcal{A}, \varphi)$ be a non-commutative probability space, and let $a \in \mathcal{A}$ be even (in the sense that $\varphi(a^n) = 0$ for odd n). The moment series of a and of a^2 will then contain precisely the same information. It is clear from the definitions that the formula which relates these two moment series is just

$$M_a = M_{a^2} \circ \mathrm{Sq}. \tag{17.19}$$

When we look at the two corresponding R-transforms, we expect them to be related by an analogous equation. This is indeed the case, only that there is an extra ingredient which appears – a boxed convolution with the Möbius series.

In order to explain how the Möbius series appears when we write the relation between R_a and R_{a^2}, let us first recall from Lecture 11 that the sequence of free cumulants of even order of the even element a is called "the determining sequence of a"(cf. Notations 11.24). In the same vein, we will then refer to the power series

$$g(z) := \sum_{n=1}^{\infty} \kappa_{2n}(a, a, \ldots, a) z^n$$

by calling it the **determining series** of the even element a. It is clear that this determining series g is connected to the R-transform R_a by the equation

$$R_a = g \circ \mathrm{Sq}. \tag{17.20}$$

Now, Proposition 11.25 which relates the cumulants of a with those of a^2 can be interpreted in our current notations to say that

$$R_{a^2} = g \boxed{\star} \mathrm{Zeta}. \tag{17.21}$$

This implies that $g = R_{a^2} \boxed{\star} \text{Möb}$, and thus gives the following formula for R_a:

$$R_a = (\, R_{a^2} \boxed{\star} \text{Möb}\,) \circ \mathrm{Sq}. \tag{17.22}$$

As announced above, Equation (17.22) is the counterpart with R-transforms for Equation (17.19) about moment series, with the extra twist brought in by the convolution with Möb. This formula can also be stated without making explicit reference to a and to $(\mathcal{A}, \varphi)$, as follows.

PROPOSITION 17.18. *For every series $f \in \Theta_1$ one has that*

$$(f \circ \mathrm{Sq}) \boxed{\star} \text{Möb} = (\, f \boxed{\star} \text{Möb} \boxed{\star} \text{Möb}\,) \circ \mathrm{Sq}. \tag{17.23}$$

PROOF. By using Exercise 16.21, one can find a non-commutative probability space $(\mathcal{A}, \varphi)$ and an even element $a \in \mathcal{A}$ such that $M_{a^2} = f$. Then the left-hand side of (17.23) is R_a, while the right-hand side of (17.23) is $(R_{a^2} \boxed{\star} \text{Möb}) \circ \mathrm{Sq}$. Thus (17.23) follows from (17.22). □

REMARKS 17.19. (1) The operations $f \mapsto f \circ \mathrm{Sq}$ and $f \mapsto f \boxed{\star} \text{Möb}$ on Θ_1 do not commute, but the preceding proposition says that they still satisfy a certain commutation relation (given by Equation (17.23)).

(2) As an application of the discussion about Sq, let us see for instance how it entails a quick solution of Exercise 11.35. Let $(\mathcal{A}, \varphi)$ be a non-commutative probability space, and let $b \in \mathcal{A}$ be a symmetric Bernoulli random variable (which means, by definition, that b is even and has $b^2 = 1_{\mathcal{A}}$). The above Equation (17.22) amounts here to

$$R_b = (R_{1_{\mathcal{A}}} \boxed{\star} \text{Möb}) \circ \text{Sq}.$$

By taking into account that

$$R_{1_{\mathcal{A}}} = \Delta = \text{ the unit for } \boxed{\star},$$

we obtain that $R_b = \text{Möb} \circ \text{Sq}$ (which was exactly the statement of Exercise 11.35).

For a second illustration of computations with Möb and Zeta we will point out a power series approach to the construction shown in Lecture 12 for free families of free Poisson elements. The construction is described as follows (cf. Example 12.19). Let $(\mathcal{A}, \varphi)$ be a $*$-probability space and let $x, e_1, \ldots, e_s \in \mathcal{A}$ be such that x is semicircular, $e_1, \ldots, e_s$ are mutually orthogonal projections, and $\{x\}$ is free from $\{e_1, \ldots, e_s\}$. Then $xe_1x, \ldots, xe_sx$ is a free family, consisting of free Poisson elements. We show here how this fact comes out effortlessly from some simple manipulations involving boxed convolutions (the resulting proof is different from the one shown in Example 12.19, though of course both proofs rely on non-crossing partitions and on free cumulants). The starting point is to identify the Zeta series as the R-transform of x^2.

PROPOSITION 17.20. *Let $(\mathcal{A}, \varphi)$ be a $*$-probability space and let $x = x^* \in \mathcal{A}$ be a standard semicircular element. Then*

$$R_{x^2, \ldots, x^2} = \text{Zeta}_s, \quad \textit{for every } s \geq 1. \tag{17.24}$$

PROOF. We first observe that

$$M_{x^2} = \text{Zeta} \boxed{\star} \text{Zeta} \tag{17.25}$$

(where here Zeta stands for $\text{Zeta}_1 \in \Theta_1$). Indeed, the coefficient of order n of Zeta $\boxed{\star}$ Zeta is

$$\sum_{\pi \in NC(n)} \text{Cf}_{(1,\ldots,1);\pi}(\text{Zeta}) \cdot \text{Cf}_{(1,\ldots,1);K(\pi)}(\text{Zeta}) = \sum_{\pi \in NC(n)} 1 = C_n,$$

the nth Catalan number. But the coefficient of order n of the moment series of x^2 is exactly the same Catalan number,

$$\varphi(\,(x^2)^n\,) = \varphi(x^{2n}) = C_n$$

(by how a standard semicircular element is defined, cf. Definition 2.16).

By convolving both sides of Equation (17.25) with Möb, we get that $M_{x^2} \boxed{\star}$ Möb $=$ Zeta, i.e. that $R_{x^2} =$ Zeta. This is the case $s = 1$ of Equation (17.24). The general case follows, since we have that

$$\begin{aligned} R_{x^2,\ldots,x^2}(z_1, \ldots, z_s) &= R_{x^2}(z_1 + \cdots + z_s) \\ &= \mathrm{Zeta}_1(z_1 + \cdots + z_s) \\ &= \mathrm{Zeta}_s(z_1, \ldots, z_s). \end{aligned}$$

□

In view of the role of the Zeta series of connecting moments with free cumulants, it then immediately follows that the left-and-right multiplication with a free semicircular element has the effect of "converting moments into free cumulants." Thus we obtain the following alternative derivation of Example 12.19.

PROPOSITION 17.21. *Let* $(\mathcal{A}, \varphi)$ *be a tracial* $*$*-probability space and let* $x, a_1, \ldots, a_s$ *be selfadjoint elements of* $\mathcal{A}$ *such that* x *is standard semicircular and such that* $\{x\}$ *is freely independent from* $\{a_1, \ldots, a_s\}$*. Then*

$$R_{xa_1x,\ldots,xa_sx} = M_{a_1,\ldots,a_s}. \tag{17.26}$$

PROOF. The traciality of φ implies that

$$\varphi(xa_{i_1}x \cdots xa_{i_n}x) = \varphi(a_{i_1}x^2 \cdots a_{i_n}x^2),$$

for every $n \geq 1$ and every $1 \leq i_1, \ldots, i_n \leq s$. At the level of series in Θ_s, this amounts to saying that

$$M_{xa_1x,\ldots,xa_sx} = M_{a_1x^2,\ldots,a_sx^2}.$$

By convolving both sides of the latter equation with $\mathrm{Möb}_s$ we obtain the equality of the corresponding R-transforms,

$$R_{xa_1x,\ldots,xa_sx} = R_{a_1x^2,\ldots,a_sx^2}. \tag{17.27}$$

But the right-hand side of (17.27) is successively equal to:

$$\begin{aligned} R_{a_1x^2,\ldots,a_sx^2} &= R_{a_1,\ldots,a_s} \boxed{\star}\, R_{x^2,\ldots,x^2} && \text{(by Proposition 17.2)} \\ &= R_{a_1,\ldots,a_s} \boxed{\star}\, \mathrm{Zeta}_s && \text{(by Proposition 17.20)} \\ &= M_{a_1,\ldots,a_s} && \text{(by Proposition 17.4).} \end{aligned}$$

□

Finally, if we want to focus strictly on free Poisson elements, we have the following.

COROLLARY 17.22. *Let $(\mathcal{A}, \varphi)$ be a $*$-probability space and let $x, e_1, \ldots, e_s$ be selfadjoint elements of $\mathcal{A}$ such that*

(i) x is semicircular of radius $r > 0$,

(ii) $e_i^2 = e_i$, for $1 \leq i \leq n$, and $e_i e_j = 0$ whenever $i \neq j$,

(iii) $\{x\}$ is freely independent from $\{e_1, \ldots, e_s\}$.

Then the elements $xe_1x, \ldots, xe_sx$ form a free family in $(\mathcal{A}, \varphi)$, and every xe_ix is a free Poisson element of parameters $\lambda = \varphi(e_i)$ and $\alpha = r^2/4$ (in the sense of Definition 12.12).

PROOF. By rescaling x we may assume without loss of generality that $r = 2$ (i.e. that x is a standard semicircular element). Also, we can assume without loss of generality that the unital $*$-algebra generated by x and $e_1, \ldots, e_s$ is all of $\mathcal{A}$ (otherwise we just replace $\mathcal{A}$ by $\text{alg}(1_{\mathcal{A}}, x, e_1, \ldots, e_s)$). Since the restrictions of φ to $\text{alg}(1_{\mathcal{A}}, x)$ and to $\text{alg}(1_{\mathcal{A}}, e_1, \ldots, e_s)$ are traces (which happens because the two algebras in question are commutative), Proposition 5.19 in Lecture 5 gives us that φ is a trace.

We are hence in a situation where we can apply Proposition 17.21; when doing this we obtain:

$$\begin{aligned} R_{xe_1x,\ldots,xe_sx}(z_1, \ldots, z_s) &= M_{e_1,\ldots,e_s}(z_1, \ldots, z_s) \\ &= \sum_{i=1}^{s} \Big(\varphi(e_i) \cdot \sum_{n=1}^{\infty} z_i^n \Big), \end{aligned}$$

where the last equality follows from the hypothesis that $e_1, \ldots, e_s$ are mutually orthogonal projections.

We thus obtained that the joint R-transform of $xe_1x, \ldots, xe_sx$ is a series in separate variables, and Theorem 16.6 now implies that $xe_1x, \ldots, xe_sx$ form a free family. At the same time we obtain that for every $1 \leq i \leq s$, the R-transform of xe_ix is

$$R_{xe_ix}(z) = \varphi(e_i) \cdot \sum_{n=1}^{\infty} z^n,$$

and this corresponds to the fact that xe_ix is free Poisson of parameters $\lambda = \varphi(e_i)$ and $\alpha = 1$. □

Exercises

EXERCISE 17.23. (1) Give an alternative proof of Proposition 17.5.1, by proceeding directly from the definition of $\boxed{\star}$, and by only using the basic combinatorics of non-crossing partitions.
[Note: it may be useful to look ahead at the concept of relative Kreweras complement for non-crossing partitions, which is discussed in the next lecture.]

(2) Give an alternative proof of Proposition 17.5.2, by using the connection of $\boxed{\star}$ with the multiplication of free s-tuples, and the fact that in a non-commutative probability space $(\mathcal{A}, \varphi)$ one always has $R_{1_{\mathcal{A}},\ldots,1_{\mathcal{A}}} = \Delta$.

EXERCISE 17.24. Let $s \geq 2$ be an integer. Determine explicitly the elements of the set:

$$\Big\{ f \in \Theta_s \mid f \text{ invertible and central with respect to } \boxed{\star}_s \Big\}.$$

The next exercise discusses the behavior of $\boxed{\star}$ in connection to the operations of scalar multiplication and of dilation for power series. We will use the following notation.

NOTATION 17.25. Let s be a positive integer, let f be a series in Θ_s, and let α be a number in $\mathbb{C} \setminus \{0\}$. We denote by $f \circ D_\alpha$ the series in Θ_s which is defined by the equation

$$(f \circ D_\alpha)(z_1, \ldots, z_s) = f(\alpha z_1, \ldots, \alpha z_s),$$

or more rigorously by the fact that

$$\text{Cf}_{(i_1,\ldots,i_n)}(f \circ D_\alpha) = \alpha^n \cdot \text{Cf}_{(i_1,\ldots,i_n)}(f),$$

for all $n \geq 1$ and $1 \leq i_1, \ldots, i_n \leq s$.

EXERCISE 17.26. Let s be a positive integer, let f, g be series in Θ_s, and let α be a number in $\mathbb{C} \setminus \{0\}$.

(1) Prove that

$$(f \circ D_\alpha) \boxed{\star}\, g = f \boxed{\star} (g \circ D_\alpha) = (f \boxed{\star}\, g) \circ D_\alpha.$$

(2) Prove that

$$\frac{1}{\alpha}(\alpha f \boxed{\star}\, \alpha g) = (f \boxed{\star}\, g) \circ D_\alpha.$$

LECTURE 18

More on the one-dimensional boxed convolution

The preceding lecture introduced the operation $\boxed{\star}$ and showed its meaning in connection to free probability, but did not detail the relation between $\boxed{\star}$ and the Möbius inversion theory of the lattices of non-crossing partitions. This relation is best observed when looking at the one-dimensional boxed convolution $\boxed{\star}_1$. Indeed, the monoid $(\Theta_1, \boxed{\star}_1)$ turns out to capture exactly the convolution of families of multiplicative functions on $NC^{(2)}$. This will be proved in the first section of the present lecture (thus completing the discussion about these multiplicative families, which was started in Lecture 10).

In the second section of this lecture we will look further at the monoid $(\Theta_1, \boxed{\star}_1)$, and we will give a precise description of the group of invertible elements in this monoid. Here one can find a transformation $\mathcal{F}$ which converts one-dimensional boxed convolution into plain multiplication of power series (cf. Theorem 18.14 below). At the level of free probabilistic interpretations, this gives us the concept of *S-transform*, a very useful tool for computing the distribution of the product of two freely independent random variables.

Relation to multiplicative functions on NC

Since in this lecture we are dealing exclusively with series of 1 variable, we will use a shortened version of the notations for coefficients introduced in Lecture 16.

NOTATION 18.1. Let $f(z) = \sum_{n=1}^{\infty} \alpha_n z^n$ be a series in Θ_1. For every $n \geq 1$ and $\pi = \{B_1, \ldots, B_r\} \in NC(n)$, we denote

$$\mathrm{Cf}_\pi(f) := \alpha_{|B_1|} \cdots \alpha_{|B_r|}. \tag{18.1}$$

REMARK 18.2. Let $f(z) = \sum_{n=1}^{\infty} \alpha_n z^n$ and $g(z) = \sum_{n=1}^{\infty} \beta_n z^n$ be two series in Θ_1, and consider their boxed convolution,

$$(f \boxed{\star}_1 g)(z) =: \sum_{n=1}^{\infty} \gamma_n z^n.$$

The explicit formula for the coefficients of $f \boxast_1 g$ (obtained from Equation (17.1) in the preceding lecture) is then

$$\gamma_n = \sum_{\substack{\pi \in NC(n) \\ \pi=\{V_1,\ldots,V_p\} \\ K(\pi)=\{W_1,\ldots,W_q\}}} \alpha_{|V_1|}\cdots\alpha_{|V_p|}\beta_{|W_1|}\cdots\beta_{|W_q|}. \tag{18.2}$$

Our starting point in this section is to observe that the very same formula appears in the framework of multiplicative functions on $NC^{(2)}$ which were considered in Lecture 10. Indeed, in the framework of Definition 10.16, let $(F_n)_{n\geq 1}$ and $(G_n)_{n\geq 1}$ be the two multiplicative families of functions in $NC^{(2)}$ which are determined by the sequences $(\alpha_n)_{n\geq 1}$ and respectively $(\beta_n)_{n\geq 1}$ Then for every $n \geq 1$ we have:

$$\begin{aligned}
(F_n * G_n)(0_n, 1_n) &= \sum_{\pi \in NC(n)} F_n(0_n,\pi)G_n(\pi,1_n) \\
&= \sum_{\pi \in NC(n)} F_n(0_n,\pi)G_n(\,0_n, K(\pi)\,) \\
&= \sum_{\substack{\pi \in NC(n) \\ \pi=\{V_1,\ldots,V_p\} \\ K(\pi)=\{W_1,\ldots,W_q\}}} \alpha_{|V_1|}\cdots\alpha_{|V_p|}\beta_{|W_1|}\cdots\beta_{|W_q|}.
\end{aligned}$$

(The second equality follows by property (ii) in Remark 10.17, applied to $(G_n)_{n\geq 1}$, the third equality by property (iii) in Remark 10.17.) In other words we have obtained that

$$(F_n * G_n)(0_n, 1_n) = \gamma_n, \quad \forall\, n \geq 1; \tag{18.3}$$

so if we knew that the family $(F_n * G_n)_{n\geq 1}$ is multiplicative, then it would follow that it is precisely the family of multiplicative functions determined by the sequence $(\gamma_n)_{n\geq 1}$.

But of course, it is not obvious that $(F_n * G_n)_{n\geq 1}$ is multiplicative. The goal of the present section is to prove this fact. We will take an approach which relies on a "relative" version of the Kreweras complementation map on $NC(n)$.

DEFINITION 18.3. Let π, σ be partitions in $NC(n)$, such that $\pi \leq \sigma$. Write explicitly $\sigma = \{V_1, \ldots, V_r\}$ and for $1 \leq q \leq r$ let J_q denote the unique order-preserving bijection from V_q onto $\{1, \ldots, |V_q|\}$. The **relative Kreweras complement** of π in σ is the partition $\rho \in NC(n)$ determined by the following two conditions:

(i) $\rho \leq \sigma$;

(ii) for every $1 \leq q \leq r$, we have that $\rho_q = K(\pi_q)$, where

$$\pi_q := J_q(\pi|V_q), \qquad \rho_q := J_q(\rho|V_q),$$

and where "K" denotes the Kreweras complementation map on $NC(|V_q|)$.

The relative Kreweras complement of π in σ will be denoted as $K_\sigma(\pi)$.

EXAMPLE 18.4. Consider the partitions $\pi, \sigma \in NC(12)$ defined by

$$\pi = \{\{1,9\},\{2,5\},\{3\},\{4\},\{6\},\{7,8\},\{10\},\{11\},\{12\}\} \text{ and}$$
$$\sigma = \{\{1,6,9,12\},\{2,4,5\},\{3\},\{7,8\},\{10,11\}\}$$

(these are the same as in Example 9.31). If we look at the block $V_1 = \{1,6,9,12\}$ of σ (which is split into the blocks $\{1,9\}$, $\{6\}$, $\{12\}$ of π) we see that

$$\pi_1 = J_1(\pi|V_1) = \{\ \{1,3\},\{2\},\{4\}\ \} \in NC(4).$$

Therefore we must have

$$\rho_1 = K(\pi_1) = \{\ \{1,2\},\{3,4\}\ \} \in NC(4),$$

which implies that V_1 is split into blocks of ρ as $\{1,6\}$, $\{9,12\}$. By doing the same kind of calculation for the other blocks of σ, we find that

$$\rho = K_\sigma(\pi) = \{\ \{1,6\},\{2,4\},\{3\},\{5\},\{7\},\{8\},\{9,12\},\{10,11\}\ \}.$$

REMARK 18.5. For a fixed $\sigma \in NC(n)$, the map $\pi \mapsto K_\sigma(\pi)$ is a bijection from $\{\pi \in NC(n) \mid \pi \leq \sigma\}$ to itself, which is order reversing and maps $0_n \mapsto \sigma$, $\sigma \mapsto 0_n$. These properties follow immediately from the corresponding properties of the Kreweras complementation map which were discussed in Lecture 9 (cf. Exercise 9.23).

Note that in fact the Kreweras complementation map K on $NC(n)$ can itself be viewed as a relative complementation, with respect to the partition with one block $1_n \in NC(n)$. Indeed, it is obvious that we have $K_{1_n}(\pi) = K(\pi)$, $\forall\ \pi \in NC(n)$.

The relevance of the relative Kreweras complementation map in connection to multiplicative functions comes from the following lemma.

LEMMA 18.6. *Let π, σ be partitions in $NC(n)$ such that $\pi \leq \sigma$. Consider the relative Kreweras complement $K_\sigma(\pi)$, and write explicitly*

$$K_\sigma(\pi) = \{W_1, \ldots, W_k\}.$$

Then the canonical factorization of the interval $[\pi, \sigma] \subset NC(n)$ (as introduced in Definition 9.30) is

$$[\pi, \sigma] \simeq NC(|W_1|) \times \cdots \times NC(|W_k|).$$

PROOF. Let us denote $K_\sigma(\pi) =: \rho$, and let us write explicitly the list of blocks of σ,

$$\sigma = \{V_1, \ldots, V_r\}.$$

Since $\rho \leq \sigma$, the explicit writing of ρ can then be put in the form

$$\rho = \{W_{1,1}, \ldots, W_{1,k_1}, \ldots, W_{r,1}, \ldots, W_{r,k_r}\},$$

with $W_{q,1} \cup \cdots \cup W_{q,k_q} = V_q$ for $1 \leq q \leq r$. With these notations, the statement to be proved is that the canonical factorization of $[\pi, \sigma]$ is

$$[\pi, \sigma] \simeq \prod_{q=1}^{k} \prod_{j=1}^{k_q} NC(\,|W_{q,j}|\,). \tag{18.4}$$

Let us re-examine the proof of Theorem 9.29, while at the same time using the notations "$\pi_1, \ldots, \pi_r, \rho_1, \ldots, \rho_r$" as in Definition 18.3. The proof of Theorem 9.29 starts by identifying $[\pi, \sigma]$ with

$$[\pi_1, 1_{|V_1|}] \times \cdots \times [\pi_r, 1_{|V_r|}].$$

This direct product is then found to be anti-isomorphic to

$$[0_{|V_1|}, K(\pi_1)] \times \cdots \times [0_{|V_r|}, K(\pi_r)],$$

i.e. to

$$[0_{|V_1|}, \rho_1] \times \cdots \times [0_{|V_r|}, \rho_r].$$

Finally every interval $[0_{|V_q|}, \rho_q)] \subset NC(\,|V_q|\,)$ is factored into a direct product by using the block structure of ρ_q,

$$[0_{|V_q|}, \rho_q] \simeq NC(\,|W_{q,1}|\,) \times \cdots \times NC(\,|W_{q,k_q}|\,)$$

(where the latter product is then noticed to be anti-isomorphic to itself). It is clear that the combination of these steps leads precisely to (18.4). □

EXAMPLE 18.7. The relative Kreweras complement $K_\sigma(\pi)$ computed in Example 18.4 has 4 blocks with 1 element and 4 blocks with 2 elements, corresponding to the factorization

$$[\pi, \sigma] \simeq NC(1)^4 \times NC(2)^4$$

found in Example 9.31.

REMARK 18.8. In Lecture 10 we considered separately the concepts of "multiplicative family of functions on NC" and "multiplicative family of functions on $NC^{(2)}$." There exists a canonical correspondence between these two kinds of families – this is clear, just from the fact that the data determining either a multiplicative family on NC or a multiplicative family on $NC^{(2)}$ is always a sequence of complex numbers $(\alpha_n)_{n\geq 1}$. The preceding lemma gives us an even better formula for moving back and forth between a multiplicative family $(f_n)_{n\geq 1}$ on NC and a multiplicative family $(F_n)_{n\geq 1}$ on $NC^{(2)}$ which are determined by the same sequence $(\alpha_n)_{n\geq 1}$. This formula does not involve the α_n in an explicit way, but rather goes as follows.

- If we know $(f_n)_{n=1}^{\infty}$ then the F_n are determined by

$$F_n(\pi, \sigma) = f_n(K_\sigma(\pi)), \quad \forall\, n \geq 1, \ \forall\, \pi \leq \sigma \text{ in } NC(n). \tag{18.5}$$

- If we know $(F_n)_{n=1}^{\infty}$ then the f_n are determined by

$$f_n(\pi) = F_n(0_n, \pi), \quad \forall\, n \geq 1, \ \forall\, \pi \in NC(n). \tag{18.6}$$

Based on this observation, the statement that we want to prove about the convolution of multiplicative families on $NC^{(2)}$ can be pulled out from the weaker result proved in Lecture 10 (Proposition 10.21), combined with a few additional facts about the relative Kreweras complementation map. These additional facts are listed in the following lemma.

LEMMA 18.9. *Let π, σ be partitions in $NC(n)$, such that $\pi \leq \sigma$.*

(1) *For every $\tau \in [\pi, \sigma]$ we have that $K_\tau(\pi) \leq K_\sigma(\pi)$. Moreover, the map $\tau \mapsto K_\tau(\pi)$ is a lattice isomorphism between the intervals $[\pi, \sigma]$ and $[0_n, K_\sigma(\pi)]$ in $NC(n)$.*

(2) *Let τ be in $[\pi, \sigma]$, and consider the partition $K_\tau(\pi) \in [0_n, K_\sigma(\pi)]$. Then we have*

$$K_{K_\sigma(\pi)} \left(K_\tau(\pi) \right) = K_\sigma(\tau). \tag{18.7}$$

REMARK 18.10. Note that the first statement of the preceding lemma is in some sense a reinforcement of Lemma 18.6. Indeed, Lemma 18.6 says essentially that $[\pi, \sigma]$ and $[0_n, K_\sigma(\pi)]$ have the same canonical factorization – so in particular these two intervals have to be isomorphic as lattices. Lemma 18.9.1 indicates explicitly an isomorphism between them.

For the second statement of the preceding lemma, it is useful to think informally of the relative Kreweras complement as a kind of "division." Indeed, if for $\pi \leq \sigma$ in $NC(n)$ we were to write σ/π instead of $K_\sigma(\pi)$, then Equation (18.7) would become

$$\frac{\sigma/\pi}{\tau/\pi} = \sigma/\tau$$

(and would thus look less mysterious). In fact the proof of (18.7) is most conveniently done by making the idea of "division of σ by π" become rigorous; this is achieved by viewing π and σ as elements of the symmetric group S_n.

In order not to divert too much from the main line of this section, we will break the proof of Lemma 18.9 into a set of exercises left to the reader (see Exercises 18.25 and 18.26 below). What we will do here is to show how Lemma 18.9 is used to complete the discussion about multiplicative families of functions on $NC^{(2)}$, and their connection to the operation of boxed convolution $\boxed{\star}_1$.

THEOREM 18.11. (1) *If $(F_n)_{n\geq 1}$ and $(G_n)_{n\geq 1}$ are multiplicative families of functions on $NC^{(2)}$, then so is $(F_n * G_n)_{n\geq 1}$. Thus the set of multiplicative families of functions on $NC^{(2)}$ has a semigroup structure, under the operation of convolution.*

(2) *Consider the map which associates to a series $f(z) = \sum_{n=1}^{\infty} \alpha_n z^n$ in Θ_1 the multiplicative family on $NC^{(2)}$ determined by the sequence $(\alpha_n)_{n\geq 1}$. Then this map is an isomorphism between the semigroup $(\Theta_1, \boxed{\star}_1)$ and the semigroup structure observed in the first part of the theorem.*

PROOF. If part (1) of the theorem is assumed to be true then part (2) follows in the way observed in Remark 18.2. Thus we only need to do part (1).

Let $(F_n)_{n\geq 1}$ and $(G_n)_{n\geq 1}$ be multiplicative families of functions on $NC^{(2)}$. Denote by $(f_n)_{n\geq 1}$ and $(g_n)_{n\geq 1}$ the multiplicative families of functions on NC which correspond to $(F_n)_{n\geq 1}$ and $(G_n)_{n\geq 1}$, respectively (as discussed in Remark 18.8). Let us moreover denote

$$h_n := f_n * G_n, \quad \forall\, n \geq 1.$$

Then $(h_n)_{n\geq 1}$ is also a multiplicative family of functions on NC, by Proposition 10.21. We will prove that

$$(F_n * G_n)(\pi, \sigma) = h_n(\, K_\sigma(\pi)\,), \tag{18.8}$$

for every $n \geq 1$ and every $\pi, \sigma \in NC(n)$ such that $\pi \leq \sigma$. This will imply that $(F_n * G_n)_{n\geq 1}$ is a multiplicative family, by Remark 18.8.

In order to establish the equality stated in (18.8), we evaluate separately its two sides. On the left-hand side:

$$\begin{aligned}(F_n * G_n)(\pi, \sigma) &= \sum_{\tau\in[\pi,\sigma]} F_n(\pi, \tau) G_n(\tau, \sigma)\\ &= \sum_{\tau\in[\pi,\sigma]} f_n(K_\tau(\pi)) g_n(K_\sigma(\tau)).\end{aligned} \tag{18.9}$$

On the right-hand side:

$$
\begin{aligned}
h_n(\, K_\sigma(\pi)\,) &= (f_n * G_n)(\, K_\sigma(\pi)\,) \\
&= \sum_{\theta \le K_\sigma(\pi)} f_n(\theta) G_n(\theta, K_\sigma(\pi)) \\
&= \sum_{\theta \in [0_n, K_\sigma(\pi)]} f_n(\theta) g_n(\, K_{K_\sigma(\pi)}(\theta)\,). \qquad (18.10)
\end{aligned}
$$

Finally, we observe that the sums appearing in (18.9) and in (18.10) are identified with each other term by term, via the bijection

$$[\pi, \sigma] \ni \tau \mapsto \theta \in [0_n, K_\sigma(\pi)]$$

put into evidence in Lemma 18.9. □

The S-transform

In this section we continue our study of the monoid $(\Theta_1, \boxed{\star}_1)$, and we prove a theorem describing the group of invertible elements of this monoid.

We start by introducing a few notations which will ease the presentation of the theorem.

NOTATIONS 18.12. (1) We will denote by $\Theta_1^{(\mathrm{inv})}$ the set of series $f \in \Theta_1$ which are invertible with respect to $\boxed{\star}_1$. As implied by Proposition 17.7, a series $f(z) = \sum_{n=1}^{\infty} \alpha_n z^n$ belongs to $\Theta_1^{(\mathrm{inv})}$ if and only if $\alpha_1 \neq 0$.

(2) We denote by Γ the set of all power series $u(z) = \sum_{n=0}^{\infty} u_n z^n$ with the property that the constant term u_0 of u is not equal to 0. On Γ we consider the group structure given by the usual multiplication of power series.

(3) For $f \in \Theta_1^{(\mathrm{inv})}$ we denote:

$$[\mathcal{F}(f)\,](z) \;=\; \frac{1}{z} f^{<-1>}(z), \qquad (18.11)$$

where $f^{<-1>}$ is the inverse of f under composition.

REMARK 18.13. Observe that $\Theta_1^{(\mathrm{inv})}$ can at the same time be described as the set of series in Θ_1 which are invertible under composition; thus the inverse "$f^{<-1>}$" on the right-hand side of (18.11) does make sense. Let us also observe that the formula (18.11) is defining a map $\mathcal{F} : \Theta_1^{(\mathrm{inv})} \to \Gamma$ (in order to check that $\mathcal{F}(f) \in \Gamma$ for every $f \in \Theta_1^{(\mathrm{inv})}$, we just have to see that the constant coefficient of $\mathcal{F}(f)$ is $\alpha_1^{-1} \neq 0$, where α_1 is the linear coefficient of f).

Then the theorem of this section is stated as follows.

THEOREM 18.14. *The function $\mathcal{F}$ defined in Equation (18.11) is a group isomorphism between $(\Theta_1^{(\mathrm{inv})}, \boxed{\star}_1)$ and $(\Gamma, \cdot)$.*

Before starting to discuss the proof of the theorem, let us record the consequence it has concerning products of free random variables – the multiplicativity property of the S-transform of Voiculescu.

DEFINITION 18.15. Let $(\mathcal{A}, \varphi)$ be a non-commutative probability space, and let $a \in \mathcal{A}$ be such that $\varphi(a) \neq 0$. The **S-transform** of a is the series in Γ defined by:

$$S_a(z) := \frac{1}{z} R_a^{<-1>}(z). \tag{18.12}$$

REMARK 18.16. The series S_a always belongs to the space Γ (indeed, it is immediate that the constant coefficient of S_a is equal to $1/\varphi(a) \neq 0$). An alternative definition for S_a could be given in terms of the moment series, namely:

$$S_a(z) = \frac{1+z}{z} M_a^{<-1>}(z). \tag{18.13}$$

The equality of the series appearing on the right-hand sides of Equations (18.12) and (18.13) was established in Remark 16.18.

COROLLARY 18.17. *Let $(\mathcal{A}, \varphi)$ be a non-commutative probability space, and let a, b be in $\mathcal{A}$. If a is free from b, then:*

$$S_{ab}(z) \;=\; S_a(z) \cdot S_b(z). \tag{18.14}$$

PROOF. It is clear from the definitions that we have

$$\mathcal{F}(\ R_x\ ) \;=\; S_x, \tag{18.15}$$

for every $x \in \mathcal{A}$ such that $\varphi(x) \neq 0$. Hence we can write:

$$\begin{aligned} S_{ab} &= \mathcal{F}(R_{ab}) \\ &= \mathcal{F}(R_a \boxed{\star} R_b) && \text{(by Proposition 17.2)} \\ &= \mathcal{F}(R_a) \cdot \mathcal{F}(R_b) && \text{(by Theorem 18.14)} \\ &= S_a \cdot S_b. \end{aligned}$$

□

REMARK 18.18. One could also introduce the S-transform and state its multiplicativity property in reference to the operation of multiplicative free convolution $\boxtimes$ which was discussed in Lecture 14.

Indeed, if μ is a compactly supported probability measure on $\mathbb{R}$ such that $\int_{\mathbb{R}} t \, d\mu(t) \neq 0$, then one can define the **S-transform** of μ by the formula

$$S(\mu) := S_a \in \Gamma, \tag{18.16}$$

where a is a random variable (in some non-commutative probability space $(\mathcal{A}, \varphi)$) which has the same moments as μ. It is immediate that this definition makes sense: first of all the S-transform S_a exists because

$$\varphi(a) = \int_{\mathbb{R}} t \, d\mu(t) \neq 0;$$

then secondly, the series S_a is defined in terms of the moments of a, hence it only depends on μ (and not on the particular choice of an a which has the same moments as μ).

Now recall (cf. Remark 14.2.3) that if μ, ν are compactly supported probability measures on $\mathbb{R}_+$, then the multiplicative free convolution $\mu \boxtimes \nu$ is also a compactly supported probability measure on $\mathbb{R}_+$, which can be described by the following condition: the moments of $\mu \boxtimes \nu$ coincide with the moments of ab, where a and b are positive random variables in some C^*-probability space $(\mathcal{A}, \varphi)$ such that a has distribution μ, b has distribution ν, and a is free from b. It is immediate that Equation (18.14) in the above corollary amounts in this case to

$$S(\mu \boxtimes \nu) = S(\mu) \cdot S(\nu), \tag{18.17}$$

as an equality of power series in Γ.

REMARK 18.19. On our way towards the proof of Theorem 18.14 a key point will be to obtain a formula which is similar in nature to the functional equation of the R-transform (in the case of 1 variable), but where we consider boxed convolution with a series different from Zeta. To be more precise, the case $s = 1$ of Theorem 16.15 says that for two series $f, g \in \Theta_1$, the relation $g = f \boxed{\star}$ Zeta is equivalent to $g = f \circ (z(1+g))$. If f is invertible under composition, we thus obtain that $f^{<-1>} \circ g = z(1+g)$, or writing only in terms of f:

$$f^{<-1>} \circ (f \boxed{\star} \text{Zeta}) \;=\; z(1 + f \boxed{\star} \text{Zeta}). \tag{18.18}$$

In the following Proposition 18.21 we will discuss a generalization of (18.18) to the case when Zeta is replaced by an arbitrary series $h \in \Theta_1$. The generalization uses a concept of *incomplete* one-dimensional boxed convolution.

DEFINITION 18.20. Let f and h be series in Θ_1. The **incomplete boxed convolution** of f and h, denoted by $f \overset{\vee}{\boxed{\star}} h$, is the series

$$(f \overset{\vee}{\boxed{\star}} h)(z) \;:=\; \sum_{n=1}^{\infty} \lambda_n z^n \in \Theta_1,$$

where for every $n \geq 1$ we set:

$$\lambda_n = \sum_{\substack{\pi \in NC(n) \text{ such} \\ \text{that } \{1\} \text{ is} \\ \text{a block of } \pi}} \mathrm{Cf}_\pi(f) \cdot \mathrm{Cf}_{K(\pi)}(h). \tag{18.19}$$

PROPOSITION 18.21. *Let f and h be series in Θ_1. We denote by α_1 the coefficient of z in f. Suppose that f is invertible under composition, i.e. that $\alpha_1 \neq 0$. Then we have that:*

$$f^{<-1>} \circ (f \boxed{\star}\, h) = \frac{1}{\alpha_1}(f \overset{\vee}{\boxed{\star}}\, h). \tag{18.20}$$

PROOF. We will show, equivalently, that

$$f \boxed{\star}\, h = f \circ \Big(\frac{1}{\alpha_1}(f \overset{\vee}{\boxed{\star}}\, h) \Big). \tag{18.21}$$

We fix a positive integer m, and we will verify the equality of the coefficients of order m in the series appearing on the two sides of Equation (18.21). The verification is similar to part (b) in the proof of Theorem 16.15; for this reason we will not give all the details.

Let us write explicitly:

$$f(z) = \sum_{n=1}^{\infty} \alpha_n z^n, \qquad (f \overset{\vee}{\boxed{\star}}\, h)(z) = \sum_{n=1}^{\infty} \lambda_n z^n.$$

The coefficient of order m in the series on the right-hand side of (18.21) is expressed in terms of the α_n and the λ_n as:

$$\sum_{n=1}^{m} \sum_{\substack{i_1,\dots,i_n \geq 1 \\ i_1+\cdots+i_n=m}} \alpha_n \alpha_1^{-n} \cdot \lambda_{i_1} \cdots \lambda_{i_n}. \tag{18.22}$$

Let us now look at the coefficient of order m on the left-hand side of (18.21). By the definition of $\boxed{\star}$, this is equal to:

$$\sum_{\pi \in NC(m)} \mathrm{Cf}_\pi(f) \cdot \mathrm{Cf}_{K(\pi)}(h).$$

The summation over $NC(m)$ can be detailed by enumerating the partitions in $NC(m)$ according to their first block. When this is done, we obtain the following expression (analogous to Equation (16.24) in

Lecture 16):

$$\sum_{n=1}^{m} \sum_{1=b_1<\cdots<b_n\leq m} \sum_{\substack{\pi\in NC(m) \text{ such} \\ \text{that } \{b_1,\ldots,b_n\} \\ \text{is a block of } \pi}} \mathrm{Cf}_{\pi}(f)\cdot \mathrm{Cf}_{K(\pi)}(h). \qquad (18.23)$$

But if a partition $\pi \in NC(m)$ is subjected to the condition of having a prescribed block $B = \{b_1, \ldots, b_n\}$, with $1 = b_1 < \cdots < b_n \leq m$, then knowing π is equivalent to knowing its restrictions to the spaces left between the consecutive elements of B. This time we proceed slightly differently from what we did in the proof of Theorem 16.15, and we denote: π_1 = the restriction of π to $\{b_1, \ldots, b_2-1\}$, π_2 = the restriction of π to $\{b_2, \ldots, b_3 - 1\}, \ldots, \pi_n$ = the restriction of π to $\{b_n, \ldots, m\}$. The difference consists in the fact that each of $\pi_1, \ldots, \pi_n$ also has (in addition to containing a union of blocks of π) a block of one element at the left end. The advantage of setting the notations this way is that we get a nice relation when we look at Kreweras complements: as is immediately checked, we have that $K(\pi)$ is just the juxtaposition of the Kreweras complements $K(\pi_1), \ldots, K(\pi_n)$. The relations between π and $\pi_1, \ldots, \pi_n$ lead us to the equations:

$$\mathrm{Cf}_{\pi}(f) = \alpha_n\alpha_1^{-n}\cdot \mathrm{Cf}_{\pi_1}(f)\cdots \mathrm{Cf}_{\pi_n}(f), \qquad (18.24)$$

and

$$\mathrm{Cf}_{K(\pi)}(h) = \mathrm{Cf}_{K(\pi_1)}(h)\cdots \mathrm{Cf}_{K(\pi_n)}(h) \qquad (18.25)$$

((18.24) is the analog of Equation (16.25) in Lecture 16, while (18.25) is an additional formula obtained by looking at Kreweras complements).

We substitute (18.24) and (18.25) into (18.23); we get that the quantity in (18.23) is thus equal to:

$$\sum_{n=1}^{m} \sum_{1=b_1<\cdots<b_n\leq m} \alpha_n\alpha_1^{-n}\cdot \sum_{\pi_1,\ldots,\pi_n} \mathrm{Cf}_{\pi_1}(f)\cdots$$
$$\cdots \mathrm{Cf}_{\pi_n}(f)\cdot \mathrm{Cf}_{K(\pi_1)}(h)\cdots \mathrm{Cf}_{K(\pi_n)}(h), \qquad (18.26)$$

where $\pi_1 \in NC(b_2-b_1), \pi_2 \in NC(b_3-b_2), \ldots, \pi_n \in NC(m-b_n+1)$ are only subjected to the condition that they start with a one-element block on the left. In (18.26) we can clearly factor out separate summations over $\pi_1, \ldots, \pi_n$. Due to the way how $\check{\boxed{\star}}$ was defined, we find on the other hand that

$$\sum_{\pi_1} \mathrm{Cf}_{\pi_1}(f)\mathrm{Cf}_{K(\pi_1)}(h) = \lambda_{b_2-b_1}, \ldots, \sum_{\pi_n} \mathrm{Cf}_{\pi_n}(f)\mathrm{Cf}_{K(\pi_n)}(h) = \lambda_{m-b_n+1}.$$

Hence (18.26) becomes:

$$\sum_{n=1}^{m} \sum_{1=b_1<\cdots<b_n\leq m} \alpha_n \alpha_1^{-n} \cdot \lambda_{b_2-b_1}\lambda_{b_3-b_2}\cdots\lambda_{m-b_n+1}, \tag{18.27}$$

and it is obvious that we have got the same quantity as in (18.22). □

EXERCISE 18.22. Verify that if in the framework of Proposition 18.21 we set $h = \text{Zeta}$, then Equation (18.20) reduces to the reformulation (18.18) of the functional equation for the R-transform.

We can now present the proof of Theorem 18.14.

PROOF. The fact that $\mathcal{F}$ is bijective is immediate, the problem is to prove that the relation

$$\mathcal{F}(f \boxed{\star} g) = \mathcal{F}(f)\cdot\mathcal{F}(g) \tag{18.28}$$

holds for every $f, g \in \Theta_1^{(\text{inv})}$. By substituting in (18.28) the explicit definition of $\mathcal{F}$ (as given in Notation 18.12.3) we see that we have to show:

$$z\cdot(f \boxed{\star} g)^{<-1>}(z) = f^{<-1>}(z)\cdot g^{<-1>}(z), \quad \forall\, f,g \in \Theta_1^{(\text{inv})}. \tag{18.29}$$

For the rest of the proof we fix two series in $\Theta_1^{(\text{inv})}$,

$$f(z) = \sum_{n=1}^{\infty} \alpha_n z^n \text{ and } g(z) = \sum_{n=1}^{\infty} \beta_n z^n,$$

for which we will show that (18.29) holds.

In order to eliminate the inverses under composition which appear in (18.29), we will compose both sides of this equation, on the right, with $f \boxed{\star} g$. The new equation obtained in this way will be equivalent to (18.29), since we can always go back by composing with $(f \boxed{\star} g)^{<-1>}$.

When composed with $f \boxed{\star} g$ on the right, the left-hand side of (18.29) becomes:

$$\begin{aligned}\left(z\cdot(f \boxed{\star} g)^{<-1>}\right)\circ(f \boxed{\star} g) &= \left(z\circ(f \boxed{\star} g)\right)\cdot\left((f \boxed{\star} g)^{<-1>}\circ(f \boxed{\star} g)\right)\\ &= (f \boxed{\star} g)(z)\cdot z.\end{aligned}$$

A similar calculation done on the right-hand side of (18.29) leads us to the series

$$\left(\, f^{<-1>}\circ(f \boxed{\star} g)\,\right)\cdot\left(\, g^{<-1>}\circ(f \boxed{\star} g)\,\right),$$

which, by Proposition 18.21, is equal to $\alpha_1^{-1}\beta_1^{-1}\cdot(f \overset{\vee}{\boxed{\star}} g)(g \overset{\vee}{\boxed{\star}} f)$. The equation equivalent to (18.29) which we obtain is thus:

$$(f \overset{\vee}{\boxed{\star}} g)(z)\cdot(g \overset{\vee}{\boxed{\star}} f)(z) = \alpha_1\beta_1 z\cdot(f \boxed{\star} g)(z). \tag{18.30}$$

In order to conclude the proof, we fix a positive integer m, for which we show that the coefficients of z^{m+1} on the two sides of (18.30) are equal. As is immediately seen, the coefficient of z^{m+1} on the left-hand side of (18.30) is equal to:

$$\sum_{n=1}^{m} \sum_{\substack{\pi \in NC(n), \\ \{1\} \text{ block of } \pi}} \sum_{\substack{\rho \in NC(m+1-n), \\ \{1\} \text{ block of } \rho}} \mathrm{Cf}_{\pi}(f) \cdot \mathrm{Cf}_{K(\pi)}(g) \cdot \mathrm{Cf}_{\rho}(g) \cdot \mathrm{Cf}_{K(\rho)}(f), \quad (18.31)$$

while the corresponding coefficient on the right-hand side of (18.30) is

$$\sum_{\sigma \in NC(m)} \alpha_1 \beta_1 \cdot \mathrm{Cf}_{\sigma}(f) \cdot \mathrm{Cf}_{K(\sigma)}(g). \quad (18.32)$$

Now, the point is that there exists a natural bijection between the index sets of the sums in (18.31) and (18.32),

$$\bigcup_{1 \leq n \leq m} \{\pi \in NC(n) \mid \{1\} \text{ is block of } \pi\} \quad (18.33)$$
$$\times \{\rho \in NC(m-n+1) \mid \{1\} \text{ is block of } \rho\} \longleftrightarrow NC(m)$$

such that whenever $(\pi, \rho) \leftrightarrow \sigma$ by this bijection, the term indexed by (π, ρ) in the sum (18.31) equals the term indexed by σ in the sum (18.32) – and even more precisely:

$$\begin{cases} \mathrm{Cf}_{\pi}(f) \cdot \mathrm{Cf}_{K(\rho)}(f) &= \alpha_1 \cdot \mathrm{Cf}_{\sigma}(f), \\ \mathrm{Cf}_{K(\pi)}(g) \cdot \mathrm{Cf}_{\rho}(g) &= \beta_1 \cdot \mathrm{Cf}_{K(\sigma)}(g). \end{cases} \quad (18.34)$$

Described from left to right, the bijection (18.33) goes as follows: start with $1 \leq n \leq m$, $\pi \in NC(n)$ such that $\{1\}$ is a block of π, and with $\rho \in NC(m+1-n)$ such that $\{1\}$ is a block of ρ. Denote by $\pi_o \in NC(n-1)$ the partition obtained by deleting the one-element block $\{1\}$ of π, and consider on the other hand the Kreweras complement $K(\rho) \in NC(m+1-n)$. Then $\sigma \in NC(m)$ which corresponds by (18.33) to (π, ρ) is obtained by simply juxtaposing π_o and $K(\rho)$, in this order.

[Numerical example: if $m = 6$, $n = 3$, $\pi = \{\{1\}, \{2,3\}\}$, $\rho = \{\{1\}, \{2,4\}, \{3\}\}$, then $\sigma = \{\{1,2\}, \{3,6\}, \{4,5\}\}$.]

If on the other hand one wants to describe the bijection (18.33) from right to left, this is done as follows: start with $\sigma \in NC(m)$, and denote by n the smallest element of the block of σ containing m. Then each of $\{1, \ldots, n-1\}$ and $\{n, \ldots, m\}$ is a union of blocks of σ, thus σ is obtained as the juxtaposition of two non-crossing partitions $\sigma_1 \in NC(n-1)$ and $\sigma_2 \in NC(m+1-n)$. We let $\pi \in NC(n)$ be the partition obtained by adding a one-element block to the left of σ_1, and we put $\rho = K^{-1}(\sigma_2) \in NC(m+1-n)$ ($K^{-1}(\sigma_2)$ has $\{1\}$ as a block – this is implied by the fact that 1 and $m+1-n$ are in the same block of σ_2). The pair (π, ρ) obtained in this way is what corresponds to σ by the map (18.33).

We leave it as an exercise to the reader to check that the bijection described in the preceding paragraph also has the following property: if $(\pi, \rho) \leftrightarrow \sigma$ by this bijection, then $K^{-1}(\sigma)$ is the juxtaposition of $K(\pi)$ and ρ_o, where ρ_o denotes the partition obtained by deleting the left-most block (of one element) of the partition ρ.

Finally, let us observe that if $(\pi, \rho) \leftrightarrow \sigma$ by the bijection (18.33), then Equations (18.34) are indeed satisfied. The first of these equations follows directly from how σ is obtained as a juxtaposition of π_o and $K(\rho)$, while the other follows from the analogous property of $K^{-1}(\sigma)$:

$$\begin{aligned} \mathrm{Cf}_{K(\sigma)}(g) &= \mathrm{Cf}_{K^{-1}(\sigma)}(g) \\ &= \mathrm{Cf}_{K(\pi)}(g) \cdot \mathrm{Cf}_{\rho_o}(g) \\ &= \frac{1}{\beta_1} \mathrm{Cf}_{K(\pi)}(g) \cdot \mathrm{Cf}_{\rho}(g) \end{aligned}$$

(The first equality is valid because $K(\sigma)$ and $K^{-1}(\sigma)$ are obtained from each other by a cyclic permutation – same argument as in Equation (17.10) in Lecture 17; the second equality follows because $K^{-1}(\sigma)$ is the juxtaposition of $K(\pi)$ and ρ_o; the third equality is due to the relation between ρ and ρ_o.) □

Exercises

EXERCISE 18.23. Prove the following generalization of the formula (9.18) which appeared in Exercise 9.23. If $\pi, \sigma \in NC(n)$ are such that $\pi \le \sigma$, then

$$|\pi| + |K_\sigma(\pi)| = n + |\sigma|. \tag{18.35}$$

NOTATIONS 18.24. Let n be a positive integer, and let S_n denote the group of all permutations of $\{1, \ldots, n\}$.

(1) For $\alpha \in S_n$ and $a, b \in \{1, \ldots, n\}$ we will say that a and b are in the same **orbit** of α if there exists an integer m such that $\alpha^m(a) = b$. The set $\{1, \ldots, n\}$ is then partitioned into orbits of α; the partition obtained in this way will be denoted by $\mathrm{Orb}(\alpha)$.

(2) Let π be a partition in $NC(n)$. We will denote by P_π the permutation $\alpha \in S_n$ which is determined by the following properties:

(i) $\mathrm{Orb}(\alpha) = \pi$;

(ii) if $V = \{a_1, a_2, \ldots, a_k\}$ is a block of π, with $1 \leq a_1 < a_2 < \cdots < a_k \leq n$, then we have $\alpha(a_1) = a_2, \ldots, \alpha(a_{k-1}) = a_k$, $\alpha(a_k) = a_1$.

The first part of the following exercise is a rigorous formulation of the fact that the relative Kreweras complement $K_\sigma(\pi)$ is in some sense "the quotient of σ by π."

EXERCISE 18.25. (1) Let π, σ be partitions in $NC(n)$, such that $\pi \leq \sigma$. Prove that

$$P_{K_\sigma(\pi)} = P_\pi^{-1} P_\sigma \tag{18.36}$$

(equality holding in the symmetric group S_n).

(2) Let π, ρ, σ be partitions in $NC(n)$ such that $\rho \leq \sigma$ and such that $P_\pi P_\rho = P_\sigma$. Prove that $\pi \leq \sigma$ and that $K_\sigma(\pi) = \rho$.

EXERCISE 18.26. Let π, σ be partitions in $NC(n)$, such that $\pi \leq \sigma$.

(1) Prove that for $\tau \in [\pi, \sigma]$ we have that $K_\tau(\pi) \leq K_\sigma(\pi)$, and that

$$K_{K_\sigma(\pi)}(\, K_\tau(\pi)\,) = K_\sigma(\tau).$$

(2) Prove that the map $\tau \mapsto K_\tau(\pi)$ is a lattice isomorphism between the intervals $[\pi, \sigma]$ and $[0_n, K_\sigma(\pi)]$ in $NC(n)$.
[Hint for (1): The partitions $K_\tau(\pi), K_\sigma(\tau), K_\sigma(\pi)$ (in this order) satisfy the hypothesis of Exercise 18.25.2. Hint for (2): $[\pi, \sigma]$ and $[0_n, K_\sigma(\pi)]$ have the same canonical factorization, hence the same cardinality.]

EXERCISE 18.27. Let $(\mathcal{A}, \varphi)$ be a $*$-probability space. Let $e, f \in \mathcal{A}$ be two selfadjoint projections such that e is free from f, and consider the element $x := efe$. By using S-transforms, determine the distribution in analytic sense of x in $(\mathcal{A}, \varphi)$ (you should get an explicit answer depending on the parameters α and β, where $\alpha := \varphi(e)$ and $\beta := \varphi(f)$).

LECTURE 19

The free commutator

The original use of the R-transform was in connection to the problem of describing the distribution of a sum of free random variables (via the formula $R_{a+b} = R_a + R_b$, which always holds when a is free from b in some non-commutative probability space – cf. Lectures 12 and 16). Similarly, the S-transform was introduced to solve the problem of multiplication of free random variables (cf. Lecture 18). By following these lines, it is natural to ask what happens when one considers the commutator $ab-ba$, or the anti-commutator $ab+ba$ of two free elements. Some remarks about this have already been made in Lecture 15. In the present lecture we will continue the discussion started there, by using the convenient language of the operation of boxed convolution $\boxed{\star}$.

The problem of the free commutator can be treated on two levels, which will be discussed separately.

First there is a level where one considers *even* random variables. At this level the problem can be solved as an application of the results on R-diagonal elements. One comes to a formula which is at the same time valid for the anti-commutator (of two free, even random variables), and which was already presented in Theorem 15.20.

Then there is the general level, where the assumption that the random variables are even is dropped. Quite surprisingly, it turns out that the free commutator (unlike the free anti-commutator) is still described in this case by the same formula as we had in the even case. This is caused by a non-trivial cancelation phenomenon: if a is free from b, then the free cumulants of odd order of a and of b simply disappear (all the terms involving them cancel out) in the process of computing the moments of $ab - ba$.

Free commutators of even elements

Recall from Lecture 11 that an element a in a non-commutative probability space $(\mathcal{A}, \varphi)$ is said to be *even* if it has the property that $\varphi(a^n) = 0$ for every odd positive integer n. For such an element we have introduced its *determining sequence* (cf. Notations 11.24) and then its *determining series* (cf. Remark 17.17). The latter series will play an important role in this lecture, so we will introduce a notation for it.

NOTATION 19.1. Let $(\mathcal{A}, \varphi)$ be a non-commutative probability space, and let a be an even element of $\mathcal{A}$. The determining series of a will be denoted by $R_a^{(\text{even})}$. That is, we have:

$$R_a^{(\text{even})}(z) := \sum_{n=1}^{\infty} \alpha_{2n} z^n, \tag{19.1}$$

where α_{2n} is the free cumulant of order $2n$ of a, for $n \geq 1$.

REMARK 19.2. If a is an even element in $(\mathcal{A}, \varphi)$, then knowing the determining series $R_a^{(\text{even})}$ is equivalent to knowing the full R-transform R_a of a (since the coefficients of R_a that are not used in $R_a^{(\text{even})}$ are all equal to 0). In fact it is clear that R_a is obtained back from $R_a^{(\text{even})}$ via the formula

$$R_a(z) = R_a^{(\text{even})}(z^2), \tag{19.2}$$

which is a restatement of the formula (17.20) in Remark 17.17. Let us also record here the equation

$$R_{a^2} = R_a^{(\text{even})} \boxed{\star} \text{Zeta}, \tag{19.3}$$

which is a copy of Equation (17.21) of the same remark, and will be used in the main result of this section.

We will next review (from Lecture 15) a few facts about R-diagonal elements, and we will rewrite the corresponding formulas in terms of power series and their boxed convolution.

It will be more convenient to consider here the general framework of a non-commutative probability space $(\mathcal{A}, \varphi)$ (where $\mathcal{A}$ does not necessarily have a $*$-operation). In this framework instead of R-diagonal elements we talk about *R-diagonal pairs*. More precisely, we have the following.

DEFINITION 19.3. Let $(\mathcal{A}, \varphi)$ be a non-commutative probability space, and let x_1, x_2 be elements of $\mathcal{A}$. We say that (x_1, x_2) is an **R-diagonal pair** if the joint R-transform R_{x_1,x_2} is of the form

$$R_{x_1,x_2}(z_1, z_2) = f(z_1 z_2) + g(z_2 z_1), \tag{19.4}$$

where $f(z)$ and $g(z)$ are series of one variable. The series f and g are called the **determining series** of (x_1, x_2).

REMARKS 19.4. (1) If in the notations of the preceding definition we write explicitly:

$$f(z) = \sum_{n=1}^{\infty} \alpha_n z^n \quad \text{and} \quad g(z) = \sum_{n=1}^{\infty} \beta_n z^n,$$

then the fact that the pair (x_1, x_2) is R-diagonal amounts thus to the following.

(i) For every $n \geq 1$ we have that:

$$\kappa_{2n}(x_1, x_2, \ldots, x_1, x_2) = \alpha_n, \text{ and } \kappa_{2n}(x_2, x_1, \ldots, x_2, x_1) = \beta_n.$$

(ii) The free cumulants of x_1 and x_2 which are not listed in (i) are all equal to zero.

(2) The connection with the R-diagonal elements of Lecture 15 is quite clear: an element x in a $*$-probability space is R-diagonal precisely when the pair (x, x^*) is R-diagonal. Most of the results about R-diagonal elements shown in Lecture 15 can in fact be extended, with the same proofs, to the framework of R-diagonal pairs. Among these results, one which is most relevant for our purposes is Theorem 15.17, about the product of two free even elements; when using the framework of R-diagonal pairs, this is stated as follows.

PROPOSITION 19.5. *Let $(\mathcal{A}, \varphi)$ be a non-commutative probability space. Let $a, b \in \mathcal{A}$ be even, and suppose that a is free from b. Then (ab, ba) is an R-diagonal pair.*

REMARK 19.6. One of the important points in Lecture 15 was to observe a basic relation which connects the determining series of an R-diagonal element x with the distributions of the elements xx^* and x^*x. Let us also review this relation, stated in the framework of R-diagonal pairs. We will look at the case of a tracial non-commutative probability space; thus what we will write is the counterpart of Proposition 15.6.2 (specifically of Equation (15.8) obtained there).

So let $(\mathcal{A}, \varphi)$ be a tracial non-commutative probability space, and let (x_1, x_2) be an R-diagonal pair in $(\mathcal{A}, \varphi)$. In this case we have that

$$\kappa_{2n}(x_1, x_2, \ldots, x_1, x_2) = \kappa_{2n}(x_2, x_1, \ldots, x_2, x_1), \quad \forall\, n \geq 1,$$

so that the two determining series f and g of Equation (19.4) are equal to each other, and Equation (19.4) is now just

$$R_{x_1,x_2}(z_1, z_2) = f(z_1 z_2) + f(z_2 z_1). \tag{19.5}$$

The traciality of φ also implies that the elements x_1x_2 and x_2x_1 have the same moment series, and hence the same R-transform. Proposition

15.6.2 (adjusted to our current framework) then simply says that the common R-transform of x_1x_2 and x_2x_1 is given by the equation:

$$R_{x_1x_2} = f \boxed{\star} \text{Zeta}. \tag{19.6}$$

Equivalent to (19.6), one can of course also write:

$$f = R_{x_1x_2} \boxed{\star} \text{Möb}. \tag{19.7}$$

Equations (19.5) and (19.7) imply together the important fact that the joint distribution of x_1 and x_2 (i.e. the values of φ on all the words in x_1 and x_2) is completely determined by the distribution of x_1x_2 (i.e. by the values of φ on just the alternating words of even length $x_1x_2 \cdots x_1x_2$).

After these preliminaries, we can now state the main result of this section.

THEOREM 19.7. *Let $(\mathcal{A}, \varphi)$ be a non-commutative probability space, and let a, b be even elements of $\mathcal{A}$, such that a is free from b. Then*

$$R_{ab-ba}(z) = 2(R_a^{(\text{even})} \boxed{\star} R_b^{(\text{even})} \boxed{\star} \text{Zeta})(-z^2). \tag{19.8}$$

PROOF. Due to the fact that a and b are free and generate commutative algebras, it follows that the restriction of φ to the unital algebra $\mathcal{A}_o$ generated by a and b is a trace. Thus by replacing $\mathcal{A}$ with $\mathcal{A}_o$, we can assume without loss of generality that φ is a trace.

Denote $ab =: x_1$ and $ba =: x_2$. As reviewed in Proposition 19.5, the pair (x_1, x_2) is R-diagonal. Let f denote its determining series. Then we have:

$$\begin{aligned}
f &= R_{x_1x_2} \boxed{\star} \text{Möb} && \text{(by Equation (19.7))} \\
&= R_{abba} \boxed{\star} \text{Möb} && \text{(by replacing } x_1 = ab,\ x_2 = ba) \\
&= R_{a^2b^2} \boxed{\star} \text{Möb} && \text{(since } R_{abba} = R_{a^2b^2}\text{, by traciality)} \\
&= R_{a^2} \boxed{\star} R_{b^2} \boxed{\star} \text{Möb} && \text{(by Proposition 17.2)} \\
&= (R_a^{(\text{even})} \boxed{\star} \text{Zeta}) \boxed{\star} (R_b^{(\text{even})} \boxed{\star} \text{Zeta}) \boxed{\star} \text{Möb} && \text{(by Eqn.(19.3))}.
\end{aligned}$$

If in the last expression obtained above we use the commutativity of the operation $\boxed{\star}$ of one variable, and the fact that Möb is the inverse of Zeta, we arrive to the formula:

$$f = R_a^{(\text{even})} \boxed{\star} R_b^{(\text{even})} \boxed{\star} \text{Zeta}. \tag{19.9}$$

On the other hand, let us note that the formula for the behavior of the R-transform under linear transformations (cf. Lecture 16, Proposition 16.12) gives us that:

$$R_{ab-ba}(z) = R_{x_1-x_2}(z) = R_{x_1,x_2}(z, -z). \tag{19.10}$$

By combining this with Equation (19.5), we thus get:

$$R_{ab-ba}(z) \;=\; f\big(z\cdot(-z)\big)+f\big((-z)\cdot z\big) \;=\; 2f(-z^2). \qquad (19.11)$$

Finally, Equations (19.9) and (19.11) together yield (19.8). □

REMARKS 19.8. Let $(\mathcal{A},\varphi)$ be a non-commutative probability space, and let $a,b\in\mathcal{A}$ be even, such that a is free from b.

(1) The commutator $ab-ba$ is again an even element. It would not be hard to prove this directly from the definitions, but at this point it is more convenient to observe that it is an immediate consequence of the formula (19.8) – indeed, the R-transform R_{ab-ba} is a series in z^2.

(2) In the discussions about the free commutator it is usually more convenient to deal with $i(ab-ba)$ instead of $ab-ba$. (For instance, if $(\mathcal{A},\varphi)$ was to be a $*$-probability space, and if a and b were to be selfadjoint, then $i(ab-ba)$ would again be a selfadjoint element.) The R-transforms of $ab-ba$ and $i(ab-ba)$ are related by

$$R_{i(ab-ba)}(z) \;=\; R_{ab-ba}(iz),$$

thus instead of Equation (19.8) of Theorem 19.7 we would now have

$$R_{i(ab-ba)}(z) \;=\; 2(\,R_a^{(\text{even})} \boxed{\star}\, R_b^{(\text{even})} \boxed{\star}\, \text{Zeta}\,)(z^2) \qquad (19.12)$$

(the minus sign disappears, since $-(iz)^2=+z^2$). Note that the latter equation can be written without making explicit use of the indeterminate z, in the form

$$R^{(\text{even})}_{i(ab-ba)} \;=\; 2(\,R_a^{(\text{even})} \boxed{\star}\, R_b^{(\text{even})} \boxed{\star}\, \text{Zeta}\,). \qquad (19.13)$$

(3) The anti-commutator $ab+ba$ has the same distribution as $i(ab-ba)$, and consequently the expression on the right-hand side of (19.12) also describes the R-transform R_{ab+ba}.

Indeed, the proof of Theorem 19.7 only involved the element $ab-ba$ at the very end, in Equation (19.10). But (in the framework of that proof) we can also write the analog of (19.10) for $ab+ba$:

$$\begin{aligned} R_{ab+ba}(z) &= R_{x_1+x_2}(z) \\ &= R_{x_1,x_2}(z,z) && \text{(by Proposition 16.12)} \\ &= 2f(z^2) && \text{(by Eqn.(19.5))}, \end{aligned}$$

hence the same expression as for $R_{i(ab-ba)}$ is obtained.

A short calculation (left as an exercise) shows that the formula for the free anti-commutator which is obtained in this way is precisely the translation in terms of R-transforms and boxed convolution of Equation (15.13) in Theorem 15.20 of Lecture 15.

EXAMPLE 19.9. In the framework of Theorem 19.7, let us suppose that the even element b is such that $b^2 = 1_{\mathcal{A}}$. Then the even free cumulants of b are the signed Catalan numbers (cf. Exercise 11.35, or Remark 17.19.2); in other words, $R_b^{(\text{even})}$ is equal to Möb, the Möbius series of one variable. The formula (19.12) becomes:

$$\begin{aligned} R_{i(ab-ba)}(z) &= 2(R_a^{(\text{even})} \boxed{\star} \text{Möb} \boxed{\star} \text{Zeta})(z^2) \\ &= 2R_a^{(\text{even})}(z^2) \\ &= 2R_a(z) \qquad\qquad \text{(by Eqn.(19.2))}. \end{aligned}$$

So we see that in this case we have

$$R_{i(ab-ba)} = 2R_a; \tag{19.14}$$

or in other words, $i(ab - ba)$ has the same distribution as the sum of two free elements a_1 and a_2, such that each of a_1 and a_2 has the same distribution as a.

REMARK 19.10. Theorem 19.7 has a nice reformulation in terms of S-transforms. More precisely, let a, b be as in Theorem 19.7, and consider the element $c = i(ab - ba)$. Since a, b, c are even, their distributions are determined by the S-transforms S_{a^2}, S_{b^2}, S_{c^2}, respectively. The version with S-transforms of Theorem 19.7 is a formula which expresses S_{c^2} in terms of S_{a^2} and S_{b^2}; this will be presented in Proposition 19.12 below. The conversion from Theorem 19.7 to Proposition 19.12 is obtained by using the isomorphism "$\mathcal{F}$" from Theorem 18.14 of the preceding lecture. We will first prove a lemma.

LEMMA 19.11. *Let $(\mathcal{A}, \varphi)$ be a non-commutative probability space, and let $a \in \mathcal{A}$ be an even element. Suppose that $\varphi(a^2) \neq 0$, so that the S-transform S_{a^2} exists. Then $R_a^{(\text{even})}$ belongs to the set $\Theta_1^{(\text{inv})}$ of series which are invertible with respect to $\boxed{\star}$ (so that $\mathcal{F}$ can be applied to $R_a^{(\text{even})}$), and we have:*

$$S_{a^2}(w) = \frac{(\mathcal{F}(R_a^{(\text{even})}))(w)}{1+w}. \tag{19.15}$$

PROOF. $R_a^{(\text{even})}$ belongs to $\Theta_1^{(\text{inv})}$ because its linear coefficient is $\varphi(a^2) \neq 0$. In order to obtain (19.15), we apply $\mathcal{F}$ to both sides of Equation (19.3). On the left-hand side we will get $\mathcal{F}(R_{a^2})$ which is S_{a^2} (cf. Equation (18.15) in Lecture 18). On the right-hand side we will get $\mathcal{F}(R_a^{(\text{even})} \boxed{\star} \text{Zeta})$, which is $\mathcal{F}(R_a^{(\text{even})}) \cdot \mathcal{F}(\text{Zeta})$, by Theorem 18.14. Hence we obtain:

$$S_{a^2} = \mathcal{F}(R_a^{(\text{even})}) \cdot \mathcal{F}(\text{Zeta}).$$

But a direct calculation using the definition of $\mathcal{F}$, and the fact that $\mathrm{Zeta}(z) = z/(1-z)$, leads to the formula:

$$\big(\mathcal{F}(\mathrm{Zeta})\big)(w) = \frac{1}{1+w}, \tag{19.16}$$

and (19.15) follows. □

PROPOSITION 19.12. *Let $(\mathcal{A}, \varphi)$ be a non-commutative probability space, and let a, b be even elements of $\mathcal{A}$, such that a is free from b. Assume that $\varphi(a^2) \neq 0 \neq \varphi(b^2)$. Denote $c := i(ab - ba)$. Then $\varphi(c^2) = 2\varphi(a^2) \cdot \varphi(b^2) \neq 0$, and:*

$$S_{c^2}(w) = \frac{1+\frac{w}{2}}{2(1+w)} \cdot S_{a^2}\Big(\frac{w}{2}\Big) \cdot S_{b^2}\Big(\frac{w}{2}\Big). \tag{19.17}$$

PROOF. The fact that $\varphi(c^2) = 2\varphi(a^2) \cdot \varphi(b^2)$ is obtained by direct computation: we have

$$c^2 = -abab - baba + ab^2a + ba^2b$$

and $\varphi(abab) = \varphi(baba) = 0$ (directly from the definition of free independence) and $\varphi(ab^2a) = \varphi(ba^2b) = \varphi(a^2)\varphi(b^2)$.

Consider now the equation:

$$R_c^{(\mathrm{even})} = 2(\, R_a^{(\mathrm{even})} \boxed{\star}\, R_b^{(\mathrm{even})} \boxed{\star}\, \mathrm{Zeta}\,), \tag{19.18}$$

which is a reformulation of Equation (19.12) in Remark 19.8.2. We apply $\mathcal{F}$ to both sides of (19.18). On the left-hand side we get:

$$S_{c^2}(w) \cdot (1+w), \tag{19.19}$$

by Lemma 19.11 applied to the even element c. For the right-hand side we use the identity

$$\big(\mathcal{F}(2f)\big)(w) = 2^{-1} \cdot \big(\mathcal{F}(f)\big)(w/2) \qquad (\forall f \in \Theta_1^{(\mathrm{inv})}),$$

which follows immediately from the definition of $\mathcal{F}$, and thus we get:

$$\begin{aligned}
&\frac{1}{2}\big(\mathcal{F}(\, R_a^{(\mathrm{even})} \boxed{\star}\, R_b^{(\mathrm{even})} \boxed{\star}\, \mathrm{Zeta}\,)\big)(w/2) \\
&= \frac{1}{2} \cdot \big(\mathcal{F}(R_a^{(\mathrm{even})})\big)(w/2) \cdot \big(\mathcal{F}(R_b^{(\mathrm{even})})\big)(w/2) \cdot \big(\mathcal{F}(\mathrm{Zeta})\big)(w/2) \\
&= \frac{1}{2} \cdot S_{a^2}(w/2) \cdot \big(1+\frac{w}{2}\big) \cdot S_{b^2}(w/2) \cdot \big(1+\frac{w}{2}\big) \cdot \frac{1}{1+\frac{w}{2}} \\
&= \frac{1}{2} \cdot \big(1+\frac{w}{2}\big) \cdot S_{a^2}(w/2) \cdot S_{b^2}(w/2)
\end{aligned}$$

(by using Theorem 18.14, Lemma 19.11 and also Equation (19.16)). The result follows by equating the last line with (19.19). □

Free commutators in the general case

We start by extending Notation 19.1 to the framework of elements that are not necessarily even.

NOTATION 19.13. Let $(\mathcal{A}, \varphi)$ be a non-commutative probability space, let a be an element of $\mathcal{A}$ (not necessarily even!), and consider the R-transform $R_a(z) := \sum_{n=1}^{\infty} \alpha_n z^n$. We denote:

$$R_a^{(\text{even})}(z) := \sum_{n=1}^{\infty} \alpha_{2n} z^n.$$

REMARK 19.14. The formula for $R_a^{(\text{even})}$ is the same as in the discussion about even elements; but it is clear that for an arbitrary $a \in \mathcal{A}$, the series $R_a^{(\text{even})}$ no longer contains the full information about the distribution of a.

On the other hand it is also clear that in order to determine the commutator of two free elements a and b, one does not need to know the full information about the distributions of a and b. For instance a trivial remark, which does not in fact depend on the freeness of a and b, is that the expectations $\varphi(a)$ and $\varphi(b)$ do not have any influence on $ab - ba$. The main point here will be that the series $R_a^{(\text{even})}$ and $R_b^{(\text{even})}$ provide exactly the partial information about a and b which is needed in order to determine the distribution of their commutator. More precisely, the main result of this section goes as follows.

THEOREM 19.15. *Let $(\mathcal{A}, \varphi)$ and $(\mathcal{A}', \varphi')$ be non-commutative probability spaces, and suppose that we have elements $a, b \in \mathcal{A}$, $a', b' \in \mathcal{A}'$, such that a is free from b in $(\mathcal{A}, \varphi)$, a' is free from b' in $(\mathcal{A}', \varphi')$, and such that:*

$$R_a^{(\text{even})} = R_{a'}^{(\text{even})}, \qquad R_b^{(\text{even})} = R_{b'}^{(\text{even})}. \tag{19.20}$$

Then the commutators $ab - ba \in \mathcal{A}$ and $a'b' - b'a' \in \mathcal{A}'$ are identically distributed.

The proof of Theorem 19.15 will be discussed in the next section. Here we will only examine the formulas which this theorem entails, and we will present a concrete example of calculation for a distribution of a free commutator.

COROLLARY 19.16. *Let $(\mathcal{A}, \varphi)$ be a non-commutative probability space, and let a, b be elements of $\mathcal{A}$, such that a is free from b. Then the relation:*

$$R_{i(ab-ba)}(z) = 2(R_a^{(\text{even})} \boxed{\star} R_b^{(\text{even})} \boxed{\star} \text{Zeta})(z^2)$$

holds (even if a and b are not even!)

PROOF. By using Exercise 16.21, one can easily construct a non-commutative probability space $(\mathcal{A}', \varphi')$ and *even* elements $a', b' \in \mathcal{A}'$ such that a' is free from b', and such that Equation (19.20) holds. But then:

$$\begin{aligned} R_{i(ab-ba)}(z) &= R_{i(a'b'-b'a')}(z) && \text{(by Theorem 19.15)} \\ &= 2(R_{a'}^{(\text{even})} \boxed{\star} R_{b'}^{(\text{even})} \boxed{\star} \text{Zeta})(z^2) && \text{(by Theorem 19.7)} \\ &= 2(R_{a}^{(\text{even})} \boxed{\star} R_{b}^{(\text{even})} \boxed{\star} \text{Zeta})(z^2) && \text{(by Eqn.(19.20)).} \end{aligned}$$

□

Let us also state the S-transform version of the above corollary.

COROLLARY 19.17. *Let $(\mathcal{A}, \varphi)$ be a non-commutative probability space, let a, b be elements of $\mathcal{A}$ such that a is free from b, and suppose that the variances $\gamma_a = \varphi(a^2) - \varphi(a)^2$ and $\gamma_b = \varphi(b^2) - \varphi(b)^2$ are different from 0. Denote $c = i(ab - ba)$. Then c is even, with $\varphi(c^2) = 2\gamma_a\gamma_b \neq 0$, and we have:*

$$\begin{aligned} S_{c^2}(w) = \frac{2}{w^2(1+w)(1+\frac{w}{2})} &\cdot (R_a^{(\text{even})})^{<-1>}(w/2) \\ &\cdot (R_b^{(\text{even})})^{<-1>}(w/2). \quad (19.21) \end{aligned}$$

PROOF. This is very similar to the proof of Proposition 19.12 in the preceding section, involving the use of the isomorphism $\mathcal{F}$ from Lecture 18. The difference is that now Lemma 19.11 can only be applied in connection to the element c (for a and b it does not apply, since these elements are not assumed to be even). So all we do now about the series $\mathcal{F}(R_a^{(\text{even})})$ and $\mathcal{F}(R_b^{(\text{even})})$ (which appear during the calculation) is to replace them by using the definition of $\mathcal{F}$ – then Equation (19.21) is obtained. □

We will conclude this section by presenting a concrete example of computation: we will determine the distribution for the free commutator of a projection and a semicircular element. For this example it feels more natural to use the framework of a $*$-probability space (even though the $*$-operation does not really play a role in the computation). We start by examining the $R^{(\text{even})}$ series of the projection.

EXAMPLE 19.18. Let $(\mathcal{A}, \varphi)$ be a $*$-probability space, and let $b \in \mathcal{A}$ be a selfadjoint projection such that $\varphi(b) = \lambda \in (0, 1)$. We will show that:

$$\left(R_b^{(\text{even})}\right)^{<-1>}(b) \;=\; \frac{w(1+w)(1+2w)^2}{(w+\lambda)(w+1-\lambda)}. \tag{19.22}$$

In order to obtain (19.22), we start from the moment series of b, which is

$$M_b(z) \;=\; \sum_{n=1}^{\infty} \lambda z^n \;=\; \frac{\lambda z}{1-z}.$$

It is then immediate that

$$M_b^{<-1>}(w) \;=\; \frac{w}{\lambda + w},$$

hence Remark 16.18 gives us:

$$R_b^{<-1>}(w) \;=\; (1+w)M_b^{<-1>}(w) \;=\; \frac{w(1+w)}{\lambda+w}.$$

This formula for $R_b^{<-1>}$ amounts to an algebraic equation satisfied by R_b, namely

$$z \;=\; \frac{R_b(z)(1+R_b(z))}{\lambda + R_b(z)}.$$

The conclusion up to this point is hence that the R-transform R_b satisfies the quadratic equation:

$$R_b(z)^2 + (1-z)R_b(z) - \lambda z \;=\; 0. \tag{19.23}$$

What we need, however, is some information concerning only the even part of the series R_b. This can be obtained as follows: write

$$R_b(z) \;=\; g(z^2) + zh(z^2),$$

where g coincides with $R_b^{(\text{even})}$, while h is a series made up by using the coefficients of odd degree of R_b. Equation (19.23) is then turned into a system of two equations in g and h:

$$\begin{cases} g + g^2 - z(h-h^2) \;=\; 0 \\ 2gh + h - g - \lambda = 0 \end{cases}$$

By solving for h in the second equation of the system, and by plugging the result into the first equation, we obtain a polynomial equation of degree 4 satisfied by g:

$$(g+g^2)(1+2g)^2 \;=\; z(g+\lambda)(g+1-\lambda). \tag{19.24}$$

Having to find g ($= R_b^{(\text{even})}$) from this equation would not be a very pleasant thing to do. Fortunately, we do not need the series $R_b^{(\text{even})}$ itself, but its inverse under composition. This means that in (19.24) we actually need to solve for z in terms of g, which is much easier and leads to the formula stated in (19.22).

EXAMPLE 19.19. Let $(\mathcal{A},\varphi)$ be a $*$-probability space, and let a, b be selfadjoint elements of $\mathcal{A}$ such that a is a standard semicircular element, b is a projection with $\varphi(b) = \lambda \in (0,1)$, and a is free from b. Denote $c = i(ab - ba)$; we want to calculate the distribution of c.

By using Corollary 19.17, we see that:

$$S_{c^2}(w) = \frac{2}{w^2(1+w)(1+\frac{w}{2})} \cdot \frac{w}{2} \cdot \frac{\frac{w}{2}(1+\frac{w}{2})(1+w)^2}{(\frac{w}{2}+\lambda)(\frac{w}{2}+1-\lambda)}$$

(where we used the fact that $R_a^{(\text{even})}(z) = z$, and we substituted $(\,R_b^{(\text{even})}\,)^{<-1>}$ from the preceding example). After simplification this becomes:

$$S_{c^2}(w) = \frac{1+w}{2(\frac{w}{2}+\lambda)(\frac{w}{2}+1-\lambda)}.$$

Then we compute the inverse of the moment series of c^2:

$$\begin{aligned} M_{c^2}^{<-1>}(w) &= \frac{w}{1+w} S_{c^2}(w) \\ &= \frac{2w}{w^2+2w+4\lambda(1-\lambda)}. \end{aligned}$$

This yields a quadratic equation for M_{c^2}. By using the obvious relation $M_c(z) = M_{c^2}(z^2)$ (which holds because we know that c is even) we obtain a quadratic equation for the moment series M_c, which reads as follows:

$$M_c(z)^2 + \Big(2 - \frac{2}{z^2}\Big) M_c(z) + 4\lambda(1-\lambda) = 0. \tag{19.25}$$

From this point on we assume that $(\mathcal{A},\varphi)$ is a C^*-probability space, with φ faithful. Then the selfadjoint element $c \in \mathcal{A}$ has a spectral distribution μ_c which is a probability measure with compact support on $\mathbb{R}$, and which can be calculated by the techniques described in Lectures 2 and 3. More precisely, if we denote by G the Cauchy transform of μ_c, then from (19.25) we get a quadratic equation for G:

$$\zeta^2 G(\zeta)^2 - 2\zeta^3 G(\zeta) + (2\zeta^2 - 1 + 4\lambda(1-\lambda)) = 0, \tag{19.26}$$

holding for all $\zeta \in \mathbb{C}$ with positive imaginary part. By solving for G in (19.26) and then by using the Stieltjes inversion formula, one can find an explicit description of μ_c. This is:

$$\mu_c = \sqrt{1-4\lambda(1-\lambda)}\,\delta_0 + \frac{1}{\pi|t|}\sqrt{4\lambda(1-\lambda)-(t^2-1)^2}\,\chi_{[-\beta,-\alpha]\cup[\alpha,\beta]}\,dt, \quad (19.27)$$

where δ_0 denotes the Dirac measure at 0, and the numbers α, β are defined by

$$\alpha := \sqrt{1-2\sqrt{\lambda(1-\lambda)}}, \qquad \beta := \sqrt{1+2\sqrt{\lambda(1-\lambda)}}.$$

The cancelation phenomenon

In this section we outline the argument proving Theorem 19.15. We first observe that there is no loss of generality if we strengthen the hypothesis of the theorem in the way indicated in the next proposition.

PROPOSITION 19.20. *Let $(\mathcal{A},\varphi)$ and $(\mathcal{A}',\varphi')$ be non-commutative probability spaces, and suppose that we have elements $a, b \in \mathcal{A}$, $a', b' \in \mathcal{A}'$, such that a is free from b in $(\mathcal{A},\varphi)$, a' is free from b' in $(\mathcal{A}',\varphi')$, and such that:*

$$R_a^{(\text{even})} = R_{a'}^{(\text{even})}, \qquad R_b = R_{b'}. \quad (19.28)$$

Then the commutators $ab - ba \in \mathcal{A}$ and $a'b' - b'a' \in \mathcal{A}'$ are identically distributed.

What is different in the statement of Proposition 19.20 is the hypothesis "$R_b = R_{b'}$," replacing the weaker hypothesis "$R_b^{(\text{even})} = R_{b'}^{(\text{even})}$" of Theorem 19.15. If we assume Proposition 19.20, then Theorem 19.15 is proved as follows.

PROOF OF THEOREM 19.15. Let $(\mathcal{A},\varphi)$, $(\mathcal{A}',\varphi')$, $a, b \in \mathcal{A}$ and $a', b' \in \mathcal{A}'$ be as in the statement of Theorem 19.15. One can construct (by appealing again to Exercise 16.21) a non-commutative probability space $(\mathcal{A}'',\varphi'')$ and elements $a'', b'' \in \mathcal{A}''$ such that a'' is free from b'' and such that

$$R_{a''} = R_{a'}, \qquad R_{b''} = R_b. \quad (19.29)$$

The first equality in (19.29) implies in particular that

$$R_{a''}^{(\text{even})} = R_{a'}^{(\text{even})} = R_a^{(\text{even})};$$

so Proposition 19.20 applies to $a, b \in \mathcal{A}$ and $a'', b'' \in \mathcal{A}''$, and gives us that $ab - ba$ and $a''b'' - b''a''$ are identically distributed. Analogously, we have that

$$R_{b'}^{(\text{even})} = R_b^{(\text{even})} = R_{b''}^{(\text{even})} \qquad \text{and} \qquad R_{a'} = R_{a''},$$

so that Proposition 19.20 applies to $b', a' \in \mathcal{A}'$ and $b'', a'' \in \mathcal{A}''$, and gives us that $b'a' - a'b'$ and $b''a'' - a''b''$ are identically distributed. It follows that both $ab - ba$ and $a'b' - b'a'$ have the same distribution as $a''b'' - b''a''$, and the conclusion follows. □

We next look at the formula for the moment of order n of a free commutator $ab - ba$, in terms of the free cumulants of a and of b. We will use the following notations.

NOTATIONS 19.21. For $\varepsilon = (\varepsilon(1), \ldots, \varepsilon(n)) \in \{1, 2\}^n$ we denote:
- $A(\varepsilon) := \{2i-1 \mid 1 \leq i \leq n,\ \varepsilon(i) = 1\} \cup \{2j \mid 1 \leq j \leq n,\ \varepsilon(j) = 2\}$;
- $B(\varepsilon) := \{2i \mid 1 \leq i \leq n,\ \varepsilon(i) = 1\} \cup \{2j-1 \mid 1 \leq j \leq n,\ \varepsilon(j) = 2\}$;
- $t(\varepsilon) := \text{card}\{j \mid 1 \leq j \leq n,\ \varepsilon(j) = 2\}$.

Thus $t(\varepsilon)$ just stands for "number of twos" in ε, while $\{A(\varepsilon), B(\varepsilon)\}$ is a partition of $\{1, \ldots, 2n\}$ into two blocks of n elements each. The significance of $A(\varepsilon)$ and $B(\varepsilon)$ is the following: if one denotes $ab =: x_1$, $ba =: x_2$ and then writes explicitly $x_{\varepsilon(1)} x_{\varepsilon(2)} \cdots x_{\varepsilon(n)}$ as a monomial of length $2n$ in a and b, then this monomial will have a on the positions indicated by $A(\varepsilon)$ and will have b on the positions indicated by $B(\varepsilon)$.

PROPOSITION 19.22. *Let $(\mathcal{A}, \varphi)$ be a non-commutative probability space, and let $a, b \in \mathcal{A}$ be such that a is free from b. Denote*

$$R_a(z) = \sum_{n=1}^{\infty} \alpha_n z^n, \ R_b(z) = \sum_{n=1}^{\infty} \beta_n z^n.$$

Then for every $n \geq 1$ we have:

$$\varphi(\,(ab - ba)^n\,) = \sum_{\varepsilon \in \{1,2\}^n} \sum_{\substack{\pi \in NC(2n), \\ \pi \leq \{A(\varepsilon), B(\varepsilon)\}}} (-1)^{t(\varepsilon)} \Big(\prod_{\substack{V \in \pi, \\ V \subset A(\varepsilon)}} \alpha_{|V|} \Big) \Big(\prod_{\substack{W \in \pi, \\ W \subset B(\varepsilon)}} \beta_{|W|} \Big). \tag{19.30}$$

(Note: the inequality $\pi \leq \{A(\varepsilon), B(\varepsilon)\}$ under the second summation sign in (19.30) is in the lattice of all partitions of $\{1, \ldots, 2n\}$ – it thus simply means that every block of π either is contained in $A(\varepsilon)$ or is contained in $B(\varepsilon)$.)

PROOF. Denote $ab =: x_1$ and $ba =: x_2$. Then we can write:

$$\begin{aligned} \varphi(\,(ab - ba)^n\,) &= \varphi(\,(x_1 - x_2)^n\,) \\ &= \sum_{\varepsilon \in \{1,2\}^n} (-1)^{t(\varepsilon)} \varphi(x_{\varepsilon(1)} \cdots x_{\varepsilon(n)}). \end{aligned}$$

Now for every $\varepsilon \in \{1,2\}^n$ let us write $x_{\varepsilon(1)} \cdots x_{\varepsilon(n)}$ as a monomial in a and b (with a and b on the positions indexed by $A(\varepsilon)$ and respectively $B(\varepsilon)$), and let us express $\varphi(x_{\varepsilon(1)} \cdots x_{\varepsilon(n)})$ in terms of the free cumulants of a and b. The mixed free cumulants of a and b will vanish (since a is free from b); so in the resulting summation over $NC(2n)$, the partitions which are not smaller than $\{A(\varepsilon), B(\varepsilon)\}$ will have zero contribution. We arrive to

$$\varphi(x_{\varepsilon(1)} \cdots x_{\varepsilon(n)}) = \sum_{\substack{\pi \in NC(2n), \\ \pi \leq \{A(\varepsilon), B(\varepsilon)\}}} \Big(\prod_{\substack{V \in \pi, \\ V \subset A(\varepsilon)}} \alpha_{|V|} \Big) \cdot \Big(\prod_{\substack{W \in \pi, \\ W \subset B(\varepsilon)}} \beta_{|W|} \Big),$$

and substituting this in the formula obtained above for $\varphi(\,(ab - ba)^n\,)$ leads to Equation (19.30). $\square$

We now arrive to the main point of the argument, which is that in the double summation on the right-hand side of (19.30), the terms which involve the cumulants of odd order of a (that is, they contain factors α_k for odd k) are *canceling each other in pairs*. In order to formalize this, it is convenient to use the following notation.

NOTATION 19.23. Let S be a totally ordered finite set, and consider the lattice $NC(S)$ of non-crossing partitions of S. We denote

$$NCE(S) := \{\pi \in NC(S) \mid \text{every block of } \pi \text{ has even cardinality }\}$$
$$NCO(S) := NC(S) \setminus NCE(S).$$

PROPOSITION 19.24. *Let n be a positive integer, and denote*

$$\mathcal{X}_n := \Big\{ (\pi, \varepsilon) \in NC(2n) \times \{1,2\}^n \;\Big|\; \begin{matrix} \pi \leq \{A(\varepsilon), B(\varepsilon)\} \text{ and} \\ \pi | A(\varepsilon) \in NCO(A(\varepsilon)) \end{matrix} \Big\}.$$

One can find a map $\Phi_n : \mathcal{X}_n \to \mathcal{X}_n$ such that $\Phi_n \circ \Phi_n$ is the identity map on $\mathcal{X}_n$ and such that for every $(\pi, \varepsilon) \in \mathcal{X}_n$ the element $\Phi_n(\pi, \varepsilon) =: (\pi', \varepsilon')$ has the following properties:

(i) *$t(\varepsilon')$ is not of the same parity as $t(\varepsilon)$.*

(ii) *$\pi' | A(\varepsilon')$ has the same block structure as $\pi | A(\varepsilon)$, and $\pi' | B(\varepsilon')$ has the same block structure as $\pi | B(\varepsilon)$ (where two partitions are said to have the same block structure when they have the same number of blocks of size k for every $k \geq 1$).*

A possible way of putting into evidence a map Φ_n with the properties listed in Proposition 19.24 is indicated in the section of exercises, at the end of the lecture. If one assumes this combinatorial statement, then the proof of Proposition 19.20 (to which Theorem 19.15 had been reduced) follows easily. Indeed, when putting together Propositions 19.22 and 19.24 we get the following corollary.

COROLLARY 19.25. *Let $(\mathcal{A}, \varphi)$ be a non-commutative probability space, and let $a, b \in \mathcal{A}$ be such that a is free from b. Denote*

$$R_a(z) = \sum_{n=1}^{\infty} \alpha_n z^n, \ R_b(z) = \sum_{n=1}^{\infty} \beta_n z^n.$$

Then for every $n \geq 1$ we have:

$$\varphi(\,(ab - ba)^n\,) \quad = \sum_{(\pi,\varepsilon)\in\mathcal{Y}_n} (-1)^{t(\varepsilon)} \Big(\prod_{\substack{V\in\pi,\\ V\subset A(\varepsilon)}} \alpha_{|V|} \Big) \Big(\prod_{\substack{W\in\pi,\\ W\subset B(\varepsilon)}} \beta_{|W|} \Big), \tag{19.31}$$

where

$$\mathcal{Y}_n := \Big\{ (\pi, \varepsilon) \in NC(2n) \times \{1,2\}^n \ \Big| \ \begin{array}{c} \pi \leq \{A(\varepsilon), B(\varepsilon)\} \text{ and} \\ \pi | A(\varepsilon) \in NCE(A(\varepsilon)) \end{array} \Big\}.$$

PROOF. Proposition 19.22 gives us a formula like the one which is required, with the difference that the sum on the right-hand side is over $\mathcal{X}_n \cup \mathcal{Y}_n$ instead of just $\mathcal{Y}_n$ (with $\mathcal{X}_n$ as defined in Proposition 19.24). But Proposition 19.24 implies that $\sum_{\mathcal{X}_n} \cdots = 0$, as the terms indexed by $\mathcal{X}_n$ can be grouped (by using the bijection Φ_n) in pairs which cancel each other. The formula (19.31) follows. □

The proof of Proposition 19.20 is now immediate.

PROOF OF PROPOSITION 19.20. Let $(\mathcal{A}, \varphi)$, $(\mathcal{A}', \varphi')$, $a, b \in \mathcal{A}$ and $a', b' \in \mathcal{A}'$ be as in the statement of Proposition 19.20. Denote

$$R_a^{(\text{even})}(z) = R_{a'}^{(\text{even})}(z) =: \sum_{n=1}^{\infty} \alpha_{2n} z^n, \qquad R_b(z) = R_{b'}(z) =: \sum_{n=1}^{\infty} \beta_n z^n.$$

For every $n \geq 1$, both $\varphi(\,(ab - ba)^n\,)$ and $\varphi'(\,(a'b' - b'a')^n\,)$ are equal to the quantity on the right-hand side of Equation (19.31). Thus $ab - ba$ and $a'b' - b'a'$ are identically distributed. □

Exercises

EXERCISE 19.26. Let $(\mathcal{A}, \varphi)$ be a $*$-probability space, and let a, b be even selfadjoint elements of $\mathcal{A}$ such that a is free from b and such that $a^2 = 1_{\mathcal{A}} = b^2$. Prove that $c := i(ab - ba)$ is a semicircular element of radius $\sqrt{2}$.

EXERCISE 19.27. Let $(\mathcal{A}, \varphi)$ be a $*$-probability space, and let $\{a_n \mid n \geq 0\}$ be a free family of elements of $\mathcal{A}$, such that every a_n is a standard semicircular element. Denote $c_1 = \frac{i}{\sqrt{2}}(a_0 a_1 - a_1 a_0)$ and

$c_n = \frac{i}{\sqrt{2}}(c_{n-1}a_n - a_n c_{n-1})$, $\forall\, n \geq 2$. Prove that the sequence $(c_n)_{n=1}^{\infty}$ converges in distribution, and describe the limit distribution.

Exercises 19.28–19.33 suggest a possible way of constructing a bijection with the properties stated in Proposition 19.24.

EXERCISE 19.28. Let S be a totally ordered set and let π be a partition in $NCO(S)$. Prove that there exists a block $V = \{v_1, \ldots, v_k\}$ of π, with $v_1 < v_2 < \cdots < v_k$, such that k is odd and such that each of the intervals $[v_1, v_2], [v_2, v_3], \ldots, [v_{k-1}, v_k]$ in S has even cardinality.

A block V with the properties described in Exercise 19.28 is sometimes said to be *parity-alternating* (because of how it looks in the particular case when $S = \{1, \ldots, n\}$ – in that case the numbers $v_1 < v_2 < \cdots < v_k$ will have alternating parities). The next exercise records an immediate consequence of the fact that NCO partitions must always have such blocks.

EXERCISE 19.29. Let S be a totally ordered set and let π be a partition in $NCO(S)$. Prove that one can find elements $p \leq q$ of S such that the following hold.

(i) The interval $[p, q]$ of S has odd cardinality.

(ii) p and q belong to the same block V of π, and moreover we have that $p = \min(V)$, $q = \max(V)$.

(iii) The block V appearing in (ii) has odd cardinality.

In order to shorten the statements of the following exercises, it is convenient to make up some names for the terminology which will be used in them.

NOTATIONS 19.30. (1) Let S be a totally ordered set and let π be a partition in $NCO(S)$. Among all the intervals $[p, q]$ of S which have the properties described in Exercise 19.29, pick the one for which the element p is as small as possible (in the order of S). That interval $[p, q]$ will be called the *odd-marked interval* of π.

(2) On the set $\cup_{n=0}^{\infty}\{1, 2\}^n$ of finite sequences of 1 and 2 we will consider the (obvious) operation of multiplication by concatenation. Moreover on this set we will also consider the $*$-operation defined as follows: for $\varepsilon = (\varepsilon(1), \ldots, \varepsilon(n)) \in \{1, 2\}^n$ we set

$$\varepsilon^* := (3 - \varepsilon(n), \ldots, 3 - \varepsilon(1))$$

(for example $(1,1,1,2,2,1)^* = (2,1,1,2,2,2)$).

(3) For $n \geq 1$ and $1 \leq a \leq b \leq n$ we will denote by $\sigma_{a,b}^{(n)}$ the permutation of $\{1,\ldots,n\}$ defined by:

$$\sigma_{a,b}^{(n)}(m) = \begin{cases} a+b-m & \text{if } a \leq m \leq b \\ m & \text{otherwise.} \end{cases}$$

In other words, $\sigma_{a,b}^{(n)}$ turns around the interval $[a,b]$ of $\{1,\ldots,n\}$, and does not change what is outside it.

EXERCISE 19.31. Let π be a partition in $NC(n)$, let W be a block of π, and denote $\min(W) =: p$, $\max(W) =: q$.

(1) Prove that $\sigma_{p,q}^{(n)} \cdot \pi \in NC(n)$ (where the action of a permutation τ on π is as discussed in Notation 9.40, $\tau \cdot \pi := \{\tau(V) \mid V \text{ block of } \pi\}$).

(2) Suppose that $p > 1$. Prove that $\sigma_{p-1,q}^{(n)} \cdot \pi \in NC(n)$.

(3) Suppose that $q < n$. Prove that $\sigma_{p,q+1}^{(n)} \cdot \pi \in NC(n)$.

EXERCISE 19.32. Let (π, ε) be an element of the set $\mathcal{X}_n$ defined in Proposition 19.24. Consider the partition $\pi|A(\varepsilon) \in NCO(A(\varepsilon))$, and let $[p,q]$ be its odd-marked interval (as defined in Notation 19.30.1).

(1) Let $i \in \{1,\ldots,n\}$ be the ceiling of $p/2$ (that is, $i = p/2$ if p is even and $i = (p+1)/2$ if p is odd); similarly, let j be the ceiling of $q/2$. Consider the (unique) factorization, in sense of multiplication by concatenation:

$$\varepsilon = \varepsilon_- \varepsilon_0 \varepsilon_+,$$

where the lengths of $\varepsilon_-, \varepsilon_0, \varepsilon_+$ are equal to $i-1, j-i+1$ and $n-j$, respectively. Then consider the n-tuple

$$\varepsilon' := \varepsilon_- \varepsilon_0^* \varepsilon_+ \in \{1,2\}^n.$$

Prove that $t(\varepsilon')$ is not of the same parity as $t(\varepsilon)$.

(2) Consider the permutation τ of $\{1,\ldots,2n\}$ defined by:

$$\tau = \begin{cases} \sigma_{p,q}^{(2n)} & \text{if } p,q \text{ have different parities} \\ \sigma_{p-1,q}^{(2n)} & \text{if } p,q \text{ are both even} \\ \sigma_{p,q+1}^{(2n)} & \text{if } p,q \text{ are both odd.} \end{cases}$$

Prove that $\tau(A(\varepsilon)) = A(\varepsilon')$ and that $\tau(B(\varepsilon)) = B(\varepsilon')$, where ε' is the n-tuple defined in the part (1) of the exercise.

(3) Let $\pi' := \tau \cdot \pi$, where τ is as in part (2) of the exercise. Prove that (π', ε') belongs to the set $\mathcal{X}_n$ defined in Proposition 19.24.

EXERCISE 19.33. Consider the set $\mathcal{X}_n$ defined in Proposition 19.24, and for every element $(\pi, \varepsilon) \in \mathcal{X}_n$ define $\Phi_n(\pi, \varepsilon) := (\pi', \varepsilon')$, with π' and ε' constructed as in Exercise 19.32. Prove that the map $\Phi_n : \mathcal{X}_n \to \mathcal{X}_n$

which is defined in this way has the properties stated in Proposition 19.24.

LECTURE 20

R-cyclic matrices

In this lecture we look at the natural operation of taking $d \times d$ matrices over a non-commutative probability space $(\mathcal{A}, \varphi)$. If $A = (a_{ij})_{1 \leq i,j \leq d}$ is a matrix in $M_d(\mathcal{A})$, then it is immediate (from the definitions) that the distribution of A in $M_d(\mathcal{A})$ is completely determined by the joint distribution of the family $\{a_{ij} \mid 1 \leq i, j \leq d\}$ in $(\mathcal{A}, \varphi)$; but, as a rule, there is no explicit formula for computing the distribution of A from the joint distribution of the a_{ij}. A notable exception to this rule was observed early on by Voiculescu, and occurs when a free family of circular/semicircular elements is used in order to build up a selfadjoint matrix A – in this case the matrix A itself turns out to have a semicircular distribution in $M_d(\mathcal{A})$.

In this lecture we introduce the concept of *R-cyclic matrix*, which is a generalization of the situation of the matrix with free circular/semicircular entries. The definition of R-cyclicity is in terms of the joint R-transform of the entries of the matrix: one requires that only the cyclic non-crossing cumulants of the entries are allowed to be different from 0. We show that for an R-cyclic matrix $A = (a_{ij})_{1 \leq i,j \leq d}$ one has an explicit $\boxed{\star}$-convolution formula for computing the distribution of A (considered in $M_d(\mathcal{A})$) in terms of the joint distribution of the a_{ij} (considered in the original space $(\mathcal{A}, \varphi)$).

The discussion extends without difficulty to the situation when one considers several matrices $A_1, \ldots, A_s \in M_d(\mathcal{A})$. Thus one has the concept of *R-cyclic family of matrices*, and a convolution formula for computing the joint distribution of such a family of matrices, in terms of the joint distribution of their entries. Several important situations of families of matrices with tractable joint distributions arise by applying this formula.

Definition and examples of R-cyclic matrices

Let us first recall what we mean when we talk about "$d \times d$ matrices over a non-commutative probability space" (cf. also Exercise 1.23).

NOTATION 20.1. Let $(\mathcal{A}, \varphi)$ be a non-commutative probability space, and let d be a positive integer. Consider the algebra $M_d(\mathcal{A})$

of $d \times d$ matrices over $\mathcal{A}$. We denote by φ_d the linear functional on $M_d(\mathcal{A})$ defined by the formula:

$$\varphi_d\big((a_{ij})_{i,j=1}^d\big) = \frac{1}{d}\sum_{i=1}^{d} \varphi(a_{ii}). \tag{20.1}$$

Then $(M_d(\mathcal{A}), \varphi_d)$ is itself a non-commutative probability space.

DEFINITION 20.2. Let $(\mathcal{A}, \varphi)$ and d be as above, and let $A = (a_{ij})_{i,j=1}^d$ be a matrix in $M_d(\mathcal{A})$.

(1) A is said to be ***R*-cyclic** if the following condition holds:

$$\kappa_n(a_{i_1 j_1}, \dots, a_{i_n j_n}) = 0 \tag{20.2}$$

for every $n \geq 1$ and every $1 \leq i_1, j_1, \dots, i_n, j_n \leq d$ for which it is not true that $j_1 = i_2, \dots, j_{n-1} = i_n, j_n = i_1$.

(2) If the matrix A is R-cyclic, then the series:

$$f(z_1, \dots, z_d) := \sum_{n=1}^{\infty} \sum_{i_1,\dots,i_n=1}^{d} \kappa_n(a_{i_n i_1}, a_{i_1 i_2}, \dots, a_{i_{n-1} i_n}) z_{i_1} z_{i_2} \cdots z_{i_n}$$

is called the **determining series of the entries** of A. (Note that $f \in \Theta_d$, where Θ_d is as in Notations 16.1.)

EXAMPLE 20.3. Consider a diagonal matrix,

$$A := \begin{pmatrix} a_1 & & 0 \\ & \ddots & \\ 0 & & a_d \end{pmatrix} \in M_d(\mathcal{A}),$$

where $(\mathcal{A}, \varphi)$ and d are as above. Then A is R-cyclic if and only if $a_1, \dots, a_d$ form a freely independent family. Indeed, the R-cyclicity condition (20.2) is spelled out here as follows: $\kappa_n(a_{i_1}, \dots, a_{i_n}) = 0$ whenever it is not true that $i_1 = i_2 = \cdots = i_n$. But in view of Theorem 11.20, the latter condition is in turn equivalent to the free independence of $a_1, \dots, a_d$.

For more elaborate examples we will use the framework of a $*$-probability space, and we will focus on selfadjoint matrices over such a space.

EXAMPLE 20.4. Let $(\mathcal{A}, \varphi)$ be a $*$-probability space, and let $\{e_{ij} \mid 1 \leq i, j \leq d\}$ be a family of matrix units, i.e. a family of elements of $\mathcal{A}$ which satisfy the following relations: $e_{ij}^* = e_{ji}$ for all $1 \leq i, j \leq d$, $e_{ij} e_{kl} = \delta_{j,k} e_{il}$ for all $1 \leq i, j, k, l \leq d$, and $\sum_{i=1}^d e_{ii} = 1_{\mathcal{A}}$. We will assume in addition that $\varphi(e_{ij}) = 0$ whenever $i \neq j$, and that

$\varphi(e_{11}) = \cdots = \varphi(e_{dd}) = 1/d$. We denote by $(\mathcal{C}, \psi)$ the compression of $(\mathcal{A}, \varphi)$ by e_{11}, i.e.

$$\mathcal{C} := e_{11}\mathcal{A}e_{11}, \quad \psi := d \cdot \varphi|\mathcal{C}.$$

Let now a be a selfadjoint element of $\mathcal{A}$, which is free from $\{e_{ij} \mid 1 \leq i, j \leq d\}$. We compress a by the matrix unit formed by the e_{ij}, and we move the compressions under the projection e_{11}; that is, we consider the family of elements

$$c_{ij} \ := \ e_{1i}ae_{j1} \in \mathcal{C}, \quad 1 \leq i, j \leq d.$$

Theorem 14.18 gives us an explicit formula for the free cumulants of the family $\{c_{ij} \mid 1 \leq i, j \leq d\}$. Namely, for every $n \geq 1$ and $1 \leq i_1, j_1, \ldots, i_n, j_n \leq d$, we have that $\kappa_n(c_{i_1j_1}, \ldots, c_{i_nj_n})$ equals:

$$\begin{cases} d^{-(n-1)}\kappa_n(a, \ldots, a) & \text{if } j_1 = i_2, \ldots, j_{n-1} = i_n, j_n = i_1 \\ \\ 0 & \text{otherwise.} \end{cases}$$

In other words, the matrix $C = (c_{ij})_{i,j=1}^d \in M_d(\mathcal{C})$ is R-cyclic, and the determining series of its entries is:

$$\begin{aligned} f(z_1, \ldots, z_d) &= \sum_{n=1}^{\infty} \sum_{i_1,\ldots,i_n=1}^{d} d^{-(n-1)}k_n(a, \ldots, a)z_{i_1} \cdots z_{i_n} \\ &= d \cdot \sum_{n=1}^{\infty} k_n(a, \ldots, a) \cdot \Big(\frac{z_1 + \cdots + z_d}{d}\Big)^n \\ &= d \cdot R_a\Big(\frac{z_1 + \cdots + z_d}{d} \Big), \end{aligned}$$

where R_a is the R-transform of a, in the original space $(\mathcal{A}, \varphi)$.

EXAMPLE 20.5. Let $(\mathcal{A}, \varphi)$ be a $*$-probability space, and let $a \in \mathcal{A}$ be an R-diagonal element (as in Lecture 15). Consider the non-commutative probability space $(M_2(\mathcal{A}), \varphi_2)$ defined as in Notation 20.1, and the selfadjoint matrix:

$$A \ = \ \begin{pmatrix} 0 & a^* \\ a & 0 \end{pmatrix} \ \in \ M_2(\mathcal{A}).$$

One immediately checks that A is R-cyclic (and in fact also that conversely, the R-cyclicity of A implies the R-diagonality of a). Moreover, the determining series of the entries of A coincides here with the joint R-transform R_{a,a^*} (as appearing for instance in Equation (16.13) of Example 16.9). A number of results about R-diagonal elements can be incorporated in the theory of R-cyclic matrices by using this trick.

EXAMPLE 20.6. On the lines of Example 20.5, one can consider the situation of a more general selfadjoint matrix with free R-diagonal entries. More precisely, let $(\mathcal{A}, \varphi)$ be a $*$-probability space, let d be a positive integer, and suppose that the elements $\{a_{ij} \mid 1 \le i, j \le d\}$ of $\mathcal{A}$ have the following properties:

(i) $a_{ij}^* = a_{ji}, \forall\, 1 \le i, j \le d$;

(ii) a_{ij} is R-diagonal whenever $i \ne j$;

(iii) the $d(d+1)/2$ families: $\{a_{ii}\}$ for $1 \le i \le d$, together with $\{a_{ij}, a_{ji}\}$ for $1 \le i < j \le d$, are free in $(\mathcal{A}, \varphi)$.

Then the matrix $A := (a_{ij})_{i,j=1}^d \in M_d(\mathcal{A})$ is R-cyclic. Indeed, the freeness condition (iii) combined with the R-diagonality of a_{ij} for $i \ne j$ implies that the only free cumulants made with the entries of A which could possibly be non-zero are:

$$\begin{cases} \kappa_n(a_{ii}, \ldots, a_{ii}) & \text{with } n \ge 1,\ 1 \le i \le d, \text{and} \\ \kappa_n(a_{ij}, a_{ji}, \ldots, a_{ij}, a_{ji}) & \text{with } n \ge 1 \text{ even},\ 1 \le i, j \le d,\ i \ne j; \end{cases}$$

all these cumulants fall within the pattern allowed by the definition of R-cyclicity.

An important particular case of the situation described above is the one when the a_{ij} are circular and semicircular. In order to describe this particular case, it is convenient to allow "rescaled" circular elements; that is, we will say that an element c in a $*$-probability space $(\mathcal{A}, \varphi)$ is *circular of radius* r if $\frac{2}{r}c$ is a circular element in the sense of Definition 11.22. (In other words, circular elements of arbitrary positive radius are defined such that the radius is scaling linearly with the element, and such that the "standard" circular element from Definition 11.22 has radius 2.) With this convention, one obtains an example of R-cyclic matrix $A = (a_{ij})_{i,j=1}^d$ by taking every a_{ii} to be semicircular (of some radius r_{ii}), by taking every a_{ij} with $i \ne j$ to be circular (of some radius r_{ij}), and by asking that the above conditions (i), (iii) are satisfied.

The convolution formula for an R-cyclic matrix

In this section we show how the distribution of an R-cyclic matrix can be obtained from the determining series of the entries of the matrix.

We start by introducing a few notations.

NOTATIONS 20.7. Let d be a positive integer. We will denote:

$$G_d(z_1, \ldots, z_d) = \frac{1}{d} \sum_{n=1}^{\infty} \sum_{i=1}^{d} z_i^n = \frac{1}{d}\Big(\frac{z_1}{1-z_1} + \cdots + \frac{z_d}{1-z_d}\Big),$$

and we will also denote

$$H_d := G_d \boxstar \text{Möb}_d$$

(where $\text{Möb}_d \in \Theta_d$ is the Möbius series in d indeterminates, as discussed in Lecture 17).

REMARKS 20.8. (1) A way of looking at the series G_d and H_d is to observe how they appear in the framework of the non-commutative probability space $(M_d(\mathbb{C}), \text{tr}_d)$, where tr_d is the normalized trace. Consider the matrices $E_1, \ldots, E_d \in M_d(\mathbb{C})$ where E_i has its (i,i)-entry equal to 1 and all the other entries equal to 0. It is obvious that the moment series $M_{E_1,\ldots,E_d}$ is equal to G_d; as a consequence, H_d has to be equal to the corresponding R-transform:

$$H_d = M_{E_1,\ldots,E_d} \boxed{\star} \text{Möb}_d \;=\; R_{E_1,\ldots,E_d}. \tag{20.3}$$

The series H_d plays a key role in the next theorem. In order to give an idea of how it looks, here is its truncation to order three:

$$H_d(z_1, \ldots, z_d) \;=\; \sum_{i=1}^{d} \frac{1}{d} z_i \;+\; \sum_{i_1,i_2=1}^{d} \frac{1}{d}\Big(\delta_{i_1,i_2} - \frac{1}{d}\Big) z_{i_1} z_{i_2}$$

$$+ \sum_{i_1,i_2,i_3=1}^{d} \frac{1}{d}\Big(\delta_{i_1,i_2,i_3} - \frac{1}{d}(\delta_{i_1,i_2} + \delta_{i_1,i_3} + \delta_{i_2,i_3}) + \frac{2}{d^2}\Big) z_{i_1} z_{i_2} z_{i_3} \;+ \cdots$$

(where δ_{i_1,i_2} is the usual Kronecker symbol, while δ_{i_1,i_2,i_3} is equal to 1 when $i_1 = i_2 = i_3$ and is equal to 0 otherwise).

(2) An application of Equation (20.3) which will be used later in the lecture is that for every $n \geq 2$, $k \in \{1, \ldots, n\}$, and for fixed indices $i_1, \ldots, i_{k-1}, i_{k+1}, \ldots, i_n \in \{1, \ldots d\}$, we have:

$$\sum_{i=1}^{d} \text{Cf}_{(i_1,\ldots,i_{k-1},i,i_{k+1},\ldots,i_n)}(H_d) = 0. \tag{20.4}$$

Indeed, the sum on the left-hand side of (20.4) can be interpreted as $\sum_{i=1}^{d} \kappa_n(E_{i_1}, \ldots, E_{i_{k-1}}, E_i, E_{i_{k+1}}, \ldots, E_{i_n})$, and is hence equal to the cumulant $\kappa_n(E_{i_1}, \ldots, E_{i_{k-1}}, I_d, E_{i_{k+1}}, \ldots, E_{i_n})$ (by the multilinearity of κ_n, and where I_d is the unit $d \times d$ matrix). But the latter quantity vanishes by Proposition 11.15 in Lecture 11.

We can now state the main result of this section.

THEOREM 20.9. *Let $(\mathcal{A}, \varphi)$ be a non-commutative probability space, let $A = (a_{ij})_{1 \leq i,j \leq d} \in M_d(\mathcal{A})$ be an R-cyclic matrix, and let $f \in \Theta_d$ be the determining series of the entries of A. Then the moment series and R-transform of A in $(M_d(\mathcal{A}), \varphi_d)$ can be computed by the formulas:*

$$M_A(z) \;=\; \frac{1}{d}(f \boxed{\star} dG_d)(\underbrace{z, \ldots, z}_{d \text{ times}}), \tag{20.5}$$

and

$$R_A(z) = \frac{1}{d}(f \boxed{\star} dH_d)(\underbrace{z, \ldots, z}_{d \text{ times}}), \tag{20.6}$$

with G_d and H_d as defined in Notations 20.7.

In the proof of the theorem, we will use the following lemma.

LEMMA 20.10. *Consider the framework of Theorem 20.9. Let n be a positive integer, let π be in $NC(n)$, and consider some indices $1 \leq i_1, \ldots, i_n \leq d$. Then we have the equality:*

$$\kappa_\pi(\, a_{i_n i_1}, a_{i_1 i_2}, \ldots, a_{i_{n-1} i_n}\,) = \text{Cf}_{(i_1,\ldots,i_n);\pi}(f) \cdot \text{Cf}_{(i_1,\ldots,i_n);K(\pi)}(dG_d). \tag{20.7}$$

PROOF. We will use cyclic notations modulo n for indices – i.e. "i_{k+1}" will mean "i_1" if $k = n$ and "i_{k-1}" will mean "i_n" if $k = 1$. Also, we will work with the permutations P_π and $P_{K(\pi)}$ which are associated to π and respectively to $K(\pi)$. These permutations are defined as in the exercise section of Lecture 18 (cf. Notations 18.24), and they satisfy the relation

$$P_{K(\pi)} = P_\pi^{-1} \cdot (1, 2, \ldots, n) \tag{20.8}$$

(which is a particular case of Exercise 18.25).

Since every coefficient of dG_d is equal either to 0 or to 1, the generalized coefficient of dG_d appearing on the right-hand side of (20.7) is also equal to 0 or 1. So we have two cases.

Case 1. $\text{Cf}_{(i_1,\ldots,i_n);K(\pi)}(dG_d) = 1$.

By writing explicitly what the generalized coefficient of dG_d is, we find that:

$$\left\{ \begin{array}{c} 1 \leq k, l \leq n, \\ k, l \text{ in the same block of } K(\pi) \end{array} \right\} \Longrightarrow i_k = i_l. \tag{20.9}$$

Under this assumption, we have to show that:

$$\kappa_\pi(a_{i_n i_1}, a_{i_1 i_2}, \ldots, a_{i_{n-1} i_n}) = \text{Cf}_{(i_1,\ldots,i_n);\pi}(f). \tag{20.10}$$

Each of the two sides of (20.10) is a product of factors indexed by the blocks of π; we will prove (20.10) by showing that actually for any given block V of π, the factor corresponding to V on the left-hand side is equal to the factor corresponding to V on the right-hand side.

So let us fix a block $V = \{k(1) < k(2) < \cdots < k(p)\}$ of π. The factor corresponding to V on the left-hand side of (20.10) is:

$$\kappa_p(a_{i_{k(1)-1} i_{k(1)}}, a_{i_{k(2)-1} i_{k(2)}}, \ldots, a_{i_{k(p)-1} i_{k(p)}}) \tag{20.11}$$

(where recall that if $k_1 = 1$, then we use i_n for "$i_{k(1)-1}$"). On the other hand, the factor corresponding to V on the right-hand side of (20.10) is $\mathrm{Cf}_{(i_{k(1)},\ldots,i_{k(p)})}(f)$, i.e.

$$\kappa_p(a_{i_{k(p)}i_{k(1)}}, a_{i_{k(1)}i_{k(2)}}, \ldots, a_{i_{k(p-1)}i_{k(p)}}). \tag{20.12}$$

But now let us notice that $k(1)$ and $k(2) - 1$ belong to the same block of $K(\pi)$, and the same for $k(2)$ and $k(3) - 1, \ldots$, the same for $k(p)$ and $k(1) - 1$. This is easily seen by looking at the permutations P_π and $P_{K(\pi)}$: we have

$$P_\pi(k(1)) = k(2), \quad \ldots, \quad P_\pi(k(p-1)) = k(p), \quad P_\pi(k(p)) = k(1),$$

so from Equation (20.8) we get that:

$$P_{K(\pi)}(k(2) - 1) = k(1), \quad \ldots, \quad P_{K(\pi)}(k(1) - 1) = k(p).$$

As a consequence of this remark and of the implication stated in (20.9) we see that the expressions appearing in (20.11) and (20.12) are in fact identical.

Case 2. $\mathrm{Cf}_{(i_1,\ldots,i_n);K(\pi)}(dG_d) = 0$.

In this case we know that (20.9) does not hold, and we have to show that the left-hand side of (20.7) is equal to 0.

It is immediate that, under the current assumption, we can find $1 \le k, l \le n$ such that:

$$P_{K(\pi)}(l) = k, \quad \text{and } i_k \ne i_l. \tag{20.13}$$

Indeed, if it were true that $i_k = i_l$ whenever $P_{K(\pi)}(l) = k$, then by moving along the cycles of $P_{K(\pi)}$ we would find that (20.9) holds.

By taking into account the fact that $P_\pi \cdot P_{K(\pi)}$ is the long cycle $(1, \ldots, n)$, we then immediately see that for k, l as in (20.13) we must also have $P_\pi(k) = l + 1$. Hence k and $l + 1$ belong to the same block V of π; and moreover, if the block V is written as $V = \{k(1) < k(2) < \cdots < k(p)\}$, then there exists an index j, $1 \le j \le p$ such that $k = k(j)$ and $l + 1 = k(j + 1)$ (with the convention that if $k = k(p)$, then $l + 1 = k(1)$). But then the fact that $i_k \ne i_l$ reads $i_{k(j)} \ne i_{k(j+1)-1}$, which in turn implies that

$$\kappa_p(a_{i_{k(1)-1}i_{k(1)}}, a_{i_{k(2)-1}i_{k(2)}}, \ldots, a_{i_{k(p)-1}i_{k(p)}}) = 0$$

(by the definition of R-cyclicity). Since the latter expression is the factor corresponding to V in the product defining $\kappa_\pi(a_{i_n i_1}, a_{i_1 i_2}, \ldots, a_{i_{n-1}i_n})$, we conclude that the left-hand side of (20.7) is indeed equal to 0. □

We can now give the proof of Theorem 20.9.

PROOF OF THEOREM 20.9. Let n be a positive integer, and consider some indices $1 \leq i_1, \ldots, i_n \leq d$. By summing over $\pi \in NC(n)$ in Equation (20.7) of Lemma 20.10 (and by taking into account the relation between moments and cumulants, and the definition of $\boxed{\star}$), we get:

$$\varphi(a_{i_n i_1} a_{i_1 i_2} \cdots a_{i_{n-1} i_n}) = \mathrm{Cf}_{(i_1,\ldots,i_n)}(f \boxed{\star}\, dG_d). \tag{20.14}$$

For every $1 \leq i \leq d$, let us denote by $E_i \in M_d(\mathcal{A})$ the matrix which has $1_{\mathcal{A}}$ on the (i,i)-entry, and has all the other entries equal to 0. It is immediately verified that

$$\varphi_d(AE_{i_1} \cdots AE_{i_n}) = \frac{1}{d}\varphi(a_{i_n i_1} a_{i_1 i_2} \cdots a_{i_{n-1} i_n}),$$

for every $n \geq 1$ and $1 \leq i_1, \ldots, i_n \leq d$. By combining this with (20.14), we get that

$$M_{AE_1,\ldots,AE_d} = \frac{1}{d}(f \boxed{\star}\, dG_d) \tag{20.15}$$

(equality of power series from Θ_d). Equation (20.5) is an immediate consequence of (20.15), since for every $n \geq 1$:

$$\begin{aligned}\varphi_d(A^n) &= \sum_{i_1,\ldots,i_n=1}^{d} \varphi_d(AE_{i_1} \cdots AE_{i_n}) \\ &= \sum_{i_1,\ldots,i_n=1}^{d} \mathrm{Cf}_{(i_1,\ldots,i_n)}(M_{AE_1,\ldots,AE_d}) \\ &= \frac{1}{d} \sum_{i_1,\ldots,i_n=1}^{d} \mathrm{Cf}_{(i_1,\ldots,i_n)}(f \boxed{\star}\, dG_d);\end{aligned}$$

and the latter quantity is easily identified as the coefficient of z^n in the series on the right-hand side of (20.5).

On the other hand let us $\boxed{\star}$-convolve with Möb_d on the right, on both sides of (20.15). On the left-hand side we get $R_{AE_1,\ldots,AE_d}$, while on the right-hand side we get $\left(\frac{1}{d}(f \boxed{\star}\, dG_d) \right) \boxed{\star}\, \text{Möb}_d$. But in view of Exercise 17.26.2 we can replace $\frac{1}{d}(f \boxed{\star}\, dG_d)$ with $\left(\frac{1}{d} f \boxed{\star}\, G_d \right) \circ D_d$, and thus we find that:

$$\begin{aligned}\Big(\frac{1}{d}(f \boxed{\star}\, dG_d)\Big) \boxed{\star}\, \text{Möb}_d &= \Big(\big(\frac{1}{d} f \boxed{\star}\, G_d\big) \circ D_d \Big) \boxed{\star}\, \text{Möb}_d \\ &= \Big(\frac{1}{d} f \boxed{\star}\, G_d \boxed{\star}\, \text{Möb}_d\Big) \circ D_d \quad \text{(by Ex. 17.26.1)} \\ &= \Big(\frac{1}{d} f \boxed{\star}\, H_d\Big) \circ D_d \quad (\text{since } G_d \boxed{\star}\, \text{Möb}_d = H_d) \\ &= \frac{1}{d}(f \boxed{\star}\, dH_d) \quad \text{(by Exercise 17.26.2)}.\end{aligned}$$

So we come to the equation:

$$R_{AE_1,\ldots,AE_d} = \frac{1}{d}(f \boxed{\star}\, dH_d), \tag{20.16}$$

out of which (20.6) is obtained in the same way as (20.5) was obtained from (20.15). $\square$

REMARK 20.11. During the proof of Theorem 20.9 we obtain Equations (20.15) and (20.16), stronger than what was originally stated, and which show better the significance of the series $f \boxed{\star}\, dG_d$ and $f \boxed{\star}\, dH_d$.

R-cyclic families of matrices

DEFINITION 20.12. Let $(\mathcal{A}, \varphi)$ be a non-commutative probability space, and let d be a positive integer. Let $A_1 = (a_{ij}^{(1)})_{i,j=1}^d, \ldots, A_s = (a_{ij}^{(s)})_{i,j=1}^d$ be matrices in $M_d(\mathcal{A})$. We say that the family $A_1, \ldots, A_s$ is ***R*-cyclic** if the following condition holds:

$$\kappa_n(a_{i_1 j_1}^{(r_1)}, \ldots, a_{i_n j_n}^{(r_n)}) \;=\; 0,$$

for every $n \geq 1$, every $1 \leq r_1, \ldots, r_n \leq s$, and every $1 \leq i_1, j_1, \ldots, i_n, j_n \leq d$ for which it is not true that $j_1 = i_2, \ldots, j_{n-1} = i_n, j_n = i_1$.

If the family $A_1, \ldots, A_s$ is R-cyclic, then the power series in ds indeterminates:

$$f(z_{1,1}, \ldots, z_{s,d}) \;:= \tag{20.17}$$

$$\sum_{n=1}^{\infty} \sum_{\substack{1 \leq i_1, \ldots, i_n \leq d \\ 1 \leq r_1, \ldots, r_n \leq s}} \kappa_n(\, a_{i_n i_1}^{(r_1)}, a_{i_1 i_2}^{(r_2)}, \ldots, a_{i_{n-1} i_n}^{(r_n)} \,) \cdot z_{r_1,i_1} z_{r_2,i_2} \cdots z_{r_n,i_n}$$

is called the **determining series of the entries** of the family.

The convolution formula presented in the preceding section extends, with minor adjustments, to the situation when we have a family of matrices. In order to state precisely how this goes, we will need to use an adjusted version of the operation $\boxed{\star}$, which is discussed in the next remark.

REMARK 20.13. Let s and d be positive integers. Consider the set Θ_{sd} of power series in sd non-commuting indeterminates $z_{1,1}, \ldots, z_{r,i}, \ldots, z_{s,d}$. The formula defining $\boxed{\star}$ in Lecture 17 (cf. Definition 17.1) can be used to define a "convolution operation," denoted in what follows by $\widetilde{\boxed{\star}}$, which gives a right action of Θ_d on Θ_{sd}. More precisely, if $f \in \Theta_{sd}$ and $g \in \Theta_d$ then we define $f \,\widetilde{\boxed{\star}}\, g \in \Theta_{sd}$ by the following formula:

$$\mathrm{Cf}_{((r_1,i_1),\ldots,(r_n,i_n))}(f \,\widetilde{\boxed{\star}}\, g) := \sum_{\pi \in NC(n)} \mathrm{Cf}_{((r_1,i_1),\ldots,(r_n,i_n));\pi}(f) \cdot \mathrm{Cf}_{(i_1,\ldots,i_n);K(\pi)}(g), \tag{20.18}$$

holding for every $n \geq 1$ and for every $1 \leq r_1, \ldots, r_n \leq s$, $1 \leq i_1, \ldots, i_n \leq d$. Some straightforward adjustments of the arguments presented for $\boxed{\star}$ in Lecture 17 show that $\widetilde{\boxed{\star}}$ is indeed a right action of Θ_d on Θ_{sd}, in the sense that the equation

$$(f \,\widetilde{\boxed{\star}}\, g) \,\widetilde{\boxed{\star}}\, h = f \,\widetilde{\boxed{\star}}\, (g \boxed{\star} h) \tag{20.19}$$

holds for every $f \in \Theta_{sd}$ and $g, h \in \Theta_d$.

Let us also record the fact that:

$$f \,\widetilde{\boxed{\star}}\, \mathrm{Zeta}_d = f \boxed{\star} \mathrm{Zeta}_{sd}, \quad \forall\, f \in \Theta_{sd} \tag{20.20}$$

(where on the right-hand side of (20.20), $\boxed{\star}$ denotes the boxed convolution operation on Θ_{sd}). This relation is obvious if one takes into account the fact that Zeta series have all the coefficients equal to 1.

From (20.19) and (20.20) it is immediate that one also has:

$$f \,\widetilde{\boxed{\star}}\, \mathrm{M\ddot{o}b}_d = f \boxed{\star} \mathrm{M\ddot{o}b}_{sd}, \quad \forall\, f \in \Theta_{sd}. \tag{20.21}$$

Note that, as a consequence, we can write the relation

$$M_{a_{1,1},\ldots,a_{r,i},\ldots,a_{s,d}} \,\widetilde{\boxed{\star}}\, \mathrm{M\ddot{o}b}_d = R_{a_{1,1},\ldots,a_{r,i},\ldots,a_{s,d}}, \tag{20.22}$$

holding for any family $\{a_{r,i} \mid 1 \leq r \leq s,\ 1 \leq i \leq d\}$ of elements in some non-commutative probability space $(\mathcal{A}, \varphi)$.

The multi-matrix version of Theorem 20.9 is then stated as follows.

THEOREM 20.14. *Let $(\mathcal{A}, \varphi)$ be a non-commutative probability space. Let $A_1, \ldots, A_s$ be an R-cyclic family of matrices in $M_d(\mathcal{A})$, and let f be the determining series of the entries of this family. Then the joint moment series and R-transform of $(A_1, \ldots, A_s)$ in the non-commutative probability space $(M_d(\mathcal{A}), \varphi_d)$ can be computed by:*

$$M_{A_1,\ldots,A_s}(z_1, \ldots, z_s) = \frac{1}{d}(f \,\widetilde{\boxed{\star}}\, dG_d)(\underbrace{z_1, \ldots, z_1}_{d\ times}, \ldots, \underbrace{z_s, \ldots, z_s}_{d\ times}), \tag{20.23}$$

and

$$R_{A_1,\ldots,A_s}(z_1, \ldots, z_s) = \frac{1}{d}(f \,\widetilde{\boxed{\star}}\, dH_d)(\underbrace{z_1, \ldots, z_1}_{d\ times}, \ldots, \underbrace{z_s, \ldots, z_s}_{d\ times}), \tag{20.24}$$

where the operation $\widetilde{\boxed{\star}}$ *is as described in the preceding remark, and where* G_d *and* H_d *are as in Notations 20.7.*

The proof of Theorem 20.14 is obtained by adjusting in a straightforward way the proof which was shown for Theorem 20.9 in the preceding section – see Exercises 20.20–20.22 at the end of the lecture.

Applications of the convolution formula

In this section we look at R-cyclic families $A_1, \ldots, A_s$ of selfadjoint $d \times d$ matrices over a $*$-probability space $(\mathcal{A}, \varphi)$; we would like to put into evidence some non-trivial situations when $A_1, \ldots, A_s$ form a free family in $(M_d(\mathcal{A}), \varphi_d)$, and where the individual R-transform of each of $A_1, \ldots, A_s$ can be explicitly determined. The main point of the section is to observe that situations of this kind appear whenever we have a "partial summation condition," as described in the next proposition.

PROPOSITION 20.15. *Let* $(\mathcal{A}, \varphi)$ *be a* $*$*-probability space, and let* d, s *be positive integers. Let* $A_1, \ldots, A_s$ *be an* R*-cyclic family of selfadjoint matrices in* $M_d(\mathcal{A})$*, and let* f *denote the determining series of the entries of this family. Suppose that for every* $n \geq 1$ *and every* $1 \leq r_1, \ldots, r_n \leq s$, $1 \leq i_1, \ldots, i_n \leq d$*, the sum*

$$\sum_{i_1,\ldots,i_{n-1}=1}^{d} \operatorname{Cf}_{((r_1,i_1),\ldots,(r_{n-1},i_{n-1}),(r_n,i_n))}(f) \;=:\; \lambda_{r_1,\ldots,r_n} \tag{20.25}$$

does not depend on i_n *(even though the sum is only over* $i_1, \ldots, i_{n-1}$*). Then:*

$$R_{A_1,\ldots,A_s}(z_1, \ldots, z_s) \;=\; \sum_{n=1}^{\infty} \sum_{r_1,\ldots,r_n=1}^{s} \lambda_{r_1,\ldots,r_n} z_{r_1} \cdots z_{r_n}. \tag{20.26}$$

PROOF. Equation (20.26) is equivalent to the fact that for every $n \geq 1$ and every $1 \leq r_1, \ldots, r_n \leq s$ we have:

$$\kappa_n(A_{r_1}, \ldots, A_{r_n}) \;=\; \lambda_{r_1,\ldots,r_n}. \tag{20.27}$$

We fix n and $r_1, \ldots, r_n$ for which we will show that (20.27) is true. The case when $n = 1$ is immediate (and left as an exercise to the reader); we will assume that $n \geq 2$.

By taking the coefficient of $z_{r_1} \cdots z_{r_n}$ on both sides of Equation (20.24) we find that the free cumulant $\kappa_n(A_{r_1}, \ldots, A_{r_n})$ is equal to

$$\frac{1}{d} \sum_{i_1,\ldots,i_n=1}^{d} \sum_{\pi \in NC(n)} \operatorname{Cf}_{((r_1,i_1),\ldots,(r_n,i_n));\pi}(f) \cdot \operatorname{Cf}_{(i_1,\ldots,i_n);K(\pi)}(dH_d).$$

We will write this in the form:

$$\kappa_n(A_{r_1}, \ldots, A_{r_n}) = \sum_{\pi \in NC(n)} T_\pi, \tag{20.28}$$

where for every $\pi \in NC(n)$ we set:

$$T_\pi := \frac{1}{d} \sum_{i_1,\ldots,i_n=1}^{d} \mathrm{Cf}_{((r_1,i_1),\ldots,(r_n,i_n));\pi}(f) \cdot \mathrm{Cf}_{(i_1,\ldots,i_n);K(\pi)}(dH_d).$$

We first consider the quantity T_π in the special case when $\pi = 1_n$, the partition of $\{1, \ldots, n\}$ which has only one block. In this case $K(\pi)$ is the partition into n blocks of one element; since all the coefficients of degree 1 of dH_d are equal to 1, it follows that

$$\mathrm{Cf}_{(i_1,\ldots,i_n);K(1_n)}(dH_d) = 1, \quad \forall\, i_1, \ldots, i_n \in \{1, \ldots, d\}.$$

We hence get:

$$T_{1_n} = \frac{1}{d} \sum_{i_1,\ldots,i_n=1}^{d} \mathrm{Cf}_{((r_1,i_1),\ldots,(r_n,i_n))}(f).$$

The partial summation property of the series f (given in Equation (20.25)) implies that the latter sum is equal to $\lambda_{r_1,\ldots,r_n}$. Thus, in view of (20.28), the proof will be over if we can show that $T_\pi = 0$ for every $\pi \neq 1_n$ in $NC(n)$.

So for the rest of the proof we fix a partition $\pi \neq 1_n$ in $NC(n)$. Moreover, we will also fix a block B_o of π which is an interval, $B_o = \{p, \ldots, q\}$ with $1 \leq p \leq q \leq n$ (every non-crossing partition has such a block). The considerations below, leading to the conclusion that $T_\pi = 0$, will be made by looking at the case when B_o has more than one element; the case when $|B_o| = 1$ (which is similar, and easier) is left as an exercise to the reader. We denote by "Rest" the set of blocks of π which are different from B_o.

Let us now look at the Kreweras complement $K(\pi)$. It is immediate that $\{p\}, \{p+1\}, \ldots, \{q-1\}$ are one-element blocks of $K(\pi)$. We denote by B_o' the block of $K(\pi)$ which contains q; observe that B_o' has more than one element – indeed, it is clear that $p-1$ also belongs to B_o' (where if $p = 1$, then "$p-1$" means "n"; even in this case we have that $p - 1 \neq q$, since it was assumed that $\pi \neq 1_n$). Let us denote by Rest$'$ the set of blocks of $K(\pi)$ (if any) which remain after $\{p\}$, $\{p+1\}, \ldots, \{q-1\}$ and B_o' are deleted.

For any $i_1, \ldots, i_n \in \{1, \ldots, d\}$ we have:

$$\begin{aligned} \mathrm{Cf}_{((r_1,i_1),\ldots,(r_n,i_n));\pi}(f) \cdot \mathrm{Cf}_{(i_1,\ldots,i_n);K(\pi)}(dH_d) &= \\ \mathrm{Cf}_{((r_p,i_p),\ldots,(r_q,i_q))}(f) \cdot \mathrm{Cf}_{(i_1,\ldots,i_n)|B_o'}(dH_d)&\cdot \end{aligned} \tag{20.29}$$

$$\cdot \prod_{B \in \text{Rest}} \text{Cf}_{((r_1,i_1),\ldots,(r_n,i_n))|B}(f) \cdot \prod_{B' \in \text{Rest}'} \text{Cf}_{(i_1,\ldots,i_n)|B'}(dH_d)$$

(we took into account that the factors $\text{Cf}_{(i_p)}(dH_d), \ldots, \text{Cf}_{(i_{q-1})}(dH_d)$, which should also appear on the right-hand side of (20.29), are all equal to 1). The indices $i_p, \ldots, i_{q-1}$ appear only in the factor "$\text{Cf}_{((r_p,i_p),\ldots,(r_q,i_q))}(f)$" of (20.29). Thus, if in (20.29) we sum over i_p, $\ldots$, i_{q-1}, and make use of the partial summation property from (20.25), then we get:

$$\lambda_{r_p,\ldots,r_q} \cdot \text{Cf}_{(i_1,\ldots,i_n)|B'_o}(dH_d) \tag{20.30}$$

$$\cdot \prod_{B \in \text{Rest}} \text{Cf}_{((r_1,i_1),\ldots,(r_n,i_n))|B}(f) \cdot \prod_{B' \in \text{Rest}'} \text{Cf}_{(i_1,\ldots,i_n)|B'}(dH_d)$$

(expression depending on some arbitrary indices $i_1, \ldots, i_{p-1}, i_q, \ldots, i_n$, chosen from $\{1, \ldots, d\}$).

Next, in (20.30) we sum over the index i_q. The only factor in (20.30) which involves i_q is "$\text{Cf}_{(i_1,\ldots,i_n)|B'_o}(dH_d)$," so as a result of this new summation we get:

$$\lambda_{r_p,\ldots,r_q} \cdot \Big\{ \sum_{i_q=1}^{d} \text{Cf}_{(i_1,\ldots,i_n)|B'_o}(dH_d) \Big\}$$

$$\cdot \prod_{B \in \text{Rest}} \text{Cf}_{((r_1,i_1),\ldots,(r_n,i_n))|B}(f) \cdot \prod_{B' \in \text{Rest}'} \text{Cf}_{(i_1,\ldots,i_n)|B'}(dH_d).$$

But, as an immediate consequence of Equation (20.4) in Remark 20.8.2 we have that $\sum_{i_q=1}^{d} \text{Cf}_{(i_1,\ldots,i_n)|B'_o}(dH_d) = 0$.

The conclusion that we draw from the preceding three paragraphs is the following: for any choice of the indices $i_1, \ldots, i_{p-1}, i_{q+1}, \ldots, i_n \in \{1, \ldots, d\}$, we have that

$$\sum_{i_p,\ldots,i_q=1}^{d} \text{Cf}_{((r_1,i_1),\ldots,(r_n,i_n));\pi}(f) \cdot \text{Cf}_{(i_1,\ldots,i_n);K(\pi)}(dH_d) = 0.$$

It only remains that we sum over $i_1, \ldots, i_{p-1}, i_{q+1}, \ldots, i_n$ in the latter equation, to obtain the desired fact that $T_\pi = 0$. □

COROLLARY 20.16. *Let $(\mathcal{A}, \varphi)$ be a $*$-probability space, let d, s be positive integers, and let $A_1 = (a_{ij}^{(1)})_{i,j=1}^{d}, \ldots, A_s = (a_{ij}^{(s)})_{i,j=1}^{d}$ be an R-cyclic family of selfadjoint matrices in $M_d(\mathcal{A})$. Suppose that the s families of entries $\{a_{ij}^{(r)} \mid 1 \le i, j \le d\}$, with $1 \le r \le s$, are free in $(\mathcal{A}, \varphi)$. Moreover, for every $1 \le r \le s$ let $f_r \in \Theta_d$ be the determining* *series of the entries of* A_r. *We assume that for every* $n \geq 1$ *and for every* $1 \leq r \leq s$, $1 \leq i \leq d$, *the sum*

$$\sum_{i_1,\ldots,i_{n-1}=1}^{d} \mathrm{Cf}_{(i_1,\ldots,i_{n-1},i)}(f_r) \;=:\; \lambda_n^{(r)} \tag{20.31}$$

does not depend on the choice of i *(but only on* n *and* r*). Then* $A_1, \ldots, A_s$ *are free in* $(M_d(\mathcal{A}), \varphi_d)$, *and have R-transforms*

$$R_{A_r}(z) \;=\; \sum_{n=1}^{\infty} \lambda_n^{(r)} z^n, \quad 1 \leq r \leq s. \tag{20.32}$$

PROOF. Let f denote the determining series of the entries of the whole R-cyclic family $A_1, \ldots, A_s$. The condition of free independence between the families of entries of $A_1, \ldots, A_s$ implies the formula:

$$f(z_{1,1}, \ldots, z_{r,i}, \ldots, z_{s,d}) \;=\; \sum_{r=1}^{s} f_r(z_{r,1}, \ldots, z_{r,i}, \ldots, z_{r,d}),$$

where f_r is (as in the statement of the corollary) the determining series for just the entries of A_r. It is immediate that f satisfies the partial summation condition described in Equation (20.25) of Proposition 20.15, where we set:

$$\lambda_{r_1,\ldots,r_n} \;=\; \begin{cases} \lambda_n^{(r)} & \text{if } r_1 = \cdots = r_n = r \\ 0 & \text{otherwise.} \end{cases}$$

Thus Proposition 20.15 can be applied, and gives us:

$$R_{A_1,\ldots,A_s}(z_1, \ldots, z_s) \;=\; \sum_{r=1}^{s} \sum_{n=1}^{\infty} \lambda_n^{(r)} z_r^n.$$

In view of the description of free independence in terms of R-transforms (cf. Theorem 16.6), this is equivalent to saying that $A_1, \ldots, A_s$ are freely independent and have individual R-transforms as described in (20.32). □

Corollary 20.16 can be in turn particularized to the situation of a family of matrices with free R-diagonal entries (on the lines of Example 20.6). The precise wording of this particular case goes as follows.

COROLLARY 20.17. *Let* $(\mathcal{A}, \varphi)$ *be a tracial* $*$*-probability space, let* d, s *be positive integers, and suppose that the elements* $\{a_{ij}^{(r)} \mid 1 \leq i, j \leq d,\ 1 \leq r \leq s\}$ *of* $\mathcal{A}$ *have the following properties.*

(i) $(\, a_{ij}^{(r)}\,)^* = a_{ji}^{(r)}$, *for every* $1 \leq i, j \leq d$ *and* $1 \leq r \leq s$.

(ii) *For every* $1 \leq i, j \leq d$ *such that* $i \neq j$, *and for every* $1 \leq r \leq s$, *the element* $a_{ij}^{(r)}$ *is R-diagonal.*

(iii) *The* $sd(d+1)/2$ *families:* $\{a_{ii}^{(r)}\}$ *for* $1 \le i \le d$, $1 \le r \le s$, *together with* $\{a_{ij}^{(r)}, a_{ji}^{(r)}\}$ *for* $1 \le i < j \le d$, $1 \le r \le s$ *are free in* $(\mathcal{A}, \varphi)$.

For $1 \le i, j \le d$, $1 \le r \le s$ *and* $n \ge 1$ *let us denote:*

$$\alpha_{ij;n}^{(r)} := \kappa_n(a_{ij}^{(r)}, a_{ji}^{(r)}, a_{ij}^{(r)}, \ldots)$$

(free cumulant of order n*, with alternating occurrences of* $a_{ij}^{(r)}$ *and* $a_{ji}^{(r)}$*; note that* $\alpha_{ij;n}^{(r)} = 0$ *when* $i \neq j$ *and* n *is odd, due to the R-diagonality of* $a_{ij}^{(r)}$*). Suppose that for every* $n \ge 1$ *and every* $1 \le r \le s$, $1 \le i \le d$, *the sum*

$$\sum_{j=1}^{d} \alpha_{ij;n}^{(r)} =: \lambda_n^{(r)} \tag{20.33}$$

does not actually depend on i*. Then the matrices* $A_1 = (a_{ij}^{(1)})_{i,j=1}^d, \ldots, A_s = (a_{ij}^{(s)})_{i,j=1}^d$ *are free in* $(M_d(\mathcal{A}), \varphi_d)$*, and have R-transforms*

$$R_{A_r}(z) = \sum_{n=1}^{\infty} \lambda_n^{(r)} z^n, \quad 1 \le r \le s.$$

REMARK 20.18. The summation conditions (20.33) become extremely simple when the elements $a_{ii}^{(r)}$ are semicircular, and the elements $a_{ij}^{(r)}$ with $i \neq j$ are circular. Indeed, in this case we have that $\alpha_{ij;n}^{(r)} = 0$ whenever $n \neq 2$, and that $\alpha_{ij;2}^{(r)}$ is one quarter of the squared radius of the circular/semicircular element $a_{ij}^{(r)}$. So if we denote the radius of $a_{ij}^{(r)}$ by $\gamma_{ij}^{(r)}$, then (20.33) amounts here to asking that, for every $1 \le r \le s$, the corresponding matrix of squared radii has constant sums along its columns:

$$\sum_{j=1}^{d} (\gamma_{1j}^{(r)})^2 = \cdots = \sum_{j=1}^{d} (\gamma_{dj}^{(r)})^2 =: \gamma_r^2.$$

The conclusion of Corollary 20.17 becomes that the matrices $A_1 = (a_{ij}^{(1)})_{i,j=1}^d, \ldots, A_s = (a_{ij}^{(s)})_{i,j=1}^d$ are free, and that A_r is semicircular of radius γ_r, for $1 \le r \le s$.

Exercises

EXERCISE 20.19. Let $(\mathcal{A}, \varphi)$ be a $*$-probability space, let d, s be positive integers, and let $A_1 = (a_{ij}^{(1)})_{i,j=1}^d, \ldots, A_s = (a_{ij}^{(s)})_{i,j=1}^d$ form an R-cyclic family of selfadjoint matrices in $M_d(\mathcal{A})$. Suppose that the cyclic cumulants of the entries of these matrices depend only on the superscript indices:

$$\kappa_n(a^{(r_1)}_{i_n,i_1}, a^{(r_2)}_{i_1,i_2}, \dots, a^{(r_n)}_{i_{n-1},i_n}) =: \alpha_{r_1,\dots,r_n}, \tag{20.34}$$

for every $n \geq 1$ and every $1 \leq r_1, \dots, r_n \leq s$, $1 \leq i_1, \dots, i_n \leq d$. Prove that:

$$R_{A_1,\dots,A_s}(z_1, \dots, z_s) = \sum_{n=1}^{\infty} \sum_{r_1,\dots,r_n=1}^{s} d^{n-1}\alpha_{r_1,\dots,r_n} z_{r_1} \cdots z_{r_n}. \tag{20.35}$$

EXERCISE 20.20. Let $(\mathcal{A}, \varphi)$ be a non-commutative probability space. Let $A_1, \dots, A_s$ be an R-cyclic family of matrices in $M_d(\mathcal{A})$, and let f denote the determining series of the entries of this family. Moreover, for every $1 \leq i, j \leq d$ and $1 \leq r \leq s$, let us denote the (i, j)-entry of A_r by $a^{(r)}_{ij}$.

Let n be a positive integer, let π be in $NC(n)$, and consider some indices $1 \leq i_1, \dots, i_n \leq d$ and $1 \leq r_1, \dots, r_n \leq s$. Prove the equality:

$$\begin{aligned} &\kappa_\pi(a^{(r_1)}_{i_n i_1}, a^{(r_2)}_{i_1 i_2}, \dots, a^{(r_n)}_{i_{n-1} i_n}) \\ &= \mathrm{Cf}_{((i_1,r_1),\dots,(i_n,r_n));\pi}(f) \cdot \mathrm{Cf}_{(i_1,\dots,i_n);K(\pi)}(dG_d). \end{aligned} \tag{20.36}$$

[Hint: This is just a multi-matrix version of Lemma 20.10, the only difference is that now one must also carry along a family of superscript indices $r_1, \dots, r_n$.]

EXERCISE 20.21. Consider the framework of the preceding exercise. For $1 \leq i \leq d$ we denote by E_i the matrix in $M_d(\mathcal{A})$ which has the (i, i)-entry equal to $1_\mathcal{A}$, and all the other entries equal to 0. Prove that the moment series and R-transform of the family of ds matrices $A_1E_1, \dots, A_rE_i, \dots, A_sE_d$ are described as follows:

$$M_{A_1E_1,\dots,A_sE_d} = \frac{1}{d}(f \mathbin{\widetilde{\boxed{\star}}} dG_d), \tag{20.37}$$

$$R_{A_1E_1,\dots,A_sE_d} = \frac{1}{d}(f \mathbin{\widetilde{\boxed{\star}}} dH_d) \tag{20.38}$$

(equalities of power series in sd non-commuting indeterminates $z_{1,1}, \dots, z_{r,i}, \dots, z_{s,d}$).

[Hint: This exercise goes in parallel with the proofs of Equations (20.15) and (20.16), obtained while working on Theorem 20.9, only that now we use Exercise 20.20 instead of Lemma 20.10.]

EXERCISE 20.22. Show how Equations (20.23) and (20.24) stated in Theorem 20.14 follow from Equations (20.37) and (20.38) obtained in the preceding exercise.

The next two exercises go toward showing that the property of R-cyclicity for a family $A_1, \dots, A_s$ of $d \times d$ matrices is in fact a property of the algebra generated by $A_1, \dots, A_s$ together with the scalar diagonal matrices.

EXERCISE 20.23. Let $(\mathcal{A}, \varphi)$ be a non-commutative probability space, let d be a positive integer, and let $A_1, \dots, A_s$ be an R-cyclic family of matrices in $M_d(\mathcal{A})$. Prove the following statements.

(1) Rearranging $A_1, \dots, A_s$ in a different order does not affect the R-cyclicity of the family.

(2) If we enlarge $A_1, \dots, A_s$ with a matrix $A \in \text{span}\{A_1, \dots, A_s\}$, then the enlarged family $A_1, \dots, A_s, A$ is still R-cyclic.

(3) If we enlarge $A_1, \dots, A_s$ with the matrix $B := A_1 A_2$, then the enlarged family $A_1, \dots, A_s, B$ is still R-cyclic.

(4) If we enlarge $A_1, \dots, A_s$ with a scalar diagonal matrix D (which has the diagonal entries of the form $\lambda_i 1_{\mathcal{A}}$, $1 \leq i \leq d$, and the off-diagonal entries equal to 0), then the enlarged family $A_1, \dots, A_s, D$ is still R-cyclic.

EXERCISE 20.24. Let $(\mathcal{A}, \varphi)$ be a non-commutative probability space, let d be a positive integer, and let $A_1, \dots, A_s$ be an R-cyclic family of matrices in $M_d(\mathcal{A})$. We denote by $\mathcal{D}$ the algebra of scalar diagonal matrices in $M_d(\mathcal{A})$, and by $\mathcal{C}$ the subalgebra of $M_d(\mathcal{A})$ which is generated by $\{A_1, \dots, A_s\} \cup \mathcal{D}$. Prove that every finite family of matrices from $\mathcal{C}$ is R-cyclic.

LECTURE 21

The full Fock space model for the R-transform

A convenient fact which was observed and then repeatedly used in the preceding lectures is that one can construct a family of non-commutative random variables with a prescribed joint distribution (where quite often the joint distribution is indicated in the guise of a prescribed joint R-transform). Sometimes there may be more than one way of doing such a construction; an example of such a situation is the one of free semicircular families, which can be obtained by an abstract free product construction, but can also be "concretely" put into evidence by using operators of creation and annihilation on the full Fock space (cf. Lecture 7).

Of course, all the different methods for constructing a family of elements with a prescribed joint distribution are ultimately equivalent, in the respect that the calculations with moments and with cumulants performed on the family give the same results, no matter how the family was constructed. Nevertheless, there can be a substantial difference in the transparency of the calculations – it may happen that the solution to the problem we are trying to solve shows up more easily if one method of construction is used over another.

So this is, in a nut-shell, the idea of "modeling": find a good way of constructing a family of non-commutative random variables with a given joint distribution, so that we are at an advantage when computing moments and cumulants of that family. In this lecture we present such a recipe of construction, which uses operators of creation and annihilation on the full Fock space (and generalizes the situation encountered in Lecture 7 for the special case of free semicircular elements). In order to illustrate how this modeling recipe works, we will show how we can use it to re-derive the formulas about free compressions which were obtained by direct combinatorial analysis in Lecture 14.

Description of the Fock space model

REMARK 21.1. We refer to the concept of full Fock space $\mathcal{F}(\mathcal{H})$ over a Hilbert space $\mathcal{H}$, which was introduced in Lecture 7 (cf. Definitions 7.13). Recall that for $\xi \in \mathcal{H}$ one has a creation operator

$l(\xi) \in B(\mathcal{F}(\mathcal{H}))$ and that one has the relations

$$l(\xi)^* l(\eta) = \langle \eta, \xi \rangle \cdot 1_{B(\mathcal{F}(\mathcal{H}))}, \quad \forall\, \xi, \eta \in \mathcal{H}$$

(cf. Remark 7.14.4). In particular, if $\xi_1, \ldots, \xi_s$ is an orthonormal system of vectors in $\mathcal{H}$, then we get that:

$$l(\xi_i)^* l(\xi_j) = \delta_{i,j} \cdot 1_{B(\mathcal{F}(\mathcal{H}))}, \quad 1 \leq i, j \leq s.$$

The latter equations are called the *Cuntz relations*. It will be more convenient to use them in an abstract framework, which we introduce next. This abstract framework will also capture, in part (2) of Definition 21.2, how the vacuum-state acts on monomials in creation and annihilation operators. (Indeed, Equation (21.2) below can be viewed as an abstract version of the formula

$$\langle l(\xi_1) \cdots l(\xi_m) l(\eta_1)^* \cdots l(\eta_n)^* \Omega, \Omega \rangle = 0,$$

which holds for all non-negative integers m, n such that $m + n \geq 1$, and for every $\xi_1, \ldots, \xi_m, \eta_1, \ldots, \eta_n \in \mathcal{H}$.)

DEFINITION 21.2. (1) Let $\mathcal{A}$ be a unital $*$-algebra, and let $l_1, \ldots, l_s$ be elements of $\mathcal{A}$. We say that $l_1, \ldots, l_s$ form a **family of Cuntz isometries** if they satisfy:

$$l_i^* l_j = \delta_{i,j} 1_{\mathcal{A}}, \quad 1 \leq i, j \leq s. \tag{21.1}$$

(2) Let $(\mathcal{A}, \varphi)$ be a $*$-probability space. We say that $l_1, \ldots, l_s \in \mathcal{A}$ form a **free** family of Cuntz isometries if (21.1) is satisfied, and if in addition we have:

$$\varphi(l_{i_1} \cdots l_{i_m} l_{j_1}^* \cdots l_{j_n}^*) = 0 \tag{21.2}$$

for all non-negative integers m, n such that $m + n \geq 1$, and for every $i_1, \ldots, i_m, j_1, \ldots, j_n \in \{1, \ldots, s\}$.

REMARKS 21.3. We record here a few comments which help explain some of the terminology used in the preceding definition.

(1) Recall that an element v of a unital $*$-algebra $\mathcal{A}$ is said to be an *isometry* if it satisfies $v^* v = 1_{\mathcal{A}}$ (cf. Lecture 7, Definition 7.21). So if $l_1, \ldots, l_s \in \mathcal{A}$ form a family of Cuntz isometries, then it is part of (21.1) that each l_i is indeed an isometry.

(2) Let $\mathcal{A}$ be a unital $*$-algebra, and let $l_1, \ldots, l_s \in \mathcal{A}$ be a family of Cuntz isometries. By repeating the argument in Remark 7.14.4, it is immediately seen that the unital $*$-subalgebra of $\mathcal{A}$ generated by $l_1, \ldots, l_s$ is:

$$*\text{-alg}(l_1, \ldots, l_s) = \tag{21.3}$$

$$\text{span}\Big(\{1_{\mathcal{A}}\} \cup \Big\{ l_{i_1} \cdots l_{i_m} l_{j_1}^* \cdots l_{j_n}^* \;\Big|\; \begin{array}{l} m, n \geq 0,\ m + n \geq 1, \\ 1 \leq i_1, \ldots, i_m, j_1, \ldots, j_n \leq s \end{array} \Big\} \Big).$$

As a consequence, if φ is a linear functional on $\mathcal{A}$, normalized by $\varphi(1_{\mathcal{A}}) = 1$, then a prescription like the one in Equation (21.2) gives us how φ acts on $*$-alg$(l_1, \ldots, l_s)$.

(3) Let $(\mathcal{A}, \varphi)$ be a $*$-probability space, and let $l_1, \ldots, l_s \in \mathcal{A}$ be a free family of Cuntz isometries. The use of the adjective "free" is justified by the fact that in this case the sets $\{l_1, l_1^*\}, \ldots, \{l_s, l_s^*\}$ are indeed freely independent with respect to φ (cf. Exercise 21.19).

We are now ready to present the "model" announced in the introduction of the lecture – that is, a canonical construction for an s-tuple with a prescribed joint R-transform. The next theorem deals with the case when the prescribed joint R-transform is a polynomial.

THEOREM 21.4. *Let $(\mathcal{A}, \varphi)$ be a $*$-probability space, and $l_1 \ldots, l_s \in \mathcal{A}$ be a free family of Cuntz isometries. Let $f \in \Theta_s$ be a polynomial, i.e. a series for which only finitely many coefficients are different from 0:*

$$f(z_1, \ldots, z_s) = \sum_{n=1}^{k} \sum_{i_1, \ldots, i_n = 1}^{s} \alpha_{(i_1, \ldots, i_n)} z_{i_1} \cdots z_{i_n}. \tag{21.4}$$

Consider the element

$$a = \sum_{n=1}^{k} \sum_{i_1, \ldots, i_n = 1}^{s} \alpha_{(i_1, \ldots, i_n)} \, l_{i_n} \cdots l_{i_1} \in \mathcal{A}, \tag{21.5}$$

and set:

$$a_i = l_i^* (1_{\mathcal{A}} + a), \qquad 1 \leq i \leq s. \tag{21.6}$$

Then $R_{a_1, \ldots, a_s} = f$.

PROOF. In view of the relation between moment series and R-transforms, the statement to be proved is equivalent to showing that $M_{a_1, \ldots, a_s} = f \boxed{\star}$ Zeta. The latter equality is in turn equivalent (upon identification of coefficients on both sides) to the fact that:

$$\varphi(a_{i_1} \cdots a_{i_n}) = \sum_{\pi \in NC(n)} \mathrm{Cf}_{(i_1, \ldots, i_n); \pi}(f), \tag{21.7}$$

for every $n \geq 1$ and every $1 \leq i_1, \ldots, i_n \leq s$. For the remainder of the proof we fix n and $i_1, \ldots, i_n$ for which we will show that (21.7) holds.

We will consider the set of all "words" of finite length (including an empty word ϕ of length 0) which can be made with letters from $\{1, \ldots, s\}$. This set of words is usually denoted as $[s]^*$:

$$[s]^* := \{\phi\} \cup \Big(\bigcup_{n=1}^{\infty} \{1, \ldots, s\}^n \Big). \tag{21.8}$$

The length (i.e. number of letters) of a word $w \in [s]^*$ will be denoted by $|w|$. For $w = (j_1, \ldots, j_m) \in [s]^*$ we will consider the element $l_w \in \mathcal{A}$ and the number $\widetilde{\alpha}(w) \in \mathbb{C}$ defined by

$$l_w := l_{j_1} \cdots l_{j_m}, \qquad \widetilde{\alpha}(w) = \mathrm{Cf}_{(j_m,\ldots,j_1)}(f)$$

(with the convention that if $w = \phi$, then we take $l_w := 1_{\mathcal{A}}$ and $\widetilde{\alpha}(w) := 1$). With these notations, the element $a \in \mathcal{A}$ from (21.5) satisfies:

$$1_{\mathcal{A}} + a \;=\; \sum_{w\in[s]^*,\ |w|\le k} \widetilde{\alpha}(w) l_w. \tag{21.9}$$

Since we have that $a_{i_1} \cdots a_{i_n} = l_{i_1}^*(1_{\mathcal{A}} + a) \cdots l_{i_n}^*(1_{\mathcal{A}} + a)$, we get (by substituting $1_{\mathcal{A}} + a$ from (21.9) and by using the linearity of φ):

$$\varphi(a_{i_1} \cdots a_{i_n}) = \sum_{\substack{w_1,\ldots,w_n\in[s]^* \\ |w_1|,\ldots,|w_n|\le k}} \widetilde{\alpha}(w_1) \cdots \widetilde{\alpha}(w_n)\ \varphi(l_{i_1}^* l_{w_1} \cdots l_{i_n}^* l_{w_n}). \tag{21.10}$$

Now, for every $w_1, \ldots, w_n \in [s]^*$, we have that either $l_{i_1}^* l_{w_1} \cdots l_{i_n}^* l_{w_n}$ is 0 or it simplifies to one of the elements listed on the right-hand side of Equation (21.3). But with the exception of $1_{\mathcal{A}}$, all those elements belong to the kernel of φ. It follows that:

$$\varphi(l_{i_1}^* l_{w_1} \cdots l_{i_n}^* l_{w_n}) \;=\; \begin{cases} 0 & \text{if } (w_1, \ldots, w_n) \notin [s]^*{}_{i_1,\ldots,i_n} \\ \\ 1 & \text{if } (w_1, \ldots, w_n) \in [s]^*{}_{i_1,\ldots,i_n}, \end{cases} \tag{21.11}$$

where we denoted

$$[s]^*{}_{i_1,\ldots,i_n} \;:=\; \left\{ (w_1, \ldots, w_n) \;\middle|\; \begin{array}{l} w_1, \ldots, w_n \in [s]^* \\ l_{i_1}^* l_{w_1} \cdots l_{i_n}^* l_{w_n} = 1_{\mathcal{A}} \end{array} \right\}. \tag{21.12}$$

The conclusion drawn from (21.10) and (21.11) is hence that:

$$\varphi(a_{i_1} \cdots a_{i_n}) \;=\; \sum_{(w_1,\ldots,w_n)\in[s]^*{}_{i_1,\ldots,i_n}} \widetilde{\alpha}(w_1) \cdots \widetilde{\alpha}(w_n). \tag{21.13}$$

(Note: in the definition of $[s]^*_{i_1,\ldots,i_n}$ we dropped the length restrictions $|w_1| \le k, \ldots, |w_n| \le k$. It is immediate that this does not introduce an error in the calculation of the moment $\varphi(a_{i_1} \cdots a_{i_n})$, due to the fact that the products of the form $\widetilde{\alpha}(w_1) \cdots \widetilde{\alpha}(w_n)$ vanish whenever it is not true that $|w_1|, \ldots, |w_n| \le k$.)

Now, the point to observe is that one has a natural bijection $\Phi : NC(n) \to [s]^*{}_{i_1,\ldots,i_n}$. For $\pi \in NC(n)$, the n-tuple $\Phi(\pi) =: (w_1, \ldots, w_n) \in [s]^*_{i_1,\ldots,i_n}$ is described as follows:

• if $m \in \{1, \ldots, n\}$ belongs to the block $V = \{b_1 < b_2 < \cdots < b_p\}$ of π, and if m is the maximal element of V (i.e. $m = b_p$), then $w_m = (i_{b_p}, \ldots, i_{b_2}, i_{b_1})$;

• if $m \in \{1, \ldots, n\}$ belongs to the block V of π, but m is not the maximal element of V, then $w_m = \phi$.

[As a numerical example, let us take $n = 5$ and consider $\pi = \{\{1,5\},\{2,3\},\{4\}\}$; then $\Phi(\pi) = (\phi, \phi, (i_3, i_2), (i_4), (i_5, i_1))$. The latter 5-tuple of words really belongs to $[s]^*_{i_1,\ldots,i_5}$, since the Cuntz relations imply that $l^*_{i_1}(1_{\mathcal{A}})l^*_{i_2}(1_{\mathcal{A}})\; l^*_{i_3}(l_{i_3}l_{i_2})\; l^*_{i_4}(l_{i_4})l^*_{i_5}(l_{i_5}l_{i_1})$ is indeed equal to $1_{\mathcal{A}}$.]

Verification of the bijectivity of Φ will be left for the exercises at the end of the lecture (cf. Exercises 21.20–21.22).

Finally, the only thing left to be noticed is that if $\pi \in NC(n)$ has $\Phi(\pi) = (w_1, \ldots, w_n) \in [s]^*$, then $\widetilde{\alpha}(w_1) \cdots \widetilde{\alpha}(w_n) = \mathrm{Cf}_{(i_1,\ldots,i_n);\pi}(f)$ (this follows directly from the explicit description of how Φ works). So if we use the bijection Φ to convert the right-hand side of Equation (21.13) into a summation over $NC(n)$, then the desired formula (21.7) is obtained. □

The applicability of Theorem 21.4 can sometimes be enhanced by using the following simple trick.

EXERCISE 21.5. Let $(\mathcal{A}, \varphi)$ be a $*$-probability space, let $l_1, \ldots, l_s \in \mathcal{A}$ be a free family of Cuntz isometries, and let $\lambda_1, \ldots, \lambda_s$ be in $\mathbb{C} \setminus \{0\}$. Then the family:

$$\lambda_1 l_1, \frac{1}{\lambda_1} l_1^*, \ldots, \lambda_s l_s, \frac{1}{\lambda_s} l_s^*$$

has the same joint distribution as $l_1, l_1^*, \ldots, l_s, l_s^*$. As a consequence, if instead of a from Equation (21.5) we use in Theorem 21.4 the element

$$a' := \sum_{n=1}^{k} \sum_{i_1,\ldots,i_n=1}^{s} \alpha_{(i_1,\ldots,i_n)} \lambda_{i_1} \cdots \lambda_{i_n} \; l_{i_n} \cdots l_{i_1} \in \mathcal{A},$$

and if we set

$$a'_i := \frac{1}{\lambda_i} l_i^*(1_{\mathcal{A}} + a'), \quad 1 \le i \le s,$$

then it will still be true that $R_{a'_1,\ldots,a'_s} = f$.

REMARK 21.6. Theorem 21.4 can be easily generalized to the case when f is an arbitrary series in Θ_s (instead of a polynomial), as long as we can give a meaning to the infinite sum in the formula:

$$a = \sum_{n=1}^{\infty} \Big(\sum_{i_1,\ldots,i_n=1}^{s} \alpha_{(i_1,\ldots,i_n)} \, l_{i_n} \cdots l_{i_1} \Big)$$

which would have to replace Equation (21.5) (with $\alpha_{(i_1,\ldots,i_n)} = \mathrm{Cf}_{(i_1,\ldots,i_n)}(f)$ for every $i_1,\ldots,i_n$). Thus we need to move to a setting where we can talk about convergent sequences (and, consequently, about convergent series). This could be the setting of a C^*-probability space – but quite a bit less structure than that is really needed. In fact, all we need is the setting of a non-commutative probability space $(\mathcal{A},\varphi)$ where on the algebra $\mathcal{A}$ we have a concept of convergence of sequences, such that the algebra operations on $\mathcal{A}$ and the linear functional $\varphi : \mathcal{A} \to \mathbb{C}$ do respect the convergence of sequences. In the next theorem we will refer to this kind of setting by the ad hoc name of *topological non-commutative probability space.*

So, the next theorem contains in particular the situation when (in the notations of the theorem) we have

$$(\mathcal{A},\varphi) = (\widetilde{\mathcal{A}},\widetilde{\varphi}) = C^*\text{-probability space.}$$

But the next theorem can also refer to a situation which is rigged in such a way that the series appearing in (21.14) below is convergent for *any* choice of the coefficients $\alpha_{(i_1,\ldots,i_n)}$. A precise description of how this can happen is given in Remark 21.8; the interest in displaying such a situation is that it completes our Fock space model for the R-transform (in the sense that the R-transform $R_{a_1,\ldots,a_s}$ which appears from the construction can now be an arbitrary series $f \in \Theta_s$).

We conclude this remark with a note for the reader who may find the framework of Theorem 21.7 a bit too intricate. The good news for this reader is that, in applications, one can often arrange the arguments so that Theorem 21.4 is applied directly, without having to be generalized. (For an illustration of this, see e.g. Remark 21.14 in the next section.)

THEOREM 21.7. *Suppose that we are given:*

(i) *a topological non-commutative probability space* $(\widetilde{\mathcal{A}},\widetilde{\varphi})$*;*

(ii) *a $*$-operation defined on a unital subalgebra $\mathcal{A} \subset \widetilde{\mathcal{A}}$, such that $(\mathcal{A},\widetilde{\varphi}|\mathcal{A})$ is a $*$-probability space;*

(iii) *a free family of Cuntz isometries $l_1,\ldots,l_s \in \mathcal{A}$;*

(iv) *a family of complex numbers $\{(\alpha_{(i_1,\ldots,i_n)} \mid n \geq 1,\ 1 \leq i_1, \ldots, i_n \leq s\}$ such that the series*

$$\sum_{n=1}^{\infty} \Big(\sum_{i_1,\ldots,i_n=1}^{s} \alpha_{(i_1,\ldots,i_n)}\, l_{i_n} \cdots l_{i_1} \Big) \tag{21.14}$$

is convergent in $\widetilde{\mathcal{A}}$.

Let $a \in \widetilde{\mathcal{A}}$ denote the sum of the convergent series in (21.14), and let us denote

$$a_i \;:=\; l_i^*(1_{\mathcal{A}} + a) \in \widetilde{\mathcal{A}} \qquad (1 \leq i \leq s).$$

Then the joint R-transform of $a_1, \ldots, a_s$ *in* $(\widetilde{\mathcal{A}}, \widetilde{\varphi})$ *is:*

$$R_{a_1,\ldots a_s} = \sum_{n=1}^{\infty} \sum_{i_1,\ldots,i_n=1}^{s} \alpha_{(i_1,\ldots,i_n)}\, z_{i_1} \cdots z_{i_n}. \tag{21.15}$$

PROOF. Let us denote the series on the right-hand side of Equation (21.15) by f, and for every $k \geq 1$ let us denote by f_k the truncation of f to terms of length up to k. On the other hand, for every $k \geq 1$ let us denote by $a^{(k)}$ the element of $\mathcal{A}$ which is obtained by truncating the first summation in (21.14) to its first k terms; and let us denote $a_i^{(k)} := l_i^*(1_{\widetilde{\mathcal{A}}} + a^{(k)})$, $1 \leq i \leq s$.

Theorem 21.4 gives us that $R_{a_1^{(k)},\ldots,a_s^{(k)}} = f_k$, $\forall\ k \geq 1$. On the other hand we have that $\lim_{k\to\infty} a^{(k)} = a$ (by how $a^{(k)}$ is defined), and consequently that $\lim_{k\to\infty} a_i^{(k)} = a_i$, $\forall\ 1 \leq i \leq s$, since the operations in $\widetilde{\mathcal{A}}$ respect the convergence of sequences. The same continuity of the operations on $\widetilde{\mathcal{A}}$ together with the continuity of $\widetilde{\varphi}$ will then imply that the series $R_{a_1^{(k)},\ldots,a_s^{(k)}}$ converge coefficientwise, as $k \to \infty$, to the series $R_{a_1,\ldots,a_s}$. The conclusion is that $R_{a_1,\ldots,a_s} = f$ (since both series can be expressed as the coefficientwise limit $\lim_{k\to\infty} f_k$). □

REMARK 21.8. We conclude this section by outlining how a construction with the properties (i), (ii), (iii) of Theorem 21.7 can be made, so that a series as in (21.14) is convergent for an arbitrary choice of the coefficients $\alpha_{(i_1,\ldots,i_n)}$. The idea is to use a certain algebra of matrices of infinite size (its definition is inspired by looking at matrices of operators on the full Fock space $\mathcal{F}(\mathbb{C}^s)$).

Let s be the positive integer which appears in (iii) of Theorem 21.7. We will use the set $[s]^*$ of words over the alphabet $\{1, \ldots, s\}$ which was introduced during the proof of Theorem 21.4 (cf. Equation (21.8)). Moreover, a family of complex numbers $(\alpha_{v,w})_{v,w\in[s]^*}$ will be called a *matrix over* $[s]^*$.

Let $\widetilde{\mathcal{A}}$ be the set of matrices $(\alpha_{v,w})_{v,w}$ over $[s]^*$ which have the property that

$$\sup\{|w| - |v| \mid v, w \in [s]^*,\ \alpha_{v,w} \neq 0\} < \infty, \tag{21.16}$$

and let $\widetilde{\varphi} : \widetilde{\mathcal{A}} \to \mathbb{C}$ be the linear functional defined by the formula

$$\widetilde{\varphi}(\ (\alpha_{v,w})_{v,w\in[s]^*}\ ) = \alpha_{\phi,\phi}. \tag{21.17}$$

On $\widetilde{\mathcal{A}}$ we consider the vector space operations defined entrywise, and the multiplication defined by the formula:

$$\begin{cases} (\alpha_{v,w})_{v,w} \cdot (\beta_{v,w})_{v,w} = (\gamma_{v,w})_{v,w}, \\ \text{where } \gamma_{v,w} := \sum_{u\in[s]^*} \alpha_{v,u}\beta_{u,w}, \quad \forall\ v, w \in [s]^*. \end{cases} \tag{21.18}$$

(Note that the sum appearing in (21.18) has only finitely many non-zero terms, and hence presents no convergence problems, due to the finiteness of $\sup\{|u| - |v| \mid \alpha_{v,u} \neq 0\}$.) If on $\widetilde{\mathcal{A}}$ we also consider the topology of entrywise convergence for matrices, it is immediately verified that $(\widetilde{\mathcal{A}}, \widetilde{\varphi})$ becomes a topological non-commutative probability space, in the sense discussed in Remark 21.6.

Let now $\mathcal{A} \subset \widetilde{\mathcal{A}}$ be the subset consisting of the matrices $(\alpha_{v,w})_{v,w}$ which (in addition to (21.16)) also have the property that:

$$\inf\{|w| - |v| \mid v, w \in [s]^*,\ \alpha_{v,w} \neq 0\} > -\infty. \tag{21.19}$$

It is immediate that $\mathcal{A}$ is a unital subalgebra of $\widetilde{\mathcal{A}}$, on which a natural $*$-operation is defined by the formula:

$$\big((\alpha_{v,w})_{v,w} \big)^* = (\overline{\alpha_{w,v}})_{v,w}.$$

Denoting $\varphi = \widetilde{\varphi}|\mathcal{A}$, we have that $(\mathcal{A}, \varphi)$ is a $*$-probability space. (This statement also contains the assertion that φ is positive on $\mathcal{A}$. The reader should have no difficulty in checking that this follows immediately from the formula (21.17) – indeed, φ simply selects a diagonal entry of the considered matrix, and such a functional is always positive, when considered on a $*$-algebra of matrices.)

Finally, for $1 \leq i \leq s$, let $l_i \in \mathcal{A}$ be the matrix $(\lambda^{(i)}_{v,w})_{v,w}$ with entries defined by the formula:

$$\lambda^{(i)}_{v,w} = \begin{cases} 1 & \text{if } v = (i) \cdot w, \text{ i.e. } v \text{ is obtained from } w \\ & \text{by adding the letter } i \text{ on its left} \\ 0 & \text{otherwise.} \end{cases}$$

Direct calculations show that $l_1, \ldots, l_s$ form a free family of Cuntz isometries in $(\mathcal{A}, \varphi)$.

So now we have constructed objects as indicated in (i), (ii), (iii) of Theorem 21.7. It is easy to verify that in the framework of this construction, a series like the one in (21.14) is convergent in $\widetilde{\mathcal{A}}$, for any choice of the coefficients $\alpha_{(i_1,\ldots,i_n)}$. The sum of the series is the matrix $(\beta_{v,w})_{v,w} \in \widetilde{\mathcal{A}}$ which has entries:

$$\beta_{v,w} = \begin{cases} \alpha_{(i_1,\ldots,i_n)} & \text{if } v = (i_n, \ldots, i_1) \cdot w \\ 0 & \text{otherwise.} \end{cases}$$

An application: revisiting free compressions

In order to illustrate how the full Fock space model can be put to work, we present in this section how modeling can be used to derive the free compression formulas which were obtained by direct combinatorial analysis in Lecture 14.

We start by reviewing the framework where our compressions are considered (cf. also the sections about compressions in Lecture 14).

NOTATIONS 21.9. Consider a non-commutative probability space $(\mathcal{A}, \varphi)$, and suppose that in $\mathcal{A}$ we have a $d \times d$ matrix unit – i.e. a family $\{e_{ij} \mid 1 \leq i, j \leq d\}$ such that $e_{ij}e_{kl} = \delta_{jk}e_{il}$, for every $1 \leq i, j, k, l \leq d$. We do not assume that $\sum_{i=1}^{d} e_{ii} = 1_{\mathcal{A}}$. On the other hand, we make the following assumptions on the values $\varphi(e_{ij})$, $1 \leq i, j \leq d$:

$$\begin{cases} \varphi(e_{ij}) = 0, & \forall i \neq j \text{ in } \{1, \dots, d\} \\ \varphi(e_{ii}) =: \lambda_i > 0,\ 1 \leq i \leq d, & \text{where } \lambda_1 + \dots + \lambda_d \leq 1. \end{cases} \tag{21.20}$$

We will denote by $(\mathcal{C}, \tau)$ the non-commutative probability space obtained by compressing $(\mathcal{A}, \varphi)$ with the projection e_{11}; that is,

$$\mathcal{C} := e_{11}\mathcal{A}e_{11}, \quad \tau := \frac{1}{\lambda_1}\varphi|\mathcal{C}.$$

An s-tuple $a_1, \dots, a_s$ of elements from $\mathcal{A}$ can then be compressed to an d^2s-tuple $\{c_{ij;r} \mid 1 \leq i, j \leq d,\ 1 \leq r \leq s\}$ of elements in $\mathcal{C}$, where:

$$c_{ij;r} := e_{1i}a_r e_{j1}, \quad 1 \leq i, j \leq d,\ 1 \leq r \leq s.$$

REMARKS 21.10. (1) If in Notations 21.9 the functional φ were assumed to be a trace, then the fact that $\varphi(e_{ij}) = 0$ for $i \neq j$ would be automatic, since we could write

$$\varphi(e_{ij}) = \varphi(e_{ii}e_{ij}) = \varphi(e_{ij}e_{ii}) = \varphi(0) = 0.$$

In the tracial case we would also obtain that $\lambda_1 = \dots = \lambda_d$, because:

$$\lambda_i = \varphi(e_i) = \varphi(e_{ij}e_{ji}) = \varphi(e_{ji}e_{ij}) = \varphi(e_j) = \lambda_j, \quad \forall\, 1 \leq i, j \leq d.$$

(2) In Notations 21.9, let us assume that $\{a_1, \dots, a_s\}$ is free from $\{e_{ij} \mid 1 \leq i, j \leq d\}$. It is then immediate from general freeness considerations that the joint distribution of the compressed (d^2s)-tuple $\{c_{ij;r} \mid 1 \leq i, j \leq d,\ 1 \leq r \leq s\}$ depends only on the joint distribution of $a_1, \dots a_s$ and on the numbers $\lambda_1, \dots, \lambda_d$ appearing in Equation (21.20). Our goal is to make this dependence precise, by indicating an exact formula for the R-transform of $\{c_{ij;r} \mid 1 \leq i, j \leq d,\ 1 \leq r \leq s\}$, in terms of $R_{a_1,\dots,a_s}$ and of $\lambda_1, \dots, \lambda_d$. This formula is given in the next theorem.

THEOREM 21.11. *Let $(\mathcal{A}, \varphi)$, $\{e_{ij} \mid 1 \leq i, j \leq d\}$, $(\mathcal{C}, \tau)$, $\{a_1, \dots, a_s\}$, and $\{c_{ij;r} \mid 1 \leq i, j \leq d,\ 1 \leq r \leq s\}$ be as in Notations 21.9, and let us denote:*

$$R_{a_1,\dots,a_s}(z_1, \dots, z_s) =: \sum_{n=1}^{\infty} \sum_{r_1,\dots,r_n=1}^{s} \alpha_{(r_1,\dots,r_n)} z_{r_1} \cdots z_{r_n}. \tag{21.21}$$

We suppose in addition that $\{a_1, \ldots, a_s\}$ *is free from* $\{e_{ij} \mid 1 \le i, j \le d\}$ *in* $(\mathcal{A}, \varphi)$. *Then the R-transform of the family* $\{c_{ij;r} \mid 1 \le i, j \le d,\ 1 \le r \le s\}$, *calculated in the compressed space* $(\mathcal{C}, \tau)$, *has the formula:*

$$R_{c_{11;1},\ldots,c_{ij;r},\ldots,c_{dd;s}}(z_{11;1}, \ldots, z_{ij;r}, \ldots, z_{dd;s}) =$$
$$\sum_{n=1}^{\infty} \sum_{\substack{1 \le r_1,\ldots,r_n \le s \\ 1 \le i_1,\ldots,i_n \le d}} \lambda_{i_1} \cdots \lambda_{i_{n-1}} \alpha_{(r_1,\ldots,r_n)} z_{i_n i_1;r_1} z_{i_1 i_2;r_2} \cdots z_{i_{n-1} i_n;r_n}. \tag{21.22}$$

REMARK 21.12. One sees without difficulty that Equation (21.22) of Theorem 21.11 contains in a concentrated form the results of Theorems 14.10 and 14.18.

Indeed, if $d = 1$ then the matrix unit $\{e_{ij} \mid 1 \le i, j \le d\}$ reduces to the projection $e := e_{11}$, having $\varphi(e) = \lambda := \lambda_1$. In this case Equation (21.22) takes the form:

$$R_{c_1,\ldots,c_s}(z_1, \ldots, z_s) = \frac{1}{\lambda} R_{a_1,\ldots,a_s}(\lambda z_1, \ldots, \lambda z_s), \tag{21.23}$$

where $c_r := ea_re \in e\mathcal{A}e$, $1 \le r \le s$, and where the R-transform on the right-hand side of (21.23) is calculated in the compressed space $(e\mathcal{A}e, \lambda^{-1}\varphi|e\mathcal{A}e)$. In terms of free cumulants, (21.23) means that:

$$\kappa_n(c_{r_1}, \ldots, c_{r_n}) = \lambda^{n-1}\kappa_n(a_{r_1}, \ldots, a_{r_n}),$$

for every $n \ge 1$ and $1 \le r_1, \ldots, r_n \le s$, as stated in Theorem 14.10.

On the other hand, if $d \ge 1$ is arbitrary, and we ask that $\lambda_1 = \cdots = \lambda_d =: \lambda > 0$, then by extracting coefficients in Equation (21.22) we obtain that:

$$\kappa_n(c_{i_1 j_1;r_1}, \ldots, c_{i_n j_n;r_n}) = \begin{cases} \lambda^{n-1}\kappa_n(a_{r_1}, \ldots, a_{r_n}) & \text{if } j_1 = i_2, \ldots, j_{n-1} = i_n, j_n = i_1 \\ 0 & \text{otherwise,} \end{cases}$$

exactly as stated in Theorem 14.18.

REMARK 21.13. A concise version of Equation (21.22) in Theorem 21.11 can be obtained if we use "matrix notations" for the variables, in the way described as follows.

Let the family of d^2s indeterminates $\{z_{i,j;r} \mid 1 \le i, j \le d,\ 1 \le r \le s\}$ and the numbers $\lambda_1, \ldots, \lambda_d > 0$ be given. We will use the notation "Θ_{d^2s}" for the set of series of the type introduced in Lecture 16, and which act in the indeterminates $z_{ij;r}$. Note that we will use Θ_{d^2s} in parallel with Θ_s (which is, as in the preceding lectures, a set of series acting in indeterminates $z_1, \ldots, z_s$).

Now, for every $1 \leq r \leq s$ let us denote by Z_r the matrix of indeterminates defined as follows:

$$Z_r := (\lambda_j z_{ij;r})_{i,j=1}^d. \tag{21.24}$$

Note that for every $n \geq 1$ and $1 \leq r_1, \ldots, r_n \leq s$ it makes sense to form the product $Z_{r_1} \cdots Z_{r_n}$ which is a $d \times d$ matrix having as entries some homogeneous polynomials of degree n in the indeterminates $z_{ij;r}$. And moreover, for a series

$$f(z_1, \ldots, z_s) = \sum_{n=1}^{\infty} \sum_{r_1,\ldots,r_n=1}^{s} \alpha_{(r_1,\ldots,r_n)} z_{r_1} \cdots z_{r_n} \in \Theta_s,$$

it makes sense to look at

$$f(Z_1, \ldots, Z_s) := \sum_{n=1}^{\infty} \sum_{r_1,\ldots,r_n=1}^{s} \alpha_{(r_1,\ldots,r_n)} Z_{r_1} \cdots Z_{r_n}, \tag{21.25}$$

which is a $d \times d$ matrix with entries from $\Theta_{d^2 s}$. (The expression on the right-hand side of (21.25) does not pose any convergence problems, precisely because its part "$\sum_{r_1,\ldots,r_n=1}^{s} \alpha_{(r_1,\ldots,r_n)} Z_{r_1} \cdots Z_{r_n}$" is homogeneous of degree n, for every $n \geq 1$.)

With these notations, the right-hand side of Equation (21.22) can be given the form:

$$\sum_{i=1}^{d} \frac{1}{\lambda_i} \cdot [\ (i,i)\text{-entry of } R_{a_1,\ldots,a_s}(Z_1, \ldots, Z_s)\ ]. \tag{21.26}$$

Or even more concisely, one can just write:

$$R_{c_{11;1},\ldots,c_{dd;s}}(z_{11;1}, \ldots, z_{dd;s}) = \mathrm{Tr}\big(\Lambda^{-1} \cdot R_{a_1,\ldots,a_s}(Z_1, \ldots, Z_s)\big), \tag{21.27}$$

where

$$\Lambda := \begin{pmatrix} \lambda_1 & & \\ & \ddots & \\ & & \lambda_d \end{pmatrix} \in M_d(\mathbb{C}).$$

REMARK 21.14. As announced at the beginning of this section, we will show a proof of Theorem 21.11 which is obtained by modeling on the full Fock space (in the sense discussed in the first part of the lecture). In order to fall back on the simpler setting of Theorem 21.4, we first observe the following reduction:

"It suffices to prove the Theorem 21.11 in the case when the R-transform $R_{a_1,\ldots,a_s}$ is a polynomial."

Indeed, if this is proved, then the general case of the theorem is obtained as follows. For every $k \geq 1$, one can construct a non-commutative probability space $(\mathcal{A}_k, \varphi_k)$ and families of elements $\{e_{ij}^{(k)} \mid 1 \leq i,j \leq d\}$, $\{a_1^{(k)}, \ldots, a_s^{(k)}\}$ in $\mathcal{A}_k$, such that:

(i) $\{e_{ij}^{(k)} \mid 1 \leq i,j \leq d\}$ is freely independent from $\{a_1^{(k)}, \ldots, a_s^{(k)}\}$;

(ii) $\{e_{ij}^{(k)} \mid 1 \leq i,j \leq d\}$ is a matrix unit with the same distribution as $\{e_{ij} \mid 1 \leq i,j \leq d\}$;

(iii) the R-transform $R_{a_1^{(k)},\ldots,a_s^{(k)}}$ is the truncation to terms of length not exceeding k of the R-transform $R_{a_1,\ldots,a_s}$.

(The possibility of finding such elements is a particular case of the fact that one can construct families of elements with any prescribed joint R-transform, which was seen in Exercise 16.21 and then was repeatedly used in subsequent lectures.)

But now, the particular case of Theorem 21.11 which is assumed here to be true can be applied to the compressions $\{c_{ij;r}^{(k)} \mid 1 \leq i,j \leq d,\ 1 \leq r \leq s\}$ of $a_1^{(k)}, \ldots, a_s^{(k)}$ by the matrix unit $\{e_{ij}^{(k)} \mid 1 \leq i,j \leq d\}$, and gives us that:

$$R_{c_{11;1}^{(k)},\ldots,c_{dd;s}^{(k)}}(z_{11;1}, \ldots, z_{dd;s}) = \sum_{n=1}^{k} \sum_{\substack{1 \leq r_1,\ldots,r_n \leq s \\ 1 \leq i_1,\ldots,i_n \leq d}} \lambda_{i_1} \cdots \lambda_{i_{n-1}} \alpha_{(r_1,\ldots,r_n)} z_{i_n i_1;r_1} z_{i_1 i_2;r_2} \cdots z_{i_{n-1} i_n;r_n}. \tag{21.28}$$

And finally, straightforward calculations show that the R-transforms $R_{c_{11;1}^{(k)},\ldots,c_{dd;s}^{(k)}}$ converge coefficientwise to $R_{c_{11;1},\ldots,c_{dd;s}}$, as $k \to \infty$. So (21.22) is obtained from (21.28) by letting $k \to \infty$.

REMARK 21.15. Before starting the proof of the reduced statement of Theorem 21.11, we would like to comment on the modeling framework which will be used. Our basic modeling pattern says that if $R_{a_1,\ldots,a_s}$ is a polynomial, then $a_1, \ldots, a_s$ can be replaced in a canonical way by $*$-polynomials in a free family of Cuntz isometries (as shown in Theorem 21.4), such that the R-transform of the s-tuple does not change. But note that in the case at hand this is not useful by itself, because it is not clear how one could also arrange to replace – or in other words "model" – the matrix unit $\{e_{ij} \mid 1 \leq i,j \leq d\}$ by a matrix unit which lives in the same space as the Cuntz isometries. It was observed by Shlyakhtenko that the simultaneous modeling of $a_1, \ldots, a_s$ and of $\{e_{ij} \mid 1 \leq i,j \leq d\}$ can be carried over if the Cuntz isometries that we start with are themselves $d \times d$ matrices. We will follow this idea, and thus we will use the modeling framework described as follows.

NOTATIONS 21.16. $(\mathcal{B}, \psi)$ will be a $*$-probability space which contains a free family of $(d+1)^2 s$ Cuntz isometries $\{l_{ij;r} \mid 1 \leq i, j \leq d+1,\ 1 \leq r \leq s\}$. We form the matrix algebra $\widetilde{\mathcal{B}} = M_{d+1}(\mathcal{B})$. On $\widetilde{\mathcal{B}}$ we consider the state $\widetilde{\psi} : \widetilde{\mathcal{B}} \to \mathbb{C}$ defined by:

$$\widetilde{\psi}(\ (y_{ij})_{ij=1}^{d+1}\ ) := \sum_{i=1}^{d+1} \lambda_i \psi(y_{ii}), \quad (y_{ij})_{ij=1}^{d+1} \in \widetilde{\mathcal{B}}, \tag{21.29}$$

where the numbers $\lambda_1, \ldots, \lambda_d > 0$ are taken from Equation (21.20) of Notations 21.9, and where we take $\lambda_{d+1} := 1 - (\lambda_1 + \cdots + \lambda_d) \geq 0$. Note that the choice of λ_{d+1} ensures that $\widetilde{\psi}(1_{\widetilde{\mathcal{B}}}) = 1$, where $1_{\widetilde{\mathcal{B}}}$ denotes the unit of $\widetilde{\mathcal{B}}$ (that is, the diagonal entries of $1_{\widetilde{\mathcal{B}}}$ are all equal to the unit $1_{\mathcal{B}}$ of $\mathcal{B}$, while the off-diagonal entries are all equal to 0).

We will denote by $M_{d+1}(\mathbb{C}1_{\mathcal{B}}) \subset \widetilde{\mathcal{B}}$ the subalgebra consisting of matrices which have all their entries in $\mathbb{C}1_{\mathcal{B}}$.

Moreover, for every $1 \leq r \leq s$ we will denote:

$$L_r := (\ \sqrt{\lambda_i} l_{ji;r}\ )_{i,j=1}^{d+1} \ \in\ \widetilde{\mathcal{B}}. \tag{21.30}$$

The next lemma and proposition refer to the framework which has just been introduced.

LEMMA 21.17. *For $Y \in M_{d+1}(\mathbb{C}1_{\mathcal{B}})$ we have that*

$$L_p^* Y L_q = \delta_{p,q} \cdot \widetilde{\psi}(Y) \cdot 1_{\widetilde{\mathcal{B}}}, \quad \forall\ 1 \leq p, q \leq s. \tag{21.31}$$

PROOF. This is done by direct calculation of the entries of $L_p^* Y L_q$ (which is straightforward, and left as an exercise to the reader). □

PROPOSITION 21.18. (1) *$L_1, \ldots, L_s$ is a free family of Cuntz isometries in $(\widetilde{\mathcal{B}}, \widetilde{\psi})$.*

(2) *$\{L_1, L_1^*, \ldots, L_s, L_s^*\}$ is free from $M_{d+1}(\mathbb{C}1_{\mathcal{B}})$ in $(\widetilde{\mathcal{B}}, \widetilde{\psi})$.*

PROOF. (1) The Cuntz relations for $L_1, \ldots, L_s$, i.e.

$$L_p^* L_q = \delta_{p,q} \cdot 1_{\widetilde{\mathcal{B}}}, \quad 1 \leq p, q \leq s,$$

follow by setting $Y = 1_{\widetilde{\mathcal{B}}}$ in Lemma 21.17. The other condition which we need to check is that:

$$\widetilde{\psi}(\ L_{p_1} \cdots L_{p_m} L_{q_1}^* \cdots L_{q_n}^*\ ) = 0,$$

for every $m, n \geq 0$ such that $m + n \geq 1$, and for every $1 \leq p_1, \ldots, p_m, q_1, \ldots, q_n \leq s$ (cf. Definition 21.2.2). This is a direct consequence of the corresponding property of the Cuntz family $\{l_{ij;r} \mid 1 \leq i, j \leq d+1,\ 1 \leq r \leq s\}$. Indeed, we have:

$$\widetilde{\psi}(\, L_{p_1} \cdots L_{p_m} L^*_{q_1} \cdots L^*_{q_n}\,) = \sum_{i=1}^{d+1} \lambda_i \psi(\,(i,i)\text{-entry of } L_{p_1} \cdots L_{p_m} L^*_{q_1} \cdots L^*_{q_n}\,),$$

and every term of the latter sum is found to be equal to zero (by writing explicitly the (i,i)-entry which appears, and by using the property stated in Definition 21.2.2, applied to the family of Cuntz isometries $l_{ij;r}$).

(2) We will appeal directly to the definition of free independence. We will verify that $\widetilde{\psi}(W) = 0$ whenever W is a word of the form

$$W = X_1 Y_1 \cdots X_k Y_k, \quad k \geq 1, \tag{21.32}$$

where $X_1, \ldots, X_k \in *\text{-alg}(L_1, \ldots, L_s)$, $Y_1, \ldots, Y_k \in M_{d+1}(\mathbb{C}1_{\mathcal{B}})$, and we have that:

$$\widetilde{\psi}(X_1) = \cdots = \widetilde{\psi}(X_k) = 0, \qquad \widetilde{\psi}(Y_1) = \cdots = \widetilde{\psi}(Y_k) = 0. \tag{21.33}$$

(Note: in order to complete this freeness verification, one would also need to consider words analogous to the above W, but which start with a "Y" letter instead of an "X" letter, and/or end with an "X" letter instead of a "Y" letter. We leave it as an immediate exercise to the reader to check that the same argument as presented below applies to these other types of words as well.)

By taking into account how $*\text{-alg}(L_1, \ldots, L_s)$ is concretely described (cf. Remark 21.3.2), we can assume without loss of generality that every X_h, $1 \leq h \leq k$, is of the form:

$$X_h = L_{p_1} \cdots L_{p_{m(h)}} L^*_{q_1} \cdots L^*_{q_{n(h)}} \tag{21.34}$$

for some $m(h), n(h) \geq 0$ with $m(h) + n(h) \geq 1$, and for some $1 \leq p_1, \ldots, p_{m(h)}, q_1, \ldots, q_{n(h)} \leq s$.

We now complete the verification that $\widetilde{\psi}(W) = 0$ by discussing two possible cases.

Case 1. There exists $h \in \{2, \ldots, k\}$ such that $n(h-1) \neq 0 \neq m(h)$. Then when we substitute the expressions of $X_1, \ldots, X_k$ from (21.34) into W of (21.32), we find a factorization of W which contains three consecutive factors $L^*_{q_{n(h-1)}}, Y_{h-1}, L_{p_1}$ with product:

$$\begin{aligned} L^*_{q_{n(h-1)}} Y_{h-1} L_{p_1} &= \delta_{q_{n(h-1)}, p_1} \cdot \widetilde{\psi}(Y_{h-1}) \cdot 1_{\widetilde{\mathcal{B}}} && \text{(by Lemma 21.17)} \\ &= 0 && \text{(by (21.33))}. \end{aligned}$$

So in this case we actually find that $W = 0$, which of course implies that $\widetilde{\psi}(W) = 0$.

Case 2. There exists no $h \in \{2, \ldots, k\}$ such that $n(h-1) \neq 0 \neq m(h)$. Then it is immediately seen that there exists $h \in \{1, \ldots, k+1\}$ such that $n(1) = \cdots = n(h-1) = 0$ and $m(h) = \cdots = m(k) = 0$ ("$h = 1$" corresponds to the extreme case when $m(1) = \cdots = m(k) = 0$, while "$h = k+1$" corresponds to the case when $n(1) = \cdots = n(k) = 0$). Let us substitute the expressions of $X_1, \ldots, X_k$ from (21.34)) into W of (21.32), and let us insert by force a $1_{\widetilde{\mathcal{B}}}$ between every two consecutive factors from the products of (21.34). We arrive to a factorization of W in the form:

$$W = C_0 L_{p_1} C_1 \cdots L_{p_M} C_M L^*_{q_1} C_{M+1} \cdots L^*_{q_N} C_{M+N}, \qquad (21.35)$$

where $M, N \geq 0$ and $M + N \geq 1$, $1 \leq p_1, \ldots, p_M, q_1, \ldots, q_N \leq s$, and $C_0, C_1, \ldots, C_{M+N} \in M_{d+1}(\mathbb{C}1_{\mathcal{B}})$. From (21.35) it follows that $\widetilde{\psi}(W) = 0$ by exactly the same argument which concluded the proof of statement (1) of this proposition. □

After all these preparations, we can now present the proof of Theorem 21.11.

Proof of Theorem 21.11. Based on Remark 21.14, we will assume that $R_{a_1,\ldots,a_s}$ is a polynomial of degree at most k (that is, we will assume that in Equation (21.21) we have $\alpha_{(r_1,\ldots,r_n)} = 0$ whenever $n > k$). Consider our modeling framework, as in Notations 21.16, and for $1 \leq r \leq s$ let us set:

$$A_r = L^*_r \Big(1_{\widetilde{\mathcal{B}}} + \sum_{n=1}^{k} \sum_{r_1,\ldots,r_n=1}^{s} \alpha_{(r_1,\ldots,r_n)} L_{r_n} \cdots L_{r_1} \Big) \in \widetilde{\mathcal{B}}. \qquad (21.36)$$

By Theorem 21.4, the R-transform of the s-tuple $A_1, \ldots, A_s$ coincides with $R_{a_1,\ldots,a_s}$. On the other hand, for every $1 \leq i, j \leq d$ let us denote by $E_{ij} \in \widetilde{\mathcal{B}}$ the matrix which has its (i,j)-entry equal to $1_{\mathcal{B}}$, and all its other entries equal to 0. It is immediate that $\{E_{ij} \mid 1 \leq i, j \leq d\}$ is a matrix unit which has (with respect to $\widetilde{\psi}$) the same distribution as the matrix unit $\{e_{ij} \mid 1 \leq i, j \leq d\}$ appearing in the hypothesis of Theorem 21.11. Since Proposition 21.18.2 gives us that $\{A_1, \ldots, A_s\}$

is free from $\{E_{ij} \mid 1 \leq i,j \leq d\}$, it follows that the joint distribution of

$$\{A_1, \ldots, A_s\} \cup \{E_{ij} \mid 1 \leq i,j \leq d\} \quad (\text{in } (\widetilde{\mathcal{B}}, \widetilde{\psi}))$$

coincides with that of

$$\{a_1, \ldots, a_s\} \cup \{e_{ij} \mid 1 \leq i,j \leq d\} \quad (\text{in } (\mathcal{A}, \varphi)).$$

But then let us consider the compression $E_{11}\widetilde{\mathcal{B}}E_{11}$ of $\widetilde{\mathcal{B}}$ and the family $\{C_{ij;r} \mid 1 \leq i,j \leq d,\ 1 \leq r \leq s\} \subset E_{11}\widetilde{\mathcal{B}}E_{11}$, where:

$$C_{ij;r} := E_{1i}A_rE_{j1}, \ \forall\, 1 \leq i,j \leq d, \ \forall\, 1 \leq r \leq s. \tag{21.37}$$

From the conclusion of the preceding paragraph it is immediate that the family in (21.37) has the same joint distribution as the family $\{c_{ij;r} \mid 1 \leq i,j \leq d,\ 1 \leq r \leq s\}$ from the statement of Theorem 21.11. So in order to prove the theorem, it will suffice to calculate the joint R-transform of $\{C_{ij;r} \mid 1 \leq i,j \leq d,\ 1 \leq r \leq s\}$ (in $E_{11}\widetilde{\mathcal{B}}E_{11}$, with respect to the functional $\lambda_1^{-1}\widetilde{\psi}|E_{11}\widetilde{\mathcal{B}}E_{11}$), and verify that this R-transform has the required form.

The compressed space $E_{11}\widetilde{\mathcal{B}}E_{11}$ is naturally identified to $\mathcal{B}$; in this identification the linear functional $\lambda_1^{-1}\widetilde{\psi}|E_{11}\widetilde{\mathcal{B}}E_{11}$ becomes ψ, and the element $C_{ij;r} \in \widetilde{\mathcal{B}}$ becomes:

$$b_{ij;r} := [\ (i,j)\text{-entry of } A_r\ ] \in \mathcal{B}, \ \ 1 \leq i,j \leq d, \ \ 1 \leq r \leq s. \tag{21.38}$$

Thus our task becomes to check that:

$$R_{b_{11;1},\ldots,b_{ij;r},\ldots,b_{dd;s}}(z_{11;1}, \ldots, z_{ij;r}, \ldots, z_{dd;s}) = \sum_{n=1}^{k} \sum_{\substack{1\leq r_1,\ldots,r_n\leq s\\ 1\leq i_1,\ldots,i_n\leq d}} \lambda_{i_1}\cdots\lambda_{i_{n-1}}\alpha_{(r_1,\ldots,r_n)} z_{i_n i_1;r_1} z_{i_1 i_2;r_2} \cdots z_{i_{n-1} i_n;r_n}. \tag{21.39}$$

Let us write the elements $b_{ij;r}$ explicitly. First, A_r from (21.36) can also be written as

$$A_r = L_r^* + \alpha_{(r)}1_{\widetilde{\mathcal{B}}} + \sum_{r_1=1}^{s} \alpha_{(r_1,r)}L_{r_1} + \sum_{n=3}^{k} \sum_{r_1,\ldots r_{n-1}=1}^{s} \alpha_{(r_1,\ldots,r_{n-1},r)} L_{r_{n-1}} \cdots L_{r_1};$$

hence, by identifying the (i,j)-entry:

$$
\begin{aligned}
b_{ij;r} &= \sqrt{\lambda_j} l^*_{ij;r} + \alpha_{(r)}\delta_{i,j}1_{\mathcal{B}} + \sum_{r_1=1}^{s} \sqrt{\lambda_i} l_{ji;r_1} \\
&\quad + \sum_{n=3}^{k} \sum_{\substack{1\le r_1,\dots r_{n-1}\le s \\ 1\le i_1,\dots,i_{n-2}\le d}} \alpha_{(r_1,\dots,r_{n-1},r)}(\sqrt{\lambda_i} l_{i_{n-2}i;r_{n-1}}) \\
&\quad \cdot (\sqrt{\lambda_{i_{n-2}}} l_{i_{n-3}i_{n-2};r_{n-2}}) \cdots (\sqrt{\lambda_{i_2}} l_{i_1i_2;r_2})(\sqrt{\lambda_{i_1}} l_{ji_1;r_1}). \qquad (21.40)
\end{aligned}
$$

At this point we use the trick described in the above Exercise 21.5, which tells us that the joint distribution of the $b_{ij;r}$ will not change if in (21.40) we replace every $l_{hk;r}$ by $\lambda_k^{1/2} l_{hk;r}$ and every $l^*_{hk;r}$ by $\lambda_k^{-1/2} l^*_{hk;r}$. We obtain that $R_{b_{11;1},\dots,b_{ij;r},\dots,b_{dd;s}} = R_{b'_{11;1},\dots,b'_{ij;r},\dots,b'_{dd;s}}$, where:

$$
\begin{aligned}
b'_{ij;r} &= l^*_{ij;r} + \alpha_{(r)}\delta_{i,j}1_{\mathcal{B}} + \sum_{r_1=1}^{s} \lambda_i l_{ji;r_1} \\
&\quad + \sum_{n=3}^{k} \sum_{\substack{1\le r_1,\dots r_{n-1}\le s \\ 1\le i_1,\dots,i_{n-2}\le d}} \alpha_{(r_1,\dots,r_{n-1},r)} \lambda_i \lambda_{i_1} \cdots \lambda_{i_{n-2}}\, l_{i_{n-2}i;r_{n-1}} \\
&\quad \cdot l_{i_{n-3}i_{n-2};r_{n-2}} \cdots l_{i_1i_2;r_2} l_{ji_1;r_1}. \qquad (21.41)
\end{aligned}
$$

In Equation (21.41) let us now factor $l^*_{ij;r}$ on the left. We get:

$$
\begin{aligned}
b'_{ij;r} &= l^*_{ij;r}\Big(1_{\mathcal{B}} + \alpha_{(r)}\delta_{i,j} l_{ij;r} + \\
&\sum_{n=2}^{N} \sum_{\substack{1\le r_1,\dots r_n\le s,\ r_n=r \\ 1\le i_1,\dots i_n\le d,\ i_{n-1}=i,\ i_n=j}} \alpha_{(r_1,\dots,r_n)} \lambda_{i_1} \cdots \lambda_{i_{n-1}}\, l_{i_{n-1}i_n;r_n} \cdots l_{i_1i_2;r_2} l_{i_ni_1;r_1}\Big).
\end{aligned}
\qquad (21.42)
$$

Dropping the restrictions "$r_n = r$" and "$i_{n-1} = i,\ i_n = j$" in the summations on the right-hand side of (21.42) will not change that expression, because the extra terms added to the summations are multiplied to 0 by the factor $l^*_{ij;r}$ on the left of the parentheses. Hence we arrive to the formula:

$$
b'_{ij;r} = l^*_{ij;r}(1_{\mathcal{B}} + x), \quad 1 \le i,j \le d,\ 1 \le r \le s, \qquad (21.43)
$$

where

$$x = \sum_{r_1=1}^{s}\sum_{i_1=1}^{d} \alpha_{(r_1)} l_{i_1 i_1; r_1} + \sum_{n=2}^{k} \sum_{\substack{1\leq r_1,\ldots r_n\leq s \\ 1\leq i_1,\ldots i_n\leq d}} \alpha_{(r_1,\ldots,r_n)} \lambda_{i_1}\cdots\lambda_{i_{n-1}}\, l_{i_{n-1}i_n;r_n}\cdots l_{i_1 i_2;r_2} l_{i_n i_1;r_1}. \tag{21.44}$$

But Theorem 21.4 allows us to read explicitly, from Equations (21.43) and (21.44), what is the R-transform of the family $\{b'_{ij;r} \mid 1 \leq i,j \leq d,\ 1 \leq r \leq s\}$; in this way (21.39) is obtained. □

Exercises

EXERCISE 21.19. Let $(\mathcal{A}, \varphi)$ be a $*$-probability space, and let $l_1, \ldots, l_s \in \mathcal{A}$ be a free family of Cuntz isometries.

(1) Show that the elements listed on the right-hand side of Equation (21.3) in Remark 21.3 are linearly independent, hence form a linear basis for $*$-alg$(l_1, \ldots, l_s)$.

(2) By appealing directly to the definition of free independence, show that $\{l_1, l_1^*\}, \ldots, \{l_s, l_s^*\}$ are freely independent in $(\mathcal{A}, \varphi)$.

Exercises 21.20–21.22 fill in the details left about the bijection $\Phi : NC(n) \to [s]^*_{i_1,\ldots,i_n}$ which was used in the proof of Theorem 21.4. We refer to that proof for the definition of the set $[s]^*_{i_1,\ldots,i_n}$, and for the description of how Φ works.

EXERCISE 21.20. Verify that Φ indeed takes values into the set $[s]^*_{i_1,\ldots,i_n}$.

EXERCISE 21.21. Consider the set $Luk(n)$ of Lukasiewicz paths with n steps, which was introduced in Lecture 9 (cf. Definition 9.6.2).

(1) Prove that if $(w_1, \ldots, w_n) \in [s]^*_{i_1,\ldots,i_n}$, then the n-tuple $(|w_n| - 1, \ldots, |w_1| - 1)$ is the rise-vector of a path in $Luk(n)$.

(2) Consider the map $\Psi : [s]^*_{i_1,\ldots,i_n} \to Luk(n)$ which associates to $(w_1, \ldots, w_n) \in [s]^*_{i_1,\ldots,i_n}$ the unique path $\gamma \in Luk(n)$ with rise-vector equal to $(|w_n| - 1, \ldots, |w_1| - 1)$. Prove that Ψ is injective.

[Hint for (2): Proceed by induction on n. More precisely, use the fact that if $(w_1, \ldots, w_n) \in [s]^*_{i_1,\ldots,i_n}$, then the word w_n must be of the form $(i_n) \cdot w'$, with $|w'| = |w| - 1$, and where $(w_1, \ldots, w_{n-2}, w_{n-1}w' \in [s]^*_{i_1,\ldots,i_{n-1}}$.]

EXERCISE 21.22. Consider the functions $\Phi : NC(n) \to [s]^*_{i_1,\dots,i_n}$ and $\Psi : [s]^*_{i_1,\dots,i_n} \to Luk(n)$ which appeared in the preceding two exercises. Consider also the bijection $\Lambda : NC(n) \to Luk(n)$ which was put into evidence in Proposition 9.8. Verify that $\Psi \circ \Phi = \Lambda$ and then (by using the facts that Λ is bijective and Ψ is injective) prove that Φ is a bijection.

REMARK 21.23. Consider the particular case $s = 1$ of Theorem 21.4, where we thus deal with only one isometry in a $*$-probability space $(\mathcal{A}, \varphi)$. We will denote this isometry by l (rather than by l_1, as it is denoted in Theorem 21.4). So we have that $l^*l = 1_{\mathcal{A}}$ and that $\varphi(l^m l^{*n}) = 0$ for every non-negative integers m, n with $m+n \geq 1$. Theorem 21.4 says that for a polynomial $f(z) = \alpha_1 z + \alpha_2 z^2 + \cdots + \alpha_k z^k$, the element

$$a := l^*(1_{\mathcal{A}} + f(l)) \in \mathcal{A}$$

has R-transform $R_a = f$.

The next exercise presents an analogous construction, observed by Haagerup, which involves an S-transform instead of an R-transform.

EXERCISE 21.24. Let $(\mathcal{A}, \varphi)$ be a $*$-probability space, and let $l \in \mathcal{A}$ be an isometry such that $\varphi(l^m l^{*n}) = 0$ for every non-negative integers m, n with $m+n \geq 1$. Let $g(z) = \beta_0 + \beta_1 z + \cdots + \beta_k z^k$ be a polynomial in $\mathbb{C}[z]$ such that $\beta_0 \neq 0$. Consider the element

$$b := g(l)(1_{\mathcal{A}} + l^*) \in \mathcal{A}.$$

Prove that the S-transform of b is

$$S_b(z) = \frac{1}{g(z)}$$

(where the reciprocal $1/g(z)$ is considered in the algebra of formal power series in z).

[Hint: Let $\widetilde{b} := (1_{\mathcal{A}} + l^*)g(l) \in \mathcal{A}$. Observe that the distributions of b and $\widetilde{b}$ are related by the formula $\varphi(b^n) = \beta_0 \varphi(\widetilde{b}^{n-1})$, $n \geq 1$. Use Theorem 21.4 to compute the R-transform of $\widetilde{b}$, then use the connections between moment series, R-transforms and S-transforms which were presented in the section on S-transforms of Lecture 18.]

LECTURE 22

Gaussian random matrices

In the final two lectures we want to treat one of the most important and inspiring realizations of free independence. Canonical examples for free random variables appeared in the context of group algebras of free products of groups and in the context of creation and annihilation operators on full Fock spaces. These are two (closely related) examples where the occurrence of free independence is not very surprising, because its definition was just modeled according to the situation on the group (or von Neumann) algebra of the free group.

But there are objects from a quite different mathematical universe which are also free (at least asymptotically), namely special *random matrices*. A priori, random matrices have nothing to do with free independence and this surprising connection is one of the key results in free probability theory. It establishes links between quite different fields.

We will present in this and the next lecture the fundamental results of Voiculescu on the asymptotic free independence of special random matrices. Our approach will be quite combinatorial and fits well with our combinatorial description of free independence. In a sense, we will show that the combinatorics of free probability theory arises as the limit $N \to \infty$ of the combinatorics of the considered $N \times N$ random matrices.

Moments of Gaussian random variables

Random matrices are matrices whose entries are classical random variables, and the most important class of random matrices are the so-called Gaussian random matrices whose entries form a Gaussian family of classical random variables. So, before we talk about random matrices, we should recall the basic properties of Gaussian families.

A Gaussian family is a collection of classical random variables whose joint density has a very special form, as given in the following definition. In this and the next lecture we will denote the states corresponding to classical probability spaces usually by E, i.e.

$$E[a] := \int_\Omega a(\omega)\, dP(\omega)$$

for classical random variables $a \in L^{\infty-}(\Omega, P)$.

DEFINITION 22.1. (1) A family of selfadjoint random variables $x_1, \dots, x_n$ living in some $*$-probability space $(L^{\infty-}(\Omega, P), E)$ is called a **(centered) Gaussian family** if its joint density is of a Gaussian form, i.e. if there exists a non-singular positive $n \times n$ matrix C such that we have for all $k \in \mathbb{N}$ and all $1 \leq i(1), \dots, i(k) \leq n$ that

$$E[x_{i(1)} \cdots x_{i(k)}] = (2\pi)^{-n/2}(\det C)^{-1/2} \int_{\mathbb{R}^n} t_{i(1)} \cdots t_{i(k)} e^{-\frac{1}{2}\langle t, C^{-1}t\rangle} \, dt_1 \cdots dt_n, \quad (22.1)$$

where $t = (t_1, \dots, t_n) \in \mathbb{R}^n$ and $\langle \cdot, \cdot \rangle$ denotes the standard inner product in $\mathbb{R}^n$. We call C the **covariance matrix** of the Gaussian family. ((2) A family of classical complex-valued random variables $a_1, \dots, a_n$ is a **complex Gaussian family** if the collection of their real and imaginary parts $\Re a_1, \Im a_1, \dots, \Re a_n, \Im a_n$ is a Gaussian family.

REMARK 22.2. The definition via Equation (22.1) is equivalent to saying that the characteristic function of the random vector $x = (x_1, \dots, x_n)$ is of the form

$$E[e^{i\langle t, x\rangle}] = \exp\left\{-\frac{1}{2}\langle t, Ct\rangle\right\}.$$

Of course, we expect that a Gaussian family should be the limit distribution appearing in a multivariate version of the classical central limit theorem. In Remarks 8.18 we pointed out that such a limit distribution has a very nice combinatorial description of its joint moments in terms of summing over all pairings (as opposed to a semicircular family, where we only sum over non-crossing pairings). That this combinatorial formula is indeed the result if one evaluates the integrals in (22.1) goes usually under the name of "Wick formula." This will be the starting point for our use of Gaussian families. We will leave the proof of this to the reader.

THEOREM 22.3. **(Wick formula)**
Let $x_1, \dots, x_n$ be a Gaussian family. Then we have for all $k \in \mathbb{N}$ and all $1 \leq i(1), \dots, i(k) \leq n$ that

$$E[x_{i(1)} \cdots x_{i(k)}] = \sum_{\pi \in \mathcal{P}_2(k)} \prod_{(r,s) \in \pi} E[x_{i(r)} x_{i(s)}]. \quad (22.2)$$

Here, $\mathcal{P}_2(k)$ denotes the set of all pairings of the set $\{1, \dots, k\}$.
If $C = (c_{ij})_{i,j=1}^n$ is the covariance matrix of the Gaussian family then we have

$$E[x_i x_j] = c_{ij} \qquad (i, j = 1, \dots, n).$$

EXERCISE 22.4. Prove Theorem 22.3.
[Hint: by diagonalizing C one can reduce the proof to the case of n independent normal random variables and thus to Exercise 8.22.]

REMARKS 22.5. (1) Of course, if k is odd, there is no pairing in $\mathcal{P}_2(k)$ and thus (22.2) states that each odd moment of a Gaussian family has to vanish.

(2) Strictly speaking, the Wick formula (22.2) was for real-valued Gaussian variables $x_1, \dots, x_n$. However, the nice feature is that it remains valid for complex-valued Gaussian variables $a_1, \dots, a_n$ – not only by replacing the x_i by $\Re a_i$ or $\Im a_i$, but also by replacing them by a_i or $\bar{a}_i$. (This follows directly from the multilinear structure of the formula.) This is important for us because the entries of our Gaussian random matrices will be complex valued.

Random matrices in general

In Lecture 1 we briefly addressed random matrices as an example of a non-commutative probability space. Let us repeat here the relevant information.

Random matrices are matrices whose entries are classical random variables. As usual, in our algebraic frame, we encode a classical probability space (Ω, P) by the algebra $L^{\infty-}(\Omega, P)$ of random variables for which all moments exist and by the state E, which is given by taking the expectation with respect to P. For the matrix part, the most canonical choice of a linear functional is given by taking the trace. Note that one can identify $N \times N$ matrices over an algebra $\mathcal{A}$, which we denote according to Exercise 1.23 by $M_N(\mathcal{A})$, with the tensor product $M_N(\mathbb{C}) \otimes \mathcal{A}$, and so we will denote the corresponding linear functional on $M_N\big(L^{\infty-}(\Omega, P)\big)$ by $\mathrm{tr} \otimes E$.

DEFINITION 22.6. A $*$-probability space of $\boldsymbol{N \times N}$ **random matrices** is given by $(M_N(L^{\infty-}(\Omega, P)), \mathrm{tr} \otimes E)$, where (Ω, P) is a classical probability space,

$$L^{\infty-}(\Omega, P) := \bigcap_{1 \le p < \infty} L^p(\Omega, P),$$

$M_N(\mathcal{A})$ denotes $N \times N$ matrices with entries from $\mathcal{A}$, E denotes the expectation with respect to P and tr denotes the normalized trace on $M_N(\mathbb{C})$. More concretely, this means elements in our probability space are of the form

$$A = (a_{ij})_{i,j=1}^N, \qquad \text{with} \qquad a_{ij} \in L^{\infty-}(\Omega, P)$$

and

$$(\mathrm{tr} \otimes E)(A) = E[\mathrm{tr}(A)] = \frac{1}{N} \sum_{i=1}^{N} E[a_{ii}].$$

The $*$-operation is given by

$$A^* = (\hat{a}_{ij})_{i,j=1}^{N} \qquad \text{with} \qquad \hat{a}_{ij} := \bar{a}_{ji}.$$

The choice of the trace as the state for matrices might look a bit arbitrary, so let us recall in the following remark the relevance of this.

REMARKS 22.7. (1) If one is dealing with matrices, then the most important information is contained in their eigenvalues and the most prominent analytical object is the **eigenvalue distribution**. This is, by definition, a probability measure which puts mass $1/N$ on each of the N eigenvalues (counted with multiplicity) of the $N \times N$ matrix. Assume we have a normal (e.g. a selfadjoint or a unitary) matrix A with eigenvalues $\lambda_1, \dots, \lambda_N$, then its eigenvalue distribution is the probability measure

$$\mu_A := \frac{1}{N}(\delta_{\lambda_1} + \cdots + \delta_{\lambda_N}).$$

Note that the unitary invariance of the trace shows that the $*$-moments of this measure are exactly the $*$-moments of our matrix with respect to the trace,

$$\mathrm{tr}(A^k A^{*l}) = \frac{1}{N} \sum_{i=1}^{N} \lambda_i^k \bar{\lambda}_i^l = \int_{\mathbb{C}} z^k \bar{z}^l \, d\mu_A(z) \qquad \forall\, k, l \in \mathbb{N}.$$

This says that the eigenvalue distribution μ_A is the $*$-distribution in analytical sense of the matrix A with respect to the trace. So the trace encodes exactly that kind of information in which one usually is interested when dealing with matrices.

(2) Generalizing the above argument, one sees that the $*$-distribution in analytical sense of a normal random matrix (with respect to $\mathrm{tr} \otimes E$) will usually be given by the averaged eigenvalue distribution. If $A(\omega) = (a_{ij}(\omega))_{i,j=1}^{N}$ is a normal matrix for all $\omega \in \Omega$, and if $\lambda_1(\omega), \dots, \lambda_N(\omega)$ are the eigenvalues of $A(\omega)$, then the **averaged eigenvalue distribution** of A is defined as

$$\mu_A = \frac{1}{N} \int_{\Omega} \sum_{i=1}^{N} \delta_{\lambda_i(\omega)} \, dP(\omega).$$

This definition of μ_A by a measure-valued integral just means that we have for all $k, l \in \mathbb{N}$

$$\operatorname{tr} \otimes E(A^k A^{*l}) = \frac{1}{N} \int_\Omega \sum_{i=1}^{N} \lambda_i(\omega)^k \overline{\lambda_i(\omega)}^l \, dP = \int_{\mathbb{C}} z^k \bar{z}^l \, d\mu_A(z).$$

For the usual random matrix ensembles the averaged eigenvalue distribution μ_A has no compact support; however in most interesting cases (as for Gaussian random matrices which we will consider in the next section) it is determined by its moments, and thus we can identify in such cases the averaged eigenvalue distribution of A with the $*$-distribution in analytical sense of A with respect to $\operatorname{tr} \otimes E$.

In the generality as considered up to now there is not much more interesting to say about random matrices; for concrete statements we have to specify the classical distribution P, i.e. the joint distribution of the entries of our matrices.

Selfadjoint Gaussian random matrices and genus expansion

A selfadjoint Gaussian random matrix $A = (a_{ij})_{i,j=1}^N$ is a special random matrix where the distribution of the entries is specified as follows:

- the matrix is selfadjoint, $A = A^*$, which means for its entries:
$$a_{ij} = \bar{a}_{ji} \qquad \forall\, i, j = 1, \dots, N;$$
- apart from this restriction on the entries, we assume that they are independent Gaussian random variables (real on the diagonal, complex above the diagonal) with variance $1/N$.

We can summarize this in the following form.

DEFINITION 22.8. A **selfadjoint Gaussian random matrix** is an $N \times N$ random matrix $A = (a_{ij})_{i,j=1}^N$ with $A = A^*$ and such that the entries a_{ij} $(i, j = 1, \dots, N)$ form a complex Gaussian family which is determined by the covariance

$$E[a_{ij} a_{kl}] = \frac{1}{N} \delta_{il} \delta_{jk} \qquad (i, j, k, l = 1, \dots, N). \tag{22.3}$$

REMARKS 22.9. (1) Note that (22.3) determines together with the selfadjointness conditions $a_{ij} = \bar{a}_{ji}$ the whole covariance matrix of the complex entries; namely we have, for example, that

$$E[a_{ij} \bar{a}_{kl}] = E[a_{ij} a_{lk}] = \frac{1}{N} \delta_{ik} \delta_{jl} \qquad (i, j, k, l = 1, \dots, N).$$

(2) In general, if one prescribes a covariance matrix for a Gaussian family one has to think about whether there exists a Gaussian family

with these covariances (which amounts to the positivity of the covariance matrix C in the definition of a Gaussian family). However, in the present case our concrete realization of the Gaussian family in the form preceding our Definition 22.8 ensures this.

(3) Note that we only consider matrices with complex entries; usually such an ensemble is addressed as GUE (Gaussian unitary ensemble). There exist also GOE (Gaussian orthogonal ensemble) and GSE (Gaussian symplectic ensemble) where the entries of the matrices are either real or quaternionic. Note that "unitary," "orthogonal," or "symplectic" refers here to the group under which the respective ensemble is invariant.

(4) The choice of the variance $1/N$ in our definition is just convention and will only become important when we consider the limit $N \to \infty$.

One can also consider non-selfadjoint Gaussian random matrices where all entries are independent.

DEFINITION 22.10. A **non-selfadjoint Gaussian random matrix** is an $N \times N$ random matrix $B = (b_{ij})_{i,j=1}^{N}$ such that the entries b_{ij} $(i, j = 1, \ldots, N)$ form a complex Gaussian family which is determined by the covariance $(i, j, k, l = 1, \ldots, N)$

$$\begin{aligned} E[b_{ij}\bar{b}_{kl}] &= \frac{1}{N}\delta_{ik}\delta_{jl} \\ E[b_{ij}b_{kl}] &= 0. \end{aligned} \tag{22.4}$$

We will in the following consider only the selfadjoint case; by "Gaussian random matrix" we will always mean a selfadjoint Gaussian random matrix. However, the non-selfadjoint versions are also quite interesting and we will address them in some of the exercises.

Our main goal is to calculate the distribution of a (selfadjoint!) Gaussian random matrix A. Calculating directly the eigenvalues of A is not very feasible, however moments are quite accessible. In the following we put

$$\varphi := \mathrm{tr} \otimes E.$$

The mth moment of A is then given by

$$\varphi(A^m) = \frac{1}{N} \sum_{i(1),\ldots,i(m)=1}^{N} E[a_{i(1)i(2)}a_{i(2)i(3)} \cdots a_{i(m)i(1)}].$$

Now we use the fact that the entries of our matrix form a Gaussian family with the covariance as described in (22.3). So, by using the Wick formula (22.2), we can continue as follows (where we count modulo m, i.e. we put $i(m+1) := i(1)$):

$$\begin{aligned}\varphi(A^m) &= \frac{1}{N} \sum_{i(1),\dots,i(m)=1}^{N} \sum_{\pi\in\mathcal{P}_2(m)} \prod_{(r,s)\in\pi} E\big[a_{i(r)i(r+1)}a_{i(s)i(s+1)}\big] \\ &= \frac{1}{N} \sum_{i(1),\dots,i(m)=1}^{N} \sum_{\pi\in\mathcal{P}_2(m)} \prod_{(r,s)\in\pi} \delta_{i(r)i(s+1)}\delta_{i(s)i(r+1)} \frac{1}{N^{m/2}} \\ &= \frac{1}{N^{1+m/2}} \sum_{\pi\in\mathcal{P}_2(m)} \sum_{i(1),\dots,i(m)=1}^{N} \prod_{(r,s)\in\pi} \delta_{i(r)i(s+1)}\delta_{i(s)i(r+1)}.\end{aligned}$$

It is convenient to identify a pairing $\pi \in \mathcal{P}_2(m)$ with a special permutation in S_m, just by declaring the blocks of π to be cycles; thus $(r,s) \in \pi$ means then $\pi(r) = s$ and $\pi(s) = r$. The advantage of this interpretation becomes apparent from the fact that in this language we can rewrite our last equation as

$$\begin{aligned}\varphi(A^m) &= \frac{1}{N^{1+m/2}} \sum_{\pi\in\mathcal{P}_2(m)} \sum_{i(1),\dots,i(m)=1}^{N} \prod_{r=1}^{m} \delta_{i(r)i(\pi(r)+1)} \\ &= \frac{1}{N^{1+m/2}} \sum_{\pi\in\mathcal{P}_2(m)} \sum_{i(1),\dots,i(m)=1}^{N} \prod_{r=1}^{m} \delta_{i(r)i(\gamma\pi(r))},\end{aligned}$$

where $\gamma \in S_m$ is the cyclic permutation with one cycle,

$$\gamma = (1, 2, \dots, m-1, m).$$

If we also identify an m-index tuple $i = (i(1), \dots, i(m))$ with a function $i : \{1, \dots, m\} \to \{1, \dots, N\}$, then the meaning of $\prod_{r=1}^{m} \delta_{i(r)i(\gamma\pi(r))}$ is quite obvious, namely it says that the function i must be constant on the cycles of the permutation $\gamma\pi$ in order to contribute a factor 1, otherwise its contribution will be zero. But in this interpretation

$$\sum_{i(1),\dots,i(m)=1}^{N} \prod_{r=1}^{m} \delta_{i(r)i(\gamma\pi(r))}$$

is very easy to determine: for each cycle of $\gamma\pi$ we can choose one of the numbers $1, \dots, N$ for the constant value of i on this orbit, and all these choices are independent from each other, which means

$$\sum_{i(1),\dots,i(m)=1}^{N} \prod_{r=1}^{m} \delta_{i(r)i(\gamma\pi(r))} = N^{\text{number of cycles of } \gamma\pi}.$$

NOTATION 22.11. For a permutation $\sigma \in S_m$ we put

$$\#(\sigma) := \text{number of cycles of } \sigma.$$

So we have finally derived the following theorem.

THEOREM 22.12. *For a selfadjoint Gaussian $N \times N$ random matrix we have for all $m \in \mathbb{N}$ that*

$$\varphi(A^m) = \sum_{\pi \in \mathcal{P}_2(m)} N^{\#(\gamma\pi)-1-m/2}.$$

This type of expansion for moments of random matrices is usually called a **genus expansion**, because pairings in S_m can also be identified with orientable surfaces (by gluing the edges of an m-gon together according to π) and then the corresponding exponent of N can be expressed (via Euler's formula) in terms of the genus g of the surface,

$$\#(\gamma\pi) - 1 - m/2 = -2g.$$

EXAMPLES 22.13. Let us look at some examples. Clearly, since there are no pairings of an odd number of elements, all odd moments of A are zero. So it is enough to consider the even powers $m = 2k$.

For $m = 2$, the formula just gives a contribution for the pairing $(1,2) \in S_2$,

$$\varphi(A^2) = 1.$$

This reflects our normalization with the factor $1/N$ for the variances of the entries to ensure that $\varphi(A^2)$ is equal to 1 (in particular, does not depend on N).

The first non-trivial case is $m = 4$. Then we have three pairings, and the relevant information about them is contained in the following table:

π	$\gamma\pi$	$\#(\gamma\pi) - 3$
$(1,2)(3,4)$	$(1,3)(2)(4)$	0
$(1,3)(2,4)$	$(1,4,3,2)$	-2
$(1,4)(2,3)$	$(1)(2,4)(3)$	0

so that we have

$$\varphi(A^4) = 2 \cdot N^0 + 1 \cdot N^{-2}.$$

For $m = 6, 8, 10$ an inspection of the 15, 105, 945 pairings of 6, 8, 10 elements yields in the end

$$\varphi(A^6) = 5 \cdot N^0 + 10 \cdot N^{-2}$$

$$\varphi(A^8) = 14 \cdot N^0 + 70 \cdot N^{-2} + 21 \cdot N^{-4}$$
$$\varphi(A^{10}) = 42 \cdot N^0 + 420 \cdot N^{-2} + 483 \cdot N^{-4}.$$

As one might suspect, the pairings contributing in leading order N^0 are exactly the non-crossing pairings.

EXERCISE 22.14. Check that in the case $m = 6$ the pairings contributing the leading order N^0 are exactly the non-crossing pairings

$$(1,2)(3,4)(5,6), \quad (1,2)(3,6)(4,5), \quad (1,4)(2,3)(5,6)$$
$$(1,6)(2,3)(4,5), \quad (1,6)(2,5)(3,4).$$

This fact that the leading orders correspond to non-crossing pairings is true in general. (In the geometric language of genus expansion, the non-crossing pairings correspond to genus zero or planar situations.) We leave its proof for the moment as an exercise, but we will come back to this in the next lecture in a more general context.

EXERCISE 22.15. Show that, for a pairing $\pi \in S_{2k}$, $\#(\gamma\pi)$ can at most be $1 + k$, and this upper bound is achieved if and only if π is non-crossing.

But this tells us that, although the moments of a Gaussian $N \times N$ random matrix for fixed N are quite involved, in the limit $N \to \infty$ they become much simpler and converge to something which we understand quite well, namely to the number of non-crossing pairings, which is given by the Catalan numbers,

$$\lim_{N\to\infty} \varphi(A^{2k}) = C_k = \frac{1}{k+1}\binom{2k}{k}.$$

On the other hand, the number of non-crossing pairings counts the even moments of semicircular elements. Thus we can rephrase the above, by using our notion of convergence in distribution, also in the following form.

THEOREM 22.16. **(Wigner's semicircle law)**
For each $N \in \mathbb{N}$, let A_N be a selfadjoint Gaussian $N \times N$-random matrix. Then A_N converges, for $N \to \infty$, in distribution towards a semicircular element s,

$$A_N \xrightarrow{\text{distr}} s.$$

EXERCISE 22.17. For each $N \in \mathbb{N}$, let B_N be a non-selfadjoint Gaussian $N \times N$ random matrix. Show that B_N converges in $*$-distribution towards a circular element.

EXERCISE 22.18. (1) Let B_N be a non-selfadjoint Gaussian $N \times N$ random matrix as in Definition 22.10 and put $X_N := B_N^* B_N$. Show that X_N converges in distribution towards a free Poisson element of rate $\lambda = 1$.

(2) Generalize part (1) by considering rectangular Gaussian $M \times N$ random matrices $B_{M,N}$, where we still assume the covariance as in (22.4), but now for $i, k = 1, \ldots, M$ and $j, k = 1, \ldots, N$. Assume that we send $M, N \to \infty$ such that the ratio M/N has a finite limit $\lambda > 0$. Show that $B_{M,N}^* B_{M,N}$ converges in distribution towards a free Poisson element of rate λ.

[Matrices of the form $B_{M,N}^* B_{M,N}$ are in the random matrix literature usually called *Wishart matrices*. Their limiting eigenvalue distribution (which is the same as a free Poisson distribution) was calculated by Marchenko and Pastur in 1967 and is accordingly referred to as *Marchenko–Pastur distribution.*]

(3) By using the fact that, for a matrix A, the non-zero eigenvalues of A^*A and of AA^* agree, derive a relation between the density of a free Poisson distribution of rate λ and the density of a free Poisson distribution of rate $1/\lambda$. Check your result by using the concrete form of the densities as given in (12.14) and (12.15).

Asymptotic free independence for several independent Gaussian random matrices

The fact that the eigenvalue distribution of Gaussian random matrices converges in the limit $N \to \infty$ to the semicircle distribution is one of the basic results in random matrix theory; it was proved by Wigner in 1955, and is accordingly usually termed "Wigner's semicircle law." Thus the semicircle distribution appeared as an interesting object long before semicircular elements were considered in free probability. This raises the question whether it is just a coincidence that Gaussian random matrices in the limit $N \to \infty$ and the sum of creation and annihilation operators on full Fock spaces have the same distribution or whether there is some deeper connection. Of course, our main interest is in the question whether there is also some free independence around for random matrices. As we see from the case of one Gaussian random matrix, it is only in the limit $N \to \infty$ where we can expect nice behavior for random matrices. Thus what we can hope for is "asymptotic free independence" for random matrices. Let us first make this notion precise.

DEFINITION 22.19. Let, for each $N \in \mathbb{N}$, $(\mathcal{A}_N, \varphi_N)$ be a non-commutative probability space. Let I be an index set and consider for each $i \in I$ and each $N \in \mathbb{N}$ random variables $a_i^{(N)} \in \mathcal{A}_N$. Let $I = I_1 \cup \cdots \cup I_m$ be a decomposition of I into m disjoint subsets. We say that

$$\{a_i^{(N)} \mid i \in I_1\}, \dots, \{a_i^{(N)} \mid i \in I_m\}$$

are **asymptotically free** (for $N \to \infty$), if $(a_i^{(N)})_{i \in I}$ converges in distribution towards $(a_i)_{i \in I}$ for some random variables $a_i \in \mathcal{A}$ $(i \in I)$ in some non-commutative probability space $(\mathcal{A}, \varphi)$ and if the limits $\{a_i \mid i \in I_1\}, \dots, \{a_i \mid i \in I_m\}$ are free in $(\mathcal{A}, \varphi)$.

REMARKS 22.20. (1) Thus $a^{(N)}$, $b^{(N)}$ asymptotically free means that in the limit $N \to \infty$ their mixed moments can be expressed in terms of the moments of $a^{(N)}$ and the moments of $b^{(N)}$ by the same formula which describes the corresponding mixed moment of free random variables. For example, we must have

$$\lim_{N \to \infty} \varphi_N(a^{(N)} b^{(N)}) = \lim_{N \to \infty} \varphi_N(a^{(N)}) \cdot \lim_{N \to \infty} \varphi_N(b^{(N)}).$$

(2) Asymptotic free independence of sequences $(a_i^{(N)})_{i \in I}$ can be characterized as follows by the asymptotic form of the definition of free independence: whenever we have, for some positive integer k, $i(1), \dots, i(k) \in I$ with $i(1) \neq i(2) \neq \cdots \neq i(k)$ and polynomials p_j $(j = 1, \dots, k)$ such that

$$\lim_{N \to \infty} \varphi_N\big(p_j(a_{i(j)}^{(N)})\big) = 0$$

for all $j = 1, \dots, k$, then we must also have the asymptotic vanishing of the corresponding alternating moment,

$$\lim_{N \to \infty} \varphi_N\big(p_1(a_{i(1)}^{(N)}) \cdots p_k(a_{i(k)}^{(N)})\big) = 0.$$

(3) Note that the existence of the limits of all moments is required as part of the definition of asymptotic free independence.

That we have indeed asymptotic free independence for random matrices is one of the fundamental discoveries of Voiculescu in free probability theory.

In order to have asymptotic free independence we should consider at least two random matrices. Let us try the simplest case, by taking two Gaussian random matrices

$$A^{(1)} = (a_{ij}^{(1)})_{i,j=1}^N, \qquad A^{(2)} = (a_{ij}^{(2)})_{i,j=1}^N.$$

Of course, we must also specify the relation between them, i.e. we must prescribe the joint distribution of the whole family

$$a_{11}^{(1)}, \dots, a_{NN}^{(1)}, a_{11}^{(2)}, \dots, a_{NN}^{(2)}.$$

Again we stick to the simplest possible case and assume that all entries of $A^{(1)}$ are independent from all entries of $A^{(2)}$; i.e. we consider now two independent Gaussian random matrices.

To put it more formally (and in order to use our Wick formula), the collection of all entries of our two matrices forms a complex Gaussian family with covariance

$$E[a_{ij}^{(r)} a_{kl}^{(p)}] = \frac{1}{N}\delta_{il}\delta_{jk}\delta_{rp} \qquad (i,j,k,l = 1,\dots,N;\ r,p = 1,2). \quad (22.5)$$

We will see that we can extend the genus expansion to this situation. Actually, it turns out that we can just repeat the above calculations putting superindices $p(1),\dots,p(m) \in \{1,2\}$ at our matrices. The main arguments are not affected by this:

$$\varphi(A^{(p(1))}\cdots A^{(p(m))}) = \frac{1}{N}\sum_{i(1),\dots,i(m)=1}^{N} E[a_{i(1)i(2)}^{(p(1))} a_{i(2)i(3)}^{(p(2))}\cdots a_{i(m)i(1)}^{(p(m))}]$$

$$= \frac{1}{N}\sum_{i(1),\dots,i(m)=1}^{N}\ \sum_{\pi\in\mathcal{P}_2(m)}\ \prod_{(r,s)\in\pi} E[a_{i(r)i(r+1)}^{(p(r))} a_{i(s)i(s+1)}^{(p(s))}]$$

$$= \frac{1}{N}\sum_{i(1),\dots,i(m)=1}^{N}\ \sum_{\pi\in\mathcal{P}_2(m)}\ \prod_{(r,s)\in\pi} \delta_{i(r)i(s+1)}\delta_{i(s)i(r+1)}\delta_{p(r)p(s)}\frac{1}{N^{m/2}}$$

$$= \frac{1}{N^{1+m/2}}\sum_{\pi\in\mathcal{P}_2(m)}\ \sum_{i(1),\dots,i(m)=1}^{N}\ \prod_{r=1}^{m} \delta_{i(r)i(\gamma\pi(r))}\delta_{p(r)p(\pi(r))}.$$

The only difference from the situation of one matrix is now the extra factor $\prod_{r=1}^{m}\delta_{p(r)p(\pi(r))}$, which just says that we have an extra condition on our pairings π: they must pair the same matrices, i.e. no block of π is allowed to pair $A^{(1)}$ with $A^{(2)}$. If π has this property then its contribution will be as before, otherwise it will be zero. Let us introduce for this the following notation.

NOTATION 22.21. For $p = (p(1),\dots,p(m))$, we put

$$\mathcal{P}_2^{(p)}(m) := \{\pi \in \mathcal{P}_2(m) \mid p(\pi(r)) = p(r) \text{ for all } r = 1,\dots,m\}.$$

Thinking of p as a coloring of the points $1,\dots,m$ with colors $p(1),\dots,p(m)$, we will also address elements from $\mathcal{P}_2^{(p)}(m)$ as pairings which **respect the coloring p**.

Then we can write the final conclusion of our calculation as follows.

PROPOSITION 22.22. *Let $A^{(1)}$ and $A^{(2)}$ be two independent selfadjoint Gaussian random matrices. Then we have for all choices of*

$m \in \mathbb{N}$ and $p(1), \dots, p(m) \in \{1,2\}$ that

$$\varphi(A^{(p(1))} \cdots A^{(p(m))}) = \sum_{\pi \in \mathcal{P}_2^{(p)}(m)} N^{\#(\gamma\pi)-1-m/2}. \tag{22.6}$$

EXAMPLES 22.23. Here are a few examples for $m = 4$:

$$\varphi(A^{(1)}A^{(1)}A^{(1)}A^{(1)}) = 2 \cdot N^0 + 1 \cdot N^{-2}$$
$$\varphi(A^{(1)}A^{(1)}A^{(2)}A^{(2)}) = 1 \cdot N^0 + 0 \cdot N^{-2}$$
$$\varphi(A^{(1)}A^{(2)}A^{(1)}A^{(2)}) = 0 \cdot N^0 + 1 \cdot N^{-2}.$$

As before, the leading term for $N \to \infty$, is given by contributions from non-crossing pairings, but now these non-crossing pairings must connect an $A^{(1)}$ with an $A^{(1)}$ and an $A^{(2)}$ with an $A^{(2)}$. But this is exactly the rule for calculating mixed moments in a semicircular system consisting of two free semicircular elements (compare Example 8.21). Thus we see that we indeed have asymptotic free independence between two independent Gaussian random matrices. Of course, the same is true if we consider n independent Gaussian random matrices instead of two – they are becoming asymptotically free.

THEOREM 22.24. *Let $A_N^{(1)}, \dots, A_N^{(n)}$ be, for each $N \in \mathbb{N}$, an independent family of selfadjoint Gaussian $N \times N$ random matrices. Then $(A_N^{(1)}, \dots, A_N^{(n)})$ converges in distribution to a semicircular system $(s_1, \dots, s_n)$ consisting of n free standard semicircular elements. In particular, $A_N^{(1)}, \dots, A_N^{(n)}$ are asymptotically free.*

EXERCISE 22.25. Let $B_N^{(1)}, \dots, B_N^{(n)}$ be, for each $N \in \mathbb{N}$, an independent family of non-selfadjoint Gaussian $N \times N$ random matrices. Show that $(B_N^{(1)}, \dots, B_N^{(n)})$ converges in $*$-distribution towards a family $(c_1, \dots, c_n)$, where each c_i is circular, and $c_1, \dots, c_n$ are $*$-free.

Asymptotic free independence between Gaussian random matrices and constant matrices

Theorem 22.24, Voiculescu's generalization of Wigner's semicircle law is a great step; not only do we find the semicircular distribution in random matrices, but the concept of free independence itself also shows up very canonically for random matrices. However, it might appear that in the situation considered above we find free independence only for a very restricted class of distributions, namely for semicircular elements. In classical probability theory, this would be comparable to saying that we understand the concept of independence for Gaussian families. Of course, this is only a very restricted version and we should aim at finding more general appearances of asymptotic free independence in the random matrix world.

Here is the next step in this direction. Instead of looking on the relation between two Gaussian random matrices we now replace one of them by a "constant" or "non-random" matrix. This just means ordinary matrices $M_N(\mathbb{C})$ without any randomness, where the state is given by taking the trace tr (the expectation acts trivially).

DEFINITION 22.26. For a given non-commutative probability space $(M_N(L^{\infty-}(\Omega, P)), \text{tr}\otimes E)$ of random matrices, we address matrices from

$$M_N(\mathbb{C}) \cong M_N(\mathbb{C} \cdot 1_{L^{\infty-}(\Omega,P)}) \subset M_N(L^{\infty-}(\Omega, P))$$

as **constant matrices.**

Of course, we expect free independence only asymptotically, so what we are really looking at is a sequence of constant matrices D_N, which converges in distribution for $N \to \infty$. Thus we assume the existence of all limits

$$\lim_{N\to\infty} \text{tr}(D_N^m) \qquad (m \in \mathbb{N}).$$

Let us denote this limit by an element d in some non-commutative probability space $(\mathcal{A}, \psi)$, i.e. we assume $D_N \overset{\text{distr}}{\longrightarrow} d$.

Note that we have great freedom in prescribing the wanted moments in the limit. For example, we can take diagonal matrices for the D_N and then we can approximate any fixed, let us say compactly supported, probability measure on $\mathbb{R}$ by suitably chosen matrices.

EXERCISE 22.27. Let μ be a probability measure on $\mathbb{R}$ for which all moments exist. Construct an explicit sequence of selfadjoint matrices $D_N \in M_N$, such that the distribution of D_N with respect to the trace converges to μ, i.e. such that we have

$$\lim_{N\to\infty} \text{tr}[D_N^m] = \int_{\mathbb{R}} t^m \, d\mu(t)$$

for all $m \in \mathbb{N}$.

As for the Gaussian random matrices we will usually suppress the index N at our constant matrices to lighten the notation. But one should keep in mind that we are talking about sequences of $N \times N$ matrices and that we want to take the limit $N \to \infty$ in the end.

Let us now see how far we can go with our above calculations in such a situation where we have a Gaussian $N \times N$ random matrix A and a constant $N \times N$ matrix D. What we would like to understand are mixed moments in A and D. We can put this in the form

$$\varphi(AD^{q(1)}AD^{q(2)}\cdots AD^{q(m)})$$

where each $D^{q(i)}$ is some power of the matrix D.

Let us now do the calculation of such alternating moments in D and A. We will denote the entries of the matrix $D^{q(i)}$ by $d_{ij}^{(i)}$.

$$\begin{aligned}
&\varphi(AD^{q(1)}\cdots AD^{q(m)})\\
&= \frac{1}{N}\sum_{\substack{i(1),\dots,i(m),\\ j(1),\dots,j(m)=1}}^{N} E\big[a_{i(1)j(1)}d^{(1)}_{j(1)i(2)}a_{i(2)j(2)}d^{(2)}_{j(2)i(3)}\cdots a_{i(m)j(m)}d^{(m)}_{j(m)i(1)}\big]\\
&= \frac{1}{N}\sum_{\substack{i(1),\dots,i(m),\\ j(1),\dots,j(m)=1}}^{N} E\big[a_{i(1)j(1)}a_{i(2)j(2)}\cdots a_{i(m)j(m)}\big]\cdot d^{(1)}_{j(1)i(2)}d^{(2)}_{j(2)i(3)}\cdots d^{(m)}_{j(m)i(1)}\\
&= \frac{1}{N}\sum_{\substack{i(1),\dots,i(m),\\ j(1),\dots,j(m)=1}}^{N}\sum_{\pi\in\mathcal{P}_2(m)}\prod_{(r,s)\in\pi} E\big[a_{i(r)j(r)}a_{i(s)j(s)}\big]\cdot d^{(1)}_{j(1)i(\gamma(1))}\cdots d^{(m)}_{j(m)i(\gamma(m))}\\
&= \frac{1}{N}\sum_{\substack{i(1),\dots,i(m),\\ j(1),\dots,j(m)=1}}^{N}\sum_{\pi\in\mathcal{P}_2(m)}\prod_{(r,s)\in\pi} \delta_{i(r)j(s)}\delta_{i(s)j(r)}\frac{1}{N^{m/2}}\cdot d^{(1)}_{j(1)i(\gamma(1))}\cdots d^{(m)}_{j(m)i(\gamma(m))}.
\end{aligned}$$

Again, we identify a $\pi\in\mathcal{P}_2(m)$ with a permutation in S_m and then it remains to understand, for such a fixed π, the expression

$$\begin{aligned}
&\sum_{\substack{i(1),\dots,i(m),\\ j(1),\dots,j(m)=1}}^{N}\prod_{(r,s)\in\pi}\delta_{i(r)j(s)}\delta_{i(s)j(r)}\cdot d^{(1)}_{j(1)i(\gamma(1))}\cdots d^{(m)}_{j(m)i(\gamma(m))}\\
&\qquad= \sum_{\substack{i(1),\dots,i(m),\\ j(1),\dots,j(m)=1}}^{N}\prod_{r=1}^{m}\delta_{i(r)j(\pi(r))}\cdot d^{(1)}_{j(1)i(\gamma(1))}d^{(2)}_{j(2)i(\gamma(2))}\cdots d^{(m)}_{j(m)i(\gamma(m))}\\
&\qquad= \sum_{j(1),\dots,j(m)=1}^{N} d^{(1)}_{j(1)j(\pi\gamma(1))}d^{(2)}_{j(2)j(\pi\gamma(2))}\cdots d^{(m)}_{j(m)j(\pi\gamma(m))}.
\end{aligned}$$

EXAMPLE 22.28. In order to recognize this as a quite familiar quantity, let us first look at an example. Let us denote $\pi\gamma$ by α and take $\alpha=(1,3,6)(4)(2,5)\in S_6$. Then one has

$$\sum_{j(1),\dots,j(6)=1}^{N} d^{(1)}_{j(1)j(\alpha(1))}d^{(2)}_{j(2)j(\alpha(2))}d^{(3)}_{j(3)j(\alpha(3))}d^{(4)}_{j(4)j(\alpha(4))}d^{(5)}_{j(5)j(\alpha(5))}d^{(6))}_{j(6)j(\alpha(6)}$$

$$
\begin{aligned}
&= \sum_{j(1),\dots,j(6)=1}^{N} d^{(1)}_{j(1)j(3)} d^{(2)}_{j(2)j(5)} d^{(3)}_{j(3)j(6)} d^{(4)}_{j(4)j(4)} d^{(5)}_{j(5)j(2)} d^{(6)}_{j(6)j(1)} \\
&= \sum_{j(1),\dots,j(6)=1}^{N} d^{(1)}_{j(1)j(3)} d^{(3)}_{j(3)j(6)} d^{(6)}_{j(6)j(1)} \cdot d^{(2)}_{j(2)j(5)} d^{(5)}_{j(5)j(2)} \cdot d^{(4)}_{j(4)j(4)} \\
&= \mathrm{Tr}[D^{(1)}D^{(3)}D^{(6)}] \cdot \mathrm{Tr}[D^{(2)}D^{(5)}] \cdot \mathrm{Tr}[D^{(4)}] \\
&= N^3 \cdot \mathrm{tr}[D^{(1)}D^{(3)}D^{(6)}] \cdot \mathrm{tr}[D^{(2)}D^{(5)}] \cdot \mathrm{tr}[D^{(4)}].
\end{aligned}
$$

Note that we denoted by Tr the unnormalized trace and we have also used the abbreviation $D^{(i)} := D^{q(i)}$. We see that the final expression is a product of traces along the cycles of α. It is quite suggestive to use the notation tr_α to denote this product.

Up to now we have introduced the notation φ_π only for non-crossing partitions π. The above suggests that it might be useful to define φ_α also in the case where α is a permutation.

NOTATION 22.29. Let n be a fixed positive integer and let, for all $1 \le k \le n$, multilinear functionals $\varphi_k : \mathcal{A}^k \to \mathbb{C}$ on an algebra $\mathcal{A}$ be given. Assume that each φ_k is tracial in its k arguments in the sense that

$$\varphi_k(A_1, \dots, A_k) = \varphi_k(A_k, A_1, \dots, A_{k-1})$$

for all $A_1, \dots, A_k \in \mathcal{A}$. Then we define for $\alpha \in S_n$ the expression $\varphi_\alpha[A_1, \dots, A_n]$ for $A_1, \dots, A_n \in \mathcal{A}$ as a product according to the cycle decomposition of α. Denote by $c_1, \dots, c_r$ the cycles of α, then we put

$$\varphi_\alpha[A_1, \dots, A_n] := \varphi_{c_1}[A_1, \dots, A_n] \cdots \varphi_{c_r}[A_1, \dots, A_n], \tag{22.7}$$

where, for a cycle $c = (i_1, i_2, \dots, i_p)$ we define

$$\varphi_c[A_1, \dots, A_n] := \varphi_p(A_{i_1}, \dots, A_{i_p}). \tag{22.8}$$

REMARK 22.30. Note that one can consider partitions as permutations in the following way: if π is a partition of $1, 2, \dots, n$, then we get a corresponding permutation $P_\pi \in S_n$ by declaring the blocks of π to cycles of P_π; this means that we have to choose a cyclic order on the blocks; the canonical choice for this is of course the restriction of the cyclic order on $(1, 2, \dots, n)$ to the block. (The special case of this map P for pairings was used in all our calculations in this lecture; the restriction of this map to non-crossing partitions was also introduced in Notations 18.24). Via this mapping our present notation φ_α is a generalization of the corresponding notation for the case of non-crossing partitions from Definition 11.1. Since in general, there is no natural choice for a first or last element in the cycles of a permutation, we have defined φ_α only for the case where the φ_k are tracial in their k arguments.

In our context, the multilinear functionals φ_k are given by

$$\varphi_k(D_1, \dots, D_k) = \mathrm{tr}[D_1 \cdots D_k].$$

Thus they are tracial in their k arguments and our above definition yields the notion tr_α for a product of traces along the cycles of α. That this is exactly what shows up in our above calculations can be seen from the next lemma, whose easy proof we leave to the reader.

LEMMA 22.31. *Let $D^{(1)} = (d_{ij}^{(1)})_{i,j=1}^N$, $\dots$, $D^{(m)} = (d_{i,j}^{(m)})_{i,j=1}^N$ be m $N \times N$ matrices and let $\alpha \in S_m$ be a permutation of m elements. Then we have*

$$\sum_{j(1),\dots,j(m)=1}^{N} d^{(1)}_{j(1)j(\alpha(1))} \cdots d^{(m)}_{j(m)j(\alpha(m))} = N^{\#\alpha} \cdot \mathrm{tr}_\alpha[D^{(1)}, \dots, D^{(m)}].$$

Thus we can write the conclusion of our calculation as follows:

$$\varphi(AD^{q(1)} \cdots AD^{q(m)}) = \sum_{\pi \in \mathcal{P}_2(m)} \mathrm{tr}_{\pi\gamma}[D^{q(1)}, \dots, D^{q(m)}] \cdot N^{\#(\gamma\pi)-1-m/2}.$$

Of course, we can do the same for several independent Gaussian random matrices, the only effect of this is to restrict the sum over π to pairings which respect the "color" of the matrices.

PROPOSITION 22.32. *Let $A^{(1)}$, $\dots$, $A^{(n)}$ be n independent selfadjoint Gaussian $N \times N$ random matrices, and D a constant $N \times N$ matrix. Then we have for all $m \in \mathbb{N}$, all $q(1), \dots, q(m) \in \mathbb{N}$, and all $1 \le p(1), \dots, p(m) \le n$ that*

$$\varphi(A^{(p(1))}D^{q(1)} \cdots A^{(p(m))}D^{q(m)}) = \sum_{\pi \in \mathcal{P}_2^{(p)}(m)} \mathrm{tr}_{\pi\gamma}[D^{q(1)}, \dots, D^{q(m)}] \cdot N^{\#(\gamma\pi)-1-m/2}.$$

Now let us look at the asymptotic structure of this formula. By our assumption that $D_N \xrightarrow{\text{distr}} d$ for some $d \in (\mathcal{A}, \psi)$, the quantity

$$\mathrm{tr}_{\pi\gamma}[D^{q(1)}, \dots, D^{q(m)}]$$

has a limit, namely

$$\psi_{\pi\gamma}[d^{q(1)}, \dots, d^{q(m)}].$$

Since the factor $N^{\#(\gamma\pi)-1-m/2}$ suppresses all crossing pairings in the limit $N \to \infty$ (see Exercise 22.15) we get finally the following result.

PROPOSITION 22.33. *Let, for each $N \in \mathbb{N}$, $A_N^{(1)}, \dots, A_N^{(n)}$ be n independent selfadjoint Gaussian $N \times N$ random matrices, and D_N be a constant $N \times N$ matrix such that $D_N \xrightarrow{\text{distr}} d$ for some $d \in (\mathcal{A}, \psi)$. Then we have for all $m \in \mathbb{N}$, for all $q(1), \dots, q(m) \in \mathbb{N}$, and for all $1 \le p(1), \dots, p(m) \le n$ that*

$$\lim_{N\to\infty} \varphi(A_N^{(p(1))} D_N^{q(1)} \cdots A_N^{(p(m))} D_N^{q(m)}) = \sum_{\pi \in NC_2^{(p)}(m)} \psi_{\pi\gamma}[d^{q(1)}, \dots, d^{q(m)}], \tag{22.9}$$

where $NC_2^{(p)}(m)$ denotes those pairings from $\mathcal{P}_2^{(p)}(m)$ which are non-crossing.

This resembles our formula for alternating moments in two free families of random variables, if one of them is a semicircular system. Recall from Lecture 14: if $d_1, \dots, d_m, s_1, \dots, s_n$ are elements in some non-commutative probability space $(\mathcal{B}, \phi)$ such that $s_1, \dots, s_n$ is a semicircular system, i.e. each s_i is a standard semicircular element and $s_1, \dots, s_n$ are free, and such that $\{d_1, \dots, d_m\}$ and $\{s_1, \dots, s_n\}$ are free, then we have for the alternating moments (see Equation (14.5))

$$\begin{aligned}\phi(s_{p(1)} d_1 \cdots s_{p(m)} d_m) &= \sum_{\pi \in NC(m)} \kappa_\pi[s_{p(1)}, \dots, s_{p(m)}] \cdot \phi_{K(\pi)}[d_1, \dots, d_m] \\ &= \sum_{\pi \in NC_2^{(p)}(m)} \phi_{K(\pi)}[d_1, \dots, d_m]. \end{aligned} \tag{22.10}$$

This matches the structure of our formula (22.9), the only difference is that we have $K(\pi)$ instead of $\pi\gamma$. However, it turns out that for a non-crossing pairing π this is the same; under the canonical embedding of partitions into permutations from Remark 22.30, the complement of a non-crossing pairing π corresponds to the permutation $\pi\gamma$. The reader was asked to prove this in Exercise 18.25. (Actually, in general the complement of π corresponds to $\pi^{-1}\gamma$; for pairings, however, π and π^{-1} coincide.) Here we will be satisfied with checking this for an example.

EXAMPLE 22.34. Consider the non-crossing pairing

$$\pi = \{(1,2), (3,6), (4,5), (7,8)\}.$$

Then we have

$$\pi\gamma = (1), (2,6,8), (3,5), (7).$$

That this agrees with the complement $K(\pi)$ can be seen from the graphical representation

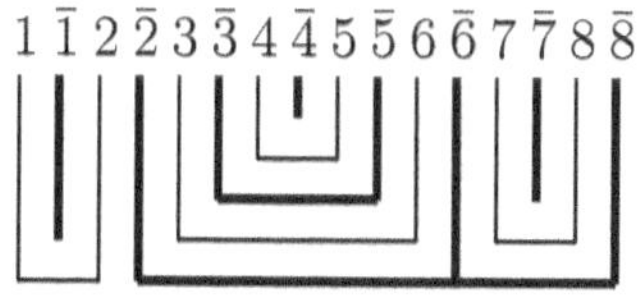

.

To conclude, let us collect what we have observed in the next theorem. Clearly, we can be a bit more general by considering not only one sequence of constant matrices, but several of them. In all our calculations, this just amounts to replacing powers of D by products of the considered constant matrices.

THEOREM 22.35. *Let, for each $N \in \mathbb{N}$, $A_N^{(1)}, \dots, A_N^{(p)}$ be p independent Gaussian random matrices and let $D_N^{(1)}, \dots, D_N^{(q)}$ be q constant matrices which converge in distribution for $N \to \infty$, i.e.*

$$D_N^{(1)}, \dots, D_N^{(q)} \overset{\text{distr}}{\longrightarrow} d_1, \dots, d_q$$

for some $d_1, \dots, d_q \in (\mathcal{A}, \psi)$. Then

$$A_N^{(1)}, \dots, A_N^{(p)}, D_N^{(1)}, \dots, D_N^{(q)} \overset{\text{distr}}{\longrightarrow} s_1, \dots, s_p, d_1, \dots, d_q,$$

where each s_i is a standard semicircular element and where $s_1, \dots, s_p$, $\{d_1, \dots, d_q\}$ are free. In particular, the Gaussian random matrices and the constant matrices are asymptotically free.

These considerations show that random matrices allow asymptotic realizations of two free variables, if one of them has a semicircular distribution. This raises, of course, the question whether we can realize with random matrices free independence between any two distributions without one of them being semicircular. That this is indeed the case relies on having similar asymptotic free independence statements for unitary random matrices instead of Gaussian ones. This will be addressed in our final lecture.

LECTURE 23

Unitary random matrices

Another important random matrix ensemble is given by Haar unitary random matrices – these are unitary matrices equipped with the Haar measure as corresponding probability measure. We will see that one can get asymptotic freeness results for Haar unitary random matrices similar to those we derived for Gaussian random matrices in the last lecture. We will also see that we have asymptotic freeness between constant matrices which are randomly rotated by a Haar unitary random matrix. (This will follow from the fact that conjugation by a free Haar unitary can be used to make general random variables free.)

Our calculations for the unitary random matrices will be of a similar kind to those from the last lecture. The main ingredient is a Wick type formula for correlations of the entries of the Haar unitary random matrices.

Haar unitary random matrices

REMARK 23.1. A fundamental fact in abstract harmonic analysis is that any compact group has an analog of the Lebesgue measure, the so-called Haar measure, which is characterized by the fact that it is invariant under translations by group elements. This Haar measure is finite and unique up to multiplication with a constant, thus we can normalize it to a probability measure – the unique **Haar probability measure** on the compact group. We will use this Haar probability measure for the case of $\mathcal{U}(N)$ – the compact group of unitary $N \times N$ matrices. It is characterized by the fact that it is a probability measure on $\mathcal{U}(N)$, and invariant under multiplication from the right and multiplication from the left with any arbitrary unitary $N \times N$ matrix.

DEFINITION 23.2. We equip the compact group $\mathcal{U}(N)$ of unitary $N \times N$ matrices with its Haar probability measure. Random matrices distributed according to this measure will be called **Haar unitary random matrices**. Thus the expectation E over this ensemble is given by integrating with respect to the Haar probability measure.

REMARKS 23.3. (1) It is not directly clear from this definition how to generate Haar unitary random matrices. An important method is to get them by polar decomposition from Gaussian random matrices. If X is a non-selfadjoint Gaussian $N \times N$ random matrix, and $X = U|X|$ is its polar decomposition, then U is almost surely a unitary matrix and the induced measure on $\mathcal{U}(N)$ by this decomposition is the Haar measure. Another possibility is to take the eigenvectors of a selfadjoint Gaussian random matrix. Again, they form a Haar unitary random matrix. These remarks show that it should be possible to transfer asymptotic freeness results from Gaussian random matrices via polar decomposition to Haar unitary random matrices. These ideas can indeed be worked out rigorously and this was actually the approach of Voiculescu to unitary random matrices.

(2) We prefer here another approach that is more combinatorial in nature and fits very well with our general combinatorial methods. In principle, we are going to imitate in the unitary case the calculations we did for Gaussian random matrices. Clearly, we need for this the expectations of products of entries of a Haar unitary random matrix U. In contrast to the Gaussian case, these are now not explicitly given. In particular, entries of U are in general not independent from each other. However, one expects that the unitary condition $U^*U = 1 = UU^*$ and the invariance of the Haar measure under multiplication from right or from left with an arbitrary unitary matrix should allow determination of all mixed moments of the entries of U. This is indeed the case; the calculation, however, is not trivial and we only describe in the following the final result, which has the flavor of the Wick formula for the Gaussian case.

The expectation of products of entries of Haar distributed unitary random matrices can be described in terms of a special function Wg on the permutation group.

NOTATION 23.4. For $\alpha \in S_n$ and $N \geq n$ we put

$$\mathrm{Wg}(N, \alpha) = E\big[u_{11} \cdots u_{nn} \overline{u}_{1\alpha(1)} \cdots \overline{u}_{n\alpha(n)}\big],$$

where $U = (u_{ij})_{i,j=1}^N$ is a Haar unitary $N \times N$ random matrix. We call Wg the **Weingarten function**.

This $\mathrm{Wg}(N, \alpha)$ depends on the permutation α only through its conjugacy class. The relevance of the Weingarten function for our purposes lies in the fact that general matrix integrals over the unitary groups can be reduced to the knowledge of Wg. This well-known "Wick type" formula is the starting point for our calculations.

LEMMA 23.5. *Let $U = (u_{ij})_{i,j=1}^N$ be a Haar unitary $N \times N$ random matrix. Then we have for all $N \geq n$ and all $1 \leq i(1), \dots, i(n) \leq N$, $1 \leq i'(1), \dots, i'(n) \leq N$, $1 \leq j(1), \dots, j(n) \leq N$, $1 \leq j'(1), \dots, j'(n) \leq N$ that*

$$\begin{aligned} &E\left[u_{i(1)j(1)} \cdots u_{i(n)j(n)} \overline{u}_{i'(1)j'(1)} \cdots \overline{u}_{i'(n)j'(n)}\right] \\ &= \sum_{\alpha,\beta \in S_n} \delta_{i(1)i'(\beta(1))} \cdots \delta_{i(n)i'(\beta(n))} \delta_{j(1)j'(\alpha(1))} \cdots \delta_{j(n)j'(\alpha(n))} \mathrm{Wg}(N, \beta\alpha^{-1}) \\ &= \sum_{\alpha,\beta \in S_n} \delta_{i(\beta(1))i'(1)} \cdots \delta_{i(\beta(n))i'(n)} \delta_{j(\alpha(1))j'(1)} \cdots \delta_{j(\alpha(n))j'(n)} \mathrm{Wg}(N, \beta\alpha^{-1}). \end{aligned}$$

REMARKS 23.6. (1) Note that corresponding integrals for which the number of u and $\bar{u}$ is different vanish, by the invariance of such an expression under the replacement $U \mapsto \lambda U$, where $\lambda \in \mathbb{C}$ with $|\lambda| = 1$.

(2) The preceding remark also shows that we have for any Haar unitary random matrix

$$\mathrm{tr} \otimes E(U^k) = 0 \qquad \text{if } k \in \mathbb{Z} \backslash \{0\},$$

and thus a Haar unitary random matrix is a Haar unitary element in the sense of our Definition 1.12.

(3) The Weingarten function is a quite complicated object. For our purposes, however, only the asymptotics for $N \to \infty$ is important. One knows that the leading order of $\mathrm{Wg}(N, \alpha)$ in $1/N$ is given by $2n - \#(\alpha)$ $(\alpha \in S_n)$ and increases in steps of 2

$$\mathrm{Wg}(N, \alpha) = \phi(\alpha) N^{\#(\alpha) - 2n} + O(N^{\#(\alpha) - 2n - 2}). \tag{23.1}$$

One also knows the function ϕ, however, for our purposes this knowledge is not needed. Actually, we will be able to determine this ϕ from our results and the fact that ϕ is multiplicative (i.e. $\phi(\alpha)$ factorizes according to the cycles of π) and we will see that it is connected with the Möbius function on non-crossing partitions.

(4) Note that we only consider unitary matrices. Similar statements are true for orthogonal and symplectic matrices; however, the behavior of subleading terms in the Weingarten function then becomes more complicated (and the decrease is not in steps of 2 any more, but in steps of 1).

The length function on permutations

In our calculations around asymptotic freeness for Haar unitary random matrices it will be important to control the appearing orders, which are given in terms of number of cycles of permutations. Let us collect here some of the basic notations and properties for later use.

NOTATIONS 23.7. Let S_n denote the symmetric group of permutations of $\{1,\dots,n\}$.

(1) The product of two permutations $\alpha,\beta \in S_n$ is taken in the "natural" order, $\alpha\beta(k) := \alpha(\beta(k))$ $(k = 1,\dots,n)$.

(2) The identity element of S_n will be denoted by e.

(3) Usually, we write a permutation in its cycle decomposition,

$$\alpha = c_1 \cdots c_k,$$

where α restricted to a cycle $c_j = (i_1, i_2, \dots, i_p)$ acts like

$$\alpha(i_1) = i_2, \quad \alpha(i_2) = i_3, \quad \dots, \qquad \alpha(i_p) = i_1.$$

Cycles of length 1, i.e. fixed points, will usually be omitted when writing a permutation as the product of its cycles.

(4) A **transposition** is a permutation of the form $\alpha = (ij)$ for $i \neq j$, i.e. all points with the exception of i and j are fixed points and i and j get exchanged under α.

(5) By $\#(\alpha)$ we denote the number of cycles of the permutation α (also counting fixed points).

(6) We consider on S_n also the **length function** $|\cdot|$, where $|\alpha|$ $(\alpha \in S_n)$ is the minimal non-negative integer k such that α can be written as product of k transpositions,

$$|\alpha| := \min\{k \in \mathbb{N} \mid \alpha = \tau_1 \cdots \tau_k \text{ for some transpositions } \tau_1, \dots, \tau_k\}.$$

By convention, $|e| = 0$.

(7) In the following, γ_n will always denote the cyclic permutation

$$\gamma_n = (1, 2, 3, \dots, n) \in S_n$$

of order n.

REMARKS 23.8. (1) We have $\#\gamma_n = 1$ and $|\gamma_n| = n - 1$.

(2) $|\cdot|$ is clearly invariant under conjugation, i.e.

$$|\beta^{-1}\alpha\beta| = |\alpha| \qquad \text{for all } \alpha, \beta \in S_n.$$

This means that

$$|\alpha\beta| = |\beta(\alpha\beta)\beta^{-1}| = |\beta\alpha|,$$

i.e. $|\cdot|$ is tracial on S_n.

Note that $|\cdot|$ is actually a length function, in particular, it satisfies a triangle inequality.

PROPOSITION 23.9. (1) *We have that* $|\alpha| = 0$ *for* $\alpha \in S_n$ *if and only if* $\alpha = e$.

(2) *We have for all* $\alpha, \beta \in S_n$ *that*

$$|\alpha\beta| \leq |\alpha| + |\beta|. \tag{23.2}$$

PROOF. This is clear by the definition of $|\cdot|$. □

Actually, our two quantities $\#$ and $|\cdot|$ are just two sides of the same coin; this follows easily from the fact that we can control very precisely what happens to the number of cycles when we multiply with a transposition τ. Namely, the number of cycles will either increase by 1 or decrease by 1, corresponding to whether τ cuts a cycle into two or glues two cycles together.

LEMMA 23.10. *Let $\alpha \in S_n$ be an arbitrary permutation and $\tau \in S_n$ a transposition, i.e. $\tau = (ij)$ with $1 \leq i, j \leq n$, $i \neq j$. Then we have*

$$\#(\alpha\tau) = \begin{cases} \#\alpha + 1 & \text{if } i \text{ and } j \text{ belong to the same cycle of } \alpha \\ \#\alpha - 1 & \text{if } i \text{ and } j \text{ belong to different cycles of } \alpha. \end{cases} \tag{23.3}$$

PROOF. We only have to look at the cycles of α containing i and j; the other cycles are not affected by the multiplication with τ.

Let us first assume that i and j are both contained in the same cycle c of α, say

$$c = (i, i_1, \dots, i_k, j, j_1, \dots, j_l).$$

Then we have

$$c \cdot (i, j) = (i, j_1, \dots, j_l)(j, i_1, \dots, i_k),$$

and thus the multiplication by τ increases the number of cycles by 1.

If on the other hand i and j are contained in different cycles, say c_1 and c_2, of α,

$$c_1 c_2 = (i, i_1, \dots, i_k)(j, j_1, \dots, j_l)$$

then we have

$$c_1 c_2 \cdot (ij) = (i, j_1, \dots, j_l, j, i_1, \dots, i_k),$$

and thus the multiplication by τ reduces the number of cycles by 1. □

PROPOSITION 23.11. *For any $\alpha \in S_n$ we have*

$$|\alpha| = n - \#(\alpha).$$

PROOF. If $|\alpha| = k$, then we can write $\alpha = \tau_1 \cdots \tau_k$ for transpositions $\tau_1, \dots, \tau_k$. By the above lemma, the multiplication of k transpositions can reduce the number of cycles, starting from the identity e with n cycles, at most by k, thus we have

$$\#(\alpha) \geq n - k = n - |\alpha|.$$

On the other hand, we can write each cycle of length k as a product of $k - 1$ transpositions,

$$(i_1, i_2, \dots, i_k) = (i_1, i_2) \cdot (i_2, i_3) \cdots (i_{k-1}, i_k),$$

and thus we can write any permutation $\alpha \in S_n$ with r cycles as a product of $n - r$ transpositions, thus

$$|\alpha| \leq n - r = n - \#(\alpha).$$

□

REMARK 23.12. By Proposition 23.11, the leading order (23.1) of the Weingarten function can be written in terms of $|\cdot|$ as

$$\mathrm{Wg}(N, \alpha) = \phi(\alpha) N^{-|\alpha|-n} + O(N^{-|\alpha|-n-2}).$$

Asymptotic freeness for Haar unitary random matrices

By using Lemma 23.5 instead of the Wick formula (22.2), one can show that one has the same kind of asymptotic freeness results for Haar unitary random matrices as for Gaussian random matrices.

The main result is the analog of Theorem 22.35

THEOREM 23.13. *Let, for each $N \in \mathbb{N}$, $U_N^{(1)}, \dots, U_N^{(p)}$ be p independent Haar unitary random matrices and let $D_N^{(1)}, \dots, D_N^{(q)}$ be q constant matrices which converge in $*$-distribution (with respect to* tr*) for $N \to \infty$, i.e.*

$$D_N^{(1)}, \dots, D_N^{(q)} \overset{*-\mathrm{distr}}{\longrightarrow} d_1, \dots, d_q$$

for some $d_1, \dots, d_q \in (\mathcal{A}, \psi)$. Then

$$U_N^{(1)}, \dots, U_N^{(p)}, D_N^{(1)}, \dots, D_N^{(q)} \overset{*-\mathrm{distr}}{\longrightarrow} u_1, \dots, u_p, d_1, \dots, d_q,$$

where $u_1, \dots, u_p, \{d_1, \dots, d_q\}$ are $*$-free and where each u_i is a Haar unitary element. In particular, the Haar unitary random matrices are asymptotically $*$-free from the constant matrices.

We will not prove the theorem in this generality, but will restrict it to a special case. In order to motivate this, let us return to our question whether we can model by random matrices free independence between any two distributions. The above theorem does not seem to provide an answer to this question, since it tells us that we get asymptotic freeness between a constant matrix (which can have an arbitrary distribution) and a Haar unitary random matrix (which has a very special distribution). However, one has to notice that one can use conjugation by a free Haar unitary to make general random variables free. (The reader was asked to prove this in Exercise 5.24.) Thus, the above Theorem 23.13 contains in particular the statement that if we have constant matrices A_N and B_N such that A_N, B_N have a limit distribution for $N \to \infty$, then $U_N A_N U_N^*$ and B_N are asymptotically free. In this formulation we can prescribe the limiting distributions of A_N (which is the same as the distribution of $U_NA_NU_N^*$) and of B_N quite arbitrarily and we thus get a random matrix realization for free independence between general distributions. In what follows we will focus on this latter consequence about randomly rotated matrices. The proof of the more general Theorem 23.13 is along the same lines and will be left to the reader.

Note also, that we have to ask for the existence of all mixed moments in A_N and B_N in the limit $N \to \infty$ in order to be able to apply Theorem 23.13 to the situation of randomly rotated matrices. One could satisfy this by a more careful asymptotic realization of two given distributions by a pair of diagonal constant matrices; however, as we will see in the following section, this assumption is actually not needed.

Asymptotic freeness between randomly rotated constant matrices

What we want to prove is the asymptotic freeness between two sequences of constant matrices which are randomly rotated against each other with the help of a Haar unitary random matrix. Here is the precise statement.

THEOREM 23.14. *Let $(A_N)_{N\in\mathbb{N}}$ and $(B_N)_{N\in\mathbb{N}}$ be sequences of $N\times N$ matrices such that A_N converges in distribution (with respect to* tr*) for $N\to\infty$, and such that B_N converges in distribution (with respect to* tr*) for $N\to\infty$. Furthermore, let $(U_N)_{N\in\mathbb{N}}$ be a sequence of Haar unitary $N\times N$ random matrices. Then, $U_NA_NU_N^*$ and B_N are asymptotically free for $N\to\infty$.*

As in the case of Gaussian random matrices the proof of this consists mainly in calculating mixed moments in our random matrices and realizing that in the limit $N\to\infty$ this converges to an expression which we recognize as corresponding to a free situation.

As usual, we suppress in the following calculations the subindex N. Thus we have to look at expressions like

$$\mathrm{tr}\otimes E(UA^{q(1)}U^*B^{p(1)}\cdots UA^{q(n)}U^*B^{p(n)}),$$

involving some powers (which might also be zero) of our matrices A and B. We will denote these powers by

$$A^{(k)} = A^{q(k)} = (a_{ij}^{(k)})_{i,j=1}^N, \qquad B^{(k)} = B^{p(k)} = (b_{ij}^{(k)})_{i,j=1}^N$$

and also use the abbreviation

$$\gamma := \gamma_n = (1,2,\ldots,n)\in S_n$$

in the following. Then we have (by using Lemma 23.5 for the expectation of products of entries of our Haar unitary matrix) that

$$\begin{aligned}
&\mathrm{tr}\otimes E\big(UA^{(1)}U^*B^{(1)}\cdots UA^{(n)}U^*B^{(n)}\big)\\
&=\frac{1}{N}\sum_{\substack{i(1),\dots,i(n),i'(1),\dots,i'(n),\\ j(1),\dots,j(n),j'(1),\dots,j'(n)=1}}^{N} E\big[u_{i(1)j(1)}a^{(1)}_{j(1)j'(1)}\bar{u}_{i'(1)j'(1)}b^{(1)}_{i'(1)i(2)}\cdots\\
&\qquad\qquad\cdots u_{i(n)j(n)}a^{(n)}_{j(n)j'(n)}u_{i'(n)j'(n)}b^{(n)}_{i'(n)i(1)}\big]\\
&=\frac{1}{N}\sum_{\substack{i(1),\dots,i(n),i'(1),\dots,i'(n),\\ j(1),\dots,j(n),j'(1),\dots,j'(n)=1}}^{N} a^{(1)}_{j(1)j'(1)}\cdots a^{(n)}_{j(n)j'(n)}b^{(1)}_{i'(1)i(2)}\cdots b^{(n)}_{i'(n)i(1)}\\
&\qquad\qquad\cdot E\big[u_{i(1)j(1)}\bar{u}_{i'(1)j'(1)}\cdots u_{i(n)j(n)}\bar{u}_{i'(n)j'(n)}\big]\\
&=\frac{1}{N}\sum_{\substack{i(1),\dots,i(n),i'(1),\dots,i'(n),\\ j(1),\dots,j(n),j'(1),\dots,j'(n)=1}}^{N} a^{(1)}_{j(1)j'(1)}\cdots a^{(n)}_{j(n)j'(n)}b^{(1)}_{i'(1)i(2)}\cdots b^{(n)}_{i'(n)i(1)}\\
&\quad\cdot\sum_{\alpha,\beta\in S_n}\delta_{i(\beta(1))i'(1)}\cdots\delta_{i(\beta(n))i'(n)}\delta_{j(\alpha(1))j'(1)}\cdots\delta_{j(\alpha(n))j'(n)}\mathrm{Wg}(N,\alpha^{-1}\beta)\\
&=\frac{1}{N}\sum_{\beta,\alpha\in S_n}\mathrm{Wg}(N,\alpha^{-1}\beta)\\
&\qquad\cdot\sum_{\substack{i'(1),\dots,i'(n),\\ j(1),\dots,j(n)=1}}^{N} a^{(1)}_{j(1)j(\alpha(1))}\cdots a^{(n)}_{j(n)j(\alpha(n))}b^{(1)}_{i'(1)i'(\beta^{-1}\gamma(1))}\cdots b^{(n)}_{i'(n)i'(\beta^{-1}\gamma(n))}\\
&=\frac{1}{N}\sum_{\alpha,\beta\in S_n}\mathrm{Wg}(N,\alpha^{-1}\beta)\cdot\mathrm{Tr}_\alpha[A^{(1)},\dots,A^{(n)}]\cdot\mathrm{Tr}_{\beta^{-1}\gamma}[B^{(1)},\dots,B^{(n)}]\\
&=\sum_{\alpha,\beta\in S_n}\mathrm{Wg}(N,\alpha^{-1}\beta)N^{\#(\alpha)+\#(\beta^{-1}\gamma)-1}\cdot\mathrm{tr}_\alpha[A^{(1)},\dots,A^{(n)}]\\
&\qquad\qquad\cdot\mathrm{tr}_{\beta^{-1}\gamma}[B^{(1)},\dots,B^{(n)}].
\end{aligned}$$

According to our assumption on the existence of a limit distribution for the A and for the B, the expressions

$$\mathrm{tr}_\alpha[A^{(1)},\dots,A^{(n)}]\qquad\text{and}\qquad\mathrm{tr}_{\beta^{-1}\gamma}[B^{(1)},\dots,B^{(n)}]$$

have a limit for $N\to\infty$. Furthermore, we know that the Weingarten function has the asymptotics

$$\mathrm{Wg}(N,\alpha^{-1}\beta)=\phi(\alpha^{-1}\beta)N^{\#(\alpha^{-1}\beta)-2n}+O(N^{\#(\alpha\beta^{-1})-2n-2}).$$

Thus the leading order in our above calculation is of the form

$$N^{\#(\alpha^{-1}\beta)+\#(\alpha)+\#(\beta^{-1}\gamma)-2n-1} = N^{n-1-|\alpha^{-1}\beta|-|\alpha|-|\beta^{-1}\gamma|}. \tag{23.4}$$

Now note that by the triangle inequality (23.2) we have

$$n-1 = |\gamma| = |\alpha(\alpha^{-1}\beta)\beta^{-1}\gamma| \le |\alpha| + |\alpha^{-1}\beta| + |\beta^{-1}\gamma|,$$

hence the highest possible order in (23.4) is N^0 and in the limit $N \to \infty$ we will remain exactly with those pairs (α, β) for which we have equality $|\alpha^{-1}\beta| + |\alpha| + |\beta^{-1}\gamma| = n-1$. So we have proved the following proposition.

PROPOSITION 23.15. *Let, for each $N \in \mathbb{N}$, $N \times N$ matrices $A_N^{(1)}, \ldots, A_N^{(n)}$ and $N \times N$ matrices $B_N^{(1)}, \ldots, B_N^{(n)}$ in $(M_N(\mathbb{C}), \mathrm{tr})$ be given such that*

$$A_N^{(1)}, \ldots, A_N^{(n)} \xrightarrow{\text{distr}} a_1, \ldots, a_n$$

for $a_1, \ldots, a_n$ in some probability space $(\mathcal{A}, \varphi)$ and such that

$$B_N^{(1)}, \ldots, B_N^{(n)} \xrightarrow{\text{distr}} b_1, \ldots, b_n$$

for $b_1, \ldots, b_n$ in some probability space $(\mathcal{B}, \psi)$. Let, for each $N \in \mathbb{N}$, U_N be a Haar unitary $N \times N$ random matrix. Then we have

$$\begin{aligned}&\lim_{N\to\infty} \mathrm{tr} \otimes E[U_N A_N^{(1)} U_N^* B_N^{(1)} \cdots U_N A_N^{(n)} U_N^* B_N^{(n)}] \\ &\quad = \sum_{\substack{\alpha,\beta \in S_n \\ |\alpha^{-1}\beta|+|\alpha|+|\beta^{-1}\gamma|=n-1}} \varphi_\alpha[a_1, \ldots, a_n] \cdot \psi_{\beta^{-1}\gamma}[b_1, \ldots, b_n] \cdot \phi(\alpha^{-1}\beta).\end{aligned} \tag{23.5}$$

In order to come back to the proof of our Theorem 23.14, we extend, by multilinearity, the formula (23.5) from powers to general polynomials in our matrices A_N and B_N.

COROLLARY 23.16. *Let $(A_N)_{N\in\mathbb{N}}$ be a sequence of $N \times N$ matrices A_N, which converges in distribution (with respect to* tr*) for $N \to \infty$,*

$$A_N \xrightarrow{\text{distr}} a \qquad \text{for } a \text{ in some probability space } (\mathcal{A}, \varphi),$$

and let $(B_N)_{N\in\mathbb{N}}$ be a sequence of $N \times N$ matrices B_N which converges (with respect to tr*) in distribution for $N \to \infty$,*

$$B_N \xrightarrow{\text{distr}} b \qquad \text{for } b \text{ in some probability space } (\mathcal{B}, \psi).$$

Furthermore, let $(U_N)_{N\in\mathbb{N}}$ be a sequence of Haar unitary $N \times N$ random matrices U_N. Then we have for all $n \in \mathbb{N}$ and all polynomials

$f_1, \dots, f_n, g_1, \dots, g_n$ *that*

$$\lim_{N\to\infty} \operatorname{tr} \otimes E[U_N f_1(A_N) U_N^* g_1(B_N) \cdots U_N f_n(A_N) U_N^* g_n(B_N)] = \sum_{\substack{\alpha,\beta\in S_n \\ |\alpha^{-1}\beta|+|\alpha|+|\beta^{-1}\gamma|=n-1}} \varphi_\alpha[f_1(a), \dots, f_n(a)] \cdot \psi_{\beta^{-1}\gamma}[g_1(b), \dots, g_n(b)] \cdot \phi(\alpha^{-1}\beta). \tag{23.6}$$

This explicit formula for the limit of the mixed moments in our randomly rotated matrices gives us directly the required asymptotic freeness, if we also make the following small observation.

LEMMA 23.17. *Consider $\alpha, \beta \in S_n$. If $|\alpha^{-1}\beta| + |\alpha| + |\beta^{-1}\gamma| = n-1$, then at least one of the permutations α and $\beta^{-1}\gamma$ must have a fixed point.*

PROOF. The assumption implies that we have $|\alpha| \leq (n-1)/2$ or $|\beta^{-1}\gamma| \leq (n-1)/2$. But the product of at most $(n-1)/2$ transpositions can move at most $n-1$ elements, hence we get the existence of at least one fixed point as asserted. □

PROOF OF THEOREM 23.14. We will show the asymptotic freeness directly by verifying the asymptotic form of free independence, as in Remarks 22.20. So assume that we have polynomials $f_1, \dots, f_n$ and $g_1, \dots, g_n$ such that for $k = 1, \dots, n$

$$\lim_{N\to\infty} \operatorname{tr} \otimes E[f_j(U_N A_N U_N^*)] = 0 \tag{23.7}$$

and

$$\lim_{N\to\infty} \operatorname{tr} \otimes E[g_j(B_N)] = 0. \tag{23.8}$$

We have to show that the corresponding alternating products tend to zero, too. Since we are working with respect to a trace, we can always assume that our alternating product starts with an f. We have to distinguish the cases that it ends with an f or with an g, i.e. we have to show that

$$\operatorname{tr} \otimes E[f_1(U_N A_N U_N^*) g_1(B_N) \cdots f_n(U_N A_N U_N^*) g_n(B_N)] \tag{23.9}$$

and

$$\operatorname{tr} \otimes E[f_1(U_N A_N U_N^*) g_1(B_N) \cdots g_{n-1}(B_N) f_n(U_N A_N U_N^*)] \tag{23.10}$$

tend to zero. We will only prove (23.9). The other case is similar and will be left to the reader, see Exercise 23.25.

In order to prove (23.9) note that

$$\lim_{N\to\infty} \operatorname{tr}[f_j(A_N)] = \lim_{N\to\infty} \operatorname{tr} \otimes E[f_j(U_N A_N U_N)] = 0$$

and

$$\lim_{N\to\infty} \operatorname{tr}[g_j(B_N)] = \lim_{N\to\infty} \operatorname{tr}\otimes E[g_j(B_N)] = 0.$$

Thus our assumptions (23.7) and (23.8) on the f_j and g_j mean that $\varphi(f_j(a)) = 0$ and $\psi(g_j(b)) = 0$ for all $j = 1, \ldots, n$. Furthermore, we have

$$\begin{aligned} &\operatorname{tr}\otimes E\big[f_1(U_N A_N U_N^*)g_1(B_N)\cdots f_n(U_N A_N U_N^*)g_n(B_N)\big] \\ &\qquad = \operatorname{tr}\otimes E\big[U_N f_1(A_N)U_N^* g_1(B_N)\cdots U_N f_n(A_N)U_N^* g_n(B_N)\big]. \end{aligned}$$

But according to Corollary 23.16 we know how to calculate the limit of the latter quantity. We have to sum in (23.6) over all pairs (α, β) in S_n which fulfill $|\alpha^{-1}\beta| + |\alpha| + |\beta^{-1}\gamma| = n - 1$. Thus, by Lemma 23.17, at least one of the partitions α and $\beta^{-1}\gamma$ has a fixed point, implying that at least one of the factors $\varphi_\alpha[f_1(a), \ldots, f_n(a)]$ and $\psi_{\beta^{-1}\gamma}[g_1(b), \ldots, g_n(b)]$ is equal to zero. Hence the whole sum (23.6) vanishes and we get the assertion. □

REMARK 23.18. By our combinatorial description of freeness we know that we can describe the mixed moments of two free sets $\{a_1, \ldots, a_n\}$, $\{b_1, \ldots, b_n\}$ as follows:

$$\begin{aligned} \varphi(a_1 b_1 \cdots a_n b_n) &= \sum_{\pi\in NC(n)} \kappa_\pi[a_1, \ldots, a_n] \cdot \varphi_{K(\pi)}[b_1, \ldots, b_n] \\ &= \sum_{\substack{\pi,\sigma\in NC(n)\\ \sigma\le\pi}} \varphi_\sigma[a_1, \ldots, a_n] \cdot \varphi_{K(\pi)}[b_1, \ldots, b_n] \cdot \mu(\sigma, \pi). \end{aligned} \tag{23.11}$$

According to formula (23.5) in Proposition 23.15 (and the knowledge that these randomly rotated matrices are asymptotically free) we now also have the formula

$$\begin{aligned} &\varphi(a_1 b_1 \cdots a_n b_n) \\ &\quad = \sum_{\substack{\alpha,\beta\in S_n\\ |\alpha^{-1}\beta|+|\alpha|+|\beta^{-1}\gamma|=n-1}} \varphi_\alpha[a_1, \ldots, a_n] \cdot \varphi_{\beta^{-1}\gamma}[b_1, \ldots, b_n] \cdot \phi(\alpha\beta^{-1}). \end{aligned} \tag{23.12}$$

Of course, these two formulas should coincide. That this is indeed the case, relies on the embedding P of non-crossing partitions $NC(n)$ into the symmetric group S_n, which we introduced in Notations 18.24. In order to see that this mapping P indeed transforms (23.11) into (23.12), we still have to recognize the image of $NC(n)$ in S_n under P. This will be done in the next section.

Embedding of non-crossing partitions into permutations

In order to understand better the condition

$$|\alpha^{-1}\beta| + |\alpha| + |\beta^{-1}\gamma| = n - 1,$$

which appears in (23.12), we will discuss it in connection with a natural distance function on the symmetric group.

DEFINITION 23.19. With the help of the length function $|\cdot|$ we can define a **distance** d on S_n in the canonical way

$$d(\alpha,\beta) := |\alpha^{-1}\beta| = |\beta\alpha^{-1}| \qquad (\alpha,\beta \in S_n).$$

PROPOSITION 23.20. *The function d on S_n is a distance function, i.e. we have*

(1) $d(\alpha,\beta) = 0$ *if and only if* $\alpha = \beta$,

(2) $d(\alpha,\beta) = d(\beta,\alpha)$ *for all* $\alpha,\beta \in S_n$,

(3) d *satisfies the triangle inequality, i.e. for all* $\alpha_1,\alpha_2,\alpha_3 \in S_n$ *we have*

$$d(\alpha_1,\alpha_3) \leq d(\alpha_1,\alpha_2) + d(\alpha_2,\alpha_3).$$

PROOF. The first and third parts follow directly from the corresponding properties of $|\cdot|$, see Proposition 23.9. The second part follows from the fact that the number of cycles of α agrees with the number of cycles of α^{-1}. □

One sees now that the condition $|\alpha^{-1}\beta| + |\alpha| + |\beta^{-1}\gamma_n| = n - 1$ on α and β in (23.12) is actually the geodesic condition

$$d(e,\alpha) + d(\alpha,\beta) + d(\beta,\gamma_n) = d(e,\gamma_n).$$

Thus the permutations α and β appearing in (23.12) are those lying on the same geodesic from the identity element e to the cycle γ_n, such that α lies before β on this geodesic. Let us introduce some notation to describe these properties.

NOTATION 23.21. (1) We denote the permutations from S_n which lie on a geodesic from e to γ_n by

$$\begin{aligned} S_{NC}(\gamma_n) :&= \{\alpha \in S_n \mid d(e,\alpha) + d(\alpha,\gamma_n) = d(e,\gamma_n)\} \\ &= \{\alpha \in S_n \mid |\alpha| + |\alpha^{-1}\gamma_n| = n - 1\}. \end{aligned} \tag{23.13}$$

(2) For $\alpha,\beta \in S_{NC}(\gamma_n)$ we say that $\alpha \leq \beta$ if α and β lie on the same geodesic and if α comes before β; this means $d(e,\alpha)+d(\alpha,\beta) = d(\alpha,\beta)$, i.e.

$$|\alpha| + |\alpha^{-1}\beta| = |\beta|.$$

PROPOSITION 23.22. (1) *If* $\alpha \in S_{NC}(\gamma_n)$, *then* $\alpha^{-1}\gamma_n, \gamma_n\alpha^{-1} \in S_{NC}(\gamma_n)$, *too.*

(2) *The set* $S_{NC}(\gamma_n)$ *consists exactly of those* $\alpha \in S_n$ *for which there exist* $n-1$ *transpositions* $\tau_1, \dots, \tau_{n-1} \in S_n$ *such that*

$$\gamma_n = \tau_{n-1} \cdots \tau_1 \qquad \text{and} \qquad \alpha = \tau_k \cdots \tau_1$$

for some $1 \le k \le n-1$ *(and then necessarily,* $k = |\alpha|$*).*

(3) *The condition* $\alpha \le \beta$ *is equivalent to: there exist* $n-1$ *transpositions* $\tau_1, \dots, \tau_{n-1}$ *such that* $\gamma_n = \tau_{n-1} \cdots \tau_1$ *and*

$$\alpha = \tau_k \cdots \tau_1 \qquad \text{and} \qquad \beta = \tau_l \cdots \tau_{k+1} \underbrace{\tau_k \cdots \tau_1}_{\alpha}$$

for some $1 \le k \le l \le n-1$.

PROOF. (1) Put $\hat{\alpha} := \alpha^{-1}\gamma_n$. Then we have $\alpha = \gamma_n \hat{\alpha}^{-1}$, so that the condition $|\alpha| + |\alpha^{-1}\gamma_n| = n-1$ reads as $|\gamma_n\hat{\alpha}^{-1}| + |\hat{\alpha}| = n-1$ in terms of $\hat{\alpha}$. The assertion follows now from Remarks 23.8 that $|\cdot|$ is a trace, i.e. that $|\gamma_n\hat{\alpha}^{-1}| = |\hat{\alpha}^{-1}\gamma_n|$.

(2) Let $\gamma_n = \tau_{n-1} \cdots \tau_1$ and $\alpha = \tau_k \cdots \tau_1$. Then $|\alpha| \le k$ and

$$|\alpha^{-1}\gamma_n| = |\gamma_n\alpha^{-1}| = |\tau_{n-1} \cdots \tau_{k+1}| \le n-k.$$

Since $|\gamma_n| = n-1$, this implies, by the triangle inequality, that we actually have equality in both these estimates and thus $|\alpha| + |\alpha^{-1}\gamma_n| = n-1$.

The other way around, consider an $\alpha \in S_n$ with $|\alpha| + |\alpha^{-1}\gamma_n| = n-1$. Let us put $k := |\alpha|$, then we have

$$|\gamma_n\alpha^{-1}| = |\alpha^{-1}\gamma_n| = n-k.$$

Thus we can write α as a product of k transpositions and $\gamma_n\alpha^{-1}$ as a product of $n-k$ transpositions, let us say,

$$\alpha = \tau_k \cdots \tau_1 \qquad \text{and} \qquad \gamma_n\alpha^{-1} = \tau_{n-1} \cdots \tau_{k+1}.$$

But then we also get

$$\gamma_n = (\gamma_n\alpha^{-1})\alpha = \tau_{n-1} \cdots \tau_1.$$

(3) The third part is proved in the same way as the second one. □

With the relation "$\le$", $S_{NC}(\gamma_n)$ becomes a poset, and in the light of the correspondence between (23.11) and (23.12) we expect that this poset should be isomorphic to $NC(n)$. This is indeed the case, and the isomorphism is given by our embedding

$$P : NC(n) \to S_n$$
$$\pi \mapsto P_\pi,$$

which we introduced in Notations 18.24. Recall that P was defined by declaring the blocks of $\pi \in NC(n)$ to become cycles of $P_\pi \in S_n$. The only non-trivial point is to choose an order on the block to change it to a cycle, and for this we just take the induced order coming from the cycle $\gamma_n = (1, 2, \ldots, n) \in S_n$.

PROPOSITION 23.23. *We have* $P(NC(n)) = S_{NC}(\gamma_n)$ *and*

$$P : NC(n) \to S_{NC}(\gamma_n)$$

is an isomorphism of posets.

PROOF. Note that, by Proposition 23.22, any $\alpha \in S_{NC}(\gamma_n)$ is, for some k and some transpositions $\tau_1, \ldots, \tau_k$, of the form $\tau_k \cdots \tau_1 \gamma_n$, such that each multiplication with a transposition reduces the length by 1,

$$|\tau_l(\tau_{l-1} \cdots \tau_1 \gamma_n)| = |\tau_{l-1} \cdots \tau_1 \gamma_n)| - 1.$$

However, by Lemma 23.10 together with Proposition 23.11, this means that in each multiplication step, starting at $\gamma_n = (1, \ldots, n)$, we split exactly one of the present cycles into two. But this is exactly the way one can produce non-crossing partitions by successively dividing blocks into two, starting from $1_n = \{(1, 2, \ldots, n)\}$. Clearly, every non-crossing partition can arise in this way. The fact that P preserves the order is also clear from this picture. □

REMARK 23.24. The bijection P between the posets $NC(n)$ and $S_{NC}(\gamma_n)$ now transforms Equation (23.11) term by term into Equation (23.12), according to the following observations.

(i) A pair σ, π in $NC(n)$ with $\sigma \leq \pi$ is mapped to a pair

$$\alpha := P_\sigma, \qquad \beta := P_\pi$$

in $S_{NC}(n)$ with $\alpha \leq \beta$ (i.e. $|\alpha^{-1}\beta| + |\alpha| + |\beta^{-1}\gamma_n| = n - 1$).

(ii) Under this mapping, φ_σ goes over to φ_α.

(iii) By Exercise 18.25 we have that

$$P_{K(\pi)} = P_\pi^{-1}\gamma_n,$$

and thus $\varphi_{K(\pi)}$ goes over to $\varphi_{\beta^{-1}\gamma_n}$.

(iv) By Lemma 18.9, the interval $[\sigma, \pi]$ is isomorphic to the interval $[0, K_\pi(\sigma)]$ and thus

$$\mu(\sigma, \pi) = \mu(0_n, K_\pi(\sigma)].$$

Furthermore, by Exercise 18.25 we have

$$P_\sigma^{-1} P_\pi = P_{K_\pi(\sigma)}.$$

(v) Define on $NC(n)$ the function $\tilde{\mu}$ by

$$\tilde{\mu}(\pi) := \phi(P_\pi).$$

Since ϕ is multiplicative, this $\tilde{\mu}$ is a multiplicative function on NC. Then (23.12) is mapped under P into the equation

$$\varphi(a_1 b_1 \cdots a_n b_n) = \sum_{\substack{\pi,\sigma\in NC(n)\\ \sigma\leq\pi}} \varphi_\sigma[a_1,\dots,a_n]\cdot\varphi_{K(\pi)}[b_1,\dots,b_n]\cdot\tilde{\mu}(K_\pi(\sigma)). \tag{23.14}$$

Note that by running over all natural n and putting $a_1 = \cdots = a_n = a$ and $b_1 = \dots b_n = b$ this can also, by using our boxed convolution and denoting by Wg the formal power series corresponding to $\tilde{\mu}$, be written as

$$M_{ab} = M_a \boxed{\star} \mathrm{Wg} \boxed{\star} M_b = M_a \boxed{\star} M_b \boxed{\star} \mathrm{Wg}.$$

But (23.11), on the other hand, tells us that this also has to be equal to

$$M_{ab} = M_a \boxed{\star} M_b \boxed{\star} \text{Möb}.$$

The equality

$$M_a \boxed{\star} M_b \boxed{\star} \text{Möb} = M_a \boxed{\star} M_b \boxed{\star} \mathrm{Wg}$$

for all moment series M_a and M_b however implies that Wg = Möb and thus we see that leading order in the Weingarten function is given by the Möbius function on non-crossing partitions,

$$\phi(P_\pi) = \mu(0,\pi).$$

(vi) Having identified ϕ with μ under P, we see that everything fits together and also the third factors in our sums (23.11) and (23.12) are mapped to each other,

$$\phi(\alpha\beta^{-1}) = \phi(P_\sigma^{-1}P_\pi) = \phi(P_{K_\pi(\sigma)}) = \mu(0_n, K_\pi(\sigma)) = \mu(\sigma,\pi).$$

Exercises

EXERCISE 23.25. Prove Equation (23.10) in the proof of Theorem 23.14.
[It might be helpful to strengthen Lemma 23.17 to the statement that α and $\beta^{-1}\gamma$ have together at least two fixed points.]

EXERCISE 23.26. Prove Theorem 23.13.

Notes and comments

Our emphasis in these lectures is on the combinatorial side of free probability theory. We only touch lightly the operator algebraic side and say nothing about operator-valued free probability or free entropy. For more information about these topics we refer to the survey articles [**84, 85, 86**] and the monographs [**38, 73, 87**]. For applications of free probability in wireless communications we refer to [**77**].

Lecture 1. The idea of "non-commutative analogs" is a recurring theme in operator algebras, and goes back all the way to the beginnings of quantum physics. The particular direction of developing a systematic free non-commutative analog for results in classical probability was initiated by Voiculescu's seminal paper [**78**].

In this lecture (and throughout the book) the framework used most of the time is an algebraic one. The basic measure-theoretic background invoked in the section on $*$-distributions for normal elements is covered for instance by the first two chapters of [**64**].

The name "Haar unitary" (cf. Definition 1.12) was coined by Voiculescu in [**81**]. Haar unitaries play an important role in free probability (which is why they keep reappearing time and again in this book).

Lecture 2. The Toeplitz algebra is a fundamental example in operator algebras, and has a very well-developed theory – see for example Chapter 7 of [**24**].

For the bicyclic semigroup, see for example the monograph [**60**].

Dyck paths are a fundamental example in the theory of lattice paths, see for example Chapter 5 in [**32**] or Chapter 6 in [**76**]. The sequence of Catalan numbers is also of fundamental importance in enumerative combinatorics – see Exercise 6.19 in [**76**], which gives more than 50 ways of how Catalan numbers can occur in enumeration problems.

The fact that the real part of the one-sided shift S has semicircular distribution with respect to the vacuum-state was observed in Section

4.5 of [**78**]. The method used there does not rely on computations of moments, but rather on an analytic formula of Helton and Howe.

For more details on Cauchy transforms and on the Stieltjes inversion formula see for example [**1**].

Lecture 3. The general facts about C^*-algebras presented in this lecture can be found in any introductory textbook on C^*-algebras, and are covered for instance by the first three chapters of [**49**]. For more details about the reduced C^*-algebra of a discrete group see Chapter VII of [**22**]. The purpose of the lecture is to collect a few of these basic facts, and streamline them so that the emphasis is on C^*-probability spaces and on random variables in such spaces.

Lecture 4. The main point of this lecture is that one can define a C^*-algebra by giving a family of generators which have a prescribed $*$-distribution (with respect to a faithful state). This fact and its counterpart taking place in the von Neumann algebra framework are well known and lie at the basis of many of the applications of free probability to operator algebras. When one looks at freely independent generators, these facts are contained in the reduced free product constructions from [**78**].

The formula (4.9) for the number of closed walks on the free group with two generators goes back to Kesten [**41**].

For more details on the rotation C^*-algebra appearing in Example 4.13, see for example the survey by Rieffel [**62**].

Lecture 5. The study of free independence was initiated in [**78**]. A standard reference for the derivation of the basic properties of free independence is Section 2.5 of the monograph [**87**].

The last section of this lecture follows the paper [**72**]; for a more axiomatic treatment see [**7**]. The additional example of universal product referred to in Exercise 5.26 is called "Boolean product," and can be traced back to the work of Bożejko [**16**].

Lecture 6. Free product constructions are studied systematically in Chapter 1 of [**87**]. In that approach, the construction of the free product functional $*_{i\in I}\varphi_i$ is preceded by studying free products of representations. In this lecture we use a direct approach to the construction of $*_{i\in I}\varphi_i$, which follows [**73**].

Lecture 7. GNS is a fundamental construction in the C^*-algebra theory. For more details see for instance Section 3.4 of [**49**].

In the section about free products of C^*-probability spaces we only discuss the technically simpler case of a free product of C^*-probability spaces $(\mathcal{A}_i, \varphi_i)$, $i \in I$, where every φ_i is a faithful trace. The C^*-reduced free product $*_{i\in I}(\mathcal{A}_i, \varphi_i)$ can be constructed without these extra assumptions on the φ_i, see [**87**], Section 1.5. In [**25**] it is proved that if one assumes every φ_i to be faithful (without assuming that it is a trace), then the free product state $*_{i\in I}\varphi_i$ is faithful on the reduced free product C^*-algebra $*_{i\in I}\mathcal{A}_i$.

The important example of how free independence appears in the framework of the full Fock space was observed in [**78**], where it is shown that we a have a free analog of the so-called "second quantization functor" from mathematical physics. For the description of this functor see Section 3 of [**78**], or the presentation made in Section 2.6 of the monograph [**87**]. The name "semicircular system" comes from the paper [**81**]. For an "incomplete" version of a semicircular system, see [**61**].

Lecture 8. The free central limit theorem, Theorem 8.10, was proved by Voiculescu [**78**] (under the more general assumptions as addressed in Remark 8.11.2) with the help of the R-transform (which was introduced in that paper for that purpose). The combinatorial proof as presented here is due to Speicher [**70**] (where "non-crossing partitions" were called "admissible partitions") – there also the multi-dimensional case, Theorem 8.17, was treated for the first time.

For other examples of non-commutative central limit theorems and a general frame to treat them, see [**75**].

For more information about the method of moments in classical probability theory we refer to the book of Billingsley [**15**], Section 30. In particular, the two statements in our Remark 8.4.2 can be found there in Example 30.1 and Theorem 30.2.

Lecture 9. The systematic study of the lattices of non-crossing partitions was initiated by the paper of Kreweras [**43**]. For a survey of the range of problems about these lattices which are of interest to combinatorialists, see Simion [**68**].

For Lukasiewicz paths see Chapter 6 of [**76**], or Chapter 11 of [**48**].

The enumeration formula (9.19) given without proof in Remark 9.24 appears (with proof) as Theorem 2.2 in the paper [**33**] by Goulden and Jackson, in an equivalent formulation given in terms of factorizations of a long cycle in a symmetric group. (The relevant relation between $NC(n)$ and the symmetric group S_n is exactly the one presented in the last section of Lecture 23 of this book.)

The canonical factorization of intervals in $NC(n)$ was done in [**71**], where its applications to free probability were also started (cf. also the notes to Lecture 11).

The uniqueness of the factorization presented in Proposition 9.38 was observed by Bormashenko, Li (undergraduate students at the University of Waterloo) and Nica, while reviewing a preliminary version of this book.

Lecture 10. The theory of Möbius inversion in lattices was started by Rota in the 1960s, see Part 1 in the collection of papers [**63**]. The computation of the Möbius function for the lattice $NC(n)$ goes back to the original paper of Kreweras [**43**]. A very good introduction to Möbius inversion on posets can be found in the monograph by Stanley, [**76**], Chapter 3.

Multiplicative functions on non-crossing partitions were introduced in [**71**]. The functional equation in Theorem 10.23 and its application to counting multi-chains in $NC(n)$ are also from [**71**]. For a direct bijective method of counting multi-chains in $NC(n)$, see the paper of Edelman [**28**].

Lecture 11. Multiplicative functionals on NC and the free cumulants were introduced by Speicher in [**71**]. There the relation between free independence and the vanishing of mixed cumulants, Theorem 11.16, was also established. For this, only a special case of the formula for free cumulants with products as entries was needed. The general form of that formula, Theorem 11.12, was found by Krawczyk and Speicher [**42**]. Simpler proofs of that result were given in [**18, 74**].

Circular elements (and, more generally, circular families) were introduced by Voiculescu in [**81**].

Proposition 11.25 on the form of cumulants of squares of even elements was proved by Nica and Speicher [**55**] in a more general context, for so-called "diagonally balanced pairs." (It is puzzling how much more complicated such calculations get if one tries to treat non-even distributions; even for the simple case $(\text{constant} + \text{even})^2$ there is no general nice formula; for the case $(\text{constant} + \text{semicircle})^2$ see [**39**].)

For more details on classical cumulants, in particular their definition, and relations with Fourier transform, classical independence and partitions we refer to Chapter II, Section 12 of the book of Shiryaev [**65**]. (Note that cumulants are there addressed as semi-invariants.) Theorem 11.30 on classical cumulants with products as entries is due to Leonov and Shiryaev [**47**]. In [**46**], Lehner develops a general theory of non-commutative cumulants which allows many aspects of classical and free (and some other variants of non-commutative) cumulants to be treated in a uniform way. In this paper one can also find more references for classical cumulants.

Exercise 11.34 for the case of pairings was essentially treated in [**17**]. The general case for all partitions can be found in [**45**], in the form of a relation between free and classical cumulants of a random variable.

Lecture 12. The free convolution and the $\mathcal{R}$-transform were introduced by Voiculescu in [**78**]; in [**79**] he proved Theorem 12.7 by analytic methods, relying on Toeplitz operator constructions. Our combinatorial approach to the $\mathcal{R}$-transform via Theorem 12.5 is due to Speicher [**71**].

The free Poisson distribution appeared in [**70, 82**].

The compound free Poisson distribution was introduced in a more general, operator-valued, context in [**73**]; Proposition 12.18 and Example 12.19 are from [**53**].

The "master equation" (12.20) was derived in [**50**].

The Kesten measures from Exercise 12.21 appeared in [**41**], in the context of random walks on free groups.

Lecture 13. The general multi-dimensional limit theorem in 13.1 was proved in [**70**]

There exists by now a well-developed theory on limit theorems and infinitely divisible distributions for free convolution. Our presentation only covers compactly supported probability measures, but by using analytic tools around the Cauchy transform (which exists for any probability measure on $\mathbb{R}$) one can extend the definition of and most results on free convolution to all probability measures on $\mathbb{R}$; for more details and references see [**8, 10, 19**].

The Levy–Khintchine type characterization of the $\mathcal{R}$-transform of infinitely divisible distributions for compactly supported distributions was proved by Voiculescu in [**79**] with analytical methods; our approach using conditionally positive definite sequences and realizations of infinitely divisible distributions on a full Fock space follows [**31**].

The relations between compound free Poisson distributions and infinitely divisible distributions, as addressed in Exercise 13.18, were treated in [**73**].

For more information on free Levy processes (Exercise 13.19), we refer to [**2, 14**].

Lecture 14. The multiplicative free convolution was introduced by Voiculescu in [**78**]. Again, by analytical methods this can be extended to a binary operation on all probability measure supported on the positive real line. Similar to the additive free convolution, there exists an extensive literature around $\boxtimes$, see [**9, 19**] for details and further references.

Our combinatorial approach to the free multiplication, as well as the applications to free compressions are due to Nica and Speicher [**53**]. The existence of the general convolution semigroup $(\mu^{\boxplus t})_{t\geq 1}$ was shown in [**53**]. Earlier, Bercovici and Voiculescu [**11**] had shown by analytic methods that for each μ there is a real T such that $\mu^{\boxplus t}$ exists for $t \geq T$. In [**17**], the semigroup for $\mu = 1/2(\delta_{-1} + \delta_{+1})$ was constructed explicitly by Fock space like constructions. For analytic properties of the semigroup $(\mu^{\boxplus t})_{t\geq 1}$ see [**5**]; for a version for multiplicative free convolution see [**6**].

The result stated in Exercise 14.22 was observed by Shlyakhtenko in [**67**].

Lecture 15. The cumulants of a Haar unitary were calculated in [**73**].

R-diagonal elements in the tracial case were introduced and investigated by Nica and Speicher in [**55**], the general case was treated in [**42, 57**]. Our presentation here mostly follows [**42**].

The polar decomposition of a circular element, Example 15.15, was proved by Voiculescu in [**81**] using random matrix approximations. An elementary combinatorial proof was given by Banica in [**3**].

Polar decomposition results for R-diagonal elements with non-trivial kernel, as addressed in Remarks 15.16, were obtained in [**4**].

The result on the product of two free even elements and the anti-commutator is from [**56**].

The result on powers of R-diagonal elements, Proposition 15.22, is due to Haagerup and Larsen [**36**].

For analytic properties of R-diagonal operators, see also [**27, 69**].

Exercise 15.26 covers results of Oravecz [**59**] (on moments) and Larsen [**44**] (on norm estimates) for powers of circular elements.

Haar partial isometries from Exercise 15.27 were introduced in [**57**]. The generalized circular element and its polar decomposition from Exercise 15.28 is due to Shlyakhtenko [**66**].

Lecture 16. The approach of considering multi-variable R-transforms in the space Θ_s of power series in several non-commuting indeterminates was started in the paper [**51**] by Nica. The behavior of the R-transform under linear transformations and its consequence stated in Exercise 16.23 were also discussed in that paper. For the classical probability statement which is paralleled by Exercise 16.23, see for example the treatise of Feller [**30**], Section III.4.

The functional equation presented in Theorem 16.15 is a generalization of Theorem 10.23 from Lecture 10, using the same kind of idea for proof. The first occurrence of this multi-variable generalization was in a preliminary version of this book.

For the proof of the Lagrange inversion formula via Lukasiewicz paths see for example [**48**], Chapter 11. As mentioned in the lecture, this is very close to the proof shown here for Proposition 16.20.

Lecture 17. The operation of boxed convolution $\boxed{\star}$ was introduced by Nica and Speicher in [**53**]. That paper also studies the basic properties of $\boxed{\star}$, and gives a number of applications to free probability, including Proposition 17.21, and a derivation based on $\boxed{\star}$ for Theorem 14.10 of Lecture 14.

Lecture 18. The considerations on relative Kreweras complements used in the first section of this lecture are taken from Section 2 of the paper [**53**].

The S-transform was introduced by Voiculescu in [**80**]. The multiplicativity of the S-transform is proved there by studying a Lie group structure on $\mathbb{R}^n$, which formalizes how the first n moments of ab are expressed in terms of the first n moments of a and of b, where a is free from b (in some non-commutative probability space).

The "combinatorial Fourier transform" $\mathcal{F}$ for the operation of boxed convolution $\boxed{\star}_1$ was introduced in [**54**]. The proof for the multiplicativity of the S-transform which is shown in this lecture also follows the arguments from [**54**].

Another proof for the multiplicativity of the S-transform is due to Haagerup [**35**]; this relies on an approach to the S-transform as outlined in Exercise 21.24.

Combinatorial interpretations of the coefficients of $1/S$ were provided in recent work of Dykema [**26**].

Lecture 19. The results presented in this lecture are from the paper [**56**]. The argument shown in the section about the cancelation phenomenon is simpler than the one originally given in [**56**], and avoids the concept of "generalized complementation map on $NC(n)$" which is used in that paper.

Lecture 20. R-cyclic matrices were introduced in [**58**], in an attempt to understand better the fundamental example of matrices with free circular/semicircular entries from [**81**]. The results about R-cyclic matrices presented in the lecture are all from [**58**].

The paper [**58**] also has a part concerning freeness with amalgamation, which is not covered by this book. Roughly, this goes as follows. Exercise 20.24 shows that the R-cyclicity of a family $A_1, \dots, A_s$ is really a property of the unital algebra $\mathcal{C}$ generated by $\{A_1, \dots, A_s\} \cup \mathcal{D}$, where $\mathcal{D}$ is the algebra of scalar diagonal matrices. In [**58**] this property of $\mathcal{C}$ is identified precisely: it is the property of being free from the algebra $M_d(\mathbb{C})$ of all scalar $d \times d$ matrices, with amalgamation over $\mathcal{D}$.

Lecture 21. The full Fock space model for the multivariable R-transform was introduced by Nica in [**51**]. This is a direct extension of how the one-dimensional R-transform was introduced by Voiculescu in [**79**].

The idea of how to use modeling on the full Fock space for computing R-transforms of free compressions appeared in the paper of Shlyakhtenko [**67**]. The applications shown in that paper are Theorem 14.10, and the particular case of Theorem 21.11 which is stated as Exercise 14.22. Another particular case of Theorem 21.11 is derived via the same method in [**52**]. The full statement of this theorem does not seem to have appeared in a research paper (but all the ideas required for the proof are present in [**67**]).

The approach to the S-transform outlined in Exercise 21.24 is from the paper of Haagerup [**35**].

Lecture 22. Random matrices have been studied in statistics and in physics since the influential papers of Wishart [**89**] and Wigner [**90**], respectively. Random matrices appear nowadays in different fields of mathematics and physics (such as combinatorics, probability theory, statistics, operator theory, number theory, quantum field theory) or applied fields (such as electrical engineering). For more information and references we refer to the recent surveys [**29, 34, 37, 77**].

For more information about Gaussian families and a proof of the Wick formula, see [**40**].

The genus expansion for Gaussian random matrices is a folklore result in physics; for a mathematical exposition see, for example, [**92**].

The notion of "asymptotic free independence" was introduced by Voiculescu in [**82**]. Our presentation of the asymptotic freeness results for Gaussian random matrices follows essentially the ideas of Voiculescu's original proofs in [**82, 83**]; however, our presentation is more streamlined by using the Wick formula and the genus expansion to make contact with our combinatorial description of free independence.

Lecture 23. For more information and references about Haar unitary random matrices, we refer to [**23, 29, 34**].

The asymptotic freeness results on Haar unitary random matrices from this lecture are due to Voiculescu [**81**]. His proof used polar decomposition of non-selfadjoint Gaussian random matrices to transfer asymptotic freeness results from Gaussian to unitary matrices. The idea of a more direct proof, by using the Wick type formula for correlations of the entries, goes back to Xu [**91**]. Our presentation here follows quite closely the work of Biane [**13**], who also considered in [**12**] the embedding of non-crossing partitions into the symmetric group.

Asymptotic evaluation of integrals for classical groups were obtained by Weingarten [**88**]. The full proof in the unitary group case, our Wick type Lemma 23.5, can be found in [**91**]. The term "Weingarten function" was coined by Collins in [**20**]. We refer to [**20, 21**] for more information about the Weingarten function.

References

[1] N. I. Akhiezer: *The Classical Moment Problem*, Oliver & Boyd, 1965.

[2] M. Anshelevich: Free stochastic measures via noncrossing partitions, *Advances in Mathematics* **155** (2000), 154–179.

[3] T. Banica: On the polar decomposition of circular variables, *Integral Equations and Operator Theory* **24** (1996), 372–377.

[4] F. Benaych-Georges: Rectangular random matrices, related free entropy and free Fisher information, Preprint, 2005, arxiv: math 0512081.

[5] S. T. Belinschi and H. Bercovici: Atoms and regularity for measures in a partially defined free convolution semigroup, *Mathematische Zeitschrift* **248** (2004), 665-674.

[6] S. T. Belinschi and H. Bercovici: Partially defined semigroups relative to multiplicative free convolution, *International Mathematics Research Notices* (2005), 65–101.

[7] A. Ben Ghorbal and M. Schürmann: Non-commutative notions of stochastic independence, *Mathematical Proceedings of the Cambridge Philosophical Society* **133** (2002), 531–561.

[8] H. Bercovici and V. Pata: Stable laws and domains of attraction in free probability theory. With an Appendix by P. Biane: The density of free stable distributions, *Annals of Mathematics* **149** (1999), 1023–1060.

[9] H. Bercovici and D. V. Voiculescu: Levy-Hincin type theorems for multiplicative and additive free convolution, *Pacific Journal of Mathematics* **153** (1992), 217–248.

[10] H. Bercovici and D. V. Voiculescu: Free convolution of measures with unbounded support, *Indiana University Mathematics Journal* **42** (1993), 733–773.

[11] H. Bercovici and D. Voiculescu: Superconvergence to the central limit and failure of the Cramer Theorem for free random variables, *Probability Theory and Related Fields* **102** (1995), 215–222.

[12] P. Biane: Some properties of crossings and partitions, *Discrete Mathematics* **175** (1997), 41–53.

[13] P. Biane: Representations of symmetric groups and free probability, *Advances in Mathematics* **138** (1998), 126–181.

[14] P. Biane: Processes with free increments, *Mathematische Zeitschrift* **227** (1998), 143–174.

[15] P. Billingsley: *Probability and Measure*, Third Edition, Wiley, 1995.

[16] M. Bożejko: Positive definite functions on the free group and the noncommutative Riesz product, *Unione Matematica Italiana, Bolletino A, Serie VI* **5** (1986), 13–21.

[17] M. Bożejko and R. Speicher: Interpolations between bosonic and fermionic relations, given by generalized Brownian motions, *Mathematische Zeitschrift* **222** (1996), 135–160.

[18] T. Cabanal-Duvillard: Non-crossing partitions for conditional freeness, Preprint, 1999, available at http://www.math-info.univ-paris5.fr/~cabanal.

[19] G. Chistyakov and F. Götze: The arithmetic of distributions in free probability theory, Preprint, 2005, arxiv: math/0508245.

[20] B. Collins: Moments and cumulants of polynomial random variables of unitary groups, the Itzykson-Zuber integral, and free probability, *International Mathematics Research Notes* (2003), 953–982.

[21] B. Collins and P. Śniady: Integration with respect to the Haar measure on unitary, orthogonal and symplectic group, *Communications in Mathematical Physics* **264** (2006), 773–795.

[22] K. R. Davidson: *C^*-Algebras by Example*, Fields Institute Monographs 6, American Mathematical Society, 1996.

[23] P. Diaconis: Patterns in eigenvalues: the 70th Josiah Willard Gibbs Lecture, *Bulletin of the American Mathematical Society* **40** (2003), 155–178.

[24] R. G. Douglas: *Banach Algebra Techniques in Operator Theory*, Second Edition, Springer Verlag, 1998.

[25] K. J. Dykema: Faithfulness of free product states, *Journal of Functional Analysis* **154** (1998), 323–329.

[26] K. Dykema: Multilinear function series and transforms in free probability theory, Preprint, 2005, arxiv: math/0504361, to appear in *Advances in Mathematics.*

[27] K. Dykema and U. Haagerup: Invariant subspaces of Voiculescu's circular operator, *Geometric and Functional Analysis* **11** (2001), 693–741.

[28] P. H. Edelman: Multichains, non-crossing partitions and trees, *Discrete Mathematics* **40** (1982), 171–179.

[29] A. Edelman and N. R. Rao: Random matrix theory. *Acta Numerica* (2005), 233–297.

[30] W. Feller: *An Introduction to Probability Theory and Its Applications*, Volume 2, Wiley, 1966.

[31] P, Glockner, M. Schürmann, and R. Speicher: Realization of free white noises, *Archiv der Mathematik* **58** (1992), 407–416.

[32] I. P. Goulden and D. M. Jackson: *Combinatorial Enumeration*, Wiley, 1983.

[33] I. P. Goulden and D. M. Jackson: The combinatorial relationship between trees, cacti and certain connection coefficients of the symmetric group, *European Journal of Combinatorics* **13** (1992), 357–365.

[34] A. Guionnet: Large deviations and stochastic calculus for large random matrices, *Probability Surveys* **1** (2004), 72–172.

[35] U. Haagerup: On Voiculescu's R- and S-transforms for free non-commuting random variables, in *Free Probability Theory* (D. Voiculescu, editor), *Fields Institute Communications* **12** (1997), 127–148.

[36] U. Haagerup and F. Larsen: Brown's spectral distribution measure for *R*-diagonal elements in finite von Neumann algebras, *Journal of Functional Analysis* **176** (2000), 331–367.

[37] U. Haagerup and S. Thorbjornsen: Random matrices with complex Gaussian entries, *Expositiones Mathematicae* **21** (2003), 293–337.

[38] F. Hiai and D. Petz: *The Semicircle Law, Free Random Variables and Entropy*, American Mathematical Society, 2000.

[39] O. Hiwatashi, T. Kuroda, M. Nagisa, and H. Yoshida: The free analogue of noncentral chi-square distributions and symmetric quadratic forms in free random variables, *Mathematische Zeitschrift* **230** (1999), 63–77.

[40] S. Janson: *Gaussian Hilbert Spaces*, Cambridge University Press, 1997.

[41] H. Kesten: Symmetric random walks on groups, *Transactions of the American Mathematical Society* **92** (1959), 336–359.

[42] B. Krawczyk and R. Speicher: Combinatorics of free cumulants, *Journal of Combinatorial Theory, Series A* **90** (2000), 267–292.

[43] G. Kreweras: Sur les partitions non-croisées d'un cycle, *Discrete Mathematics* **1** (1972), 333–350.

[44] F. Larsen: Powers of R-diagonal elements, *Journal of Operator Theory* **47** (2002), 197–212.

[45] F. Lehner: Free cumulants and enumeration of connected partitions, *European Journal of Combinatorics* **23** (2002), 1025–1031.

[46] F. Lehner: Cumulants in noncommutative probability theory. I. Noncommutative exchangeability systems, *Mathematische Zeitschrift* **248** (2004), 67–100.

[47] V. P. Leonov and A. N. Shiryaev: On a method of semi-invariants, *Theory of Probability and its Applications* **4** (1959), 319–329.

[48] M. Lothaire: *Combinatorics on Words*, Addison-Wesley, 1982.

[49] G. J. Murphy: *C^*-Algebras and Operator Theory*, Academic Press, 1990.

[50] P. Neu and R. Speicher: A self-consistent master equation and a new form of cumulants, *Zeitschrift für Physik B* **92** (1993), 399–407.

[51] A. Nica: R-transforms of free joint distributions and non-crossing partitions, *Journal of Functional Analysis* **135** (1996), 271–296.

[52] A. Nica: R-diagonal pairs arising as free off-diagonal compressions, *Indiana University Mathematics Journal* **45** (1996), 529–544.

[53] A. Nica and R. Speicher: On the multiplication of free n-tuples of noncommutative random variables. With an Appendix by D.-V. Voiculescu: Alternative proofs for the type II free Poisson variables and for the compression results, *American Journal of Mathematics* **118** (1996), 799–837.

[54] A. Nica and R. Speicher: A "Fourier transform" for multiplicative functions on non-crossing partitions, *Journal of Algebraic Combinatorics* **6** (1997), 141–160.

[55] A. Nica and R. Speicher: R-diagonal pairs – a common approach to Haar unitaries and circular elements, in *Free Probability Theory* (D. Voiculescu, editor), *Fields Institute Communications* **12** (1997), 149–188.

[56] A. Nica and R. Speicher: Commutators of free random variables, *Duke Mathematical Journal* **92** (1998), 553–592.

[57] A. Nica, D. Shlyakhtenko, and R. Speicher: R-diagonal elements and freeness with amalgamation, *Canadian Journal of Mathematics* **53** (2001), 355–381.

[58] A. Nica, D. Shlyakhtenko, and R. Speicher: R-cyclic families of matrices in free probability, *Journal of Functional Analysis* **188** (2002), 227–271.

[59] F. Oravecz: On the powers of Voiculescu's circular elements, *Studia Mathematica* **145** (2001), 85–95.

[60] M. Petrich: *Inverse Semigroups*, Wiley, 1984.

[61] F. Rădulescu: Stable equivalence of the weak closures of free group convolution algebras, *Communications in Mathematical Physics* **156** (1993), 17–36.

[62] M. A. Rieffel: Non-commutative tori – a case of study of non-commutative differentiable manifolds, in *Geometrical and Topological Invariants of Elliptic Operators* (J. Kaminker, editor), Contemporary Mathematics 105, American Mathematical Society, 1990.

[63] G.-C. Rota: *Gian-Carlo Rota on Combinatorics*, Introductory Papers and Commentaries (J. P. S. Kung, editor), Birkhäuser, 1995.

[64] W. Rudin: *Real and Complex Analysis*, Third Edition, McGraw-Hill, 1987.

[65] A. N. Shiryaev: *Probability*, Second Edition, Springer, 1995.

[66] D. Shlyakhtenko: Free quasi-free states, *Pacific Journal of Mathematics* **177** (1997), 329–368.

[67] D. Shlyakhtenko: R-transforms of certain joint distributions, in *Free Probability Theory* (D.-V. Voiculescu, editor), *Fields Institute Communications* **12** (1997), 253–256.

[68] R. Simion: Noncrossing partitions, *Discrete Mathematics* **217** (2000), 367–409.

[69] P. Śniady and R. Speicher: Continuous family of invariant subspaces for R-diagonal operators, *Inventiones Mathematicae* **146** (2001), 329–363.

[70] R. Speicher: A new example of independence and white noise, *Probability Theory and Related Fields* **84** (1990), 141–159.

[71] R. Speicher: Multiplicative functions on the lattice of non-crossing partitions and free convolution, *Mathematische Annalen* **298** (1994), 611–628.

[72] R. Speicher: On universal products, in *Free Probability Theory* (D.-V. Voiculescu, editor), *Fields Institute Communications* **12** (1997), 257–266.

[73] R. Speicher: Combinatorial theory of the free product with amalgamation and operator-valued free probability theory, *Memoirs of the American Mathematical Society* **132** (1998), no. 627.

[74] R. Speicher: A conceptual proof of a basic result in the combinatorial approach to freeness, *Infinite Dimensional Analysis, Quantum Probability, and Related Fields* **3** (2000), 213–222.

[75] R. Speicher and W. von Waldenfels: A general central limit theorem and invariance principle, *Quantum Probability and Related Topics* **IX** (1994), 371–387.

[76] R. P. Stanley: *Enumerative Combinatorics*, Cambridge Studies in Advanced Mathematics 49 (two volumes), Cambridge University Press, 1997.

[77] A. Tulino and S. Verdú: Random matrix theory and wireless communications, *Foundations and Trends in Communications and Information Theory* **1** (2004), 1–186.

[78] D. Voiculescu: Symmetries of some reduced free product C^*-algebras, in *Operator Algebras and Their Connection with Topology and Ergodic Theory* (H. Araki, C. C. Moore, S. Stratila and D. Voiculescu, editors), Springer Lecture Notes in Mathematics Volume 1132, Springer Verlag, 1985, pp. 556–588.

[79] D. Voiculescu: Addition of certain noncommuting random variables, *Journal of Functional Analysis* **66** (1986), 323–346.

[80] D. Voiculescu: Multiplication of certain non-commutative random variables, *Journal of Operator Theory* **18** (1987), 223–235.

[81] D. Voiculescu: Circular and semicircular systems and free product factors, in *Operator Algebras, Unitary Representations, Enveloping Algebras, and Invariant Theory*, Progress in Mathematics, Volume 92, Birkhäuser, 1990, pp. 45–60.

[82] D. Voiculescu: Limit laws for random matrices and free products, *Inventiones Mathematicae* **104** (1991), 201–220.

[83] D. Voiculescu: A strengthened asymptotic freeness result for random matrices with applications to free entropy, *International Mathematics Research Notices* (1998), 41–63.

[84] D. Voiculescu: Lectures on free probability theory, *Lectures on Probabiltiy Theory and Statistics (Saint-Flour, 1998)*, Lecture Notes in Mathematics 1738, Springer, 2000, pp. 279–349.

[85] D. Voiculescu: Free entropy, *Bulletin of the London Mathematical Society* **34** (2002), 257–278.

[86] D. Voiculescu: Free probability and the von Neumann algebras of free groups, *Reports on Mathematical Physics* **55** (2005), 127–133.

[87] D. Voiculescu, K. J. Dykema, and A. Nica, *Free Random Variables*, CRM Monograph Series 1, American Mathematical Society, 1992.

[88] D. Weingarten: Asymptotic behaviour of group integrals in the limit of infinite rank, *Journal of Mathematical Physics* **19** (1978), 999–1001.

[89] J. Wishart: The generalized product moment distribution in samples from a normal multivariate population, *Biometrika* **20** (1928), 35–52.

[90] E. Wigner: Characteristic vectors of bordered matrices with infinite dimensions, *Annals of Mathematics* **62** (1955), 548–564.

[91] F. Xu: A random matrix model from two-dimensional Yang-Mills theory, *Communications in Mathematical Physics* **190** (1997), 287–307.

[92] A. Zvonkin: Matrix integrals and map enumeration; an accessible introduction, *Mathematical and Computer Modelling* **26** (1997), 281–304.

Index

$\sim_\pi$, 135
0_n, 144
1_n, 144
0_P, 145
1_P, 145
$\wedge$, 144
$\vee$, 144

$\xrightarrow{\text{distr}}$, 116, 126
$\xrightarrow{*-\text{distr}}$, 126

$\boxplus$, 195
$\boxtimes$, 224

$\boxed{\star}$, 273
$\overset{\vee}{\boxed{\star}}$, 295
$\widetilde{\boxed{\star}}$, 330

$*_{i\in I}G_i$, 67
$*_{i\in I}\mathcal{A}_i$, 82
$*_{i\in I}\ (\mathcal{A}_i, \varphi_i)$, 84, 101

$|\pi|$, 135
$\#(\sigma)$, 366
$\#(\alpha)$, 382
$|\alpha|$, 382

$A(\varepsilon)$, 315
a^o, 70
$|\alpha|$, 382
$\alpha \leq \beta$, 390
α_n, 188
$(\mathcal{A}, \varphi)$, 3
$*_{i\in I}\ (\mathcal{A}_i, \varphi_i)$, 84, 101
$*_{i\in I}\mathcal{A}_i$, 82
$\mathcal{A}^+$, 38
$B(\varepsilon)$, 315
C_n, 24
$\mathbb{C}\langle X_1, X_1^*, \dots, X_s, X_s^*\rangle$, 53
$\mathbb{C}\langle X_1, \dots, X_s\rangle$, 49
$\mathbb{C}\langle X, X^*\rangle$, 14
$\mathbb{C}G$, 6
$C_n^{(k)}$, 151
$\mathrm{Cf}_{(i_1,\dots,i_n)}(f)$, 259
$\mathrm{Cf}_{(i_1,\dots,i_n);\pi}(f)$, 260
$\mathrm{Cf}_\pi(f)$, 287
$C^*_{\mathrm{red}}(G)$, 41
$d(\alpha, \beta)$, 390
Δ, 275
$\delta(\pi, \sigma)$, 157
$E[\cdot]$, 359
ε, 315
$\mathbb{F}_n$, 66
$\mathcal{F}(f)$, 293
$\mathcal{F}(\mathcal{H})$, 103
$f \circ D_\alpha$, 286
$f \circ L$, 264
$f \circ \mathrm{Sq}$, 281
$F * G$, 156
$f * G$, 156
G_μ, 30
G_d, 325
$*_{i\in I}G_i$, 67
Γ, 293
γ_n, 382
H_d, 325
$(i_1, \dots, i_n)|V$, 260
J_q, 288
κ_n^a, 185
κ_π, 118, 175
$K(\pi)$, 147
$K_\sigma(\pi)$, 289
$\Lambda(T)$, 110

$\lambda(g)$, 41
$L^{\infty-}(\Omega, P)$, 5
$L^2(\mathcal{A}, \varphi)$, 95
L^t, 265
$Luk(n)$, 138
$l(\xi)$, 103
$l(\xi)^*$, 103
$M_{a_1,\ldots,a_s}$, 260
$M_d(\mathbb{C})$, 5
$M_d(\mathcal{A})$, 16, 322
Möb, 280
$\mu^{\boxplus t}$, 232
$\mu(\pi, \sigma)$, 158
μ_P, 158
$NCE(2n)$, 153
$NCE(S)$, 316
$NC(n)$, 135
$NCO(S)$, 316
$NC(S)$, 135
$NC_2(n)$, 122
Ω, 103
$\mathrm{Orb}(\alpha)$, 300
$\mathcal{P}$, 190
$\mathcal{P}(n)$, 135
$\mathcal{P}(S)$, 135
$\Phi(f)$, 36
ϕ, 381
φ_π, 175
φ_α, 374
φ_n, 175
$|\pi|$, 135
$\pi \leq \sigma$, 144
$\pi_1 \cup \pi_2$, 137
$\pi|_W$, 137
$[\pi, \sigma]$, 148
$\pi \vee \sigma$, 144
$\pi \wedge \sigma$, 144
$P^{(k)}$, 151
P_π, 300, 375, 392
$R_a^{(\mathrm{even})}$, 304, 310
$R_{a_1,\ldots,a_s}$, 260
$\mathcal{R}_\mu$, 199
S_a, 294
S_μ, 294
S_n, 382
$S_{NC}(\gamma_n)$, 390
$\mathrm{Sp}(a)$, 35
$[s]^*$, 341
$\sigma_{a,b}^{(n)}$, 319
Θ_s, 259
$\Theta_1^{(\mathrm{inv})}$, 293
τ_G, 6
$\tau_{\mathcal{H}}$, 103
$\hat{\tau}$, 179
$t(\varepsilon)$, 315
T_F, 156
tr, 5
$\mathrm{tr} \otimes E$, 361
$\mathcal{U}(N)$, 379
Var, 15
$\mathrm{Wg}(N, \alpha)$, 380
Wg, 393
$\mathcal{W}_{i_1,\ldots,i_n}$, 82
Zeta, 274
$\zeta(\pi, \sigma)$, 158
ζ_P, 158

almost-rising path, 138
alternating (arguments), 240
annihilation operator
  left, 103
arcsine law, 13, 201
asymptotically free, 369
averaged eigenvalue distribution, 10, 362

Bell number, 206, 210
Bernoulli variable, 193, 251
bicyclic semigroup, 20
block (of partition), 135
Boolean independence, 64
boxed convolution, 273

canonical factorization, 149
  uniqueness, 152
Catalan number
  and Möbius function of NC, 163
  as moment of semicircle, 28
  counting
    Dyck paths, 24
    non-crossing pairings, 123
    non-crossing partitions, 137
  definition, 24
  recurrence relation, 24
Cauchy–Schwarz inequality, 4, 15
Cauchy transform, 198
  definition, 30

Stieltjes inversion, 31
c.c.w., 234
centering of random variable, 70
central limit theorem
classical, 120
free, 124
multi-dimensional version, 129
general version, 119
chain in poset, 171
circular element, 187
generalized, 255
of radius r, 324
classical cumulant, 191
coloring, 370
complement (Kreweras), 147
complementation map, 147
complex Gaussian family, 360
compound free Poisson distribution, 206
compression, 229
conditionally positive definite, 216
constant matrix, 372
convergence
in distribution, 116, 126
in $*$-distribution, 126
convolution
boxed, 273
free, 195
incomplete boxed, 295
multiplicative free, 224
of functions in poset, 156
couple in a cyclic way, 234
couple the blocks, 181
covariance matrix
of Gaussian family, 360
of semicircular family, 128
covers, 159
creation operator
left, 103
crossing partition, 135
C^*-algebra
definition, 35
functional calculus, 36
of semicircular system, 102
positive element, 38
reduced C^*-algebra of group, 41
reduced free product, 102
rotation algebra, 58
spectral mapping theorem, 37
spectrum, 35
cumulant
classical, 191
free, 173
vanishing of mixed cumulants, 183
Cuntz relations, 340
cycle decomposition of permutation, 382
cyclic permutation trick, 141
cyclic vector, 97

determined by moments, 117
determining sequence
of even element, 188
of R-diagonal element, 241
determining series
of entries of R-cyclic
family of matrices, 329
matrix, 322
of even element, 282, 304
of R-diagonal element, 263
of R-diagonal pair, 304
direct product of posets, 149
distance on S_n, 390
distribution
convergence in, 116, 126
joint, 50
$*$-distribution
and norm, 45
convergence in, 126
in algebraic sense, 14
in analytic sense, 14
joint, 54
of classical random variable, 9
of matrix, 10
of normal element, 8
in C^*-algebra, 43
of random variable, 14
of selfadjoint element, 9
Dyck path, 22
irreducible, 25

eigenvalue distribution
of matrix, 10, 362
of random matrix, 10, 362
element, *see also* random variable
even element, 188
determining sequence, 188

faithful, 5
family
  Gaussian, 129, 360
  semicircular, 128
family of Cuntz isometries, 340
family of matrix units, 233
Fourier transform, 194
free, 66
  asymptotically, 369
$*$-free, 66
free binomial distribution, 201, 209
free convolution, 195
free cumulant, 173, 175
  of random variables, 185
free family of Cuntz isometries, 340
free group, 66
free independence, 66
free Levy process, 222
free limit theorem
  central limit theorem, 124
    multi-dimensional version, 129
  compound Poisson limit, 206
  for triangular array, 211
  Poisson limit, 203
free Poisson distribution, 204
  compound, 206
free product
  of $*$-probability spaces, 88
  of C^*-probability spaces, 101
  of groups, 67
  of non-commutative probability spaces, 84
  of unital algebras, 81
  reduced (of C^*-algebras), 102
freely independent, 64
$*$-freely independent, 65
freeness, 66
  of subgroups, 66
full Fock space, 103
functional
  faithful, 5
  positive, 4
  selfadjoint, 4
  vacuum-state, 21
  vector-state, 7
functional calculus in C^*-algebra, 36
Fuss–Catalan number, 151

gauge operator, 110
Gaussian family, 129, 360
  complex, 360
  covariance matrix, 360
  moments of, 360
  Wick formula, 129, 360
Gaussian random matrix
  non-selfadjoint, 364
  orthogonal ensemble, 364
  selfadjoint, 363
    genus expansion, 366
    moments of, 366
  symplectic ensemble, 364
  unitary ensemble, 364
genus expansion, 366
geodesic, 390
GNS representation, 96
GOE, 364
graded poset, 171
group algebra, 6
grow faster than exponentially, 218
GSE, 364
GUE, 364

Haar measure, 379
(α, β)-Haar partial isometry, 255
Haar unitary, 11
Haar unitary random matrix, 379

incomplete boxed convolution, 295
independence
  Boolean, 64
  classical, 63
  concept of, 75
  free, 66
  tensor, 63
infinitely divisible, 215
interval in poset, 148
interval partition, 154
isometry, 19, 109

join, 144
joint
  distribution, 50
  $*$-distribution, 54
  moment, 50
  $*$-moment, 53
jump distribution (of compound free Poisson), 206
jump size (of free Poisson), 204

Kreweras complement, 147
  relative, 288

Lagrange inversion formula, 270
lattice, 144
left annihilation operator, 103
left creation operator, 103
left regular representation, 41
length function on S_n, 382
length of inverval in poset, 159
Levy–Khintchine formula, 220
Lukasiewicz path, 138

Marchenko–Pastur distribution, 368
matrices over non-commutative probability space, 321
matrix
  constant, 372
  random, 361
matrix units, 233
mean, 15
meet, 144
Möbius function, 155, 158
  of $\mathcal{P}(n)$, 172
  of $NC(n)$, 163
Möbius inversion, 155, 158
  partial, 160
Möbius series, 280
moment, 15
  determined by moments, 117
  joint, 50
$*$-moment
  joint, 53
  of random variable, 13
moment-cumulant formula, 173, 176
moment series, 260
morphism, 7
multi-chain in poset, 151
multiplicative family
  of functionals on NC, 174
  of functionals on $\mathcal{P}$, 190
  of functions on NC, 166
  of functions on $NC^{(2)}$, 165
multiplicative free convolution, 224

Narayana number, 144
NE-SE path, 22
non-commutative
  probability space, 3
  random variable, 3
non-crossing
  pairing, 122
  partition, 135
normal, 4
not grow faster than exponentially, 218
number
  Catalan, 24
  Fuss–Catalan, 151
  Narayana, 144
number of
  multi-chains in $NC(n)$, 151
  non-crossing pair partitions, 123
  non-crossing partitions, 137
    with given block sizes, 143
    with k blocks, 144
  pairings, 120, 206
  partitions, 206, 210

odd-marked interval, 318
one-sided shift, 21
orbit, 300

pair partition, 119
pairing, 119
  non-crossing, 122
parity-alternating block, 318
partial isometry, 109
  (α, β)-Haar, 255
partial Möbius inversion, 160
partial summation condition, 331
partially ordered set, 144, *see also* poset
partition, 118, 135
  interval, 154
  non-crossing, 135
path
  almost-rising, 138
    rise-vector, 138
  Dyck, 22
  Lukasiewicz, 138
  NE-SE, 22
p-Haar unitary, 11
Poisson limit theorem
  classical, 209
  free, 203
polar decomposition, 247
  of circular element, 248

of R-diagonal element, 248
poset, 144
chain in, 171
length of, 171
saturated, 171
direct product of, 149
graded, 171
interval in, 148
length of, 159
Möbius function, 158
multi-chain in, 151
rank in, 171
zeta function, 158
positive
element of C^*-algebra, 38
functional, 4
matrix, 87
positive definite, 216
probability space
non-commutative, 3
matrices over, 321
topological, 344
of random matrices, 361
C^*-probability space, 39
*-probability space, 4
process with free increments, 222
projection, 109

quarter-circular distribution, 248

radial series, 277
random matrix, 6, 361
Gaussian
non-selfadjoint, 364
selfadjoint, 363
Haar unitary, 379
Wishart, 368
random variable
p-Haar unitary, 11
circular, 187
of radius r, 324
classical, 5
compound free Poisson, 206
even, 188
free cumulant, 185
free $*$-cumulant, 185
free Poisson, 204
Haar unitary, 11
isometry, 109
matrix, 5
non-commutative, 3
normal, 4
partial isometry, 109
projection, 109
quarter-circular, 248
random matrix, 6, 361
R-diagonal, 241
selfadjoint, 4
semicircular, 29
symmetric Bernoulli, 193, 251
unitary, 4
random variables
free, 66
R-cyclic matrix, 322
R-diagonal pair, 304
Raney's lemma, 141
rank in poset, 171
rate (of compound free Poisson), 206
rate (of free Poisson), 204
R-cyclic family of matrices, 329
R-cyclic matrix, 322
R-diagonal pair, 304
R-diagonal element, 241
tracial, 241
refinement order, 144
relative Kreweras complement, 288
representation, 7
GNS, 96
respect the coloring, 370
reversed refinement order, 144
right regular representation, 42
rise vector, 138
rotation algebra, 54, 58
R-transform, 260
$\mathcal{R}$-transform, 199

saturated chain in poset, 171
Schur product, 87
selfadjoint
functional, 4
Gaussian random matrix, 363
random variable, 4
self-duality of $NC(n)$, 147
semicircle law, 368
semicircular
family, 128
random variable, 29
system

  C^*-algebra of, 102
  definition, 102
semi-invariant, 191
separating vector, 98
spectral mapping theorem, 37
spectrum, 35
Stieltjes inversion formula, 31
stochastic process, 209
S-transform, 294

tensor independence, 63
Toeplitz algebra, 20
topological non-commutative
  probability space, 344
trace, 3
tracial R-diagonal element, 241
transform
  Cauchy, 198
  Fourier, 194
  $\mathcal{F}$-, 293
  R-, 260
  $\mathcal{R}$-, 199
  S-, 294
transposition, 382
triangle inequality for permutations
  for distance, 390
  for length function, 382

unitary, 4
  Haar, 11
  p-Haar, 11
unitary random matrix, 379

vacuum
  expectation state, 103
  vector, 103
vacuum-state, 21
vanishing of mixed cumulants, 183
variance, 15
vector
  cyclic, 97
  separating, 98
vector-state, 7

Weingarten function, 380
  asymptotics, 381, 384
Wick formula, 129, 360
  analog for unitary matrices, 380
Wigner's semicircle law, 367, 368
Wishart matrix, 368

zeta function, 158
zeta series, 274

For EU product safety concerns, contact us at Calle de José Abascal, 56–1°,
28003 Madrid, Spain or eugpsr@cambridge.org.

www.ingramcontent.com/pod-product-compliance
Lightning Source LLC
LaVergne TN
LVHW090300210926
848480LV00001B/81

* 9 7 8 0 5 2 1 8 5 8 5 2 6 *